The Launching of La Belle Epoque of High Energy Physics & Cosmology

Proceedings of the
32nd Coral Gables Conference

The Launching of La Belle Epoque of High Energy Physics & Cosmology

A festschrift for Paul Frampton in his 60th year and memorial tributes to Behram Kursunoglu

Fort Lauderdale, Florida 17– 21 December 2003

Editors

Thomas Curtright
University of Miami, Coral Gables, Florida

Stephan Mintz
Florida International University, Miami, Florida

Arnold Perlmutter
University of Miami, Coral Gables, Florida

World Scientific

NEW JERSEY • LONDON • SINGAPORE • BEIJING • SHANGHAI • HONG KONG • TAIPEI • CHENNAI

Published by

World Scientific Publishing Co. Pte. Ltd.

5 Toh Tuck Link, Singapore 596224

USA office: 27 Warren Street, Suite 401-402, Hackensack, NJ 07601

UK office: 57 Shelton Street, Covent Garden, London WC2H 9HE

British Library Cataloguing-in-Publication Data
A catalogue record for this book is available from the British Library.

Cover "La Belle" paintings by Jessica Curtright.
"Orbiton symbol" courtesy of the Global Foundation.

THE LAUNCHING OF LA BELLE EPOQUE OF HIGH ENERGY PHYSICS AND COSMOLOGY
A Festschrift for Paul Frampton in His 60th Year and Memorial Tributes to Behram Kursunoglu (1922–2003)
Proceedings of the 32nd Coral Gables Conference

ISBN 981-256-202-8

This book is printed on acid-free paper.

Printed in Singapore by Mainland Press

PREFACE

The thirty-second Coral Gables Conference on High Energy Physics and Cosmology was originally slated to have a light-hearted motif. It was to have been the celebration of the 60th Birthday of Professor Paul Frampton, a leading theorist in high energy physics. A group of distinguished speakers were assembled which included three Nobel Laureates, and the conference seemed poised to be the strongest in recent memory. However, on the 25th of October, the founder and director of these conferences, Behram Kursunoglu, passed away at the age of 81.

The leadership of the conference and the family of Professor Kursunoglu decided that the conference should be held as a memorial to its founder and director. This was done and a memorial service was held during the conference at which members of the Kursunoglu family as well as Behram's friends and long-time associates spoke. A retrospective session on the scientific contributions of Professor Kursunoglu was also held at which the lead seminar was given by Professor Phillip Mannheim.

Nevertheless, it was felt that the original theme of the conference – the celebration of the birthday of Paul Frampton – should remain an integral part of the conference. Thus a festive session with a succession of speakers, including Professors S. Glashow, G. 't Hooft, and M. Veltman, was held on the evening of the conference banquet. This session provided some relief from the otherwise somber nature of the meeting. Within this volume are various photographs of the conference participants and of Paul Frampton, from his earliest days to the present, which epitomize the celebration.

Behram Kursunoglu was the heart and soul of the Coral Gables Conferences. He had founded them in 1964, with the support of the high energy physics community (including his University of Miami colleague, Arnold Perlmutter, and also the well- known Director of the Institute for Advanced Study, Robert Oppenheimer) and except for a hiatus from 1984 through 1992, Behram kept them going, even in retirement. He and his wife, Sevda, within the framework of their Global Foundation, provided the organization and obtained the financing necessary for these conferences with only modest help from the local organizing committee. Much of the responsibility for the 2003 Conference passed to Ismet Kursunoglu, MD, the new president of the Global Foundation. His efforts made possible the meeting which we so enjoyed.

The resulting conference was the most successful in the series since the early 1980s. All speakers were invited to have their talks reproduced here, and most agreed to contribute, with only a few exceptions in those cases where the work was already published elsewhere. Even for these few exceptional cases, talk summaries and bibliographies are included below.

The meeting organizers agreed that the conference series should go forward as a living memorial to Behram Kursunoglu, in the form of a high quality conference of modest size. We hope these conferences will continue long into the future.

We are grateful to Dr. Edward Bacinich for his generous support to have these proceedings published.

Thomas Curtright
Stephan Mintz
Arnold Perlmutter

Editors

ABOUT THE GLOBAL FOUNDATION

The Global Foundation was established in 1977 and utilizes the world's most important resource ... people. The Foundation consists of senior men and women in science and learning, outstanding achievers and entrepreneurs from industry, governments, and international organizations, and promising, enthusiastic young people. These people form a unique and distinguished interdisciplinary entity, and the Foundation is dedicated to assembling all the resources necessary for them to work together. The distinguished senior members of the Foundation convey their expertise and accumulated experience, knowledge, and wisdom to the younger membership on important global issues and frontier problems in science.

CONTENTS

Preface v

About the Global Foundation vii

1 Frampton Festschrift Contributions 1

Glueballs and the Universal Energy Spectrum of Tight Knots and Links 3
R. V. Buniy and T. W. Kephart

Aspects of Neutrino Mass Matrices 11
P. Frampton

Hierarchy Problem and a Bound State of $6t$ and $6\bar{t}$ 20
C. D. Froggatt, H. B. Nielsen and L. V. Laperashvili

Neutrinos – The Soft Underbelly of the Standard Model 28
S. L. Glashow

Unconventional Fermions: The Price of Quark-lepton Unification at TeV Scales 29
P. Q. Hung

On Extended Electroweak Symmetries 39
O. C. W. Kong

Spacetime Foam, Holographic Principle, and Black Hole Quantum Computers 47
Y. J. Ng and H. van Dam

Proton Decay, Neutrino Oscillations, etc. ... from IMB, Through Super-K & Antares 55
L. R. Sulak

Renormalization Without Infinities 65
G. 't Hooft

Problems and Difficulties in Standard Model and Gravitation 75
M. Veltman

Application of the OPE to the Jet Physics 80
M. B. Wise

The publications of Paul Frampton 82

2 Astrophysics and Cosmology 103

Geometric Explanation for an Accelerating Universe 105
E. J. Bacinich

A Theoretical Value for the Newton Gravitational Constant from the GEM Theory of Field Unification and the Kursunoglu–Brandenburg Hypothesis of Massive Gamma-ray Bursters 112
J. E. Brandenburg

Eccentric Inflation and WMAP 120
R. V. Buniy

Dilaton Signatures in the Electromagnetic Spectra of Stars 128
R. Casadio, S. Fabi and B. Harms

Big Crunch/Big Bang String Cosmologies 138
B. Craps and B. Ovrut

Direct Determinations of the Redshift Behavior of the Pressure, Energy Density, and Equation of State of the Dark Energy and the Acceleration of the Universe 139
R. A. Daly and S. G. Djorgovski

Cosmology with Clusters of Galaxies 147
M. Donahue

TeV Antineutrinos from Cygnus OB2 158
H. Goldberg

New Physics at Low Energies and Dark Matter – Dark Energy Transmutation 166
E. I. Guendelman and A. B. Kaganovich

Pulsar Kicks and Dark Matter from a Sterile Neutrino 174
A. Kusenko

LIGO 181
S. Meshkov

On the Mass of a Universe with a Quantum String-like Beginning 182
A. J. Meyer, II

Variable Speed of Light Cosmology and Bimetric Gravity:
An Alternative to Standard Inflation 194
J. W. Moffat

Cosmic Neutrino Physics and Astrophysics 202
T. J. Weiler

3 Lepton Experiments 215

The Status of MINOS 217
T. Bergfeld

IceCube: The Cubic Kilometer Neutrino Telescope at the
South Pole 224
A. R. Fazely

Plans for Experiments to Measure θ_{13} 232
M. Goodman

Limits on the Neutrino Magnetic Moment Using 1496 Days
of Super-Kamiokande-I Solar Neutrino Data 240
D. W. Liu

First Results from the Salt Phase of SNO 248
J. C. Maneira

Precision Measurement of the Anomalous Magnetic Moment of
the Muon 256
J. M. Paley

Measurement of the Structure Functions g_1^p and g_1^d with the
CLAS at Jefferson Lab 264
Y. Prok

Neutrino Experiments with Super-Kamiokande 265
M. B. Smy

Measurement of $\sin^2 2\theta_{13}$ by Reactor Experiments and
its Sensitivity 273
O. Yasuda

4 Neutrino Theory 281

Neutrinos, Lepton Flavor, and CP Violation in a Predictive $SO(10)$ Model 283
B. Dutta, Y. Mimura and R. N. Mohapatra

Testing the Leptogenesis Mechanism of the Seesaw Model 291
E. Jenkins

Tracing Very High Energy Tau Neutrinos from Cosmological Sources in Ice 298
J. Jones, I. Mocioiu, I. Sarcevic and M. H. Reno

The Strangeness Changing Antineutrino Reaction $\bar{\nu}_L + P \longrightarrow \Lambda + L^+$ 306
S. L. Mintz

A Natural Framework for Bi-large Neutrino Mixing 315
S. Raby

Chasing CHOOZ 328
P. Ramond

5 Lorentz Violation, Field Theory and Branes 339

Problems with Gauge Invariance in Lorentz-violating QED 341
B. Altschul

New Implications of Lorentz Violation 349
D. Colladay

Evolving Branes with Generalized Dynamics 357
T. Curtright

Hidden Symmetries in M-theory 373
M. J. Duff

Gauge/String-gravity Duality and Froissart Bound 382
K. Kang

Lorentz Violation as a Quantum-gravity Signature 391
R. Lehnert

Generalized Holonomy in M-theory 399
J. T. Liu

Soft Breaking in SUSY, String, and Intersecting D Brane Models 408
P. Nath

Some Relations Between Twisted K-theory and E8 Gauge Theory 416
H. Sati

Matter Fields on a Two-sheeted Space-time 417
K. C. Wali

6 Kurşunoğlu Memorial Tributes 425

Memorial Session for Behram Kursunoglu 427
Paul Frampton

For My Father 428
I. B. Kursunoglu

The Work of Behram Kursunoglu 434
P. D. Mannheim

The Life and Work of Behram Kursunoglu 442
A. Perlmutter

The Kursunoglus 445
B. Rosenstiel

1 Frampton Festschrift Contributions

GLUEBALLS AND THE UNIVERSAL ENERGY SPECTRUM OF TIGHT KNOTS AND LINKS

ROMAN V. BUNIY* AND THOMAS W. KEPHART†‡

Department of Physics and Astronomy
Vanderbilt University
Nashville, TN 37235
USA
**E-mail: roman.buniy@vanderbilt.edu,*
†*E-mail: thomas.w.kephart@vanderbilt.edu*
‡ *Talk at Coral Gables 2003 in celebration of Paul Frampton's 60th birthday*

Systems of tightly knotted, linked, or braided flux tubes will have a universal mass-energy spectrum if the flux is quantized. We focus on a model of glueballs as knotted QCD flux tubes.

Plasma physics informs us that linked magnetic flux tubes are much more stable than an unknotted single loop [1]. Linked and knotted flux tubes carry topological quantum numbers, and one can think of a knot as a self-linked loop. Similar comments apply to braids.

Our interest will be in tubes carrying quantized flux in tight knot and link configurations. If the loops (tubes) have fixed uniform thickness and circular cross-section, then each knot and link has a completely specified length if the configuration is tight, i.e., is of the shortest length with the tubes non-overlapping and their cross-sections undistorted. If tubes have uniform cross sections, as can be approximately the case for many physical systems, then the length of the tight knot is proportional to the mass (or energy) of the knot. This, we claim, generates a universal mass (energy) spectrum for knotted/linked configurations of objects of this type. The lengths of tight knots were not studied until the mid-1990s [2], and only recently have accurate calculations of large numbers of tight knots [3] and links [4] become available. These results now make it possible to examine physical systems and compare them with the knot spectrum. We have examined the glueball spectrum of QCD [5, 6]. Glueballs [7] are likely to be solitonic states (See Ref. [5] for detailed references.) that are solutions to the QCD field equations. While QCD will be our main focus, there are many more systems where tight knots may play a role.

In order to decide if a system of flux tubes falls into the universal class of having a tight knot energy spectrum, we must first investigate the time scales involved. These are the lifetime of the soliton τ_s and the relaxation time τ_r necessary to reach the ground state of a knotted configuration (i.e., the tight knot state). The soliton lifetime (or the corresponding decay width $\Gamma_s = 1/\tau_s$) can depend on several factors. These include the effects of flux tube breaking, rearrangement, and reconnection. The partial width for flux tube breaking is non-zero if the production of particle/anti-particle pairs at the break point is energetically allowed, for example quark/anti-quark ($q\bar{q}$) pairs for color electric flux tubes. The partial widths can vary widely depending on the particle masses. Rearrangement is a quantum effect where, for example, in a linked double donut arrangement, the loops can tunnel free of each other. Finally, reconnection is an effect where tubes break and re-attach in a different configuration. Such behavior has been seen in plasma physics and is of major importance in understanding a variety of astrophysical systems. All these processes change topological charge, and their partial widths compete more or less favorably with each other depending on the parameters that describe the system.

While no knot lengths have been calculated exactly, it is possible to calculate the exact lengths of an infinite number of links and braids [4]. For links, these calculations are possible in the case where individual elements of the link lie in planes. For braids, exact calculations are possible when the elements of the braid are either straight sections or where their centerlines follow helical paths. The shortest of all links, the double donut, is exactly calculable. The two elements lie in perpendicular planes and are tori of equal length. The shortest non–trivial braid is a helically twisted pair. "Weyl's tube formula" [8, 9] states that for a tube of constant cross-section σ normal to a path of length l, the volume of the tube in flat 3D is just $V_T = l\sigma$. If we have an analytic form for the path and a circular cross-section we can find V_T. This leads to the class of exactly calculable links and braids, but since there are no known analytic forms for the path of tight knots, their volumes can only be calculated numerically. We can then calculate or estimate the volume and therefore the energy for a corresponding physical system. As with knots, the volumes of topologically non-trivial tight braids (those where the elements are woven together) can only be found approximately. While the simple helically twisted braid has a volume that depends on the pitch angle which can potentially be adjusted by experimental conditions, tight knots and links have no such adjustable parameter.

Let us begin with a discussion of tight links of flux in electromagnetic plasma. Movement of fluids can exhibit topological properties. Interrelation between hydro- and magnetic dynamics may cause magnetic fields, to exhibit topological properties as well. For example, for a perfectly conducting fluid, the (abelian) magnetic helicity $\mathcal{H} = \int d^3x\, \epsilon^{ijk} A_i \partial_j A_k$ is an invariant of the motion [10], and this quantity can be interpreted in terms of knottedness of magnetic flux lines [11]. (The helicity for two linked flux tubes with fluxes Φ_1 and Φ_2 is $\mathcal{H} = 2n\Phi_1\Phi_2$, where n is the Gauss linking number of the two tubes. It is straightforward to generalize this to the case of linked and/or self-linked thick flux tubes.) A perfectly conducting, non-viscous, incompressible fluid relaxes to a state of magnetic equilibrium without a change in topology [1]. The system approaches a state of magnetic equilibrium by decreasing its magnetic energy by contraction of the magnetic field lines. In the case of trivial topology, closed curves contract to a point without crossing each other. The relaxation eventually leads to a state with zero fields (vacuum). If, however, the topology of the initial magnetic fields is non-trivial, the relaxation stops when flux tubes are tightly knotted or linked. This happens because the "freeze-in" condition forces topological restrictions on possible changes in field configurations and so any initial knots and links of field lines remain topologically unchanged during relaxation. The energy of a final (equilibrium) state is determined by topology.

By analogy with the abelian case, for a conserved non-abelian helicity [12], we choose the corresponding expression for helicity with topological properties, $\mathcal{H} = \int_V \left(AdA + \frac{2}{3}A^3\right)$. In a perfectly conductive relativistic non-abelian plasma, the electric field vanishes in the local frame moving with the plasma. Details are analogous to the abelian case. With these facts in mind, we are now in position to ask if one could hope to find knotted/linked flux tubes in a physical system.

For several reasons we believe the ideal physical system in which to discover and study tight knots and links is Quantum Chromodynamics (QCD). These include: (1) QCD is a solidly based part of the standard model of particle physics, and much about color confinement and the quark model is already well understood in this context, making much previous work transferable to the problem of tightly knotted flux tubes in this theory. (2) Unlike plasmas, fluids or condensed matter systems where flux tubes are excitations of some media with many parameters that could hide universal behavior, flux knots in QCD can exist in the vacuum. Thus continuum states are absent and there are no media parameters to vary and obscure the universality. Hence, the results in QCD can be far less ambiguous. (3) The hadronic energy spectrum has been measured over a large range of energies (140 MeV to 10 GeV) and already many hundreds of states are known. We expect that among these, a few dozen can be classified as tightly knotted/linked flux tubes states. These states must have no valance quarks (i.e., no flavor quantum numbers) in order to be classified as glueballs. (4) Knotted solitons in QFT are already known to exist. (5) One can efficiently search for new glueball states at accelerators. (Also, data from older experiments still exist and can be reanalyzed to check the predictions of new states described below.)

Consider a hadronic collision that produces some number of baryons and mesons plus a gluonic state in the form of a closed QCD flux tube (or a set of tubes). From an initial state, the fields in the flux tubes quickly relax to an equilibrium configuration, which is topologically equivalent to the initial state. (We assume topological quantum numbers are conserved during this rapid process.) The relaxation proceeds through minimization of the field energy. Flux conservation and energy minimization force the fields to be homogeneous across the tube cross sections. This process occurs via shrinking the tube length, and halts to form a "tight" knot or link. The radial scale will be set by $\Lambda_{\text{QCD}}^{-1}$. The energy of the final state depends only on the topology of the initial state and can be estimated as follows. An arbitrarily knotted tube of radius a and length l has the volume $\pi a^2 l$. Using conservation of flux Φ_E, the energy becomes $\propto l(\text{tr}\Phi_E^2)/(\pi a^2)$. Fixing the radius of the tube (to be proportional to $\Lambda_{\text{QCD}}^{-1}$), we find that the energy is proportional to the length l. The dimensionless ratio $\varepsilon(K) = l/(2a)$ is a topological invariant and the simplest definition of the "knot energy" [13], and can be used to fit the correspondence between knot/link energies and glueball masses.

In our model, the chromoelectric fields F_{0i} are confined to knotted/linked tubes. After an initial time evolution, the system reaches a static equilibrium state which is described by the energy density $\mathcal{E}_E = \frac{1}{2}\text{tr}F_{0i}F^{0i} - V$. Similar to the bag model, we have included a constant potential energy V needed to keep the tubes at a fixed cross-section. The chromoelectric flux Φ_E is conserved and we assume flux tubes carry one flux quantum. To account for conservation of the flux, we add the term $\text{tr}\lambda\{\Phi_E/(\pi a^2) - n^i F_{0i}\}$ to the energy density, where n^i is the normal vector to a section of the tube and λ is a Lagrange multiplier. The energy density should be constant under variations of the degrees of freedom, the gauge potentials A_μ. This leads to a constant field solution, $F_{0i} = (\Phi_E/\pi a^2)n_i$. With this solution, the energy is positive and proportional to l and thus the minimum of the energy is achieved by shortening l, i.e., tightening the knot.

Lattice calculations, QCD sum rules, electric flux tube models, and constituent glue models agree that the lightest non–$q\bar{q}$ states are glueballs with quantum numbers $J^{++} = 0^{++}$ and 2^{++} [7]. We will model all J^{++} states (i.e., all f_J and f'_J states listed by the PDG [7]), some of which will be identified with rotational excitations, as knotted/linked chromoelectric QCD flux tubes. We proceed to identify knotted and linked QCD flux tubes with glueballs, where we include all f_J and f'_J states. The lightest candidate is the $f_0(600)$, which we identify with the shortest knot/link, i.e., the 2^2_1 link; the $f_0(980)$ is identified with the next shortest knot, the 3_1 trefoil knot, and so forth. All knot and link energies have been calculated for states with energies less then 1680 MeV. Above 1680 MeV the number of knots and links grows rapidly, and few of their energies have been calculated (see alternatively, Ref. [4]). However, we do find knot energies corresponding to all known f_J and f'_J states, and so can make preliminary identifications in this region. Our detailed results are collected in Table 1, where we list f_J and f'_J masses, our identifications of these states with knots and the corresponding knot energies. In Fig. 1 we compare the mass spectrum of f_J states with the identified knot and link energies. Since errors for the knot energies in Ref. [2] were not reported, we conservatively assumed the error to be 1%. A least squares fit to the most reliable data (below 1680 MeV) gives $E(G) = (23.4 \pm 46.1) + (59.1 \pm 2.1)\varepsilon(K)$ [MeV], with $\chi^2 = 9.1$. The data used in this fit is the first seven f_J states (filled circles in Fig. 1) in the PDG summary tables. Inclusion of the remaining seven (non-excitation) states (unfilled circles in Fig. 1) in Table 1, where either the glueball or knot energies are less reliable, does not significantly alter the fit and leads to $E(G) = (26.9 \pm 24.9) + (58.9 \pm 1.0)\varepsilon(K)$ [MeV], with $\chi^2 = 10.1$. Both fits are in good agreement with our model, where $E(G)$ is proportional to $\varepsilon(K)$. Better HEP data and the calculation of more knot energies will provide further tests of the model and improve the high mass identification.

Knot complexity can be reduced (or increased) by unknotting (knotting) operations. In terms of flux tubes, these moves are equivalent to reconnection events. Hence, a metastable glueball may decay via reconnection. Once all topological charge is lost, metastability is lost, and the decay proceeds to completion. Two other glueball decay processes are: flux

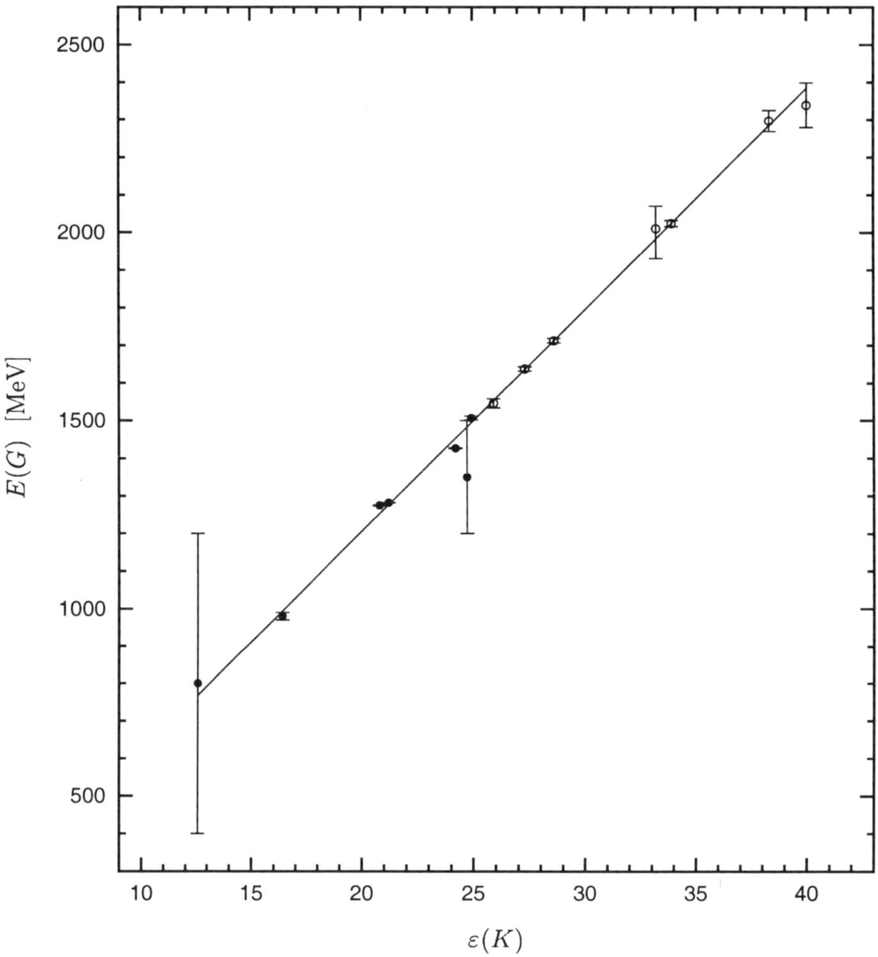

Figure 1. Relationship between the glueball spectrum $E(G)$ and knot energies $\varepsilon(K)$. Each point in this figure represents a glueball identified with a knot or link. The straight line is our model and is drawn for the fit $E(G) = (23.4 \pm 46.1) + (59.1 \pm 2.1)\varepsilon(K)$ [MeV].

tube (string) breaking, which favors large decay widths for configurations with long flux tube components; and quantum fluctuations that unlink flux tubes, which tends to broaden states with short flux tube components. As yet we are not able to go beyond providing a phenomenological fit to these qualitative observations [5], but hope to be able to do so in the future.

In conclusion, let us return to continuum physics and consider a slab of material that can support flux tubes. We have in mind a super-fluid or

Table 1. Comparison between the glueball mass spectrum and knot energies.

State	Mass	K [a]	$\varepsilon(K)$ [b]	$E(G)$ [c]
$f_0(600)$	$400-1200$	2_1^2	12.6 $[4\pi]$	768 [766]
$f_0(980)$	980 ± 10	3_1	16.4	993
$f_2(1270)$	1275.4 ± 1.2	$2_1^2*0_1$	$[6\pi+2]$	[1256]
$f_1(1285)$	1281.9 ± 0.6	4_1	21.2	1277
		4_1^2	(21.4)	(1289)
$f_1(1420)$	1426.3 ± 1.1	5_1	24.2	1454
$\{f_2(1430)$	$\approx 1430\}$ [d]	5_1	24.2	$1454+\delta'$
$f_0(1370)$	$1200-1500$	3_1*0_1	(24.7)	(1484)
$f_0(1500)$	1507 ± 5	5_2	24.9	1496
$\{f_1(1510)$	$1518\pm5\}$	5_2	24.9	$1496+\delta$
$f_2'(1525)$	1525 ± 5	5_2	24.9	$1496+3\delta$
$\{f_2(1565)$	$1546\pm12\}$	5_1^2	(25.9)	(1555)
$\{f_2(1640)$	$1638\pm6\}$	6_3^3	((27.3))	((1638))
		$(2_1^2*0_1)*0_1$ [e]	$[8\pi+3]$	[1686] [f]
$f_0(1710)$	1713 ± 6	6_2^3	((28.6))	((1714))
		$3_1\#3_1^*$	28.9 (30.5)	1732 (1827)
		$3_1\#3_1$	29.1 (30.5)	1744 (1827)
		$2_1^2*2_1^2$	$[8\pi+4]$	[1745]
		6_2	29.2	1750
		6_1	29.3	1756
		6_3	30.5	1827
		7_1	30.9	1850
		8_{19}	31.0	1856
		8_{20}	32.7	1957
$f_2(2010)$	2011^{+60}_{-80}	7_2	33.2	1986
$f_4(2050)$	2025 ± 8	8_{21}	33.9	2028
		8_1	37.0	2211
		$10_{161,162}$	37.6	2247
$f_2(2300)$	2297 ± 28	8_{18}, 9_1	38.3	2288
$f_2(2340)$	2339 ± 60	9_2	40.0	2389
		10_1	44.8	2672
		11_1	47.0	2802

[a]: Notation n_k^l means a link of l components with n crossings, and occurring in the standard table of links (see e.g. Ref. [?]) on the kth place. $K\#K'$ stands for the knot product (connected sum) of knots K and K' and $K*K'$ is the link of the knots K and K'.

[b]: Values are from Ref. [2] except for our exact calculations of 2_1^2, $2_1^2*0_1$, and $(2_1^2*0_1)*0_1$ in square brackets, our analytic estimates given in parentheses, and our rough estimates given in double parentheses.

[c]: $E(G)$ is obtained from $\varepsilon(K)$ using the fit (??).

[d]: States in braces are not in the Particle Data Group (PDG) summary tables.

[e]: This is the link product that is not $2_1^2*2_1^2$.

[f]: Resonances have been seen in this region, but are unconfirmed [7].

superconductor, but are not limited to these possibilities. Assume further that the flux tubes carry one and only one unit of flux. Next consider manipulating these flux tubes. For instance, consider a hypothetical superconductor where the flux tubes are pinned at the bottom of the slab, say by being attracted to the poles of some magnetic material, and at the top of the slab they are each associated with the pole of a movable permanent magnet, perhaps a magnetic whisker, or fine solenoid. Assuming the tubes have time to relax to tight configurations, the energy released should correspond to the universal energy spectrum described above. Another collection of physical systems of potential interest are the atomic Bose-Einstein condensates. For example, laser stirring of dilute ^{87}Rb atoms at 80 nK has produced vortices [14], which could lead to knots and links.

This work is supported in part by U.S. DoE grant # DE-FG05-85ER40226.

References

1. H. K. Moffatt, J. Fluid Mech. **159**, 359 (1985).
2. V. Katritch, et al., Nature **384**, 142 (1996); V. Katritch, et al., Nature **388**, 148 (1997).
3. K. Millett and E. Rawdon, J. of Comp. Phys. **186**, 426 (2003).
4. R. V. Buniy, T. W. Kephart, M. Piatek and E. Rawdon, in preparation.
5. R. V. Buniy and T. W. Kephart, Phys. Lett. B **576**, 127 (2003).
6. R. V. Buniy and T. W. Kephart, in combined Proceedings of "Knot Theory and its Applications," AMS Western Section Meeting (San Francisco, CA, May 2003), and "Knots, Random Walks and Biomolecules," (Les Diablerets, Switzerland, July 2003), to appear in the "Series on Knots and Everything," eds. J. Calvo, K. Millett, E. Rawdon, and A. Stasiak, World Scientific.
7. K. Hagiwara et al., Phys. Rev. D **66**, 010001 (2002).
8. H. Weyl, Am. J. Math. **61**, 461 (1939).
9. A. Gray, *Tubes*, Addison-Wesley, 1990.
10. L. Woltier, Proc. Nat. Acad. Sci. **44**, 489 (1958).
11. H. K. Moffatt, J. Fluid Mech. **35**, 117 (1969).
12. R. Jackiw, V. P. Nair, and So-Young Pi, Phys. Rev. D **62**, 085018 (2000).
13. H. K. Moffatt, Nature **347**, 367 (1990); S. G. Whittington, D. W. Sumners, and T. Lodge, editors, *Topology and Geometry in Polymer Science*, Springer, 1998; R. A. Litherland, J. Simon, O. Durumeric, and E. Rawdon, Topology Appl. **91**, 233 (1999); G. Buck and J. Simon, Topology Appl. **91**, 245 (1999).
14. M. Matthews *at al.*, Phys. Rev. Lett. **83**, 2498 (1999); K. Madison *at al.*, Phys. Rev. Lett. **84**, 806 (2000).

ASPECTS OF NEUTRINO MASS MATRICES

P.H. FRAMPTON
Department of Physics and Astronomy,
University of North Carolina at Chapel Hill,
Chapel Hill, NC 27599-3255.
E-mail: frampton@physics.unc.edu

After an Introduction briefly describing the rise and fall of the three-zero texture of the Zee model, we discuss still-allowed two-zero textures for the Majorana three-neutrino mass matrix. Finally, a model with two right-handed neutrinos and two Dirac texture zeros is described (FGY model) which can relate CP violation in leptogenesis to CP violation in long-baseline neutrino oscillations.

1. Introduction

The minimal standard model involves three chiral neutrino states, but it does not admit renormalizable interactions that can generate neutrino masses. Nevertheless, experimental evidence show that both solar and atmospheric neutrinos display flavor oscillations, and hence that neutrinos do have mass.

Among the many renormalizable and gauge-invariant extensions of the standard model that can provide neutrino masses are:

- The introduction of a complex triplet of mesons (T^{++}, T^{+}, T^{0}) coupled bilinearly to pairs of lepton doublets. They must also couple bilinearly to the Higgs doublet(s) so as to avoid spontaneous $B - L$ violation and the appearance of a massless and experimentally excluded majoron. This mechanism can generate an arbitrary complex symmetric Majorana mass matrix for neutrinos.
- The introduction of singlet counterparts to the neutrinos with very large Majorana masses. The interplay between these mass terms and those generated by the Higgs boson—the so-called see-saw mechanism—yields an arbitrary but naturally small Majorana neutrino mass matrix.
- The introduction of a charged singlet meson f^{+} coupled antisym-

metrically to pairs of lepton doublets, *and* a doubly-charged singlet meson g^{++} coupled bilinearly both to pairs of lepton singlets and to pairs of f-mesons. An arbitrary Majorana neutrino mass matrix is generated in two loops.

- The introduction of a charged singlet meson f^+ coupled antisymmetrically to pairs of lepton doublets *and* (also antisymmetrically) to a pair of Higgs doublets. This simple mechanism was first proposed by Zee [1], and results (at one loop) in a Majorana mass matrix in the flavor basis (e, μ, τ) of a special form:

$$\mathbf{M} = \begin{pmatrix} 0 & m_{e\mu} & m_{e\tau} \\ m_{e\mu} & 0 & m_{\mu\tau} \\ m_{e\tau} & m_{\mu\tau} & 0 \end{pmatrix} \tag{1}$$

  Because the diagonal entries of $\mathbf{M}$ are zero, the amplitude for no-neutrino double beta decay vanishes at lowest order and this process cannot proceed at an observable rate. Furthermore, the parameters $m_{e\mu}$, $m_{e\tau}$ and $m_{\mu\tau}$ may be taken real and non-negative without loss of generality, whence $\mathbf{M}$ becomes real as well as traceless and symmetric. With this convention, the analog to the Kobayashi-Maskawa matrix becomes orthogonal: $\mathbf{M}$ is explicitly CP invariant and $\delta = 0$.

The sum of the neutrino masses (the eigenvalues of $\mathbf{M}$) vanishes: $m_1 + m_2 + m_3 = 0$. When the squared-mass hierarchy is taken into account, that the solar squared mass difference is very much smaller than the atmospheric squared mass difference, there are two possibilities. In case A, we have $m_1 + m_2 = 0$ and $m_3 = 0$. This case arises iff at least one of the three parameters in $\mathbf{M}$ vanishes. In case B, we have $m_1 = m_2$ and $m_3 = -2m_2 > 0$. It has been shown[2,3] that neither case is consistent with the SuperKamiokande and SNO data so the the Zee model is ruled out by these experiments.

2. Two-Zero Textures

Having excluded one three-zero texture we now study two-zero possibilities.

Quark masses and mixings are described by two 3×3 matrices which involve a total of ten convention-independent real quantities: four Kobayashi-Maskawa parameters and six positive quark masses. The lepton sector requires only three additional parameters (the charged lepton masses) in the minimal—and manifestly incomplete—standard model wherein neutrinos are massless.

We assume that neutrino phenomenology can be formulated in terms of three left-handed neutrino states with a complex symmetric Majorana mass matrix $\mathbf{M}$. We do not commit ourselves to any particular mechanism by which neutrino masses are generated. Of the twelve real parameters characterizing $\mathbf{M}$, three are arbitrary phases of the three flavor eigenstates. Thus the neutrino sector involves nine convention-independent parameters. The two squared-mass differences and the four Kobayashi-Maskawa analogs are likely to be measureable in practice. So also is the quantity $|\mathbf{M}_{ee}|$ through the search for neutrinoless double beta decay. There remain two parameters that are measureable in principle but apparently not easily in practice.

It will be useful to define the ratio of squared-mass differences, whose estimated value is: $R_\nu \equiv \frac{\Delta_s}{\Delta_a} \approx 2 \times 10^{-2}$.

We turn to the question of which two independent entries of $\mathbf{M}$ can vanish in the basis wherein the charged lepton mass matrix is diagonal. Of the fifteen logical possibilities, we find just seven to be in accord with our empirical hypotheses. We discuss them individually, with the non-vanishing entries in each case denoted by X's. Our results are presented to leading order in the small parameter s_2. We begin with a texture in which $\mathbf{M}_{ee} = \mathbf{M}_{e\mu} = 0$:

$$\text{Case } A_1: \quad \begin{pmatrix} 0 & 0 & X \\ 0 & X & X \\ X & X & X \end{pmatrix}$$

For this texture, $\mathbf{M}_{ee} = 0$ so that the amplitude for no-neutrino double beta decay vanishes to lowest order in neutrino masses. If Case A_1 is realized in nature, the neutrinoless process simply cannot be detected. There is even more to say. Two of the squared neutrino masses are suppressed relative to the third by a factor of s_2^2. As a result we find that s_2 can lie close to its present experimental upper limit. This prediction will become more precise when θ_3 is better measured. However, the CP-violating parameter δ is entirely unconstrained. For case A_1 the subdominant angle θ_2 is likely to be measureable and neutrinos may display observable CP violation.

$$\text{Case } A_2: \quad \begin{pmatrix} 0 & X & 0 \\ X & X & X \\ 0 & X & X \end{pmatrix}$$

This texture has $\mathbf{M}_{ee} = \mathbf{M}_{e\tau} = 0$. Its phenomenological consequences are nearly the same as those of Case A_1.

Case B_1: $$\begin{pmatrix} X & X & 0 \\ X & 0 & X \\ 0 & X & X \end{pmatrix}$$

With $\mathbf{M}_{\mu\mu} = \mathbf{M}_{e\tau} = 0$, we find an acceptable solution if and only if $|s_2 \cos\delta \tan 2\theta_1| \ll 1$, in which case the three neutrinos are nearly degenerate in magnitude because $t_1^2 \simeq 1$. The texture is also promising in regard to the search for neutrinoless double beta decay. We obtain $\mathbf{M}_{ee} \simeq -t_1^2\sqrt{\Delta_a/|1-t_1^4|}$ likely to exceed 100 meV. The rate of neutrinoless double beta decay may approach its current experimental upper limit.

Case B_2: $$\begin{pmatrix} X & 0 & X \\ 0 & X & X \\ X & X & 0 \end{pmatrix}$$

This texture, with $\mathbf{M}_{\tau\tau} = \mathbf{M}_{e\mu} = 0$, is as in B_1 with t_1 replaced by $-1/t_1$. Its phenomenological consequences are nearly the same as those of Case B_1.

Case B_3: $$\begin{pmatrix} X & 0 & X \\ 0 & 0 & X \\ X & X & X \end{pmatrix}$$ and Case B_4: $$\begin{pmatrix} X & X & 0 \\ X & X & X \\ 0 & X & 0 \end{pmatrix}$$

The phenomenological implications of all four cases B_i (i = 1, 2, 3, 4) are substantially the same.

Case C: $$\begin{pmatrix} X & X & X \\ X & 0 & X \\ X & X & 0 \end{pmatrix}$$

Our seventh and last allowed texture has $\mathbf{M}_{\mu\mu} = \mathbf{M}_{\tau\tau} = 0$.

No other two-zero texture of the neutrino mass matrix is compatible with our empirical hypotheses. It is easily verified that no two of our allowed two-zero textures can be simultaneously satisfied while remaining consistent with our empirical hypotheses. It follows that there is no tolerable three-zero texture.

The seven allowed two-zero neutrino textures fall into three classes: A (with two members), B (with four members), and C. The textures within each class are difficult or impossible to distinguish experimentally, but each of the three classes has radically different implications. For class A, the subdominant angle θ_2 is expected to be relatively large, but no-neutrino $\beta\beta$ decay is forbidden. For class B, the latter process should be measureable by the next generation of double beta decay experiments, whilst s_2 may or may not be large enough to be detected. However, if s_2 is comparable to its experimental upper limit, CP violation must be nearly maximal and should be readily detectable by proposed experiments. For class C, no-neutrino $\beta\beta$ decay is likely to be observable and θ_2 ought to be large enough to measure and to permit a search for CP violation. It is surprising that such a great variety of textures of the neutrino mass matrix can fit what is presently known about neutrino masses and oscillations. Future data should reveal which, if any, of these textures should serve as a guide to the model builder.

3. FGY Model

One of the most profound ideas is[5] that baryon number asymmetry arises in the early universe because of processes which violate CP symmetry and that terrestrial experiments on CP violation could therefore inform us of the details of such cosmological baryogenesis.

The early discussions of baryogenesis focused on the violation of baryon number and its possible relation to proton decay. In the light of present evidence for neutrino masses and oscillations it is more fruitful to associate the baryon number of the universe with violation of lepton number[6]. In the present Letter we shall show how, in one class of models, the sign of the baryon number of the universe correlates with the results of CP violation in neutrino oscillation experiments which will be performed in the forseeable future.

Present data on atmospheric and solar neutrinos suggest that there are respective squared mass differences $\Delta_a \simeq 3 \times 10^{-3} eV^2$ and $\Delta_s \simeq 5 \times 10^{-5} eV^2$. The corresponding mixing angles θ_1 and θ_3 satisfy $tan^2\theta_1 \simeq 1$ and $0.6 \leq sin^2 2\theta_3 \leq 0.96$ with $sin^2\theta_3 = 0.8$ as the best fit. The third mixing angle is much smaller than the other two, since the data require $sin^2 2\theta_2 \leq 0.1$.

A first requirement is that our model[4] accommodate these experimental facts at low energy.

In the minimal standard model, neutrinos are massless. The most economical addition to the standard model which accommodates both neutrino masses and allows the violation of lepton number to underly the cosmological baryon asymmetry is two right-handed neutrinos $N_{1,2}$.

These lead to new terms in the lagrangian:

$$\mathbf{L} = \frac{1}{2}(N_1, N_2)\begin{pmatrix} M_1 & 0 \\ 0 & M_2 \end{pmatrix}\begin{pmatrix} N_1 \\ N_2 \end{pmatrix} +$$

$$+ (N_1, N_2)\begin{pmatrix} a & a' & 0 \\ 0 & b & b' \end{pmatrix}\begin{pmatrix} l_1 \\ l_2 \\ l_3 \end{pmatrix} H + h.c. \tag{2}$$

where we shall denote the rectangular Dirac mass matrix by D_{ij}. We have assumed a texture for D_{ij} in which the upper right and lower left entries vanish. The remaining parameters in our model are both necessary and sufficient to account for the data.

For the light neutrinos, the see-saw mechanism leads to the mass matrix[7]

$$\hat{L} = D^T M^{-1} D$$

$$= \begin{pmatrix} \frac{a^2}{M_1} & \frac{aa'}{M_1} & 0 \\ \frac{aa'}{M_1} & \frac{(a')^2}{M_1} + \frac{b^2}{M_2} & \frac{bb'}{M_2} \\ 0 & \frac{bb'}{M_2} & \frac{(b')^2}{M_2} \end{pmatrix} \tag{3}$$

We take a basis where a, b, b' are real and where a' is complex $a' \equiv |a'|e^{i\delta}$. To check consistency with low-energy phenomenology we temporarily take the specific values (these will be loosened later) $b' = b$ and $a' = \sqrt{2}a$ and all parameters real. In that case:

$$\hat{L} = \begin{pmatrix} \frac{a^2}{M_1} & \frac{\sqrt{2}a^2}{M_1} & 0 \\ \frac{\sqrt{2}a^2}{M_1} & \frac{2a^2}{M_1} + \frac{b^2}{M_2} & \frac{b^2}{M_2} \\ 0 & \frac{b^2}{M_2} & \frac{b^2}{M_2} \end{pmatrix} \tag{4}$$

We now diagonalize to the mass basis by writing:

$$\mathbf{L} = \frac{1}{2}\nu^T \hat{L} \nu = \frac{1}{2}\nu'^T U^T \hat{L} U \nu' \tag{5}$$

where

$$U = \begin{pmatrix} 1/\sqrt{2} & 1/\sqrt{2} & 0 \\ -1/2 & 1/2 & 1/\sqrt{2} \\ 1/2 & -1/2 & 1/\sqrt{2} \end{pmatrix} \times$$

$$\times \begin{pmatrix} 1 & 0 & 0 \\ 0 & cos\theta & sin\theta \\ 0 & -sin\theta & cos\theta \end{pmatrix} \tag{6}$$

We deduce that the mass eigenvalues and θ are given by

$$m(\nu_3') \simeq 2b^2/M_2; \quad m(\nu_2') \simeq 2a^2/M_1; \quad m(\nu_1') = 0 \tag{7}$$

and

$$\theta \simeq m(\nu_2^{'})/(\sqrt{2}m(\nu_3^{'})) \tag{8}$$

in which it was assumed that $a^2/M_1 \ll b^2/M_2$.

By examining the relation between the three mass eigenstates and the corresponding flavor eigenstates we find that for the unitary matrix relevant to neutrino oscillations that

$$U_{e3} \simeq sin\theta/\sqrt{2} \simeq m(\nu_2)/(2m(\nu_3)) \tag{9}$$

Thus the assumptions $a^{'} = \sqrt{2}a$, $b^{'} = b$ adequately fit the experimental data, but $a^{'}$ and $b^{'}$ could be varied around $\sqrt{2}a$ and b respectively to achieve better fits.

But we may conclude that

$$\begin{aligned} 2b^2/M_2 &\simeq 0.05eV = \sqrt{\Delta_a} \\ 2a^2/M_1 &\simeq 7 \times 10^{-3}eV = \sqrt{\Delta_s} \end{aligned} \tag{10}$$

It follows from these values that N_1 decay satisfies the out-of-equilibrium condition for leptogenesis (the absolute requirement is $m < 10^{-2}eV$ [8]) while N_2 decay does not. This fact enables us to predict the sign of CP violation in neutrino oscillations without ambiguity.

Let us now come to the main result. Having a model consistent with all low-energy data and with adequate texture zeros[9] in $\hat{L}$ and equivalently D we can compute the sign both of the high-energy CP violating parameter (ξ_H) appearing in leptogenesis and of the CP violation parameter which will be measured in low-energy ν oscillations (ξ_L).

We find the baryon number B of the universe produced by N_1 decay proportional to[10]

$$\begin{aligned} B \propto \xi_H &= (ImDD^{\dagger})_{12}^2 = Im(a^{'}b)^2 \\ &= +Y^2a^2b^2 sin2\delta \end{aligned} \tag{11}$$

in which B is positive by observation of the universe. Here we have loosened our assumption about $a^{'}$ to $a^{'} = Yae^{i\delta}$.

At low energy the CP violation in neutrino oscillations is governed by the quantity

$$\xi_L = Im(h_{12}h_{23}h_{31}) \tag{12}$$

where $h = \hat{L}\hat{L}^{\dagger}$.

Using Eq.(3) we find:

$$
\begin{aligned}
h_{12} &= \left(\frac{a^3 a^{'*}}{M_1^2} + \frac{a|a^{'}|^2 a^{'*}}{M_1^2}\right) + \frac{aa^{'}b^2}{M_1 M_2} \\
h_{23} &= \left(\frac{bb^{'}a^{'2}}{M_1 M_2}\right) + \left(\frac{b^3 b^{'}}{M_2^2} + \frac{bb^{'3}}{M_2^2}\right) \\
h_{31} &= \left(\frac{aa^{'*}bb^{'}}{M_1 M_2}\right)
\end{aligned}
\tag{13}
$$

from which it follows that

$$
\xi_L = -\frac{a^6 b^6}{M_1^3 M_2^3} sin2\delta[Y^2(2+Y^2)] \tag{14}
$$

Here we have taken $b = b^{'}$ because the mixing for the atmospheric neutrinos is almost maximal.

Neutrinoless double beta decay $(\beta\beta)_{0\nu}$ is predicted at a rate corresponding to $\hat{L}_{ee} \simeq 3 \times 10^{-3} eV$.

The comparison between Eq.(11) and Eq.(14) now gives a unique relation between the signs of ξ_L and ξ_H.

This fulfils in such a class of models the idea of [5] with only the small change that baryon number violation is replaced by lepton number violation.

Acknowledgement

I wish to thank Ruth Daly, Tom Kephart, the late Behram Kursunoglu, Steve Mintz, Arnold Perlmutter and Ina Sarcevic for organizing the very memorable Fest. This work was supported in part by the Department of Energy under Grant Number DE-FG02-97ER-410236.

References

1. A. Zee, Phys. Lett, **93B** 389 (1980).
2. P.H. Frampton and S.L. Glashow, Phys. Lett. **B461,** 95 (1999). hep-ph/9906375.
3. P.H. Frampton, M.C. Oh and T. Yoshikawa, Phys. Rev. **B65,** 073014 (2002). hep-ph/0110300.
4. P.H. Frampton, S.L. Glashow and T. Yanagida, Phys. Lett. **B548,** 119 (2002). hep-ph/0208157

5. A.D. Sakharov, Pis'ma Zh. Eksp. Teor. Fiz. **5,** 32 (1967) [JETP Lett. **5,** 24 (1967)]
6. M. Fukugita and Y. Yanagida, Phys. Lett. **B174,** 45 (1986).
7. T. Yanagida, in *Proceedings of the Workshop on Unified Theories and Baryon Number of the Universe* Editors:O.Sawada and A. Sugamoto, (Tsukuba, Japan, 1979), KEK Report KEK-79-18, page 95.
   S.L. Glashow, Cargese lectures (1979).
   M. Gell-Mann, P. Ramond and R. Slansky, in *Supergravity* Editors: D.Z. Freedman and P. Van Nieuwenhuizen, (North-Holland, Amsterdam, 1979).
8. W. Buchmüller, in Proceedings of the Eighth Symposium on Particles, Strings and Cosmology. Editors: P.H. Frampton and Y.J. Ng. Rinton Press, New Jersey (2001), page 97.
9. P.H. Frampton, S.L. Glashow and D. Marfatia, Phys. Lett. **B536,** 79 (2002).
10. See *e.g.* W. Buchmüller and M. Plumacher, Phys. Reps. **320,** 329 (1999), and references therein.

HIERARCHY-PROBLEM AND A BOUND STATE OF 6 t AND 6 $\bar{t}$.

C. D. FROGGATT

Department of Physics and Astronomy, University of Glasgow, Glasgow

H. B. NIELSEN

Niels Bohr Institute,
Blegdamsvej 17-21, DK 2100 Copenhagen, Denmark

L. V. LAPERASHVILI

Institute of Theoretical and Experimental Physics,
Cheremushkinskaya ulica 25, Moscow

We propose a unification of some fine-tuning problems – really in this article only the problem of why the weak scale is so small in energy compared to a presumed fundamental scale, being say the Planck scale – by postulating the zero or very small value of the cosmological constant not only for one but for several vacua. This postulate corresponds to what we have called the Multiple Point Principle, namely that there be many "vacuum" states with the same energy density. We further assume that 6 top quarks and 6 anti-top quarks can bind by Higgs exchange so strongly as to become tachyonic and form a condensate. This gives rise to the possibility of having a phase transition between vacua with and without such a condensate. The two vacua distinguished by such a condensate will have the same cosmological constant provided the top Yukawa coupling is about 1.1 ± 0.2, in good correspondence with the experimental value. The further requirement that this value of the Yukawa coupling, at the weak scale, be compatible with the existence of a third vacuum, with a Higgs field expectation value of the order of the fundamental scale, enforces a hierarchical scale ratio between the fundamental and weak scales of order $10^{16} - 10^{20}$.

1. Introduction

The present article has the purpose of suggesting the following two relatively new ideas:

1) A unified fine-tuning principle, according to which there exist *several* vacuum states having zero or approximately zero value for the cosmological constant.

2) The existence of a bound state of six top quarks and six anti-top quarks whose binding, due to Higgs particle exchange, is so strong that a condensate of such bound states could form and make up a phase in which essentially tachyonic bound states of this type fill the vacuum.

There are several fine-tuning problems in the Standard Model (SM): the tiny values of the cosmological constant and the strong CP violating parameter Θ_{QCD}, the small hierarchy problem of the Yukawa couplings and the large hierarchy problem of the weak to fundamental (Planck) scale ratio. However these are *only* fine-tuning problems and do not necessarily require a modification[a] of the SM. They could *a priori* be resolved by a general fine-tuning principle, which we take in the form of a zero cosmological constant postulate combined with our so-called Multiple Point Principle[1] of degenerate vacua. Thus, much like in supersymmetry, our idea is to postulate that there be many vacuum states all having zero or rather approximately zero cosmological constant, but without having supersymmetry! We do not speculate here on the underlying mechanism responsible for such degenerate vacua, but it seems likely that some kind of non-locality is required[1].

In this article we point out the possible existence of at least 3 degenerate vacua in the pure SM, which could be responsible for the hierarchy between the fundamental and weak scales. In particular we emphasize our second idea that one of these vacua is due to the condensation of an exotic meson consisting of 6 t and 6 $\bar{t}$ quarks[2]. The reason that such a strongly bound exotic meson has been overlooked until now is that its binding is based on the collective effect of attraction between several quarks due to Higgs exchange. The effect builds up for many (here 12) particles in an analogous way to that of the universal gravitational force of attraction, as we now describe.

2. Proposed bound state of 6 top quarks and 6 anti-top quarks

As emphasized above, the virtual exchange of the Higgs particle between two quarks, two anti-quarks or a quark anti-quark pair yields an attractive force in each case. We now consider putting more and more t and $\bar{t}$ quarks together in the lowest energy relative S-wave states. The Higgs exchange binding energy for the whole system becomes proportional to the number of pairs of constituents, rather than to the number of constituents. So *a priori*,

[a] Of course neutrino masses indicate some new physics at the see-saw scale, which requires a minor modification of the SM.

by combining sufficiently many constituents, the total binding energy could exceed the constituent mass of the system! However we can put a maximum of $6t + 6\bar{t}$ quarks into the ground state S-wave. So let us now estimate the binding energy of such a 12 particle bound state.

As a first step we consider the binding energy E_1 of one of them to the remaining 11 constituents treated as just one particle analogous to the nucleus in the hydrogen atom. We assume that the radius of the system turns out to be reasonably small, compared to the Compton wavelength of the Higgs particle, and use the well-known Bohr formula for the binding energy of a one-electron atom with atomic number $Z = 11$ to obtain the crude estimate:

$$E_1 = -\left(\frac{11g_t^2/2}{4\pi}\right)^2 \frac{11m_t}{24}. \tag{1}$$

Here g_t is the top quark Yukawa coupling constant, in a normalisation in which the top quark mass is given by $m_t = g_t\,174$ GeV.

The non-relativistic binding energy $E_{binding}$ of the 12 particle system is then obtained by multiplying by 12 and dividing by 2 to avoid double-counting the pairwise binding contributions. This estimate only takes account of the t-channel exchange of a Higgs particle between the constituents. A simple estimate of the u-channel Higgs exchange contribution[2] increases the binding energy by a further factor of $(16/11)^2$, giving:

$$E_{binding} = \left(\frac{11g_t^4}{\pi^2}\right) m_t \tag{2}$$

We have so far neglected the attraction due to the exchange of gauge particles. So let us estimate the main effect coming from gluon exchange[b] with a QCD fine structure constant $\alpha_s(M_Z) = g_s^2(M_Z)/4\pi = 0.118$, corresponding to an effective gluon $t - \bar{t}$ coupling constant squared of:

$$e_{tt}^2 = \frac{4}{3}g_s^2 \simeq \frac{4}{3}1.5 \simeq 2.0 \tag{3}$$

For definiteness, consider a t quark in the bound state; it interacts with 6 $\bar{t}$ quarks and 5 t quarks. The 6 $\bar{t}$ quarks form a colour singlet and so their combined interaction with the considered t quark vanishes. On the other hand the 5 t quarks combine to form a colour anti-triplet, which together interact like a $\bar{t}$ quark with the considered t quark. So the total gluon interaction of the considered t quark is the same as it would have with a

[b]Note that we here improve our earlier estimates[2].

single $\bar{t}$ quark. In this case the u-channel gluon contribution should equal that of the t-channel. Thus we should compare the effective gluon coupling strength $2 \times e_{tt}^2 \simeq 2 \times 2 = 4$ with $(16/11) \times Zg_t^2/2 \simeq 16 \times 1.0/2 = 8$ from the Higgs particle. This leads to an increase of $E_{binding}$ by a factor of $(\frac{4+8}{8})^2 = (3/2)^2$, giving our final result:

$$E_{binding} = \left(\frac{99 g_t^4}{4\pi^2}\right) m_t \tag{4}$$

We are now interested in the condition that this bound state should become tachyonic, $m_{bound}^2 < 0$, in order that a new vacuum phase could appear due to Bose-Einstein condensation. For this purpose we consider a Taylor expansion in g_t^2 for the mass *squared* of the bound state, crudely estimated from our non-relativistic binding energy formula:

$$m_{bound}^2 = (12m_t)^2 - 2\,(12m_t) \times E_{binding} + \ldots \tag{5}$$

$$= (12m_t)^2 \left(1 - \frac{33}{8\pi^2} g_t^4 + \ldots\right) \tag{6}$$

Assuming that this expansion can, to first approximation, be trusted even for large g_t, the condition $m_{bound}^2 = 0$ for the appearance of the above phase transition with degenerate vacua becomes to leading order:

$$g_t|_{phase\ transition} = \left(\frac{8\pi^2}{33}\right)^{1/4} \simeq 1.24 \tag{7}$$

We have of course neglected several effects, such as weak gauge boson exchange, s-channel Higgs exchange and relativistic corrections. In particular quantum fluctuations in the Higgs field could have an important effect in reducing $g_t|_{phase\ transition}$ by up to a factor of $\sqrt{2}$, as discussed in Section 4.3. It is therefore quite conceivable that the value of the top quark running Yukawa coupling constant, predicted from our vacuum degeneracy fine-tuning principle, could be in agreement with the experimental value $g_t(\mu_{weak})_{exp} \approx 0.95 \pm 0.03$. Assuming this to be the case, we now make a further application of our fine-tuning principle to postulate the existence of a third degenerate vacuum, in which the SM Higgs field has a vacuum expectation value of order the fundamental scale $\mu_{fundamental}$.

3. Three degenerate vacua and the huge scale ratio

In this section, we explain how our degenerate vacuum fine-tuning principle can be used to derive the huge ratio, $\mu_{fundamental}/\mu_{weak}$, between the fundamental and weak scales. The basic idea is to use this principle to tune the value of the running top quark Yukawa coupling $g_t(\mu)$ *both* at the weak scale, as described above, and at the fundamental scale. Since running couplings vary logarithmically with scale, the predicted values $g_t(\mu_{weak})$ and $g_t(\mu_{fundamental})$ can easily imply an exponentially large scale ratio.

In order to tune the value of $g_t(\mu_{fundamental})$ we postulate the existence of a third degenerate vacuum, in which the SM Higgs field has a vacuum expectation value of order $\mu_{fundamental}$. For large values of the SM Higgs field $\phi \sim \mu_{fundamental} \gg \mu_{weak}$, the renormalisation group improved effective potential is well approximated by

$$V_{eff}(\phi) \simeq \frac{1}{8}\lambda(\mu = |\phi|)|\phi|^4 \tag{8}$$

and the degeneracy condition means that $\lambda(\mu_{fundamental})$ should vanish to high accuracy. The effective potential V_{eff} must also have a minimum and so its derivative should vanish. Therefore the vacuum degeneracy requirement means that the Higgs self-coupling constant and its beta function should vanish near the fundamental scale:

$$\lambda(\mu_{fundamental}) = \beta_\lambda(\mu_{fundamental}) = 0 \tag{9}$$

This leads to the fine-tuning condition[3]

$$g_t^4 = \frac{1}{48}\left(9g_2^4 + 6g_2^2g_1^2 + 3g_1^4\right) \tag{10}$$

relating the top quark Yukawa coupling $g_t(\mu)$ and the electroweak gauge coupling constants $g_1(\mu)$ and $g_2(\mu)$ at $\mu = \mu_{fundamental}$. We must now input the experimental values of the electroweak gauge coupling constants, which we evaluate at the Planck scale using the SM renormalisation group equations, and obtain our prediction:

$$g_t(\mu_{fundamental}) \simeq 0.39. \tag{11}$$

However we note that this value of $g_t(\mu_{fundamental})$, determined from the right hand side of Eq. (10), is rather insensitive to the scale, varying by approximately 10% between $\mu = 246$ GeV and $\mu = 10^{19}$ GeV.

We now estimate the fundamental to weak scale ratio by using the leading order SM beta function for the top quark Yukawa coupling $g_t(\mu)$:

$$\beta_{g_t} = \frac{dg_t}{d\ln\mu} = \frac{g_t}{16\pi^2}\left(\frac{9}{2}g_t^2 - 8g_3^2 - \frac{9}{4}g_2^2 - \frac{17}{12}g_1^2\right) \tag{12}$$

where the $SU(3) \times SU(2) \times U(1)$ gauge coupling constants are considered as given. It should be noticed that, due to the relative smallness of the fine structure constants $\alpha_i = g_i^2/4\pi$ and particularly of $\alpha_3(\mu_{fundamental})$, the beta function β_{g_t} is numerically rather small at the fundamental scale. Hence we need many e-foldings between the two scales, where $g_t(\mu_{fundamental}) \simeq 0.39$ and $g_t(\mu_{weak}) \simeq 1.24$. The predicted scale ratio is quite sensitive to the input value of $\alpha_3(\mu_{fundamental})$. If we input the value of $\alpha_3 \simeq 1/54$ evaluated at the Planck scale in the SM, we predict the scale ratio to be $\mu_{fundamental}/\mu_{weak} \sim 10^{16} - 10^{20}$. We note that, as the rate of logarithmic running of $g_t(\mu)$ increases as α_3 increases, the value of the weak scale is naturally fine-tuned to be a few orders of magnitude above the QCD scale. We also predict[3] the Higgs mass $M_H = 135 \pm 9$ GeV.

4. Phenomenology of the bound state

4.1. *Rho parameter*

Strictly speaking, it is *a priori* not obvious within our scenario in which of the two degenerate vacua discussed in Section 2 we live. There is however one argument that we live in the phase without a condensate of new bound states rather than in the one with such a condensate. The reason is that such a condensate is not invariant under the $SU(2) \times U(1)$ electroweak gauge group and would contribute to the squared masses of the $W^{\pm}$ and Z^0 gauge bosons. Although these contributions are somewhat difficult to calculate, preliminary calculations indicate it is unlikely that, by some mathematical accident, they should be in the same ratio as those from the SM Higgs field. This ratio is essentially the ρ-parameter, which has been measured to be in accurate agreement with the SM value without a new bound state condensate. So we conclude that we live in a phase without a condensate of new bound states.

We have previously[2] given a weak argument in favour of the phase with a condensate emerging in the early Universe out of the Big Bang. However, even if it were valid, one could imagine that a phase transition occurred, in our part of the Universe, from a metastable phase with a bound state condensate into the present one without a condensate. A phenomenological signal for such a phase transition would be the slight variation of various coupling constants, on a very large scale, from region to region in cosmological space and time.

4.2. *Seeing a bound state of 6t + 5 $\bar{t}$?*

We expect the new bound state to be strongly bound and very long lived in our vacuum; it could only decay into a channel in which all 12 constituents disappeared together. The production cross-section of such a particle would also be expected to be very low, if it were just crudely related to the cross section for producing 6 t and 6 $\bar{t}$ quarks. It would be weakly interacting and difficult to detect. There would be a better chance of observing an effect, if we optimistically assume that the mass of the bound state is close to zero (i.e. very light compared to $12m_t \approx 2$ TeV and possibly a dark matter candidate) even in the phase in which we live. In this case the bound state obtained by removing one of the 12 quarks would also be expected to be light. These bound states with radii of order $1/m_t$ might then be smaller than or similar in size to their Compton wavelengths and so be well described by effective scalar and Dirac fields respectively. The 6 t + 6 $\bar{t}$ bound state would couple only weakly to gluons whereas the $6t$ + 5 $\bar{t}$ bound state would be a colour triplet and be produced like a fourth generation top quark at the LHC. If these 11 constituent bound states were pair produced, they would presumably decay into the lighter (undetected) 12 constituent bound states with the emission of a t and a $\bar{t}$ quark.

4.3. *Fine-tuning the top mass; Higgs field fluctuations*

The crucial phenomenological test of our fine-tuning principle is of course that it correctly predicts the experimental values of physical parameters. The predicted existence of a new phase at the weak scale and the value of the top Yukawa coupling $g_t|_{phase\ transition}$ at the phase transition provide, in principle, a very clean test, since it only involves SM physics. However, in practice, the calculation of the binding energy of the proposed 6 t + 6 $\bar{t}$ bound state is hard and indeed Eq. (7) overestimates g_t. So here we consider a potentially large correction due to quantum fluctuations in the Higgs field.

The fluctuations in the average of the Higgs field over the interior of the bound state get bigger and bigger, as the top Yukawa coupling is increased and the size of the bound state diminishes. There is then a significant chance that the average value would turn out to be negative compared to the usual vacuum value. By thinking of the top quark Dirac sea configuration in the bound state, we see that for a sign-inverted Higgs field this configuration becomes just the *vacuum* state. Such a sign-inverted configuration may perhaps best be described by saying that neither the non-relativistic kinetic term for the quarks nor their mass energy are present, both being in these situations approximated by zero. Let us denote by P_v the probability of fluctuating into such a vacuum configuration. The most primitive way to take the effect of these fluctuations into account is to correct the constituent mass in the bound state from m_t to $(1 - P_v)m_t$, and the non-relativistic kinetic term for the same constituents from $\vec{p}^2/(2m_t)$ to $(1 - P_v)\vec{p}^2/(2m_t)$.

It is the kinetic term which determines the binding energy and the above correction corresponds to increasing m_t by a factor $1/(1 - P_v)$ in the binding energy. Therefore the binding energy, which for dimensional reasons is proportional to the m_t occurring in the kinetic term (for fixed g_t), will increase by this factor $1/(1 - P_v)$. On the other hand the constituent masses are corrected the opposite way, meaning that they decrease from m_t to $(1 - P_v)m_t$. So the ratio of the binding energy to the constituent energy – the binding fraction one could say – increases by the square of the factor $1/(1 - P_v)$.

In principle we should now calculate the probability P_v of a sign fluctuation as a function of g_t. The probability P_v is expected to increase as a function of g_t for two reasons: the reduction of the Higgs field inside the bound state and its decreasing radius. We note that both these effects are more important for a bound state of 12 constituents than for, say, toponium. We can, however, not expect the fluctuation probability to go beyond $P_v = 1/2$. So, for a crude orientation, let us calculate the correction in this limiting case. In this case the ratio of the binding energy to the constituent energy, which is proportional to g_t^4, should be increased by the factor $(\frac{1}{1-P_v})^2 = 4$. Applying this correction to Eq. (7), we obtain the limiting value $g_t|_{phase\ transition} \simeq 1.24/4^{1/4} = 0.88$. This value corresponds to the largest possible correction from fluctuations and so we take:

$$g_t|_{phase\ transition} = 1.06 \pm 0.18 \qquad (13)$$

as our best estimate, which is in good agreement with the experimental value $g_t(\mu_{weak})_{exp} \simeq 0.95$ determined from the physical top quark mass.

References

1. D.L. Bennett, C.D. Froggatt and H.B. Nielsen, in *Proc. of the 27th International Conference on High Energy Physics*, p. 557, ed. P. Bussey and I. Knowles (IOP Publishing Ltd, 1995); *Perspectives in Particle Physics '94*, p. 255, ed. D. Klabučar, I. Picek and D. Tadić (World Scientific, 1995) [arXiv:hep-ph/9504294].
2. C.D. Froggatt and H.B. Nielsen, *Surv. High Energy Phys.* **18**, 55 (2003) [arXiv:hep-ph/0308144]; *Proc. to the Euroconference on Symmetries Beyond the Standard Model*, p. 73 (DMFA, Zaloznistvo, 2003) [arXiv:hep-ph/0312218].
3. C.D. Froggatt and H.B. Nielsen, *Phys. Lett.* **B368**, 96 (1996) [arXiv:hep-ph/9511371].

NEUTRINOS - THE SOFT UNDERBELLY OF THE STANDARD MODEL

SHELDON LEE GLASHOW

Physics Department, Boston University
590 Commonwealth Avenue, Boston, MA 02215
E-mail: slg@bu.edu

My talk at the Coral Gables 2003 Conference was based on the several papers listed below. Happy birthday, Paul!

References

1. P.H. Frampton, S.L. Glashow and T. Yanagida, "Cosmological Sign of Neutrino CP Violation," Phys. Lett. B548, 119-121 (2002). hep-ph/0208157
2. P.H. Frampton, S.L. Glashow and D. Marfatia, "Zeroes of the Neutrino Mass Matrix," Phys. Lett. 536B, 79-82 (2002). hep-ph/0201008
3. P.H. Frampton and S.L. Glashow, "Can the Zee Ansatz for Neutrino Masses be Correct?" Phys. Lett. B461, 95-98 (1999). hep-ph/9906375
4. Sheldon Lee Glashow, "Fact and Fancy in Neutrino Physics," Venice 2003, Neutrino telescope, vol. 2, 611-617. hep-ph/0306100
5. Sheldon Lee Glashow, "Neutrinos with Seesaw Masses and Suppressed Interactions." hep-ph/0301250

UNCONVENTIONAL FERMIONS: THE PRICE OF QUARK-LEPTON UNIFICATION AT TEV SCALES *

P. Q. HUNG

Dept. of Physics, University of Virginia,
382 McCormick Road
P. O. Box 400714, Charlottesville, Virginia 22904-4714, USA
E-mail: pqh@virginia.edu

The early petite unification (PUT) of quarks and leptons at TeV scales with $\sin^2\theta_W(M_Z^2)$ used as a constraint, necessitates the introduction of extra quarks and leptons with unconventional electric charges (up to 4/3 for the quarks and 2 for the leptons). This talk, in honor of Paul Frampton's 60th birthday, will be devoted to the motivation and construction of models of early unification and to their implications, including the issues of rare decays and unconventional fermions.

Happy Birthday, Paul!

1. Motivations

It is now believed that the Strong and Electroweak forces are very well described by $SU(3)_c \otimes SU(2)_L \otimes U(1)_Y$ at energies near or below the electroweak scale. It is also widely believed that the SM is just a *low energy manifestation* of some deeper unified theory which could explain why the three gauge couplings are so different, why the quantum numbers of the quarks are different from those of the leptons, and (wishfully thinking) why fermion masses are the way they are. Some of the most successful and popular unification scenarios are the quintessential $SU(5)$ or $SO(10)$ supplemented by supersymmetry. With 3 couplings $\alpha_3(M_Z^2)$, $\alpha_2(M_Z^2)$, $\alpha'(M_Z^2)$, one can make two predictions: $M_{GUT} \approx 10^{16}\,GeV$ and $\sin^2\theta_W(M_Z^2) = 0.233(2)$.

Are there *alternatives* to GUT that can make predictions for $\sin^2\theta_W(M_Z^2)$ and that can be tested? In particular, can these alterna-

*This work is supported in parts by the US Department of Energy under Grant No. DE-A505-89ER40518.

tives be constructed for energy scales in the TeV region instead of being close to the Planck scale? These are well motivated questions which are enhanced by recent interests in the possibility that the "fundamental scale" lies in the TeV region, in the context of large extra dimensions. The key quantity used in the search for such alternatives is

$$\sin^2\theta_W(M_Z^2) = 0.23113(15)\,, \tag{1}$$

which is very precisely measured.

Twenty two years ago, a construction of an alternative to GUT was made by Hung, Buras, and Bjorken [1] based on the data available at the time, namely $\sin^2\theta_W(M_Z^2) \sim 0.22$. The unification scale was found to be $\sim 1000\,TeV$, "small" enough to be coined the name *Petite Unification* (PUT) .

In light of the new and more precise data and of new theoretical motivations, a reexamination of PUT was performed by two of us (AB and PQH) [2] yielding *three* possible scenarios with some containing *unconventional fermions*. These three scenarios predict the PUT scale to be less than 10 TeV. What is most remarkable about two of the three scenarios is the existence of these unconventional fermions which provide a *natural* way for avoiding the severe constraint coming from the process $K_L \to \mu e$, as we shall see below.

Since the concept of large extra dimensions involves scales in the TeV range, it was natural to investigate the possibility of early unification within the LED context. This, in fact, has been done by Chacko, Hall, and Perelstein [3] and by Dimopoulos and Kaplan [4]. The model used in [3] was, in fact, one of the scenarios studied in [1] transported to five dimensions.

2. A petite review of Petite Unification

2.1. *What is Petite Unification?*

In any unification scheme, one would like to know what the predictions might be. For example, in GUT, starting from *three fundamental couplings*: g_3, g_2, g', one obtains *one fundamental coupling*: g_{GUT}, which results in two predictions: M_{GUT} and $\sin^2\theta_W(M_Z^2)$.

For Petite Unification, starting from *three fundamental couplings*: g_3, g_2, g', one obtains *two fundamental coupling*: g_S, g_W, which results in one prediction: $\sin^2\theta_W(M_Z^2)$ when the scale of Petite Unification is constrained independently. The requirement that the scale is less than 10 TeV, for example, severely constrains the PUT gauge groups as we shall see below.

We assume the PUT gauge group to be $G = G_S \otimes G_W$ with the following pattern of symmetry breaking:

$$G \xrightarrow{M} G_1 \xrightarrow{\tilde{M}} G_2 \xrightarrow{M_Z} SU(3)_c \otimes U(1)_{EM}, \tag{2}$$

where

$$G_1 = SU(3)_c(g_3) \otimes \tilde{G}_S(\tilde{g}_S) \otimes G_W(g_W), \tag{3}$$

and

$$G_2 = SU(3)_c(g_3) \otimes SU(2)_L(g_2) \otimes U(1)_Y(g'). \tag{4}$$

It turns out that the most economical choices for G_S and G_W are the Pati-Salam $SU(4)_{PS}$ [5] and $SU(N)^k$ respectively. The PUT group is now

$$G = SU(4)_{PS} \otimes SU(N)^k, \tag{5}$$

with a permutation symmetry assumed so that each $SU(N)$ of $SU(N)^k$ has the same gauge coupling.

The next task is to compute $\sin^2\theta_W(M_Z^2)$. Using $Q = T_{3L}+T_0$ and $T_0 = \sum_\alpha C_{\alpha W} T^0_{\alpha W} + C_S T_{15}$, and the matching of the electromagnetic coupling with the weak couplings at M_Z, one arrives at the following master formula for $\sin^2\theta_W(M_Z^2)$ [1]:

$$\sin^2\theta_W(M_Z^2) = \sin^2\theta_W^0[1 - C_S^2\frac{\alpha(M_Z^2)}{\alpha_S(M_Z^2)} - 8\pi \times \alpha(M_Z^2)(K \ln\frac{\tilde{M}}{M_Z} + K' \ln\frac{M}{\tilde{M}})], \tag{6}$$

where $\alpha(M_Z^2) \equiv e^2(M_Z^2)/4\pi$, $\alpha_S(M_Z^2) \equiv g_3^2(M_Z^2)/4\pi$, and

$$\sin^2\theta_W^0 = \frac{1}{1+C_W^2}, \tag{7}$$

with $C_W^2 = \sum_\alpha C_{\alpha W}^2$. Here

$$K = b_1 - C_W^2 b_2 - C_S^2 b_3, \tag{8}$$

$$K' = C_S^2(\tilde{b} - \tilde{b}_3). \tag{9}$$

$\tilde{b}$ and $\tilde{b}_3$ are the one-loop Renomalization Group coefficients, above $\tilde{M}$, of $U(1)_S$ and $SU(3)_c$ respectively. Furthermore, the following (fairly) precise inputs are used in Eq. 6: $1/\alpha(M_Z^2) = 127.934(27)$, $\alpha_S(M_Z^2) = 0.1172(20)$.

From Eq. 6, it is important to realize the following point: If we require M and $\tilde{M}$ to be at most 10 TeV, the logarithmic evolution of $\sin^2\theta_W$ is less important than in the GUT case. In our case, a term which is crucial in the determination of $\sin^2\theta_W$ is the following term in Eq. 6: $C_S^2 \frac{\alpha(M_Z^2)}{\alpha_S(M_Z^2)}$. This can easily be seen by looking at $\sin^2\theta_W(M_Z^2) = \sin^2\theta_W^0(1-0.067\, C_S^2 - \log terms)$. Indeed, $\sin^2\theta_W(M_Z^2)$ is found to be very sensitive to the value of C_S^2 which, as we shall see, depends crucially on the fermion representation under G_W. Since we have already chose G_S to be $SU(4)_{PS}$, the choices of G_W amount to their predictions for $\sin^2\theta_W^0$ and the choices of fermion representations as represented by C_S^2.

Our objectives are therefore the computations of $\sin^2\theta_W^0$ and C_S^2, and to examine the related physical consequences.

2.2. $\sin^2\theta_W^0$ *and* C_S^2

The computation of $\sin^2\theta_W^0$ was done in detail in [1] and repeated in [2]. Here, I will simply state the results.

By the definition of $\sin^2\theta_W^0$ as shown in Eq. 7, its computation requires simply the adjoint representation of G_W. [1] arrived at the following important constraint: Only weak gauge bosons (i.e. the gauge bosons of G_W) with charges $0, \pm 1$ are consistent with the data. We obtained a *very simple* formula for $\sin^2\theta_W^0$:

$$\sin^2\theta_W^0 = \frac{N}{kTr(Q_W^2)|_{adj}} = \frac{N}{kn_1} = \frac{N}{2kr_0(N-r_0)}, \tag{10}$$

with n_1 is the number of weak gauge bosons with charges ± 1 and $n_1 = 2r_0r_1$, with $[\underbrace{\tilde{Q}_W, \cdots \tilde{Q}_W}_{r_0}, \underbrace{\tilde{Q}_W - 1, \cdots \tilde{Q}_W - 1}_{r_1}]$. The results for $\sin^2\theta_W^0$ are listed in Table 1 below.

Since $\tilde{Q}_W^i = \frac{1}{4}(3Q_q^i + Q_l^i)$, one can see that only groups and representations with $\tilde{Q}_W^i = \pm\frac{1}{2}$ or $\tilde{Q}_W^i = 0, \pm 1$ can accomodate standard fermions. With this in mind, Table 2 gives the values of C_S^2 along with the corresponding quark and lepton charges.

From Table 1 and 2, and from $\sin^2\theta_W(M_Z^2) = \sin^2\theta_W^0(1 - 0.067\, C_S^2 - \log terms)$, it is clear that groups with "high" $\sin^2\theta_W^0$ need "high" C_S^2.

A close examination revealed *three* favorite candidates for G_W:

1. $[SU(2)]^4$: $C_S^2 = 2/3$; $\sin^2\theta_W^0 = 0.25$.

This group contains *only* conventionally-charged quarks and leptons since the fermion representations under G_W are of the type $(f, 1, ..)$ as one can easily infer from Tables 1 and 2.

Table 1. The values of $\sin^2\theta^0$ for the weak groups $G_W = SU(N)^k$ and different fermion representations.

G_W	r_0	$\sin^2\theta^0_W$	$(f,1)+(1,\bar{f})$ $\tilde{Q}^i_W$	$(f,\bar{f})$ $\tilde{Q}^i_W$
$[SU(2)]^3$	1	0.333	$\pm\frac{1}{2}$	0,±1
$[SU(2)]^4$	1	0.250	$\pm\frac{1}{2}$	0,±1
$[SU(3)]^2$	1	0.375	$\frac{2}{3},-\frac{1}{3}$	0,±1
$[SU(3)]^3$	1	0.250	$\frac{2}{3},-\frac{1}{3}$	0,±1
$[SU(4)]^2$	2	0.250	$\pm\frac{1}{2}$	0,±1
$[SU(5)]^2$	1	0.313	$\frac{4}{5},-\frac{1}{5}$	0,±1
$[SU(6)]^2$	1	0.300	$\frac{5}{6},-\frac{1}{6}$	0,±1
$SU(7)$	3	0.292	$\frac{4}{7},-\frac{3}{7}$	
$[SU(7)]^2$	1	0.292	$\frac{6}{7},-\frac{1}{7}$	0,±1
$SU(8)$	3	0.267	$\frac{5}{8},-\frac{3}{8}$	
$SU(8)$	4	0.250	$\pm\frac{1}{2}$	

Table 2. The values of lepton (Q^i_l) and quark (Q^i_q) electric charges and the corresponding weak charge ($\tilde{Q}^i_W$), and C^2_S.

$\tilde{Q}^i_W$	Q^i_l	Q^i_q	C^2_S
$\frac{1}{2}$	0	$\frac{2}{3}$	
$-\frac{1}{2}$	-1	$-\frac{1}{3}$	
$\frac{1}{2}$	1	$\frac{1}{3}$	$\frac{2}{3}$
$-\frac{1}{2}$	0	$-\frac{2}{3}$	
1	0	$\frac{4}{3}$	
0	-1	$\frac{1}{3}$	
-1	-2	$-\frac{2}{3}$	$\frac{8}{3}$
1	2	$\frac{2}{3}$	
0	1	$-\frac{1}{3}$	
-1	0	$-\frac{4}{3}$	

2. $[SU(2)]^3$, $[SU(3)]^2$: $C^2_S = 8/3$; $\sin^2\theta^0_W = 1/3,\ 3/8$.

These groups contain conventionally-charged quarks and leptons *as well as* unconventional quarks and leptons with higher charges $\pm 4/3$ and ± 2 as can be seen from Table 2. The G_W-fermion representations are of the types: $(f, \bar{f}, ..)$.

The use of the term "favorite" actually means that these are the three groups that can give $\sin^2\theta_W(M^2_Z)$ within the allowed experimental range for unification scales which are less than 10 TeV. We shall see however that $[SU(2)]^4$ suffers from problems with rare decays, and we will be left with

$[SU(2)]^3$, $[SU(3)]^2$ as the true favorites.

In order to calculate $\sin^2\theta_W(M_Z^2)$ or, equivalently, the unification scale M, a knowledge of at least the fermionic degrees of freedom that enter the evolution of $\sin^2\theta_W$ is necessary.

3. Unconventional Fermions

Although this section is titled "unconventional fermions", I will list the fermion contents of all three "favorite" candidates.

I) $PUT_0 = SU(4)_{\rm PS} \otimes SU(2)_L \otimes SU(2)_R \otimes \tilde{SU}(2)_L \otimes \tilde{SU}(2)_R$:

- Standard Fermions: $\Psi_L = (q_L, l_L) = (4,2,1,1,1)_L$, $\Psi_R = (q_R, l_R) = (4,1,2,1,1)_R$.
- Fermions of "Mirror Group": $\tilde{\Psi}_L = (\tilde{q}_L, \tilde{l}_L) = (4,1,1,2,1)_L$, $\tilde{\Psi}_R = (\tilde{q}_R, \tilde{l}_R) = (4,1,1,1,2)_R$.

It should be understood that the adjective "Mirror" refers to something completely different (groups instead of fermions) from its customary use in Left-Right symmetric models.

II) $PUT_1 = SU(4)_{\rm PS} \otimes SU(2)_L \otimes SU(2)_H \otimes SU(2)_R$:

Below the electric charges of the fermions are explicitely written down in parentheses next to their names.

- Standard Fermions:
  $\psi^q_{L,R} = \begin{pmatrix} u(2/3) \\ d(-1/3) \end{pmatrix}_{L,R}$; $\psi^l_{L,R} = \begin{pmatrix} \nu(0) \\ l(-1) \end{pmatrix}_{L,R}$
- Fermions with "weird" charges:
  $\tilde{Q}_{L,R} = \begin{pmatrix} \tilde{U}(4/3) \\ \tilde{D}(1/3) \end{pmatrix}_{L,R}$; $\tilde{L}_{L,R} = \begin{pmatrix} \tilde{l}_u(-1) \\ \tilde{l}_d(-2) \end{pmatrix}_{L,R}$
- Fermion Representations:
  $(4,2,2,1)_L = [(i\tau_2\psi^{q,*}_L, \tilde{Q}_L), (\tilde{L}_L, \psi^l_L)]$; $(4,1,2,2)_R = [(i\tau_2\psi^{q,*}_R, \tilde{Q}_R), (\tilde{L}_R, \psi^l_R)]$; $(4,2,1,1)_{L,R} = [\tilde{Q}'_{L,R}, \tilde{L}'_{L,R}]$; $(4,1,1,2)_{L,R} = [\tilde{Q}''_{L,R}, \tilde{L}''_{L,R}]$
- Tree-level $SU(2)_H$ transition:
  $i\tau_2\psi^{q,*}_L \rightarrow \tilde{Q}_L$
  $\tilde{L}_L \rightarrow \psi^l_L$
  $\Rightarrow$ No tree-level transition between normal quarks and leptons due to $SU(2)_H$ gauge bosons
- Tree-level $SU(4)/SU(3)_c \otimes U(1)_S$ transition:
  $i\tau_2\psi^{q,*}_L \rightarrow \tilde{L}_L$
  $\tilde{Q}_L \rightarrow \psi^l_L$

etc...

$\Rightarrow$ No tree-level transition between normal quarks and leptons due to PS gauge bosons.

III) $PUT_2 = SU(4)_{\rm PS} \otimes SU(3)_L \otimes SU(3)_H$:

- Fermion Representations of the type $(4, 3, \bar{3})$ and $(4, \bar{3}, 3)$.
- Same presence of higher charged fermions as in PUT_1!

Is the presence of quarks and leptons with unconventional charges in $PUT_{1,2}$ a boon or a bane? As we shall see below, the existence of these fermions turns into a virtue for $PUT_{1,2}$ when we look at the decay process $K_L \rightarrow \mu^{\pm} e^{\mp}$. Before discussing the virtues and defects of these three scenarios, let us do some RG analysis to see the range of values that the unification scales can take.

4. RG analysis and PUT scales

- If M and $\tilde{M}$ are of O(TeV), there is not much "running" to do starting from M_Z. This means that two-loop contributions to $\sin^2\theta_W(M_Z^2)$ are not as important as the one-loop contribution. A detailed RG analysis up to two loops within the context of PUT_1 will be presented in [6]. Here I will neglect that contribution for simplicity as we had done in [2].
- To find PUT scales from $\sin^2\theta_W(M_Z^2)$ in our RG analysis, we assume the unconventional fermions to have a mass $M_F = (250 \pm 50)\, GeV$. Furthermore, we assume all vector-like fermions (present in $PUT_{1,2}$) to have a mass of order M.
- Let us start out with $\tilde{M} = M$. We then use Eq. (1), namely $\sin^2\theta_W(M_Z^2) = 0.23113(15)$, as a constraint to obtain M. Since the scale of $SU(2)_R \otimes \tilde{SU}(2)_L \otimes \tilde{SU}(2)_R$ breaking is naturally of order M, we require (from the constraint on W_R) that $M \geq 800\, GeV$. Furthermore, we also require $M_F \geq 200\, GeV$. We obtained the following results shown in Figure 1.
- One can also look at the case where $\tilde{M} \neq M$. This is shown in Figure 2 below.

The following conclusions arise by examining Figure 1. 1) For PUT_0, one needs the number of generations $n_G \geq 9$ for the SM and $n_G \geq 4$ for MSSM. 2) For PUT_1 with $n_G = 3$, one obtains $M = (1.00 \pm 0.14)$. 3) For PUT_2 with $n_G = 3$, one obtains $M = (3.30 \pm 0.47)\, TeV$.

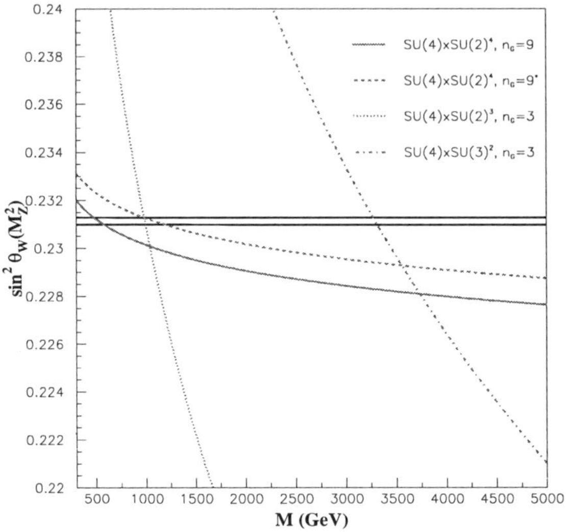

Figure 1. $\sin^2\theta_W(M_Z^2)$ versus the PUT scale M. The horizonal band represents the experimental value. The dashed curve ($n_G = 9^*$) is obtained by using $M_F = 200GeV$, while the other three curves are obtained by using $M_F = 250GeV$.

From the above analysis, one can see that the PUT scales are all below 10 TeV as promised.

5. Virtues and Defects

I will now discuss in particular the defects of PUT_0 and the virtues of the unconventional fermions in $PUT_{1,2}$.

- PUT_0:

The defects are the following:

1) Large number of generations! (On the other hand, why not?)

2) Tree-level transition between SM fermions which leads to a large $Br(K_L \to \mu e)$.

$Br(K_L \to \mu^\pm e^\mp) = 4.7 \cdot 10^{-12} \left(\frac{\alpha_S(m_G)}{0.1}\right)^2 \times \left[\frac{1.8 \cdot 10^3\, TeV}{m_G}\right]^4$

versus

$Br(K_L \to \mu e) < 4.7 \times 10^{-12}$

Since $m_G \sim M < 1\,TeV$, the bound is violated by at least 13 orders of magnitude!

Chacko, Hall and Perelstein [3] solved this problem by taking PUT_0 into

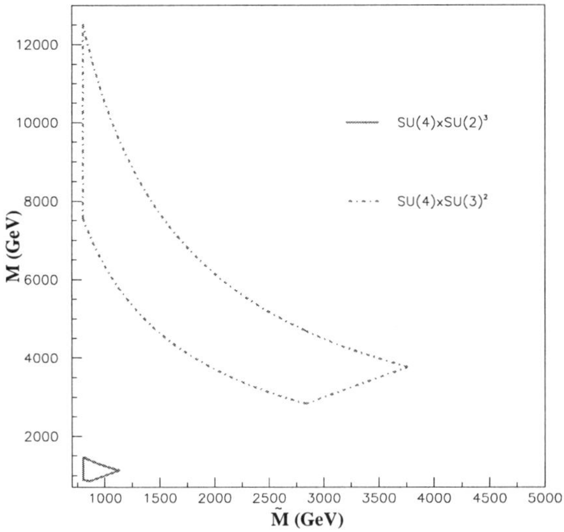

Figure 2. The allowed ranges (at 2 σ) for the $SU(2)^3$ and $SU(3)^2$ scenarios.

five dimensions.

- PUT_1 (PUT_2):

The virtues of the unconventional fermions are as follows.

1) Tree-level transitions via $SU(2)_H$ and $SU(4)/SU(3)_c \otimes U(1)_S$ gauge bosons only occur between *unconventional* and *normal* fermions. There is *no* tree-level FCNC.

The process $K_L \rightarrow \mu e$ occurs in box diagrams and can be made small! (Exactly zero when the unconventional fermions in the boxes are made degenerate.)

2) The lightest of the unconventional fermions (quark or lepton) is unstable. It can decay entirely into normal fermions since the Higgs sector for the model can mix W_L, W_R, W_H. For example, if $\tilde{l}_u(-1)$ were the lightest of such particles (still presumably having a mass larger than M_W), it can have the decay mode $\tilde{l}_u(-1) \rightarrow \nu W$. The rate will depend on the details of the mixing of the gauge bosons [6].

3) Since the lightest one is unstable, there is no cosmological constraint.

4) Fermions such as these unconventional ones can be searched for at the LHC (see [7]), especially if they are relatively 'long lived".

6. Conclusions

From our analysis, we have arrived at two favorite models: PUT_1 and PUT_2. (A detailed study of PUT_1 will appear in the very near future [6].) The correct $\sin^2\theta_W(M_Z^2)$ was obtained for a PUT scale from 1-10 TeV. These models predict an absence of tree-level FCNC because of the presence of unconventionally charged quarks and leptons: The tree-level transitions only connect these fermions to the normal ones! Heavy (less than 1 TeV) and perhaps "long lived" quarkonic or leptonic unconventional fermions are characteristic signatures of this model. Is this a heavy price to pay for early unification or an actual bonus? Last but not least, since the unification scale is in the low TeV region, one might wonder if there is any link to the physics of large extra dimensions.

Acknowledgments

I wish to thank Paul Frampton and the organizers for a wonderful conference. Our thoughts go to Behram Kursunoglu who tirelessly organized the Coral Gables series until he ultimately left us in October 2003.

References

1. P. Q. Hung, A. J. Buras and J. D. Bjorken, Phys. Rev. D **25**, 805 (1982).
2. A. J. Buras and P. Q. Hung, Phys. Rev. D **68**, 035015 (2003).
3. Z. Chacko, L. J. Hall, and M. Perelstein, JHEP **0301**, 001 (2003).
4. S. Dimopoulos and D. E. Kaplan, Phys. Lett. B **531**, 127 (2002).
5. J. C. Pati and A. Salam, Phys. Rev. D **10**, 275 (1974).
6. A. J. Buras, P. Q. Hung, N. K. Tran, A. Poschenrieder, and E. Wyszomirski, "Early $SU(4)_{\rm PS} \otimes SU(2)_L \otimes SU(2)_R \otimes SU(2)_H$ Unification of Quarks and Leptons", hep-ph/0406048.
7. P. H. Frampton, P. Q. Hung, and M. Sher, Phys. Rep. **330**, 263 (2000).

ON EXTENDED ELECTROWEAK SYMMETRIES

OTTO C. W. KONG*
Department of Physics, National Central University, Chung-li, TAIWAN 32054
E-mail: otto@phy.ncu.edu.tw

We discuss extensions of the Standard Model through extending the electroweak gauge symmetry. An extended electroweak symmetry requires a list of extra fermionic and scalar states. The former is necessary to maintain cancellation of gauge anomalies, and largely fixed by the symmetry embedding itself. The latter is usually considered quite arbitrary, so long as a vacuum structure admitting the symmetry breaking is allowed. Anomaly cancellation may be used to link the three families of quarks and leptons together, given a perspective on flavor physics. It is illustrated lately that the kind of models may also have the so-called little Higgs mechanism incorporated. This more or less fixes the scalar sector and take care of the hierarchy problem, making such models of extended electroweak symmetries quite appealing candidates as TeV scale effective field theories.

Dedicated to Paul Frampton

1. Introduction

This talk is my contribution to the event celebrating the 60th birthday of Paul Frampton. The subject here is extending electroweak symmetries, in particular, as an approach to particle physics beyond the Standard Model (SM). The focus is on my own works on the subject, which began during the time I was a student studying under Paul's supervision. I am getting back to the topic lately, with some studies more related to some of Paul's own works but in the name of little Higgs.

The SM is a model of interactions dictated by an $SU(3)_C \times SU(2)_L \times U(1)_Y$ gauge symmetry, with a anomaly free chiral fermion spectrum and a Higgs multiplet responsible for the spontaneous breaking of the electroweak (EW) symmetry $SU(2)_L \times U(1)_Y$. Extending the EW gauge symmetry

*Work partially supported by grant NSC 92-2112-M-008-044 of the National Science Council of Taiwan.

extends the SM while adding new fermions and scalars. This is to be contrasted with other approaches such as grand unification and/or supersymmetry. Compares with the latter approaches, extending EW symmetry may look like less popular or not so well motivated. Grand unification aims at providing a unified picture to all the otherwise separate parts of the gauge symmetry and their independent couplings, though only at a scale of about 10^{-16} GeV. All that can be achieved, in the case of $SU(5)$, without the need for extra fermionic states. Supersymmetry used the beautiful boson-fermion symmetry to tackle the hierarchy problem, essentially extending the chiral nature of the fermions to fix the problem for the scalar sector. Putting the two together provides a theoretical structure that promises to "explain" more or less all of particle physics. However, the large extrapolation over the many order of magnitudes of particle physics desert may certainly be taken with suspicion. Moreover, the approaches do not provide any new insight into the difficult problem of the origin of flavor structure. Why there are three families of SM fermions is still a fundamental problem that we have no credible approach to handle. On the contrary, extending the EW symmetry may provide some new perspectives to the flavor problem. It can even provide an alternative solution to the hierarchy problem, in the name of the so-called little Higgs mechanism[1].

2. Looking at the Fermionic Spectra

The spectrum of SM fermion in one family is like perfection, essentially dictated by gauge anomaly cancellation conditions. To illustrate the point of view, we recall our earlier argument[2]. Assuming that there exist a minimal multiplet carrying nontrivial quantum numbers of each of the component gauge groups, one can obtain the one-family SM spectrum as the unique solution by asking for the minimal consistent set of chiral states. Consistence here refers to the perfect cancellation of nonvanishing contributions to various gauge anomalies from individual fermionic states. A vectorlike set (or pair) is trivial but not interesting. Only chiral states are protected from heavy gauge invariant masses and relevant to physics at the relatively low energy scale.

The above suggested derivation of the one-family SM spectrum goes as follow. We are essentially starting with a quark doublet, with arbitrary hypercharge normalization. The two $SU(3)_C$ triplets require two antitriplets to cancel the anomaly. Insisting on the chiral spectrum means taking two quark singlets here, with hypercharges still to be specified. Now, $SU(2)_L$

is real, but has a global anomaly. Cancellation requires an even number of doublets, so at least one more beyond the three colored components in the quark doublet. There are still four anomaly cancellation conditions to take care of. They are the $[SU(3)_C]^2U(1)_Y$, $[SU(2)_L]^2U(1)_Y$, $[grav]^2U(1)_Y$, and $[U(1)_Y]^3$. We are however left with three relative hypercharges to fit the four equations, actually without a possible solution. A rescue comes from simply adding a $U(1)_Y$-charged singlet. But the four equation for four unknown setting is misleading. The $[U(1)_Y]^3$ anomaly cancellation equation is cubic in all the charges, with no rational solution guaranteed. The SM solution may actually be considered a beautiful surprise. Moreover, the perspective may be the best we have on understanding *why there is what there is.*

We would also like to take the opportunity here to briefly sketch the next step taken in Ref.[2], to further illustrate our perspective. The results there also may be considered a worthy comparison with our little Higgs motivated flavor/family spectrum presented below, from the point of view of the origin of the three families. The major goal of Ref.[2] is to use a similar structure with an extended symmetry to obtain the three families. For example, one can start with some $SU(4) \times SU(3) \times SU(2) \times U(1)$ gauge symmetry and try to obtain the minimal chiral spectrum contain a $(\mathbf{4}, \mathbf{3}, \mathbf{2})$ multiplet — the simplest one with nontrivial quantum number under all component groups. Having a consistent solution is not enough though. In order for the spectrum be of interest, we ask the spectrum

multiplets	X	Gauge anomalies					$U(1)_Y$ states	
		trX	44X	33X	22X	X^3		
$(\mathbf{4},\mathbf{3},\mathbf{2})$	**1**	24	6	8	12	24	3 $\mathbf{1}(Q)$	$\mathbf{-5}(Q^{'})$
$(\bar{\mathbf{4}},\bar{\mathbf{3}},\mathbf{1})$	**5**	60	15	20		1500	3 $\mathbf{-4}(\bar{u})$	$\mathbf{2}(\bar{d})$
$(\bar{\mathbf{4}},\mathbf{1},\mathbf{2})$	**3**	24	6		12	216	3 $\mathbf{-3}(L)$	$\mathbf{3}(\bar{L})$
$(\bar{\mathbf{4}},\mathbf{1},\mathbf{1})$	**9**	36	9			2916	3 $\mathbf{-6}(\bar{E})$	$\mathbf{0}(N)$
$(\mathbf{6},\mathbf{1},\mathbf{1})$	**-18**	-108	-36			-34992	3 $\mathbf{6}(E)$	3 $\mathbf{12}(S)$
$(\mathbf{1},\bar{\mathbf{3}},\mathbf{2})$	**-10**	-60		-20	-30	-6000	$\mathbf{5}(\bar{Q}^{'})$	
$(\mathbf{1},\bar{\mathbf{3}},\mathbf{1})$	**-4**	-12		-4		-192	$\mathbf{2}(\bar{d})$	
$(\mathbf{1},\bar{\mathbf{3}},\mathbf{1})$	**-4**	-12		-4		-192	$\mathbf{2}(\bar{d})$	
$(\mathbf{1},\mathbf{1},\mathbf{2})$	**6**	12			6	432	$\mathbf{-3}(L)$	
3 $(\mathbf{1},\mathbf{1},\mathbf{1})$	**24**	72				41472	3 $\mathbf{-12}(\bar{S})$	
3 $(\mathbf{1},\mathbf{1},\mathbf{1})$	**-12**	-36				-5184	3 $\mathbf{6}(E)$	
	Total	0	0	0	0	0		

Table I. An $SU(4)_A \times SU(3)_C \times SU(2)_L \times U(1)_X$ spectrum embedding three SM families.

to yield the chiral spectrum of three SM families plus a set of vectorlike states under a feasible spontaneous symmetry breaking scenario, *i.e.* when the gauge symmetry is broken to that of the SM. Ref.[2] has only partial success. A consistent group theoretical SM embedding could be obtained but only with a slight addition to the minimal chiral spectrum obtained from anomaly cancellation considerations alone. We give an example in Table I.

Next, we recall the fermionic spectrum from a simple model of extended EW symmetry, the 331 model from Paul himself[3]. The model has the EW symmetry extended to an $SU(3)_L \times U(1)_X$. To have a consistent spectrum of chiral fermions, one may first look into how the SM doublets are to be embedded into multiplets of $SU(3)_L$. It is interesting to note here that a naive family universal embedding would not work. The $SU(3)_L$ anomaly would not cancel. Instead, the model has the (t, b) doublet embedded into a $\bar{\mathbf{3}}$ while the quark doublets of the first two families into $\mathbf{3}$'s, with all leptonic doublets embedded into $\bar{\mathbf{3}}$'s. The fact that the number of color equals the number of families makes the anomaly cancellation possible. All extra quark states here are exotic, with charges $\frac{5}{3}$ and $\frac{-4}{3}$. There are no extra leptonic states though. The 331 model spectrum is given in Table II.

	$U(1)_Y$-states	
$(\mathbf{3}_\mathrm{C}, \bar{\mathbf{3}}_\mathrm{L}, \mathbf{\frac{2}{3}})$	$\frac{1}{6}[Q]$	$\frac{5}{3}(T)$
$2\ (\mathbf{3}_\mathrm{C}, \mathbf{3}_\mathrm{L}, \mathbf{\frac{-1}{3}})$	$2\ \frac{1}{6}[2\ Q]$	$2\ \frac{-4}{3}(D, S)$
$3\ (\mathbf{1}_\mathrm{C}, \bar{\mathbf{3}}_\mathrm{L}, \mathbf{0})$	$3\ \frac{-1}{2}[3\ L]$	$3\ \mathbf{0}(3\ E^+)$
$3\ (\bar{\mathbf{3}}_\mathrm{C}, \mathbf{1}_\mathrm{L}, \mathbf{\frac{-2}{3}})$	$4\ \frac{-2}{3}\ (\bar{u}, \bar{c}, \bar{t})$	
$3\ (\bar{\mathbf{3}}_\mathrm{C}, \mathbf{1}_\mathrm{L}, \mathbf{\frac{1}{3}})$	$5\ \frac{1}{3}\ (\bar{d}, \bar{s}, \bar{b})$	
$1\ (\bar{\mathbf{3}}_\mathrm{C}, \mathbf{1}_\mathrm{L}, \mathbf{\frac{-5}{3}})$	$\frac{-5}{3}\ (\bar{T})$	
$1\ (\bar{\mathbf{3}}_\mathrm{C}, \mathbf{1}_\mathrm{L}, \mathbf{\frac{4}{3}})$	$2\ \frac{4}{3}\ (\bar{D}, \bar{S})$	

Table II. A (331) model spectrum for $SU(3)_L \times U(1)_X$ extended EW symmetry.

3. Extended EW Symmetries of $SU(N)_L \times U(1)_X$

Looking at the model spectrum of Table II, one may wonder if the construction is in any sense unique, and if similar anomaly free spectra exist for a different extended EW symmetry. We look into the question lately and have the general solution. It turns out quite simple and straightforward.

For an extended EW symmetry of $SU(N)_L \times U(1)_X$, the SM doublets may be embedded into $\boldsymbol{N}$'s or $\bar{\boldsymbol{N}}$'s. Embedding one quark doublet into an $\boldsymbol{N}$ and the two others into $\bar{\boldsymbol{N}}$'s while putting all lepton doublets into $\boldsymbol{N}$'s does give a prescription with canceled $SU(N)$ anomaly. The a bit of surprising part is that no matter how one chooses to embed $U(1)_Y$ into $SU(N)_L \times U(1)_X$, simply completing the list of chiral states with appropriate $SU(N)$ singlets to ensure vectorlike matchings at the QCD and QED level does yield a completely anomaly free spectrum, essential unique for the particular symmetry embedding. The number of possible consistent model spectra of the type is then equivalent to the number of admissible symmetry embeddings. The latter can conveniently be parametrized by the choice of electric charges for the extra $N-2$ quark states sharing the $\boldsymbol{N}$ multiplet with the (t,b) doublet[4]. We have no room in this write-up to elaborate on the details though.

4. Little Higgs and Extended Electroweak Symmetries

The little Higgs mechanism[1] has been proposed as new solution to the hierarchy problem. More precisely, it alleviates the quadratic divergent quantum correction to the SM Higgs states and admits a natural little hierarchy between the EW scale and a higher scale of so-called UV-completion at around 10 TeV above which further structure would be hidden. The idea is a rather humble bottom-up approach then; but experimental hints at the existence of such a little hierarchy has been discussed[5]. What is relevant for our present discussion is that a little Higgs model necessarily has an extended gauge symmetry, EW or beyond, and extra fermion(s). The latter includes a heavy top T quark.

Simple little Higgs model(s) based on an extended EW symmetry has been introduced by Kaplan and Schmaltz[6], though the authors failed to properly address the structure of the fermionic sector. The gauge symmetry considered are $SU(3)_L \times U(1)_X$ and $SU(4)_L \times U(1)_X$. We discuss completion of the kind of models with consistent, anomaly free, fermionic spectra and the resulted implications on the flavor structure of the models in Ref.[4,7,8]. Naively, so long as one pick a model spectrum with an extra T quark in the (t,b) containing $\boldsymbol{N}$ multiplet (here $N=3$ or $N=4$, for example), one have potentially a extended EW little Higgs model. The T quark may be used to cancel the quadratic divergent contributions (only at 1-loop level) to the SM Higgs mass from the t quark, while the extra EW gauge bosons to do the same for their SM counterparts. The scalar/Higgs sector has to be explicitly constructed though, to have the SM Higgs doublet embedded as (pseudo-)Nambu-Goldstone states of some global symmetry. It is an $[SU(3)]^2/[SU(2)]^2$ symmetry for the $SU(3)_L \times U(1)_X$ case, for instance. The Higgs sector symmetry is to be explicitly violated beyond the sector, in the gauge and Yukawa couplings of the Higgs multiplets. Such a scheme can be easily achieved with pair(s) of Higgs multiplets having the right quantum number to couple to the $(T, t, b, ..)$ multiplet and a right-handed T singlet. However, there is source of further complication, related to the construction of a proper Higgs quartic coupling term[6]. We admit that, in general, the latter issue still have to be studied more carefully. We do have a definitely complete and consistent model though. This is given by the fermion spectrum of Table III, with the Higgs sector as given in Ref.[6]. Here below, we will focus on the fermionic sector and flavor physics structure.

With a specific choice of the extended EW symmetry, a little Higgs model can be built only with the inclusion of the T quark state. For the $N = 3$ case, that fixes the hypercharge embedding and hence, from our anomaly cancellation study, the unique fermionic spectrum. The spectrum can be read off from Table III, with only one set of the duplicated T, D, S, and three N states. Note that the X-charges will have to change accordingly. For the $N = 4$ case, one may consider variations of the model spectrum, essentially by choosing a different set of states beyond that of the $N = 3$ content. In particular, a spectrum with a full set of duplicated, heavy, SM fermions look very interesting[4]. However, the scalar/Higgs sector has to be explicitly constructed then. Following exactly the construction of Ref.[6], one may be restricted to the spectrum of Table III, with trivial

	Gauge anomalies					$U(1)_Y$ states
	tr$X$	LLL	LLX	CCX	$_{(144)}X^3$	
$(\mathbf{3_C}, \mathbf{4_L}, \mathbf{\frac{5}{12}})$	5	3	5/4	5/3	125	$\mathbf{\frac{1}{6}}[Q]$ 2 $\mathbf{\frac{2}{3}}(T)$
2 $(\mathbf{3_C}, \mathbf{\bar{4}_L}, \mathbf{\frac{-1}{12}})$	−2	−6	−1/2	−2/3	−2	2 $\mathbf{\frac{1}{6}}[Q]$ 4 $\mathbf{\frac{-1}{3}}(2\ D, 2\ S)$
3 $(\mathbf{1_C}, \mathbf{4_L}, \mathbf{\frac{-1}{4}})$	−3	3	−3/4		−27	3 $\mathbf{\frac{-1}{2}}[L]$ 6 $\mathbf{0}(N)$
5 $(\mathbf{\bar{3}_C}, \mathbf{1_L}, \mathbf{\frac{-2}{3}})$	−10			−10/3	−640	5 $\mathbf{\frac{-2}{3}}$ $(\bar{u}, \bar{c}, \bar{t}, 2\ \bar{T})$
7 $(\mathbf{\bar{3}_C}, \mathbf{1_L}, \mathbf{\frac{1}{3}})$	7			7/3	112	7 $\mathbf{\frac{1}{3}}$ $(\bar{d}, \bar{s}, \bar{b}, 2\ \bar{D}, 2\ \bar{S})$
3 $(\mathbf{1_C}, \mathbf{1_L}, \mathbf{1})$	3				432	3 $\mathbf{1}$ (e^+, μ^+, τ^+)
Total	0	0	0	0	0	

Table III. The $SU(3)_C \times SU(4)_L \times U(1)_X$ chiral fermionic spectrum completing the Kaplan-Schmaltz little Higgs model, with anomaly cancellation illustrated.

generalization to $N > 4$ spectra extending on the content. At the moment, one sees no motivation to go for $N > 4$.

5. Some Implications to Flavor Physics

Unlike generic models of extended EW symmetries, we do not have much freedom in picking a set of scalar multiplets with VEVs according to what mass generating Yukawa couplings we may want to include. However, a careful checking of the Higgs multiplets shows that phenomenologically acceptable mass terms for the fermions, SM ones or heavy quarks, can be obtained for the explicit models discussed above. Here, we use the $SU(3)_L \times U(1)_X$ case for the demonstration, in favor or simpler notation and expressions.

As touched on above, the little Higgs mechanism is to be implemented with two scalar multiplets having the right quantum number to couple to the chiral parts of the T quark. They are denoted by Φ_1 and Φ_2 below in the expression of which we give the Yukawa part of the Lagrangian. The latter is constructed simply by tracing the quantum numbers and admitting all terms compatible with the gauge symmetries.

$$\begin{aligned}\mathcal{L}_{\text{Yukawa}} &= \lambda_1^t\, \bar{t}'\, \Phi_1\, Q + \lambda_2^t\, \bar{T}'\, \Phi_2\, Q + \frac{1}{M} \lambda_{\alpha j}^u\, \bar{u'}_\alpha\, \Phi_1\, \Phi_2\, Q'_j \\ &+ \lambda_{\beta j}^{d1}\, \bar{d'}_\beta\, \Phi_1^\dagger\, Q'_j + \lambda_{\beta j}^{d2}\, \bar{d'}_\beta\, \Phi_2^\dagger\, Q'_j + \frac{1}{M} \lambda_\beta^b\, \bar{d'}_\beta\, \Phi_1^\dagger\, \Phi_2^\dagger\, Q\,, \end{aligned} \quad (1)$$

where Q and Q'_j denote (contrary to notation in Table III) the color triplet and antitriplets. Note that we have to include dimension five terms here. Recall that the little Higgs model actually has a high energy cut-off of only around a 10 TeV scale. The next step is to use the nonlinear sigma model expansion of the scalar multiplets in terms of the pseudo-Nambu-Goldstone states, which include the SM Higgs doublet h[6,7]. We recover

$$\begin{aligned}\mathcal{L}_{\text{Yukawa}} &= f\,(\lambda_1^t\, \bar{t}' + \lambda_2^t\, \bar{T}')\, T + f\,(\lambda_{\beta j}^{d1}\, \bar{d'}_\beta + \lambda_{\beta j}^{d2}\, \bar{d'}_\beta)\, D_j \\ &+ \frac{i}{\sqrt{2}}\,(\lambda_1^t\, \bar{t}' - \lambda_2^t\, \bar{T}')\, h \begin{pmatrix} t \\ b \end{pmatrix} - \frac{i\sqrt{2}\, f}{M} \lambda_{\alpha j}^u\, \bar{u'}_\alpha\, h \begin{pmatrix} u_j \\ d_j \end{pmatrix} \\ &- \frac{i}{\sqrt{2}}\,(\lambda_{\beta j}^{d1}\, \bar{d'}_\beta - \lambda_{\beta j}^{d2}\, \bar{d'}_\beta)\, h^\dagger \begin{pmatrix} u_j \\ d_j \end{pmatrix} + \frac{i\sqrt{2}\, f}{M} \lambda_\beta^b\, \bar{d'}_\beta\, h^\dagger \begin{pmatrix} t \\ b \end{pmatrix} + \cdots \end{aligned} \quad (2)$$

The expression shows that all the heavy quark state, T, and D_j (or D and S) get Dirac mass at scale f of the VEVs of Φ_1 and Φ_2, and standard Yukawa couplings for the SM quarks and Higgs doublet are all available. However, the expression also indicates that one has to expect mass mixings among heavy and SM quark states. The nature of the extra heavy quarks and their mass mixings with the SM counterparts dictate stringent constraints on the related couplings and interesting flavor physics.

6. Conclusions

The bottom line here is that sensible discussion of flavor physics of a little Higgs model is not possible before the full fermion spectrum is spelt out. The latter is constrained by gauge anomaly cancellation. We exhibit at least one complete model here on which detailed flavor physics still have to be studied. For the kind of models, the fermionic part has a family non-universal flavor structure just like that of the 331 model, linking the three SM families into one fully connected set. Gauge anomaly cancellation should play a major role on constructing the fermionic completion of any little Higgs model. This is, unfortunately, an issue that has been largely overlooked in the literature.

In summary, we see that studies of extended EW symmetries has arrived at the point of furnishing all round models of beyond SM physics addressing more or less all the concerns of particle physics, including the hierarchy problem. Such a model then has almost no arbitrary parts to be chosen at model-builders' discretion. It has generic appeals, but are also very humble, liable to various stringent precision EW and flavor physics constraints and begs UV-completion about an order of magnitude in energy scale above that of the electroweak theory. Building models of the kind, and studying their phenomenology in details, as well as checking the predictions experimentally should be a worthy endeavor.

References

1. N. Arkani-Hamed *et.al.*, Phys. Rev. Lett. **86**, 4757 (2001).
2. O.C.W. Kong, Mod. Phys. Lett. **A11**, 2547 (1996); Phys. Rev. **D55**, 383 (1997).
3. P.H. Frampton, Phys. Rev. Lett. **69**, 2889 (1992).
4. O.C.W. Kong, NCU-HEP-k010, hep-ph/0308148.
5. R. Barbieri, hep-ph/0312253.
6. D.E. Kaplan and M. Schmaltz, JHEP **0310**, 039 (2003).
7. O.C.W. Kong, NCU-HEP-k009, hep-ph/0307250; see also NCU-HEP-k015.
8. O.C.W. Kong, NCU-HEP-k012, hep-ph/0312060, talk given at ICFP II.

SPACETIME FOAM, HOLOGRAPHIC PRINCIPLE, AND BLACK HOLE QUANTUM COMPUTERS

Y. JACK NG AND H. VAN DAM

Institute of Field Physics, Department of Physics and Astronomy, University of North Carolina, Chapel Hill, NC 27599-3255, USA
E-mail: yjng@physics.unc.edu

Spacetime foam, also known as quantum foam, has its origin in quantum fluctuations of spacetime. Arguably it is the source of the holographic principle, which severely limits how densely information can be packed in space. Its physics is also intimately linked to that of black holes and computation. In particular, the same underlying physics is shown to govern the computational power of black hole quantum computers.

1. Introduction

Early last century, Einstein's general relativity promoted spacetime from a passive and static arena to an active and dynamical entity. Nowadays many physicists also believe that spacetime, like all matter and energy, undergoes quantum fluctuations. These quantum fluctuations make spacetime foamy on small spacetime scales. (For a discussion of the relevant phenomenology and for a more complete list of references, see Ref. 1.)

But how large are the fluctuations? How foamy is spacetime? Is there any theoretical evidence of quantum foam? In what follows, we address these questions. By analysing a gedanken experiment for spacetime measurement, we show, in section 2, that spacetime fluctuations scale as the cube root of distances or time durations. Then we argue that this cube root dependence is consistent with the holographic principle. In section 3, we discuss how quantum foam affects the physics of clocks (accuracy and lifetime) and computers (computational rate and memory space). We also show that the physics of spacetime foam is intimately connected to that of black holes, giving a poor man's derivation of the Hawking black hole lifetime and the area law of black hole entropy. Lastly a black hole computer is shown to compute at a rate linearly proportional to its mass.

2. Quantum Fluctuations of Spacetime

If spacetime indeed undergoes quantum fluctuations, the fluctuations will show up when we measure a distance (or a time duration), in the form of uncertainties in the measurement. Conversely, if in any distance (or time duration) measurement, we cannot measure the distance (or time duration) precisely, we interpret this intrinsic limitation to spacetime measurements as resulting from fluctuations of spacetime.

The question is: does spacetime undergo quantum fluctuations? And if so, how large are the fluctuations? To quantify the problem, let us consider measuring a distance l. The question now is: how accurately can we measure this distance? Let us denote by δl the accuracy with which we can measure l. We will also refer to δl as the uncertainty or fluctuation of the distance l for reasons that will become obvious shortly. We will show that δl has a lower bound and will use two ways to calculate it. Neither method is rigorous, but the fact that the two very different methods yield the same result bodes well for the robustness of the conclusion. (Furthermore, the result is also consistent with well-known semi-classical black hole physics. See section 3.)

Gedanken Experiment. In the first method, we conduct a thought experiment to measure l. The importance of carrying out spacetime measurements to find the quantum fluctuations in the fabric of spacetime cannot be over-emphasized. According to general relativity, coordinates do not have any intrinsic meaning independent of observations; a coordinate system is defined only by explicitly carrying out spacetime distance measurements. Let us measure the distance between two points. Following Wigner[2], we put a clock at one point and a mirror at the other. Then the distance l that we want to measure is given by the distance between the clock and the mirror. By sending a light signal from the clock to the mirror in a timing experiment, we can determine the distance l. However, quantum uncertainties in the positions of the clock and the mirror introduce an inaccuracy δl in the distance measurement. We expect the clock and the mirror to contribute comparable uncertainties to the measurement. Let us concentrate on the clock and denote its mass by m. Wigner argued that if it has a linear spread δl when the light signal leaves the clock, then its position spread grows to $\delta l + \hbar l (mc\delta l)^{-1}$ when the light signal returns to the clock, with the minimum at $\delta l = (\hbar l/mc)^{1/2}$. Hence one concludes that

$$\delta l^2 \gtrsim \frac{\hbar l}{mc}. \tag{1}$$

General relativity provides a complementary bound. To see this, let the clock be a light-clock consisting of a spherical cavity of diameter d, surrounded by a mirror wall of mass m, between which bounces a beam of light. For the uncertainty in distance measurement not to be greater than δl, the clock must tick off time fast enough that $d/c \lesssim \delta l/c$. But d, the size of the clock, must be larger than the Schwarzschild radius $r_S \equiv 2Gm/c^2$ of the mirror, for otherwise one cannot read the time registered on the clock. From these two requirements, it follows that

$$\delta l \gtrsim \frac{Gm}{c^2}. \tag{2}$$

The product of Eq. (2) with Eq. (1) yields

$$\delta l \gtrsim (l l_P^2)^{1/3} = l_P \left(\frac{l}{l_P} \right)^{1/3}, \tag{3}$$

where $l_P = (\hbar G/c^3)^{1/2}$ is the Planck length. (Note that the result is independent of the mass of the clock and, hence, one would hope, of the properties of the specific clock used in the measurement.) The end result is as simple as it is strange and appears to be universal: the uncertainty δl in the measurement of the distance l cannot be smaller than the cube root of $l l_P^2$.[3] Obviously the accuracy of the distance measurement is intrinsically limited by this amount of uncertainty or quantum fluctuation. We conclude that there is a limit to the accuracy with which one can measure a distance; in other words, we can never know the distance l to a better accuracy than the cube root of $l l_P^2$. Similarly one can show that we can never know a time duration τ to a better accuracy than the cube root of τt_P^2, where $t_P \equiv l_P/c$ is the Planck time. Because the Planck length is so inconceivably short, the uncertainty or intrinsic limitation to the accuracy in the measurement of any distance, though much larger than the Planck length, is still very small. For example, in the measurement of a distance of one kilometer, the uncertainty in the distance is to an atom as an atom is to a human being.

The Holographic Principle. Alternatively we can estimate δl by applying the holographic principle.[4,5] In essence, the holographic principle[6] says that although the world around us appears to have three spatial dimensions, its contents can actually be encoded on a two-dimensional surface, like a hologram. To be more precise, let us consider a spatial region measuring l by l by l. According to the holographic principle, the number of degrees of freedom that this cubic region can contain is bounded by the surface area of the region in Planck units, i.e., l^2/l_P^2, instead of by the volume of the region as one may naively expect. This principle is counterintuitive, but is supported by black hole physics in conjunction with the laws of thermodynamics, and it is embraced by both string theory and loop quantum gravity. So strange as it may be, let us now apply the holographic principle to deduce the accuracy with which one can measure a distance.

First, imagine partitioning the big cube into small cubes. The small cubes so constructed should be as small as physical laws allow so that we can associate one degree of freedom with each small cube. In other words, the number of degrees of freedom that the region can hold is given by the number of small cubes that can be put inside that region. But how small can such cubes be? A moment's thought tells us that each side of a small cube cannot be smaller than the accuracy δl with which we can measure each side l of the big cube. This can be easily shown by applying the method of contradiction: assume that we can construct small cubes each of which has sides less than δl. Then by lining up a row of such small cubes along a side of the big cube from end to end, and by counting the number of such small cubes, we would be able to measure that side (of length l) of the big cube to a better accuracy than δl. But, by definition, δl is the best accuracy with which we can measure l. The ensuing contradiction is evaded by the realization that each of the smallest cubes (that can be put inside the big cube) measures δl by δl by δl. Thus, the number of degrees of freedom in the region (measuring l by l by l) is given by $l^3/\delta l^3$, which, according to the holographic principle, is no more than l^2/l_p^2. It follows that δl is bounded (from below) by the cube root of ll_P^2, the same result as found above in the gedanken experiment argument. Thus, to the extent that the holographic principle is correct, spacetime indeed fluctuates, forming foams of size δl on the scale of l. Actually, considering the fundamental nature of spacetime and the ubiquity of quantum fluctuations, we should reverse the argument and then we will come to the conclusion that the "strange" holographic principle has its origin in quantum fluctuations of spacetime.

3. From Spacetime Foam to Black Hole Computers

So far there is no experimental evidence of spacetime foam. In view of this lack of experimental evidence, we should at least look for theoretical corroborations (aside from the "derivation" of the holographic principle discussed above). Fortunately such corroborations do exist — in the sector of black hole physics. To show that, we have to make a small detour to consider clocks and computers[7,8] first.

Clocks. Consider a clock (technically, a simple and "elementary" clock, not composed of smaller clocks that can be used to read time separately or sequentially), capable of resolving time to an accuracy of t, for a period of T (the running time or lifetime of the clock). Then bounds on the resolution time and the lifetime of the clock can be derived by following an argument very similar to that used above in the analysis of the gedanken experiment to measure distances. The two arguments are very similar; one obtains[7]

$$t^2 \gtrsim \frac{\hbar T}{mc^2}, \qquad t \gtrsim \frac{Gm}{c^3}, \tag{4}$$

the analogs of Eq. (1) and Eq. (2) respectively. One can also combine these two equations to give[7]

$$T/t^3 \lesssim t_P^{-2} = \frac{c^5}{\hbar G}, \tag{5}$$

the analog of Eq. (3), which relates clock precision to its lifetime. (For example, for a femtosecond (10^{-15} sec) precision, the bound on the lifetime of a simple clock is 10^{34} years.)

Computers. We can easily translate the above relations for clocks into useful relations for a simple computer (technically, it refers to a computer designed to perform highly serial computations, i.e., one that is not divided into subsystems computing in parallel). Since the resolution time t for clocks is the smallest time interval relevant in the problem, the fastest possible processing frequency is given by its reciprocal, i.e., $1/t$. Thus if ν denotes the clock rate of the computer, i.e., the number of operations per bit per unit time, then it is natural to identify ν with $1/t$. To identify the number I of bits of information in the memory space of a simple computer, we recall that the running time T is the longest time interval relevant in the problem. Thus, the maximum number of steps of information processing is given by the running time divided by the resolution time, i.e., T/t. It follows that one can identify the number I of bits of the computer with T/t. (One can think of a tape of length cT as the memory space, partitioned into bits each of length ct.) In other words, the translations from the case of clocks to the case of computers consist of substituting the clock rate of computation for the reciprocal of the resolution time, and substituting the number of bits for the running time divided by the resolution time. The bounds on the precision and lifetime of a clock given by Eq. (4) and Eq. (5) are now translated into bounds on the rate of computation and number of bits in the computer, yielding respectively

$$I\nu \lesssim \frac{mc^2}{\hbar}, \qquad \nu \lesssim \frac{c^3}{Gm}, \qquad I\nu^2 \lesssim \frac{c^5}{\hbar G} \sim 10^{86}/sec^2. \tag{6}$$

The first inequality shows that the speed of computation is bounded by the energy of the computer divided by Planck's constant, in agreement with the result found by Margolus and Levitin[9], and by Lloyd[10] (for the ultimate limits to computation). The last bound is perhaps even more intriguing: it requires the product of the number of bits and the square of the computation rate for *any* simple computer to be less than the square of the reciprocal of Planck time,[7] which depends on relativistic quantum gravity (involving c, $\hbar$, and G). This universal relation links together our concepts of information/computation, relativity, gravity, and quantum uncertainty. Numerically, the computation bound is about seventy-six orders of magnitude above what is available for a current lap-top computer performing ten billion operations per second on ten billion bits, for which $I\nu^2 \sim 10^{10}/s^2$.

Black Holes. Now we can apply what we have learned about clocks and computers to black holes.[7,8] Let us consider using a black hole to measure time. It is reasonable to use the light travel time around the black hole's horizon as the resolution time of the clock, i.e., $t \sim \frac{Gm}{c^3} \equiv t_{BH}$, then from the first equation in Eq. (4), one immediately finds that

$$T \sim \frac{G^2 m^3}{\hbar c^4} \equiv T_{BH}. \tag{7}$$

Thus, if we had not known of the Hawking lifetime (T_{BH}) for black hole evaporation, this remarkable result would have implied that there is a maximum lifetime for a black hole!

Finally, let us consider using a black hole to do computations. This may sound like a ridiculous proposition. But if we believe that black holes evolve according to quantum mechanical laws, it is possible[10], at least in principle, to program black holes to perform computations that can be read out of the fluctuations in the Hawking black hole radiation. How large is the memory space of a black hole computer, and how fast can it compute? Applying the results for computation derived above, we readily find the number of bits in the memory space of a black hole computer, given by the lifetime of the black hole divided by its resolution time as a clock, to be

$$I = \frac{T_{BH}}{t_{BH}} \sim \frac{m^2}{m_P^2} \sim \frac{r_S^2}{l_P^2}, \tag{8}$$

where $m_P = \hbar/(t_P c^2)$ is the Planck mass, m and r_S^2 denote the mass and event horizon area of the black hole respectively. This gives the number of bits I as the event horizon area in Planck units, in agreement with the identification of a black hole entropy. Furthermore, the number of operations per unit time for a black hole computer is given by

$$I\nu = \frac{T_{BH}}{t_{BH}} \times \frac{1}{t_{BH}} \sim \frac{mc^2}{\hbar}, \tag{9}$$

its energy divided by Planck's constant, as first found by Lloyd[10]. Note that all the bounds on computation discussed above are saturated by black hole computers. Thus one can even say that once they are programmed to do computations, black holes are the ultimate simple computers.

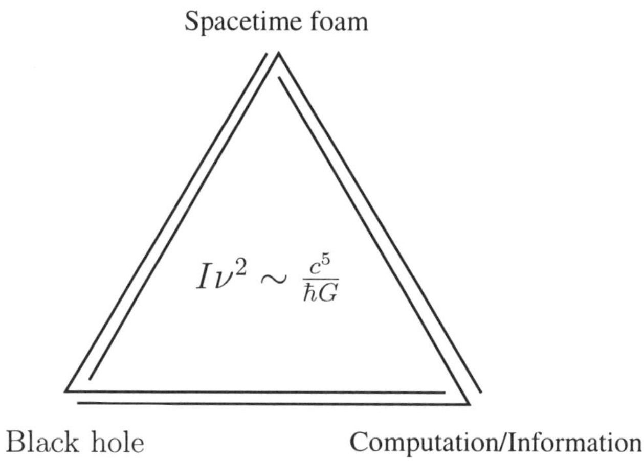

Figure 1. The *universal* relation for black hole computers, $I\nu^2 \sim c^5/\hbar G$, is a combined product of the physics behind spacetime foam, black holes, and computation/information.

All these results reinforce the conceptual interconnections of the physics underlying spacetime foam, black holes, and computation. It is intersting that these three subjects share such intimate bonds and are brought together here [see Fig. 1]. The internal consistency of the physics we have uncovered also vindicates the simple (some would say overly simple) arguments we present in section 2, in the derivation of the limits to spacetime measurements and the elucidation of the structure of spacetime foam.

4. Summary

We have analyzed a gedanken experiment for spacetime measurements to show that spacetime fluctuations scale as the cube root of distances or time durations. This cube root dependence is strange, but has been shown to be consistent with the holographic principle and with semi-classical black hole physics in general. (To us, this result for spacetime fluctuations is as beautiful as it is strange. Dare we agree with Francis Bacon in his observation: There is no excellent beauty that hath not some strangeness in the proportion.) We have also shown that the physics of spacetime foam is intimately connected to computation. The unity of physics, underlying spacetime foam, the holographic principle, black holes, and quantum computers, is hereby adequately (if not overwhelmingly) demonstrated.

Acknowledgments

One of us (YJN) thanks the organizers of the Coral Gables Conference for inviting him to present the materials contained in this paper. We dedicate this article to our colleague Paul Frampton on the occasion of his sixtieth birthday. This work was supported in part by the US Department of Energy and the Bahnson Fund of the University of North Carolina. We thank L. L. Ng and T. Takahashi for their help in the preparation of this manuscript.

References

1. Y. J. Ng, *Mod. Phys. Lett.* **A18**, 1073 (2003). See also Y.J. Ng, gr-qc/0401015.
2. E.P. Wigner, *Rev. Mod. Phys.* **29**, 255 (1957); H. Salecker and E.P. Wigner, *Phys. Rev.* **109**, 571 (1958).
3. Y.J. Ng and H. van Dam, *Mod. Phys. Lett.* **A9**, 335 (1994); **A10**, 2801 (1995); in *Proc. of Fundamental Problems in Quantum Theory*, eds. D.M. Greenberger and A. Zeilinger, Ann. New York Acad. Sci. **755**, 579 (1995). Also see F. Karolyhazy, *Nuovo Cimento* **A42**, 390 (1966); T. Padmanabhan, *Class. Quan. Grav.* **4**, L107 (1987); D.V. Ahluwalia, *Phys. Lett.* **B339**, 301 (1994); and N. Sasakura, *Prog. Theor. Phys.* **102**, 169 (1999).
4. Y. J. Ng and H. van Dam, *Found. Phys.* **30**, 795 (2000); *Phys. Lett.* **B477**, 429 (2000).
5. Y. J. Ng, *Int. J. Mod. Phys.* **D11**, 1585 (2002).
6. G. 't Hooft, in *Salamfestschrift*, edited by A. Ali et al. (World Scientific, Singapore, 1993), p. 284; L. Susskind, *J. Math. Phys.* (N.Y.) **36**, 6377 (1995). Also see J.A. Wheeler, *Int. J. Theor. Phys.* **21**, 557 (1982); J.D. Bekenstein, *Phys. Rev.* **D7**, 2333 (1973); S. Hawking, *Comm. Math. Phys.* **43**, 199 (1975).
7. Y. J. Ng, *Phys. Rev. Lett.* **86**, 2946 (2001), and (erratum) **88**, 139902-1 (2002); Y. J. Ng in *Proc. of OCPA 2000*, eds. N. P. Chang et al. (World Scientific, Singapore, 2002), p.235.
8. J.D. Barrow, *Phys. Rev.* **D54**, 6563 (1996).
9. N. Margolus and L. B. Levitin, *Physica* **D120**, 188 (1998).
10. S. Lloyd, *Nature* (London) **406**, 1047 (2000).

PROTON DECAY, NEUTRINO OSCILLATIONS, *ETC*... FROM IMB, THROUGH SUPER-K & ANTARES

LAWRENCE R. SULAK†

Department of Physics, Boston University, 590 Commonwealth Ave., Boston, MA 02215 USA, and

Centre de Physiques des Particules de Marseille, 163 Ave. de Luminy, Marseille, 13288 France

We review the early evolution of massive water Cherenkov tracking calorimeters, including the inception of the pioneering IMB and Dumand concepts and the two successor Kamiokande detectors. We also consider the ultimate facilities: a high-resolution megaton detector and the high-volume Antares detector plus its sequel, the Cubic Kilometer.

Paul Frampton played a seminal role in the scientific justification for the physics that drove IMB and its successors. We, the experimenters, pay tribute to Paul's precocious support of searches for proton decay and neutrino oscillations. Led along the garden path for the first, we found the second, as well as the first (supernova) astrophysical neutrinos. This pays homage to Paul's early days at Harvard and the decade that followed when he was a theory godfather to IMB and the detectors that followed. This includes Paul as the convener of the "Workshops on Grand Unification," where, for example, was published the first paper to advocate IMB as a neutrino oscillation search with the baseline of the earth. All of us associated with IMB thank Paul immensely.

1. The Ideas...First Recorded by "Workshops on Grand Unification"

This is a personal retrospective on what Paul and his colleagues have wrought on the experimental community, both by their theory and his "Workshops." The first publications of nascent proton decay and neutrino oscillation phenomenology and detector concepts appeared in his "journal of record." New ideas, especially for experimentalists, are not "results" and, therefore, rarely the subject for refereed journals. Hence, the conference series inaugurated in 1980 by Paul Frampton, Sheldon Glashow, and Asim Yildiz, as editors of the First Workshop on Grand Unification (FWOGU),[1] became seminal in the field.

---

† Work partially supported by grant 2-4570.5 of the Swiss National Science Foundation.

FWOGU was the first of a series of ten conferences and associated journals. After a decade, the WOGU series merged with the Coral Gables Conferences, at which we are pleased to commemorate this heritage today. My talk will draw from many presentations recorded in the ten WOGU volumes.

Amongst other topics, the WOGU series explored the theory behind the experiments shown in the Table, a summary of characteristics of the four ring-imaging Cherenkov Detectors that the author knows most about and discusses here from his personal observation of them and of Paul's early legacy.

Table. Characteristics of the Early Ring-imaging Cherenkov Detectors
World's Largest Tracking Calorimeters

Dumand	'76-'93	goal: 0.1 km^2 to 1 km^3	off Hawaii
IMB	'80-'90	10 kilotons = $(20m)^3$	salt mine, Ohio
Kamiokande	'83-'96	3 kilotons	heavy metal mine, Japan
Super-K	'96-	50 kilotons	a 2nd site at Kamioka

The scope of this review is limited to the above detectors. Other talks in this volume provide details for some of these and successor facilities. For SNO, see the presentation of Maniera; for Amanda and IceCube, see Fazely; and for Super-K, see Liu and Smy.

2. Dumand...Setting the Stage for IMB, its Successors, and Antares

Dumand was the first massive water Cherenkov detector to be conceived. Its main objective: to search for the 1-10 TeV neutrinos initially from active galactic nuclei, and later from the annihilation of supersymmetric particles from the center of our galaxy.

The author had the pleasure of convening a study group at the 1976 Dumand Summer Workshop[2] where the concepts were developed. The depth, 4 km, had to be sufficient to decrease the downward cosmic muons to a rate where they could be distinguished from upward muons induced by atmospheric neutrino interactions in the earth.

The enormous size, eventually 1 cubic kilometer, was needed to detect a minimum of a few events per year from a point source. The spacing of pms was set by the longest known light attenuation length of blue light in ocean water (60m), as well as by the length of muon induced hadronic showers. The pm size was dictated by the ability to record at least one pe at that distance, etc. An anchor at the bottom and a buoy at the top provided an ocean tolerant structure.

The conclusion of that report is true today: "The detector is entirely consistent with current technology and its cost is comparable to that of the larger accelerator experiments." This work set the stage for a massive "land-based" prototype water Cherenkov detector with Paul and his friends inventing the science motivation.

3. Proton Decay & Neutrino Oscillation Motivate IMB

In 1976 the author and collaborators were running the E1A experiment at Fermilab, a 100 Ton total-absorption liquid scintillator detector, the largest in the world at the time, which we had built to search for neutral currents. We were also running the first dedicated searches for neutrino oscillations at Brookhaven, using a 60 T liquid scintillator calorimeter we had initially constructed to discover neutrino proton elastic scattering.

The BNL neutrino oscillations experiment was frustrated: Although we had optimized the highest flux low energy neutrino beam (200 MeV) in the world, the anticipated neutrino sensitivity proved elusive. We needed an economical technology to achieve at least an order of magnitude increase in mass.

A year after, Frampton and his theoretical colleagues H. Georgi and Glashow, J. Pati and A. Salam, and S. Weinberg shocked us experimentalists with the possibility that proton decay was natural in grand unified theories, and should occur at a rate visible in a detector with a few kilotons of mass. This was intriguing physics, as it remains today, which could be achieved with a "small detector" relative to the scale of Dumand. Several of us had despaired at the enormous challenge of outfitting a cubic kilometer 4 km deep in open seawater for Dumand. But I, for one, was elated that one could conceivably search for proton decay and build a perfect staging experiment for Dumand at the same time.

During the year of 1978, my students and I at Harvard optimized an inexpensive pm and base to see single photoelectrons with ¼ pe threshold, built prototype electronics that **was** deadtimeless and had 1 ns accuracy for Cherenkov directionality and 0.1 μsec for identifying muons by their decay electrons, dunked pms in Percy Bridgman's pressure tanks to assure operation at several atmospheres, hung strings of phototubes from the ceiling of the high bay area of the former Cambridge Electron Accelerator.

All of this hardware delighted Frampton, as well as DeRujula, Glashow, Weinberg and others who saw it...that their theories might indeed be testable. By the end of 1978, all elements of a 10,000-kTon detector had been proven, and we presented our ideas at the Madison Meeting on Proton Stability.[3] The conclusion of that paper foresaw the "the proposed detector can be considered a prototype for an ultimate experiment," what is now called "the megaton detector." That paper was presented also at Neutrino '79[4] in Bergen.

The first publication of advocating a large water detector for "A Long-Baseline Neutrino Oscillation Experiment..." appeared in Paul's first edition of WOGU[5] in 1979. The abstract expands on the title with "...using as a baseline the diameter of the earth. A flux independent asymmetry in the up/down ration of the two neutrino species is the primary signal." This same paper contains a detailed description of "IMB," as it had been recently dubbed. To save page space, Paul forced me to concatenate my two talks into one!

That first WOGU was critical in obtaining funding for IMB. To help overcome the initial hesitation of DOE, Glashow, Salam, and Weinberg all wrote letters to urge their support.

4. The Conception of IMB, the First Massive Ring-Imaging Detector

IMB was the first detector, operational in 1981, with 2000 phototubes (pms), 5" in diameter, made by EMI, the only manufacturer of a hemispherical pm. Fortuitously that size was sufficient to intercept light from the far side of the tank and record it with one photoelectron (pe). The shape was essential to support the total pressure of 3 atm and to provide isochronous electron optics to achieve 1 ns timing. Before this, no experiment had attempted single photoelectron operation with a large number of tubes, which therefore had to be inexpensive.

A threshold of ¼ pe on each tube was achieved, and the trigger energy threshold was 50% efficient at 25 MeV, chosen to allow calibration by Michel electrons from the decay of stopped muons. Since the diagonal distance was 30 m, it required development of a reverse osmosis system to provide the world's purist water. The attenuation length of > 50m in the UV, 337 nm, was monitored by the first use of a fast N_2 laser, whose light was transmitted by a quartz fiber to an isotropic diffuser. Two optical filter wheels, with both linear and log attenuation, provided the energy calibration.

After a year of operation, the high purity water had unexpectedly leached sodium out of the pm glass, relieving surface tension and causing 5% of the pms to develop cracks. By this time, the engineers of Mr. Hiruma's, president of Hamamatsu, had used the IMB Poisson tracking code to design a bigger hemispherical pm, 8" in diameter, with which IMB could replace the original pms. To obtain even more light, light-collecting wavelength-shifting plates, hung around each pm, were developed by the author to enhance the light collection by an overall factor of four, the detector called IMB III in 1983.

5. IMB: Elegant SU(5) is Dead, a "Muon Anomaly," & a Super Nova

By 1983, with a year of live time, the Ph D theses of Bruce Cortez and Bill Foster set limits on proton decay into the SU(5) favored $e^+ \pi^0$ and supersymmetric μ π modes at the levels of 10^{32} and 10^{31} years respectively. Many of the most stringent limits on some 45 different possible decay modes of the nucleon are still held by the IMB detector; most of the rest have been set by Super-K. The message is that "mass matters most" in these searches, setting the stage for the megaton detector.

Within the first year of operation of IMB, it was clear that the number of cosmic ray muons was 75% lower than expected. After several years of study and the PhD thesis of Eric Shumard, IMB ruled out systematic effects except the cosmic ray flux calculations and published the result in PRL[6] in 1986 "that 34%+-1% of the events should have an identified muon decay while our data has 26%+_3%. This could be...some other as-yet-unaccounted-for physics." We were too cautious to claim a discovery.

This reticence was due in part to no muon anomaly at all in Kamioka, as recorded in the PhD thesis[7] of T. Kajita of 1986 and in several conference reports by the collaboration. J. LoSecco visited the Kamioka collaboration at the time, showed them all of are results, but could not determine why they did not see the effect at the time.

Then, in 1987, at 07:35:35 UT, three hours before an optical supernova was observed in Chile, bam...b-bam bam bam. Some 8 events in IMB and 11 in Kamioka, all within 13 sec. The world's first observation of a solar implosion in the "neutrino light" from an astrophysical explosion.

IMB searched for point neutrino sources, as anticipated by Dumand, by extrapolating back entering upward going muons. With a sample of 500 events, the neutrino sky was dark. The author's most recent graduate student, Shantanu Desai, has repeated this search with the 22 k-Ton fiducial region of Super-K. Again, the sky remains dark. We must await Antares and IceCube with their much greater surface area.

By 1990 the reservoir liner for the IMB detector had reached the lifetime limit. Further, the physics was limited by its massive, but nevertheless limiting, 3.3 kTon fiducial mass. A new detector was essential.

6. Kamiokande

In the meantime, the second detector, Kamiokande, came on line two years after IMB in 1983. It was built with an additional challenging goal, to search for neutrinos in real time coming from the direction of the sun. Even though IMB tried to do this, its depth is too shallow leading to many stopping muons: the rate of beta decays from long-lived spallation products following the absorption of negative muons was too great.

To decrease the downward cosmic ray background to acceptable levels, the detector was built at 2.7 km water equivalent (the same depth as Antares is being constructed today). To minimize cosmic muon background, it was surrounded by an active veto. To lower the threshold to nuclear levels, 7 MeV, it was built with spectacular 20" pms, which by this time Mr. Hiruma had extrapolated from the 8" pms he had developed for IMB.

Relative to IMB, the detector had 3 times less mass, half the pms (half the pixels), no fast timing on the pms, and minimal water filtering. Hence pattern recognition was significantly compromised.

Bruce Cortez, whose PhD on proton decay from IMB eliminated SU(5), joined the Kamioka collaboration as a Milliken Fellow at Cal Tech, bringing analysis packages and his knowledge of electronics for fast timing. He soon recognized that radioactive background in the water prevented lowering the energy threshold sufficient to see solar neutrinos. The deficiencies of the detector were remedied in 1986 as Kamioka II, allowing sensitivity to solar neutrinos with its unprecedented 40% photocathode coverage.

By 1988, Kamioka II clearly saw a neutrino signal that pointed back to the sun, but at only half the rate expected from the standard solar model, leading to the Nobel Prize for the experiment and T. Koshiba in 2002. However, the statistics were limited by the small size of the Kamioka fiducial mass. Finally in 1988, Kamioka III also observed a muon anomaly at about the same significance as IMB. They were bolder, assumed that the flux Monte Carlo calculations were accurate, and announced the effect as a discovery of neutrino oscillation, even though the hint was only a 3 sigma effect.

By 1989 it was clear that an ambitious new detector with a mass an order of magnitude larger was needed. (Super-K eventually had a fiducial mass 7 times larger than that of IMB.)

7. Super-Kamiokande

With the Superconducting Super Collider still under funded, the American team of IMB was strongly encouraged to join forces with the Japanese on the next generation detector that had been so strongly pushed there by T. Koshiba. The hopes were that the Japanese would reciprocate by providing international support for the SSC. The IMB collaboration offered to dismantle the detector and to redeploy it as the outer detector of Super-Kamiokande, a critical missing item in the Super-K proposal. With this impetus, Super-K, 40 m in height and diameter, turned on in 1996. Inside are an astounding 11,000 20" pms, sufficient to decrease the threshold for solar neutrino to lower than 5 MeV. The outer detector consists of the system of 2,000 8" pms and wavelength shifting light collectors recycled from IMB.

Within a short time, using the technique advocated at the first FWOGU, Super-K was observing the anticipated number of atmospheric electron neutrinos raining down as coming up, and the expected number of muon neutrinos (twice more) coming down. But 30% of those expected to be coming up in the high energy sample were missing; a similar loss of muon neutrinos occurred in six independent classes of data. The missing ratio was entirely consisted with the values reported earlier by IMB and Kamioka III. Eventually the "muon anomaly" could only be interpreted as muon neutrino oscillations.

At the same time, Super-K was observing 15 solar neutrino events a day above a threshold of 5 MeV. With 22,400 events recorded, only 47% of the number expected with the oscillation of solar neutrino, the oscillation of electron neutrinos was now unambiguous.

Recently, a beam of muon neutrinos from the KEK accelerator 300 km away has been aimed at Super-K with the same characteristic L/E (flight distance over energy) as the upward-coming atmospheric muon neutrinos. The oscillation phenomenon has been confirmed with man-made neutrinos.

8. Dumand...Almost a Reality

The lowest optical module was 100 m off the bottom to avoid the "dust" near there. An optical fiber carried processed signals from the bottom of each string to a junction box, which sent them to shore. The collaborators in IMB from Boston had some fun here too. Eric Hazen's upgraded IMB electronics evolved into the Benthos sphere of each pm.

Slowly, over a period of 15 years, Dumand proved each of these milestones, many of them helped by the parallel developments for IMB, the miniature "prototype" on the surface.

In 1987, Dumand deployed a short prototype string consisting of 7 15" pms in 17" glass pressure-tolerant (Benthos) spheres. This string was operated for 1 week down to a depth of 4.5 km. Having proven its technology with this test, Dumand was approved for first substantial funding in 1989 for a 0.1 km^2 array. Unfortunately, it was prematurely terminated in 1993 (along with the Superconducting Super Collider), in part due to budget constraints in the US

9. Antares

With the cancellation of US support for Dumand, the CEA lab at Saclay France seized the opportunity to follow-up with a very similar string and electronics structure. I had the pleasure of helping M. Spiro initiate this with L. Moscoso while on sabbatical at Saclay. The Nestor Group in Greece also attempted to continue the Dumand tradition. Since then, the CPPM lab at Marseille and other French and European countries expanded the original CEA effort into the Antares collaboration.

Antares is dedicated to the same goals and initial size (0.1 km^3) as Dumand. Many ocean campaigns have been carried out at the site over the last several years. The optical fiber and junction box are laid. Two strings, a preprototype sting and an instrumentation line, have been successfully operated and retrieved. From that experience, upgrades have been incorporated. Currently under construction are spherical optical modules housing 900 pms and electronics for 12 strings. Each string is 250 meters long with 25 stories and 3 pms at each story. Installation of Antares at 2.4 km depth in the Mediterranean 40km from the French naval base at Toulon is scheduled for 2005 to 07. It will turn on upon the placement of the first string

Nature may include supersymmetric particles. As Shelly Glashow wrote in one of the WOGUs, "Supersymmetry...has generated so many thousands of papers it must be correct." If so, Antares may well see the products of their annihilation in the sun or the galactic center before LHC. **Km^{3}NET...the Cubic Kilometer Goal**

In the meantime, the Km^{3}NET consortium of the three groups working on neutrino astrophysics in the Mediterranean, Antares, Nestor, and Nemo (an Italian group) has proposed an engineering design study to the European Community. It would extrapolate the Antares concept to the ultimate cubic kilometer detector. The study should be complete by 2007, with initial operation as early as 2010.

This Km^3 detector would complement the massive ice detector ICECUBE to be constructed at the South Pole. Km^3 will have far better pointing resolution (~0.2 degrees) since the light scattering length is ~100 m, not 1 m. Unlike a detector at the south pole, Km^3 can record neutrinos entering from below (the direction with no background) from the center of the galaxy.

11. Conclusions

Hence a quarter century of unification physics was stimulated by Paul *et al.*, espousing the search for proton decay and neutrino oscillations.

A new technology was driven by this science: pixelated, ring-imaging Cherenkov calorimetry. This required submersible, depth-tolerant pms and electronics, inexpensive single photoelectron operation, pattern recognition, fast timing for directionality, and the development of vast amounts of ultra pure water.

Unanticipated astro-neutrino physics discoveries resulted. 1) The first physics beyond standard model, the discovery of neutrino mass and the oscillation between flavors, which became the most highly cited experimental physics paper of all time. And 2) the first observation of extra-terrestrial neutrinos and the imaging the sun with them, and the first extra-galactic neutrinos, from SN 1987a.

rand unification remains the ultimate goal. We need Paul's drive and enthusiasm again, to argue for a MEGATON proton decay detector that might, if coupled to an accelerator beam, simultaneously search for CP violation in the lepton sector, and to advocate a cubic kilometer detector to search for supersymmetric particle annihilations from the sun and the galactic center and for high energy point sources. We all await the next discovery!

Acknowledgments

On behalf of his experimental friends from IMB, Super K and Antares, we wish Paul Frampton well on his continuing path of discovery...pursuing the physics of grand unification.We thank him for his pioneering theoretical work and for his enthusiastic support of our experiments in the search for proton decay and neutrino oscillations, from the very beginning of the challenge.

Appreciation is due to my hosts Elie Aslanides and Roy Aleksan at CPPM, Marseille for the sabbatical year 2003-05 during which I have had the opportunity to work both on Antares with John Carr and on Km^{3}NET. This sojourn gave me the time to remember the early days with Paul.

I thank Masayuki Nakahata for providing historical information on the Kamiokande experiment

It is essential to acknowledge the extraordinary cleverness of the young scientists of the IMB team, all of whom "cut their wings" in this effort. The realization of its inexpensive hardware was achieved by several gifted graduate students, postdoctoral fellows, and young engineers, essentially all of whom either reported at a WOGU, or know Paul from attending them.

Bill Foster and Eric Hazen invented the sensitive, low noise, deadtimeless, large dynamic range electronics. Harold Lessure and Richard Bionta developed the laser and LED ball calibration systems. John Learned wrote the pm photoelectron tracking code, still in use by Hamamatsu. Jim Stone designed the water vessel.

Eric Shumard converted the CPU of an LSI 11 into a real-time joystick driven display engine to represent for the first time events color-coded in time and in photoelectrons from the vertex. Bruce Cortez wrote the simulation and reconstruction code. John LoSecco, who was a key person on the BNL neutrino oscillation search in 1976, made the calculations to assure that we could observe the characteristic up/down muon asymmetry (first pointed out by the author) of neutrino oscillations. Mike Levi helped with this work. The perfection of the software for IMB III was the marvel produced by David Casper. Much of it continues in Super-K today.

I would like to thank Bernie Hildebrand, Bill Wallenmeyer, and P. K. Williams for their steadfast support of IMB and for establishing particle astrophysics as a showpiece of US DOE funding.

References

1. Paul H. Frampton, Sheldon L. Glashow, and Asim Yildiz, eds., Proc. of First Workshop on Grand Unification, New England Center, University of New Hampshire, April 10-12, Math Sci Press (1980).
2. L. Sulak *et al.*, "Signatures of High Energy Neutrino Interactions and their Detection via Cherenkov Light, Proc. of 1976 Dumand Summer Workshop, A. Roberts, Ed., 297, Office of Publications, Fermi National Accelerator Laboratory (1976).
3. L. Sulak, "A Test of Baryon Stability Sensitive to a Lifetime of 1033 Years," Madison Meeting on Proton Stability, D. Cline, ed., December 8, 1978.
4. L. Sulak, "A Search for Proton Decay in the IMB Detector," Proc. of the 1979 Conference on Neutrino Physics, Bergen, Norway, C. Jarlskog, ed. (1979).
5. L. Sulak, "The I-M-B Nucleon Decay Facility: Status Report on a Proton Decay Experiment Sensitive to a Lifetime of 1033 Years and A Long-Baseline Neutrino Oscillation Experiment Sensitive to Mass Differences of Hundredths of an Electron Volt," in ref. 1 (1980). (At that time, one referred to mass differences, not mass differences squared, as today.)
6. T. J. Haines, *et al.* "Calculation of Atmospheric Neutrino-Induced Background in a Nucleon-Decay Search," Phys. Rev. Let. **57**, No. 16 (1986).
7. T. Kajita, Ph. D. Thesis, University of Tokyo (1986). The number of cosmic ray muons observed in Kamioka are shown to be in complete agreement with expectations.

RENORMALIZATION WITHOUT INFINITIES

GERARD 'T HOOFT

Institute for Theoretical Physics
Utrecht University, Leuvenlaan 4
3584 CC Utrecht, the Netherlands
and
Spinoza Institute
Postbox 80.195
3508 TD Utrecht, the Netherlands
e-mail: g.thooft@phys.uu.nl
internet: http://www.phys.uu.nl/~thooft

Abstract. Most renormalizable quantum field theories can be rephrased in terms of Feynman diagrams that only contain dressed irreducible 2-, 3-, and 4-point vertices. These irreducible vertices in turn can be solved from equations that also only contain dressed irreducible vertices. The diagrams and equations that one ends up with do not contain any ultraviolet divergences. The original bare Lagrangian of the theory only enters in terms of freely adjustable integration constants. It is explained how the procedure proposed here is related to the renormalization group equations. The procedure requires the identification of unambiguous "paths" in a Feynman diagrams, and it is shown how to define such paths in most of the quantum field theories that are in use today. We do not claim to have a more convenient calculational scheme here, but rather a scheme that allows for a better conceptual understanding of ultraviolet infinities.

Congratulations

This contribution is written at the occasion of Paul Frampton's 60$^{\underline{\text{th}}}$ birthday.

1. Introduction. Rearranging Feynman Diagrams

Usually, a quantized field theory is defined through its bare Lagrangian. From this Lagrangian, one derives Feynman diagrams to represent contributions to the amplitudes that one wishes to compute. Many of the resulting expressions are found to contain ultraviolet divergences, which are subsequently neutralized by adding new counter terms to the original Lagrangian[1]. In practice, this works so well that refinements and caveats are not thought worth-while for consideration, and indeed one can formulate precise justifications of this procedure. If the total number of physically distinguishable freely adjustable parameters stays finite and fixed in the course of the perturbative expansion, the theory is called 'renormalizable'.

In many cases, however, there exists an alternative way to address renormalizable theories, such that never in the intermediate results UV divergent expressions enter. This short communication aims at explaining it.

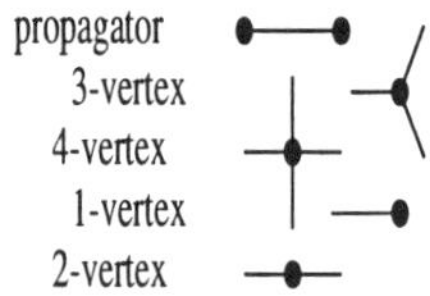

Figure 1. Bare propagators and vertices

We start with the original formulation of the Feynman rules. In general, there are three-point vertices and four-point vertices. Higher vertices will rend the theory unrenormalizable, with the exception of some theories in lower space-time dimensions which we shall not consider.

Vertices with one or two external lines may occur, but they can quickly be eliminated by shifting and renormalizing the field definitions, and hence we will ignore them, although with a little more effort one can accommodate for them in the formulation below (see Fig. 1).

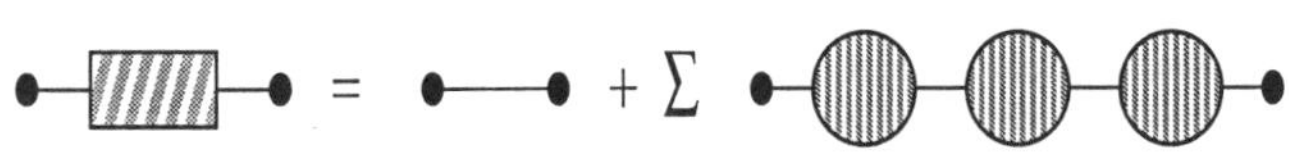

Figure 2. The dressed propagator.

Our procedure for rearranging Feynman diagrams begins with dressing all propagators. We use the notation illustrated in Fig. 2. Thus, from now on, all propagators in a diagram are assumed to include the one-particle irreducible 2-point diagrams, which form a geometric series. A generic diagram will look as illustrated in Fig. 3.

Next, we consider all one-particle irreducible 3-point diagrams. They can also be added, once and for all, to all bare 3-point vertices, to form the so-called dressed 3-point vertices. Similarly, we can collect all subdia-

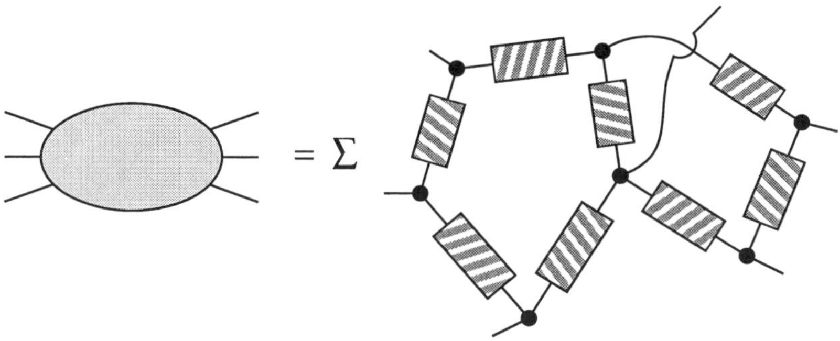

Figure 3. Example of a diagram with dressed propagator insertions

grams needed to turn all 4-vertices into dressed 4-vertices, see Fig. 4. It is important that diagrams, where the propagators and vertices are replaced by dressed ones, themselves should not contain any other subgraphs with three or four external lines.

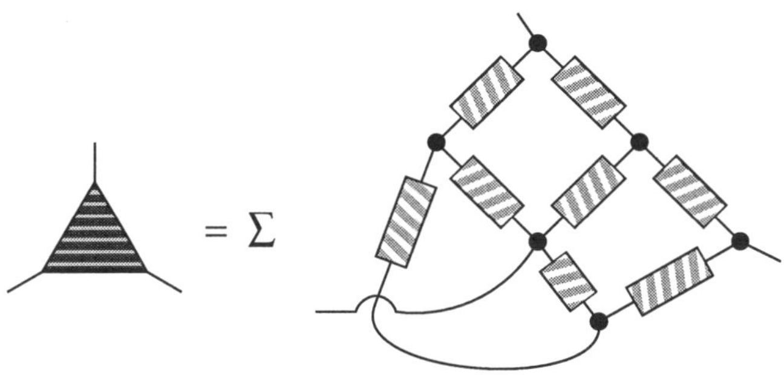

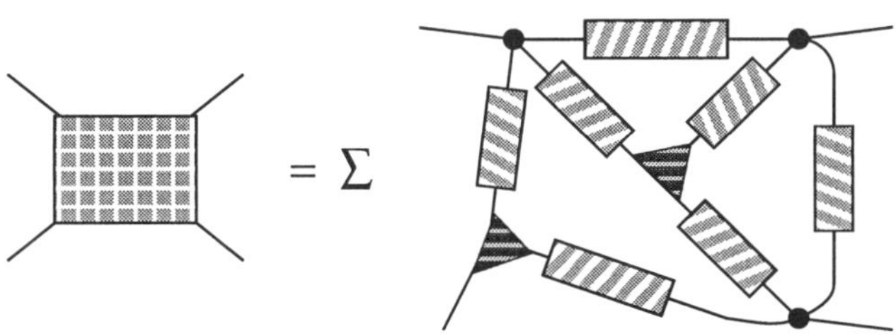

Figure 4. The dressed 3-point vertices formed from irrdeucible subdiagrams, and the dressed 4-point vertices from irrdeucible 4-point diagrams

A generic diagram then looks as in Fig. 5.

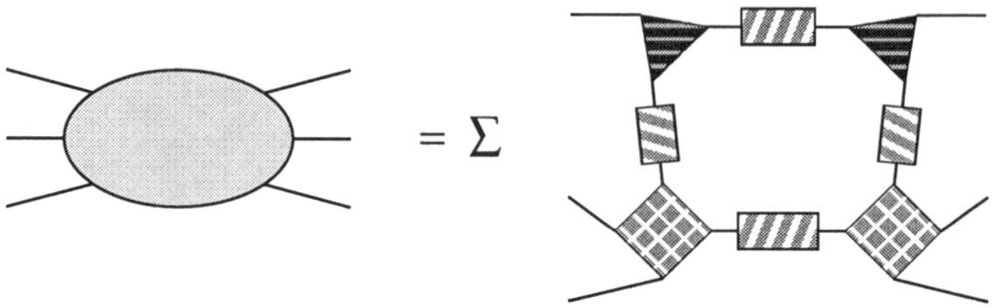

Figure 5. Diagrams with more than 4 external lines are built exclusively of dressed propagators, dressed 3-point vertices and dressed 4-point vertices

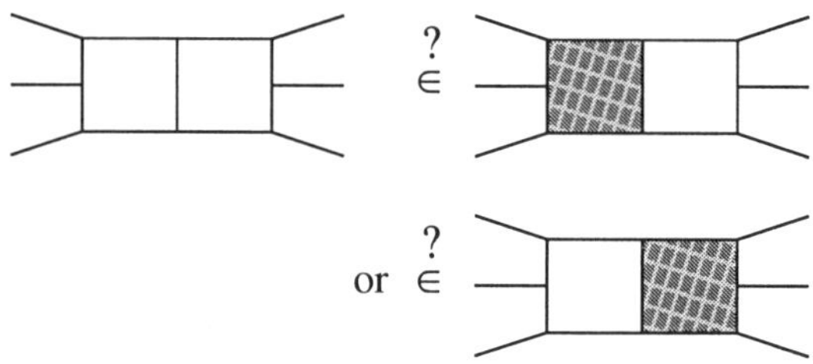

Figure 6. Ambiguity for irreducible 5-point vertices: the diagram indicated has six external lines and can be written in two ways using an irreducible 5-vertex. It would be counted twice, which is incorrect unless further correction procedures are introduced

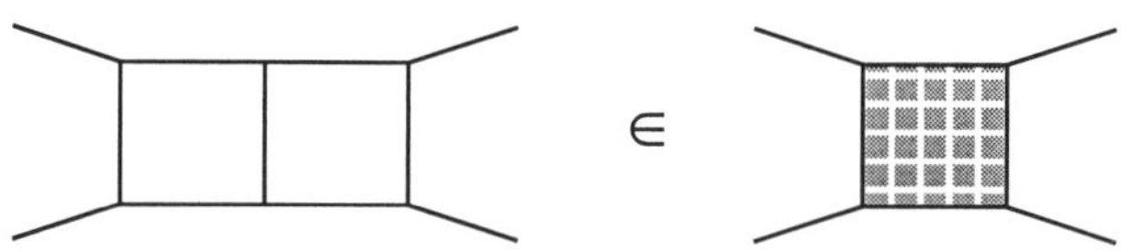

Figure 7. Subgraphs with 4 external lines do not lead to counting errors. The diagram depicted here contributes just once to the irreducble dressed 4-vertex

It is important to check here that rearranging diagrams using this prescription does not lead to omissions of any diagrams or to overcounting of diagrams. Indeed, if we were to continue the procedure towards irreducible diagrams with *five* external lines, overcounting would occur. This is illustrated in Fig. 6. Such an ambiguity cannot occur in the case of 4-vertices; cf. Fig. 7. The diagram of this Figure is counted correctly as a single contribution to the dressed 4-vertices.

We conclude from this section that all diagrams with five or more external lines can be seen to be built up in an unambiguous way from irreducible dressed propagators, 3-point functions and 4-point functions. These dressed diagrams themselves should not contain any irreducible subgraph with less than five external lines. Consequently, the integrations over any of the momenta in these dressed diagrams do not lead to any ultraviolet divergence. In particular, there are no overlapping ultraviolet divergences.

However, the dressed 2-, 3- and 4-point functions themselves cannot be reduced to convergent integrals along such lines; they themselves still seem to be built out of *bare* propagators and vertices. They will be considered in the next sections.

2. The Ariadne Procedure

The dressed 2-, 3- and 4-point vertices may be divergent[a], but if we introduce *subtractions*, more convergent expressions may arise. We claim that, if a divergent, irreducible diagram with $n < 5$ external lines is considered, then we can take the *difference* between that diagram and the same diagram at some different values of its external momenta, and rewrite that as a new irreducible Feynman diagram with $n+1$ external lines, whose degree of convergence is improved by at least one power of k.

In order to introduce unambiguous rules for these *difference diagrams*, we need the notion of a guiding path inside a diagram.[b] A guiding path is a sequence of propagators inside a diagram that form a single uninterrupted line from one external line to another, see Fig. 8. If an external line is a fermion, such as in QED, we can use this fermion as a guiding path. In $U(N)$ (gauge) theories, we can often use index lines as guiding paths, provided that not all index lines lead from one external line back to the same one; since such diagrams do not contribute in $SU(N)$ theories

[a]Diagrams with *fermions* are less divergent; one may decide to count external fermions with weight 3/2 in the procedure that follows.

[b]In planar diagram theories, the guiding path is simply the edge of a diagram. In fact, this was used in Ref[2].

(fields in the adjoint representation are traceless), index lines are assured to be useful in these theories. However, also in $SO(N)$ theories, index lines cannot run from one external line back to the same line, since here also the adjoint representation is traceless. This means that also $U(1)$ theories can sometimes be handled. If, however, an external line has two (or more) units of $U(1)$ charge, it means that it has two index lines in terms of which the representation is symmetric, hence not traceless. In that case, we have to use some other guiding line. In the electro-weak case, this appears to be possible: all our bosonic fields have several quantum numbers, so that we do not have to resort on the unreliable $U(1)$ indices.

Thus, it seems that in most cases of interest one can find a guiding path. This will be referred to as the Ariadne principle. It is the one restriction that we will assume, besides the more familiar restriction that our theory should not contain any chiral anomalies[3] (more about the anomalies later).

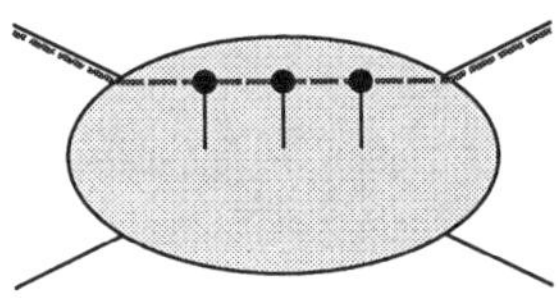

Figure 8. The introduction of a guiding path (dotted line)

Consider the sequence of (bare) propagators $P(k_i)$ and (bare) vertices $V(k_i)$ along a guiding path that leads from an external line with momentum k_0 to another line with momentum k_N.

$$F(k_0,\,k_N) = V_1(k_1)P_1(k_1)\,\cdots\,P(k_{N-1})V_N(k_N)\,; \tag{2.1}$$

Here, and k_i is the momentum in the ith propagator or vertex, i running from 1 to N. Now substitute all these momenta by the same values plus an additional, fixed, momentum q, and compute the difference between these two amplitudes:

$$\begin{aligned} F(k_0+q,\,k_N+q) - F(k_0,\,k_N) \;&= \sum_i \Big\{ V_1(k_1)\cdots P_{i-1}(k_{i-1}) \\ &\Big(\Big(V_i(k_i+q) - V_i(k_i)\Big)P_i(k_i+q) + V_i(k_i)\Big(P_i(k_i+q) - P_i(k_i)\Big)\Big) \\ &V_{i+1}(k_{i+1}+q)\,P_{i+1}(k_{i+1}+q)\,\cdots\,V_N(k_N+q)\Big\}\,. \end{aligned} \tag{2.2}$$

We see that this expression contains bare propagators and vertices that again form dressed propagators and vertices when summed. In particular, parts of this expression refer to the difference between two dressed propagators, which obey

$$P^{\,\mathrm{dr}}(k) = \Big(k^2 + m^2 - \Gamma(k) - i\varepsilon\Big)^{-1} \quad ;$$
$$P^{\,\mathrm{dr}}(k+q) - P^{\,\mathrm{dr}}(k) = P^{\,\mathrm{dr}}(k)\Big[\Gamma(k+q) - \Gamma(k) - (k+q)^2 + k^2\Big] \times P^{\,\mathrm{dr}}(k+q)\,. \tag{2.3}$$

Writing

$$k^2 - (k+q)^2 = -q_\mu(2k^\mu + q^\mu) \qquad \Gamma(k+q) - \Gamma(k) = q_\mu \Gamma_1^\mu(k,q) \tag{2.4}$$

we see that the expression between square brackets here can be regarded as an effective three-point diagram Γ_1^μ, multiplied with a factor q_μ. Eq. (2.3) is depicted diagrammatically in Fig. 9.

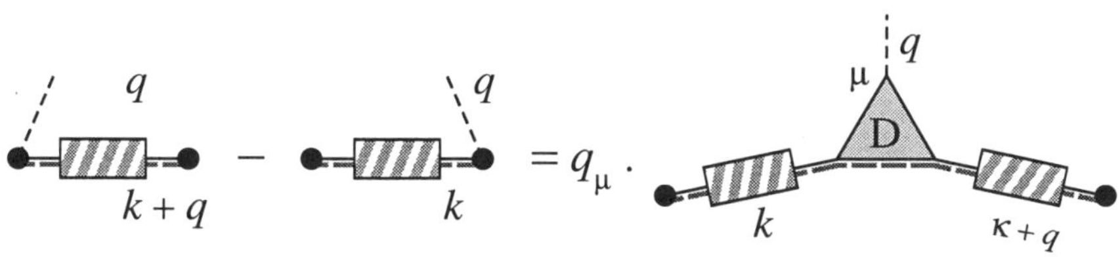

Figure 9. Difference between two dressed propagators in diagrammatic notation

3. Integrating the Equations. Conclusion

Similarly, we can handle the difference between three-point diagrams, $V^{[3]}(k+q) - V^{[3]}(k)$, at different external momenta as four-point diagrams, see Fig. 10. The difference between two four-point diagrams is a sum of convergent diagrams, see Fig. 11. The difference equations used here can be used either for the original irreducible diagrams for the theory or for the diagrams obtained after a previous differenciation. In all cases, the irreducible diagrams of five or more external lines only contain convergent expressions. As far as the ultra-violet divergences is concerned, the situation is the same as if we had differentiated with respect to the momenta rather than taking finite differences (*i.e.*, if q had been taken infinitesimal.

A disadvantage of infinitesimal q, however, is the emergence of higher order poles and the associated infra-red divergences in the propagators. Our difference procedure avoids infra-red divergences.

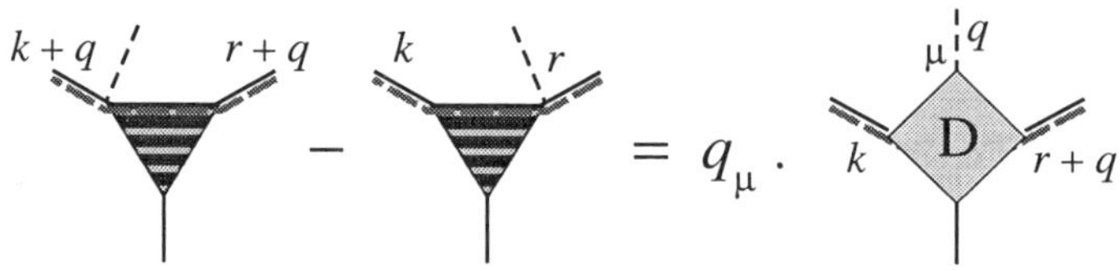

Figure 10. The difference between the 3-point functions at different values of the external momenta is a four-point diagram

Fig. 11, whose equation reads as:

$$V^{[4]}(k+q) - V^{[4]}(k) = q_\mu \sum_i V_i^{\mu\,[5]}(k, q, \cdots)\ , \tag{3.1}$$

where $V_i^{\mu[5]}$ do not contain any divergences, in many respects is to be regarded as a renormalization group equation. Since $V_i^{\mu[5]}$ are all linearly (or better) convergent, the equation can be symbolized as

$$\frac{d}{dk}\Gamma^{[4]}(k) = \frac{1}{k}\beta(\Gamma^{[4]})(k)\,, \tag{3.2}$$

the r.h.s. being essentially a beta function. The difference equations for the 2- and 3-point functions, in short-hand, are

$$\Gamma^{[2]}(k+q) - \Gamma^{[2]}(k) = q^\mu\, k\,(D\Gamma)_\mu^{[3]}(k, q)\ , \tag{3.3}$$

$$V^{[3]}(k+q) - V^{[3]}(k) = q^\mu\,(DV)_\mu^{[4]}(k, q)\ , \tag{3.4}$$

where we explicitly indicated the k-dependence apart from logarithms. Thus, we see that Eqs. (3.3) and (3.4) converge in the infrared when integrated, whereas (3.2) has the infra-red structure of the renormalization group.[4]

Our set of equations appears to be particularly elegant because no direct reference is made to the *bare* Lagrangian of the theory! All bare coupling parameters are generated by the integration constants when integrating these difference equations.

Thus, Quantum Field Theory has been recast into a self-consistent set of equations, which can be integrated to obtain the desired amplitudes. The four-point amplitudes – more precisely, the canonically dimensionless irreducible n-point functions – follow from solving the renormalization group

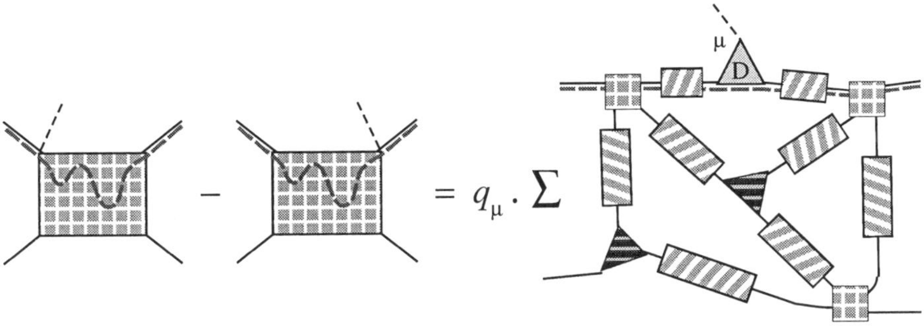

Figure 11. The difference between the 4-point functions at different values of the external momenta is a sum of convergent diagrams

equation Eq. (3.1). The required integration constant(s) replace the original free parameters of the theory.

A difficulty may arise from the fact that the 'guiding lines' may be chosen in many alternative ways. Indeed, it is in integrating the equations that one might encounter anomalies[3]: the integration constants cannot be reconciled with all symmetries of the theory.

Also, the lower irreducible Green functions may generate integration constants, which would correspond to dimensionful parameters of the theory. The usual questions concerning "naturalness" are not affected by our procedure; if the integration constants lead to small amplitudes in the far infra-red, this may be considered as 'unnatural', but there is no objection to that from a purely mathematical point of view.

Another fundamental difficulty not addressed by our procedure is the divergence of the perturbative expansion for the diagrams in the r.h.s. of Eq. (3.1), depicted in Fig. 11. In general, such expansions diverge factorially. In the planar $N \to \infty$ limit, the number of diagrams increases by calculable power laws, but individual diagrams may grow factorially, so that there still is no guarantee for a finite radius of convergence.[2] In practice, it seems to be not unreasonable to simply cut the series off at some given order.

References

1. C. Itzykson and J.-B. Zuber, Quantum Field Theory, McGraw-Hill 1985; ISBN 0-07-032071-3; 0-07-066353-X.
2. G, 't Hooft, *Planar diagram field theories*, in *Progrss in Gauge Field Theory, NATO Adv. Study Inst. Series*, eds. G. 't Hooft *et al.*, Plenum, 1984, 271, reprinted in G. 't Hooft, *Under the Spell of the Gauge principle*, World Scientific, 1994, p. 378. See also: *Nucl. Phys.* **B 72** (1974) 461.
3. S.L. Adler, *Phys. Rev.* **177** (1969) 2426; J.S. Bell and R. Jackiw, *Nuovo Cim.* **60A** (1969) 47; S.L. Adler and W.A. Bardeen, *Phys. Rev.* **182** (1969) 1517; W.A. Bardeen, *Phys. Rev.* **184** (1969) 1848.
4. C.G. Callan, *Phys. Rev.* **D2** (1970) 1541; K. Symanzik, Commun. Math. Phys. **16** (1970) 48; *ibid.* **18** (1970) 227.

PROBLEMS AND DIFFICULTIES IN STANDARD MODEL AND GRAVITATION

M. VELTMAN
Emeritus MacArthur Professor
University of Michigan

Dedicated to Paul Frampton
on the occasion of his sixtieth birthday

The Standard Model provides us with a beautiful structure of forces and particles, yet at the same time presents a bewildering set of problems. The foremost problem, in my mind is this: why are there three generations of quarks and leptons? Next there comes the large amount of free parameters in the model, such as the particle masses, quark mixing matrix and recently in addition neutrino mixing. Then there are the more profound questions, such as why the symmetry SU(3)×SU(2)×U(1) etc.

On the theoretical side there are further problems, of which we would like to mention the strong CP problem and the Cosmological Constant problem. The strong CP problem gave rise to the solution involving an axion, which however has not been seen. A somewhat unsatisfying solution called the invisible axion exists.

Concerning the Cosmological Constant, it can of course be considered as a formal problem, because a Cosmological Constant is a free parameter, and the fact that there are many corrections to this constant, including quantum corrections requiring fine tuning, is mainly unsatisfactory and ugly. The problem has been aggrevated by the recent experimental discovery that the Universe seems to be perfectly flat.

On this occasion, as an amusing exercise, we would like to think the unthinkable and speculate the outrageous. The observed flatness of the Universe appears to be a real blow to Einstein's theory of gravitation. That theory is known to work quite correctly on the scale of a solar system, in lowest non-relativistic approximation in the form of Newton's law, and furthermore giving the correct value for the perihelion rotation of Mercury. Also the gravitational radiation as observed by Taylor c.s. is from a system of relatively small size. There can be no doubt about the general theory of relativity on the scale of a solar system. However, for larger distances the situation is far from clear. There is of course standard cosmology, but no astro-scepticus would really consider that as a proof of the correctness of general relativity. The same goes for the multitude of black holes discovered almost daily by astronomers, mainly on the premise "What else could it be?"

Let us now take the following position. Imagine that a group of scientists, born and living in a spaceship hurtling through the Universe and not bound to any solar system, tries to establish the laws of gravitation on the basis of their observation of galaxies. Would they come up with the general theory of relativity? Not very likely. Not even Newtons law!

The first thing our scientists would do is to make a visual inspection of some galaxies. They would of course immediately note the spirals, or the bar in many galaxies. Here are two typical pictures.

Obviously, the stars seen in these galaxies are bound in some way. As a next step the star scientists would surely try to establish the velocity whith which these objects move around the center, and typically a star near the end of a spiral arm would be observed to move with a velocity of something like 200 km/sec (like our sun). From this they would (we assume that they had developed methods to estimate the size of the galaxy) derive that the matter in the arms would have been around some 30 times. Thus, through about 30 revolutions the arm structure would stay approximately unchanged. It follows that matter in the arms rotates with a velocity that is proportional to the distance of the center, thus

$$v \propto r$$

Assuming now a force between the center and the stars rotating around it they would derive that the force would have to behave according to the law:

$$F \propto r$$

with C some constant. It is not anywhere near Newton's law which is

$$F \propto \frac{1}{r^2}$$

That law would have produced the velocity distribution

$$v \propto \frac{1}{\sqrt{r}}$$

Of course, our scientists would be quite upset with the force law $F \propto r$, because their study of quantum field theory would certainly have revealed that long range forces, associated with particles of mass zero, always produce a Newtonian type force law. This assuming perturbation theory is valid for such a theory. Some subset of the star scientists would try assuming a force law as suggested by perturbative field theory (i.e. a Newtonian law). However, they would then face the almost impossible burden of constructing complicated mechanisms to explain the (invariant) spiral structure. Another group would assume that perturbation theory breaks down and that perturbative field theory does not apply. Yet a third group would have noted that there is a remarkable similarity to what happens inside a cyclotron. Inside a cyclotron with its homogenous magnetic field the revolution time of the charged particles is independent of their distance from the origin. Thus if there is a vector force acting on the stars, and if there was a strong homogenous field (analogous to a magnetic field) perpendicular to the plane of the galaxy then the observed pattern would follow. That vector force could for example couple to baryon number B, or to lepton number L or even B-L. In the latter case hydrogen would not be subject to that force. The main problem would of course be how to understand such a homogenous field.

At this point the confusion would be great. The scientists believing in perturbative field theory would probably develop gravitation with a spin 2 particle and come up with Einstein's theory. However, many would be unwilling to accept their fancy explanations of the observed invariant spiral structure, and if we consider modern astronomy one must conclude that there is no natural general explanation for the observed facts. Astronomers speak of the winding problem, because with a Newtonian type velocity distribution the spiral arms would wind up tightly after a few revolutions. It is not enough to produce a mechanism that shows behaviour different from the Newton structure, one must actually produce something that preserves strictly the spiral arms.

At this point the star scientists turn to the matter outside of the visible matter region as they have developed radio astronomy and can measure velocities outside the luminous regions. Velocity distributions of this type can be found in the thesis of A. Broeils (September 1992) from the University of Groningen, the Netherlands. See also K.G. Begeman, A. Broeils and R. Sanders, Mon. Not. R. astr. Soc. 249 (1991) 523. Consider first a typical plot for some galaxy.

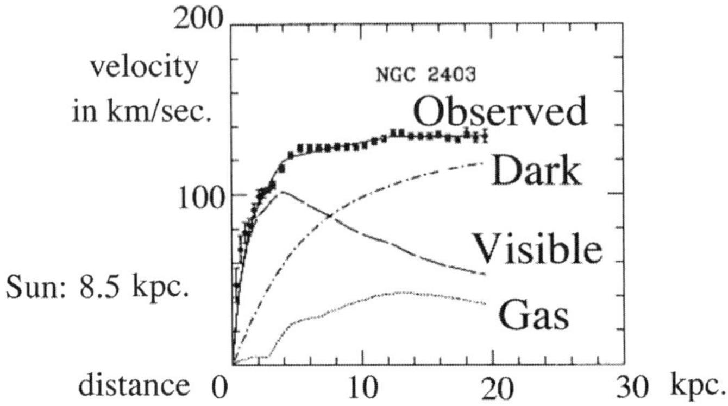

This shows one such velocity distribution. The curves give the observed velocity and those obtained using Newtons law for visible matter etc. Dark matter is needed to produce the observed result. Here are six cases, including the previous one.

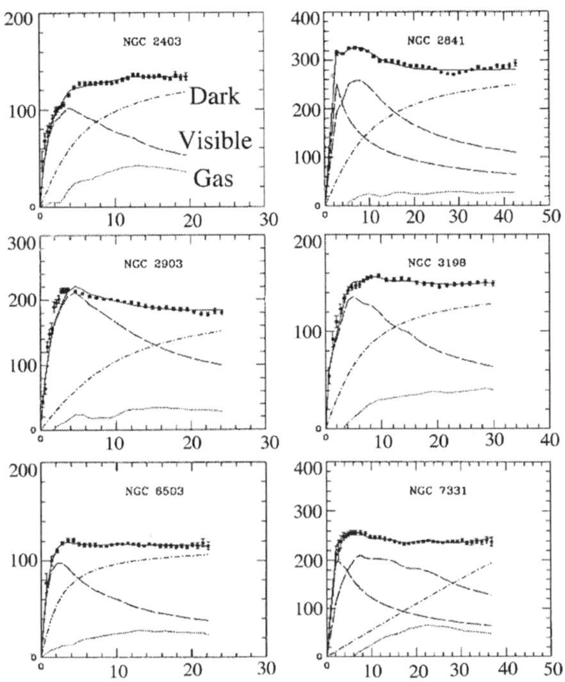

Again, the observed velocity distribution is not at all as expected from Newton's law applied to the observed visible matter and interstellar gas. In fact, one sees a velocity distribution that is nearly constant for large distances rather than something $\propto 1/\sqrt{r}$ as expected from Newton's law. The observed distribution in fact suggest a power law $F \propto 1/r$ (as compared to Newton's $F \propto 1/r^2$). The standard astronomers solution is to assume Dark Matter such that the observed curve is obtained. This leaves us with the mystery as to why at large distances the observed curves appear quite flat. And of course, it is a one time hypothesis for one problem. No further verification of this hypothesis exists.

The star scientists are now more confused than ever. Of course, the group hanging on to the perturbative spin 2 field theory will undoubtedly come up with dark matter. Those believing in a magnetic type field must now let this field decrease at a rate roughly proportional to the distance. Those advocating a breakdown of perturbation theory must now allow two different regions with different behaviour.

To be complete, there are astronomers who have tried to explain the velocity distribution without using Dark Matter. The best known attempt, due to Milgrom, is commonly called MOND (MOdified Newtonian Dynamics). This is a one-paramter modification of the law $F = ma$. The modification makes no sense from a field theoretic point of view, but one can see it as a phenomenological one parameter model. It is quite successful. For a review see R. Sanders and S. McGaugh, Annu. Rev. Astron. Astrophys. 2002, Vol. 40 (2002) 263.

Given that the star scientists do not know about planetary systems they will for sure stay utterly divided. Nobody can claim any success. The zero mass spin two theory appears to need such fancy repairs that very few would be willing to accept it. The experimental observation (from background radiation) of a seemingly perfectly flat Universe would be a hard blow. In addition, the fact that the theory is non-renormalizable does not help either. The introduction of a cut-off, perhaps in the form of a string theory, would not convince anybody.

However, the other groups among the star scientists have not much to offer in the way of proof either. From our advantage point, knowing planetary systems with the near perfect fit using Newton's law, the spin 2 theory seems preferable, but again, the difficulties are big. One may argue for a spin 2 theory with a breakdown of perturbation theory, somehow producing the observed results. That would lead to a fundamental change in our understanding of the theory (see M. Veltman, *Perturbation theory and relative space,* Acta Phys. Pol. B25 (1994) 1399). This may explain the observed flatness of the Universe but to date no satisfactory explanation of the other effects mentioned above has been found.

APPLICATION OF THE OPE TO JET PHYSICS

MARK B. WISE

California Institute of Technology, Pasadena CA 91125

Summary

Jet cross sections and shape variables in Z-decay to hadrons have a perturbative expansion in powers of the strong coupling $\alpha_s(M_Z)$. In addition non-perturbative effects suppressed by powers of Λ_{QCD}/M_Z contribute to these variables. In my talk, I reviewed how the non-perturbative contributions are calculated using the operator product expansion (OPE) and Soft-Collinear Effective field theory. Relations between the nonperturbative corrections to different shape variables were discussed.

References

[1] R. Barate et al. [ALEPH Collaboration], Eur. Phys. J. C **17**, 1 (2000).

[2] P. Abreu et al. [DELPHI Collaboration], Eur. Phys. J. C **14**, 557 (2000).

[3] M. Acciarri et al. [L3 Collaboration], Phys. Lett. B **489**, 65 (2000).

[4] G. Abbiendi et al. [OPAL Collaboration], Eur. Phys. J. C **16**, 185 (2000).

[5] K. Abe et al. [SLD Collaboration], Phys. Rev. D **51**, 962 (1995).

[6] J. Abdallah et al. [DELPHI Collaboration], Eur. Phys. J. C **29**, 285 (2003) [arXiv:hep-ex/0307048].

[7] B. R. Webber, Phys. Lett. B **339**, 148 (1994).

[8] A. V. Manohar and M. B. Wise, Phys. Lett. B **344**, 407 (1995).

[9] G. P. Korchemsky and G. Sterman, Nucl. Phys. B **555**, 335 (1999).

[10] G. P. Korchemsky and S. Tafat, JHEP **0010**, 010 (2000).

[11] Y. L. Dokshitzer and B. R. Webber, Phys. Lett. B **352**, 451 (1995).

[12] Y. L. Dokshitzer and B. R. Webber, Phys. Lett. B **404**, 321 (1997).

[13] Y. L. Dokshitzer, Invited talk at the International Conference "Frontiers of Matter," Blois, France, June 1999, arXiv:hep-ph/9911299.

[14] C. W. Bauer, A. V. Manohar and M. B. Wise, arXiv:hep-ph/0212255; C. W. Bauer, C. Lee, A. V. Manohar and M. B. Wise, arXiv:hep-ph/0309278.

[15] For reviews on factorization theorems, see J. C. Collins and D. E. Soper, Ann. Rev. Nucl. Part. Sci. **37**, 383 (1987).

[16] G. Sterman, TASI lectures, arXiv:hep-ph/9606312.

[17] G. Sterman and S. Weinberg, Phys. Rev. Lett. **39**, 1436 (1977).

[18] C. W. Bauer, S. Fleming and M. E. Luke, Phys. Rev. D **63**, 014006 (2001).

[19] [19] C. W. Bauer, S. Fleming, D. Pirjol and I. W. Stewart, Phys. Rev. D **63**, 114020 (2001).

[20] C. W. Bauer and I. W. Stewart, Phys. Lett. B **516**, 134 (2001).

[21] C. W. Bauer, D. Pirjol and I. W. Stewart, Phys. Rev. D **65**, 054022 (2002).

[22] C. W. Bauer, S. Fleming, D. Pirjol, I. Z. Rothstein and I. W. Stewart, Phys. Rev. D **66**, 014017 (2002).

[23] C. L. Basham, L. S. Brown, S. D. Ellis and S. T. Love, Phys. Rev. Lett. **41**, 1585 (1978); C. L. Basham, L. S. Brown, S. D. Ellis and S. T. Love, Phys. Rev. D **19**, 2018 (1979).

PAUL FRAMPTON'S PUBLICATIONS

1. Chirality Commutator and Vector Mesons, Nuov. Cim. 48A, 703-708 (1967)

2. Superconvergence Sum Rules in pi-rho Scattering (with J. C. Taylor), Nuov. Cim. 49A, 152-156 (1967)

3. Superconvergence in pi-N* Elastic Scattering, Nucl. Phys. B2, 518-524 (1967)

4. Factorization, Kinematic Factors of the Regge Residue Function, and Conspiracy, Nucl. Phys. B7, 507-526 (1968)

5. Crossing Symmetric Regge Model and Isovector Meson Decays, Phys. Rev. 180, 1392-1394 (1969)

6. Dispersion Sum Rules for the Baryon Spectrum in an O(3, 1) Dynamical Group Model (with B. Hamprecht), Phys. Lett. 28B, 664-667 (1969)

7. Dispersion Sum Rules for the Non-Strange Baryon Spectrum (with B. Hamprecht), Phys. Rev. 183, 1416-1424 (1969)

8. Asymptotic Behavior of Partial Widths in the Veneziano Model of Scattering Amplitudes (with Y. Nambu), in Quanta, Festschrift for G. Wentzel, Eds: P. G. O. Freund, C. J. Goebel and Y. Nambu, Univ. Chicago Press (1970) pp. 403-414

9. SU(3) Selection Rule from the Veneziano Model (with C. Michael), Nuov. Cim. Lett. 2, 81-83 (1969)

10. Regge Trajectories and Nucleon Structure, Phys. Rev. 186, 1419-1421 (1969)

11. Generalized Vector Dominance Theory and Vertex Functions, Phys. Rev. D1, 3141-3151 (1970)

12. Form Factor Models and Chiral Symmetry, Nuov. Cim. Lett. 3, 229-234 (1970)

13. Factorization of Satellite Terms in the N Particle Dual Amplitude, Phys. Lett. 32B, 195-198 (1970)

14. Factorization of Satellite Terms in the N Particle Dual Amplitude II- The Single Planar Dual Loop (with R. J. Rivers), Phys. Rev. D2, 691-694 (1970)

15. Conditions for Regge Behavior of an Absolutely Convergent Veneziano Series (with C. W. Gardiner), Phys. Rev. D2, 2378-2388 (1970)

16. No-Go Theorem for Dual Models (with A. P. Balachandran and L. N. Chang), Nuov. Cim. 1A, 545-552 (1971)

17. Quark Focusing at Hadronic Vertices (with P. G. O. Freund), Nucl. Phys. B24, 453-464 (1970)

18. Hadrodynamics and Quark Structure (with S. Ellis, P. G. O. Freund, and D. Gordon), Nucl. Phys. B24, 465-489 (1970)

19. Hyperon Beta Decay (with W. -K. Tung), Phys. Rev. D3, 1114-1121 (1971)

20. Perturbative Unitarity of Dual Loops (with P. Goddard and D. A. Wray), Nuov. Cim. 3A, 755-762 (1971)

21. No-Go Theorem for Dual Models, Addendum. Nuov. Cim. 3A, 763-765 (1971)

22. Duality and Inclusive Phenomenology (with J. Ellis, J. Finkelstein, and M. Jacob), Phys. Lett. 35B, 227-230 (1971)

23. Dipion Photoproduction. In International Conference on Meson Resonances and Related Phenomena, Eds. R. H. Dalitz and A. Zichichi. Editrice Compositori, Bologna ITALY (1972) 543-555.

24. Why is (rho)' Photoproduction Suppressed? (with K. Schilling and C. Schmid), Phys. Lett. 36B, 591-592 (1971)

25. Relative Importance of Dual Loops, Phys. Lett. 36B, 591-592 (1971)

26. Phenomenological Duality for Pomeron-Proton Scattering (with P. V. Ruuskanen), Phys. Lett. 38B, 78-80 (1972)

27. Duality and The Deck Effect (with N. Tornquist), Nuov. Cim. Lett. 4, 233-236 (1972)

28. Dual Model Spectrum Analysis, Using Virasoro-Algebra Representations (with H. B. Nielsen), Nucl. Phys. B45, 318-332 (1972)

29. O(n) Relations For Coupling Constants and Space-Time Dimensions In Dual Models, Phys. Lett. 41B, 364-370 (1972)

30. Duality and Single Particle Inclusive Spectra Involving Baryons, Nuov. Cim. Lett. 6, 58-62 (1973)

31. A Study of the Hilbert Space Properties of the Veneziano Model Operator Formalism (with L. Brink and H. B. Nielsen), J. Math. Phys. 15, 1599-1605 (1974)

32. Symmetric Group and Meson Born Amplitudes, Phys. Rev. D7, 3077-3091 (1973)

33. Regge Slope Expansion of Meson Born Amplitudes (with K. C. Wali), Phys. Rev. D8, 1879-1886 (1973)

34. Recent Developments in Dual Models, in International School of Subnuclear Physics, 11th. Course, Ettore Majorana Centre, Laws of Hadronic Matter. Edited by A. Zichichi, Academic Press (1975) p. 112-136.

35. Possible Born Amplitude for Pions, Nuov. Cim. Lett. 8, 525-530 (1973)

36. Symmetric Group Construction of a New Multiparticle Amplitude for Physical Pions, Phys. Rev. D9, 487-497 (1974)

37. Factorization Properties of Multipion Amplitude, Phys. Rev. D9, 2861-2864 (1974)

38. Dual Resonance Models (BOOK), Addison-Wesley/ Benjamin (1974). Frontiers in Physics. pages i-xxiii and 1-451.

39. An Investigation of Four-Meson Symmetric Group Dual Models. (with K. Geer), Phys. Rev. D10, 1284-1288 (1974)

40. Spontaneous Symmetry Breaking and Regge Intercepts (with R. Cahalan), Phys. Lett. 50B, 475-478 (1974)

41. Lectures on Dual Resonance Models, Part Six: Symmetric Group. Syracuse University Report SU-4205-30, COO-3533-30. January 1974. 57 pages.

42. Evidence for Critical Space-Time Dimensionality Equals Four in Symmetric Group Model, Phys. Rev. D11, 953-955 (1975)

43. Predictions of Quark and Dual Approaches to Meson Classification, Phys. Rev. Lett. 34, 840-842 (1975)

44. N(bar)-N into 3(pi) Amplitudes and the Symmetric Group Dual Resonance Model (with A. Nandy and T. Kalogeropoulos), Phys. Rev D11, 1040-1042 (1975)

45. String Approaches to Hadron Structure, Phys. Rev D12, 538-545 (1975)

46. Consequences of Direct (Psi)-Lepton Coupling (with V. Rabl), Nuov. Cim. Lett. 13, 293-296 (1975)

47. Meson Mass Formula from Duality, Phys. Rev. D12, 2164-2165 (1975)

48. Dual Models and Strings, EPS Conference on High Energy Physics, Ed. A. Zichichi, Editrice Compositore (1976) pp. 1250-1272

49. Dual Resonance Model Incorporating Two Different Parent Regge Trajectories In The Same Channel (with V. Rabl), Phys. Rev. D12, 3214-3220 (1975)

50. Model of Leptons Incorporating the New Spin 1 Particles as Gauge Bosons, Syracuse University Report SU-4206-51, COO-3533-51. April 1975. 10 pages.

51. Renormalizable Model of Leptons with New Particles as Neutral Gauge Bosons, Syracuse University Report SU-4206-52, COO-3533-52. April 1975. 19 pages.

52. Relationship Between Gauge Field Theories and Dual Resonance Models, in International School on Subnuclear Physics, 13th. Course, Ettore Majorana Centre, New Phenomena in Subnuclear Physics. Ed. A. Zichichi, Plenum Press (1977) pp. 493-532.

53. On the Multi-Regge Behavior of a Certain Class of Dual Resonance Models (with D. Knight), Nuov. Cim. Lett. 15, 125-128 (1976)

54. Uniqueness of S-Wave Non-Abelian Magnetic Monopole, Phys. Rev. D14, 528-530 (1976)

55. Parton Model Derivation of Generalized Scaling Variable, Nuov. Cim. Lett. 17, 499-500 (1976)

56. Vacuum Instability and Higgs Scalar Mass, Phys. Rev. Lett. 37, 1378-1380 (1976)

57. Weak Neutral Current and Decay of Positronium (with E.S. Abers), UCLA Report ACLA/76/TEP/5. April 1976. 7 pages.

58. Charm and Strange Particle Production in Electron-Positron Annihilation, UCLA Report UCLA/76/TEP/9. June 1976. 10 pages.

59. Lectures on Gauge Field Theories. Part One. Gauge Invariance. UCLA Report UCLA/76/TEP/15. August 1976. 104 pages.

60. 't Hooft Monopoles and Singular Gauge Transformations, UCLA Report UCLA/76/TEP/12. July 1976. 9 pages.

61. Consequences of Vacuum Instability in Quantum Field Theory, Phys. Rev. D15, 2922-2928 (1977)

62. Model for Scaling Violation (with J. J. Sakurai), Phys. Rev. D16, 572-574 (1977)

63. Vacuum Bubble Instantons, in Significance of Nonlinearity in the Natural Sciences, Editors: B. Kursonoglu, A. Perlmutter, and L. F. Scott, Plenum Press (1977) pages 57-65.

64. Lecture on Gauge Field Theories. Part Two. Quantization. UCLA Report UCLA/77/TEP/21. December 1977. 64 pages.

65. Yang-Mills Vacua in Landau Gauge (with T. Hagiwara, W. F. Palmer, and S. S. Pinsky), Phys. Lett. 76B, 455-457 (1978)

66. Concerning Gribov Copies in Nonabelian Gauge Theory (with W.F. Palmer and S.S. Pinsky), Ohio State University Report COO-1545-229-mc (microfiche). May 1978. 11 pages.

67. Lectures on Gauge Field Theories. Part Three. Renormalization. Ohio State University Report COO-1545-245. November 1978. 106 pages.

68. New Anomaly in Axial Vector Ward Identity. Ohio State University Report COO-1545-248-mc (microfiche). December 1978. 13 pages.

69. Higgs Scalar in Heavy Vector Meson Decays (with W. W. Wada), Phys. Rev. D19, 271-273 (1979)

70. Scale Invariance and Solitons (with L. Tyburski), Nuov. Cim. Lett. 25, 13-16 (1979)

71. Conditions for Renormalizability of Quantum Flavor Dynamics, Phys. Rev. D20, 3372-3377 (1979)

72. Anomalies, Unitarity, and Renormalization, in High Energy Physics in the Einstein Centennial Year, Editors: F. Krausz, A. Perlmutter, and L. F. Scott, Plenum Press (1979) pages 133-137.

73. Estimate of Flavor Number From SU(5) Grand Unification (with S. Nandi and J. J. G. Scanio), Phys. Lett. 85B, 225-228 (1979)

74. SU(N) Grand Unification with Several Quark-Lepton Generations, Phys. Lett. 88B, 299-301 (1979)

75. Gauge Ambiguities, Zero Modes, and Classification of Field Size (with T. Hagiwara), Phys. Lett. 86B, 337-339 (1979)

76. SU(9) Grand Unification of Flavor with Three Generations (with S. Nandi), Phys. Rev. Lett. 43, 1460-1462 (1979)

77. Reply to "Noncontribution of n=4 Axial Vector Anomaly", Phys. Rev. D21, 1704 (1980)

78. Reply to "Comment on a Proposed New Anomaly", Phys. Rev. D21, 1709 (1980)

79. No-Go Theorem for SU(N) Unification of Extra-Strong Interactions, Phys. Rev. Lett. 43, 1912-1914 (1979)

80. Unification of Flavor, Phys. Lett. 89B, 352-354 (1980)

81. Masses of Quarks in Grand Unification (with K. Kang), Had. J. 3, 814-823 (1980)

82. Unstable Heavy Particles (with S. L. Glashow), Phys. Rev. Lett. 44, 1481-1483 (1980)

83. First Workshop on Grand Unification, Editor with S. L. Glashow and A. Yildiz, Math Sci Press, Brookline (1980)

84. The Limits on Unstable Heavy Particles, in First Workshop on Grand Unification, Editors: P. H. Frampton, S. L. Glashow and A. Yildiz, Math Sci Press, Brookline (1980) pages 33-37.

85. Connection Between Quark-Lepton Mass Ratios in SU(5) and Renormalization (with M. D. Scadron), Harvard University Report HUTP-80/A015. June 1980. 7 pages.

86. Family Unification in SO(8) Supergravity, Harvard University Report HUTP-80/A050. July 1980. 8 pages.

87. Mass Relations Between Quarks and Leptons in Grand Unification (with M. Goldhaber and K. Kang), Brookhaven Lab. Report BNL-26459. January

1980. 11 pages.

88. Is Flavor Proliferation Explicable by Supergravity? In High Energy Physics-1980 (XX Int. Conf., Madison, WI) Editors: L. Durand and L. G. Pondrom, American Institute of Physics (1981) page 970.

89. Grand Unification in N=8 Extended Supergravity, Phys. Rev. Lett. 46, 881-883 (1981)

90. Tests of SU(7) Unification (with P. H. Cox and A. Yildiz), Phys. Rev. Lett. 46, 1051-1053 (1981)

91. Natural Axion Invisibility and SU(N) Flavor Unification (with S. P. De Alwis), Phys. Rev. D24, 3345-3347 (1981)

92. Massive Neutrinos (with P. Vogel), Phys. Reports 82, 339-357 (1982)

93. Simultaneous Solution of Strong CP and Flavor Problems, Phys. Rev. D25, 294-296 (1982)

94. Exceptionally Simple E(6) Theory (with T. W. Kephart), Phys. Rev. D25, 1459-1461 (1982)

95. Time-Averaged Neutrino Oscillations (with S. L. Glashow), Phys. Rev. D25, 1982-1984 (1982)

96. Massive Neutrinos in E(6) Unified Theory with Harmless Axion (with T. W. Kephart, Y. J. Ng, and H. Van Dam), Phys. Lett. 112B, 50-52 (1982)

97. Dynkin Weights and Global Supersymmetry in Grand Unification (with T. W. Kephart), Phys. Rev. Lett. 48, 1237-1241 (1982)

98. Supersymmetric Family Unification (with T. W. Kephart), Nucl. Phys. B211, 239-268 (1983)

99. Fractionally Charged Particles as Evidence for Supersymmetry (with T. W. Kephart), Phys. Rev. Lett. 49, 1310-1313 (1982)

100. Third Workshop on Grand Unification (BOOK), Editor with S. L. Glashow and H. Van Dam, Birkhauser (1982) pages i-ix and 1-374.

101. Gauge Symmetry Breaking Which Preserves Supersymmetry, in Third Workshop on Grand Unification [see 100] Pages 22-29

102. A Useful Formula for Witten's Mass Heirarchy (with H. Georgi and J. E. Kim), Phys. Lett. 116B, 346-348 (1982)

103. Automatic Invisible Axion Without Domain Walls (with S. Dimopoulos, H. Georgi, and M. B. Wise), Phys. Lett. 117B, 185-188 (1982)

104. Lectures on Gauge Field Theories. Part Five. Renormalization Group. UNC-Chapel Hill Report IFP-176-UNC. 1982. 71 pages.

105. Higher Anomalies and Fermion Families (with T.W. Kephart), UNC-Chapel Hill Report IFP-184-UNC. September 1982. 8 pages.

106. Towards a Realistic Gauge Hierarchy (with J. E. Kim), Phys. Lett. 117B, 349-352 (1983)

107. N=8 Supergravity and Kaluza-Klein Axial Anomalies, Phys. Lett. 122B, 351-354 (1983)

108. Invisible Axions in Supersymmetric SU(N) (with T. Matsuki and W. W. Wada), Phys. Rev. D27, 2204-2210 (1983)

109. Automaticity, Axion Invisibility, and Cosmological Baryogenesis (with G. B. Lipton), Nucl. Phys. B219, 41-60 (1983)

110. Anomalies and Fermion Masses in D Dimensions (with J. Preskill and H. Van Dam), Phys. Lett. 124B, 209-211 (1983)

111. Explicit Evaluation of Anomalies in Higher Dimensions (with T. W. Kephart), Phys. Rev. Lett. 50, 1343-1346 (1983)

112. Consistency Conditions for Kaluza-Klein Axial Anomalies (with T. W. Kephart), Phys. Rev. Lett. 50, 1347-1349 (1983)

113. Analysis of Anomalies in Higher Space-Time Dimensions (with T. W. Kephart), Phys. Rev. D28, 1010-1023 (1983)

114. Cancelling the Hexagon Anomaly (with T. W. Kephart), Phys. Lett. 131B, 80-82 (1983)

115. Staying Alive with SU(5) (with S. L. Glashow), Phys. Lett. 131B, 340-342 (1983) . Erratum 135B, 515(1984)

116. Supersymmetric Subconstituents and the Problem of Mass (with G. Mandelbaum), Phys. Lett. 133B, 311-314 (1983)

117. Analysis of Anomalies in Higher Spacetime Dimensions: Addendum (with T.W. Kephart), UT-Austin Report UTTG-9-83-Add. UNC-Chapel Hill Report IFP-210-UNC. Purdue University Report PURD-TH-83-17-Add. September 1983. 10 pages.

118. Age and Flatness of Inflationary Universe with Cosmological Constant (with G.B. Lipton), UNC-Chapel Hill Report IFP-203-UNC. July 1983. 8 pages.

119. Anomalies of Gravity-Coupled SU(N) Yang-Mills Theory, Phys. Lett. 140B, 313-316 (1984)

120. Unitary Flavor Unification Through Higher Dimensions (with K. Yamamoto), Phys. Rev. Lett. 52, 2016-2018 (1984)

121. Results on Stability of Instanton-Induced Compactification in Eight Dimensions (with P. Moxhay and K. Yamamoto), Phys. Lett. 144B, 354-356 (1984)

122. Left-Right Asymmetry from the Eight-Sphere (with T. W. Kephart), Phys. Rev. Lett. 53, 867-870 (1984)

123. Fifth Workshop on Grand Unification (BOOK), Editor with H. Fried and K. Kang, World Scientific (1984) pages i-xiii and 1-538.

124. Chiral Fermions in Kaluza-Klein Theories, in Fifth Workshop on Grand Unification [see 123] pages 494-507.

125. Lecture on Gauge Field Theories. Chapter Seven. Grand Unification, UNC-Chapel Hill Report IFP-238-UNC. December 1984. 68 pages.

126. Spontaneous Compactification and Chiral Fermions (with K. Yamamoto), Nucl. Phys. 254B, 349-366 (1985)

127. Chiral Fermions from Compactification of O(32) and E(8)xE(8) String Theories (with H. Van Dam and K. Yamamoto), Phys. Rev. Lett. 54, 1114-1117 (1985)

128. Systematics of Quark Mass Matrices in the Standard Electroweak Model (with C. Jarlskog), Phys. Lett. 154B, 421-424 (1985)

129. Weak Anomaly Cancellation in Even Dimensions (with K. Yamamoto), Phys. Lett. 156B, 345-347 (1985)

130. The Chiral Anomaly in Conformal and Ordinary Simple Supergravity in Fujikawa's Approach (with D. R. T. Jones, P. Van Nieuwenhuizen, and S. C. Zhang), in Third Seminar on Quantum Gravity, Editors: M. A. Markov , V. A. Berezin and V. P. Frolov), World Scientific (1985) pages 393-409.

131. Superstring Phenomenology, in Symposium on Anomalies, Geometry, and Topology, Editors:W. A. Bardeen and A. R. White, World Scientific (1985) pages 413-414.

132. Chiral Fermions in String Theories, in Sixth Workshop on Grand Unification, Editors:S. Rudaz and T. Walsh, World Scientific (1986) pages 120-136.

133. Kaluza-Klein and Superstring Theories, in Superstrings, Supergravity, and Unified Theories, Editors:G. Furlan, R. Jengo, J. C. Pati, D. W. Sciama, and Q. Shafi, World Scientific (1986) pages 182-229.

134. Gauge Field Theories (BOOK), Benjamin-Cummings, Frontiers In Physics (1986) pages i-xx and 1-584.

135. One-Loop Finiteness in O(32) Open Superstring Theory (with Y. J. Ng and P. Moxhay), Phys. Rev. Lett. 55, 2107-2110 (1985)

136. Lectures on Gauge Field Theories. Chapter Six. Quantum Chromodynamics. UNC-Chapel Hill Report IFP-257-UNC. 1985. 75 pages.

137. Dual Resonance Models and Superstrings (BOOK) [contains reprints of 38 and 111 above]

World Scientific (1986) pages i-xvi and 1-538.

138. Infinity Cancellation for O(32) Open Strings, in Oregon Meeting, Editor:R. W. Hwa, World Scientific (1986) pages 797-804.

139. Introduction to Superstrings, in Lewes String Theory Workshop, Editors:L. Clavelli and A. Halprin, World Scientific (1986) pages 21-68.

140. Modular Invariance in Closed Superstrings (with K. Kikuchi and Y. J. Ng), Phys. Lett. 174B, 262-266 (1986)

141. Limitations of Heterotic Superstring Phenomenology (with H. W. Braden, T. W. Kephart, and A. K. Kshirsagar), Phys. Rev. Lett. 56, 2668-2671 (1986)

142. Regularization of Open Superstring from Orientable Closed Surface (with A. K. Kshirsagar and Y. J. Ng), Phys. Rev. D34, 2532-2534 (1986)

143. Explicit Evaluation of Pentagon Diagram for Open Superstring (with P. Moxhay and Y. J. Ng), Nucl. Phys. B276, 599-616 (1986)

144. Worldsheet Supersymmetry (with H. W. Braden), Phys. Rev. Lett. 57, 2112-2114 (1986)

145. Topics in Superstring Theory, in Superstrings, Unified Theories, and Cosmology, Editors:G. Furlan, R. Jengo, J. Pati, D. Sciama, E. Sezgin, and Q. Shafi, World Scientific (1987) pages 167-175.

146. Naturalness of Discrete Symmetries in Superstrings (with A. K. Kshirsagar), Phys. Lett. 188B, 321-323 (1987)

147. Chiral Color: An Alternative to the Standard Model (with S. L. Glashow), Phys. Lett. 190B, 157-161 (1987)

148. Unifiable Chiral Color with Natural GIM Mechanism (with S. L. Glashow), Phys. Rev. Lett. 58, 2168-2170 (1987)

149. Two-Loop Chiral Anomalies in Open Superstrings (with T. W. Kephart and T. C. Yuan), Phys. Rev. Lett. 59, 1802-1804 (1987)

150. Superstring Anomalies, in Superstrings, Editors: P. G. O. Freund, and K. T. Mahanthappa, Plenum Publishing (1988) pages 243-249.

151. The P-Adic String N-Point Function (with Y. Okada), Phys. Rev. Lett. 60, 484-486 (1988)

152. Lower Limit on the Axigluon Mass from Upsilon Decay (with F. Cuypers), Phys. Rev. Lett. 60, 1237-1238 (1988)

153. Effective Scalar Field Theory of P-Adic String (with Y. Okada), Phys. Rev. D37, 3077-3079 (1988)

154. Four-Dimensional Parafermionic String (with M. Ubriaco), Phys. Rev. D38 1341-1343 (1988)

155. The P-Adic Approach to String Theory, in Ninth Workshop on Grand Unification, Editor: R. Barloutaud, World Scientific (1988) pages 360-372.

156. On Adelic Formulas for P-Adic Strings (with Y. Okada and M. R. Ubriaco), Phys. Lett. 213B, 260-262 (1988)

157. New P-Adic Strings from Old Dual Models (with Y. Okada and M. R. Ubriaco), Phys. Rev. D39, 1152-1157 (1989)

158. P-Adic Number Fields and String Tree Amplitudes, in Problems On High Energy Physics and Field Theory, Academy of Sciences of the USSR, Editor: V. A. Petrov, Moscow Nauka (1989) pages 49-61.

159. Theory of P-Adic Closed Strings (with H. Nishino), Phys. Rev. Lett. 62, 1960-1963 (1989)

160. Analysis of Bound States of Charged Fermion and Abelian Dyon (with Y. C. Qi and J. Z. Zhang), Phys. Rev. D40, 3533-3536 (1989)

161. Lower Bound on Axigluon Mass from Electron-Positron Annihilation (with F. Cuypers), Phys. Rev. Lett. 63, 125-127 (1989)

162. P-Adic Superstrings and P-Adic Quantum Cosmology, Editors: R.Arnowitt, R.Bryan, M.J.Duff, D. Nanopoulos, and C.N.Pope, World Scientific (1990) pages 13 -20.

163. Last Workshop on Grand Unification (BOOK), Editor, World Scientific (1989) pages i-xi and 1-362.

164. Chiral Color Extension of the Standard Model, in Last Workshop on Grand Uniication [see 163] pages 155-163.

165. SU(3) x SU(3) Chiral Theory, Solution of Strong CP and Wormholes (with J.-Z. Zhang). UNC-Chapel Hill Report IFP-355-UNC. June 1989. 11 pages.

166. Primary Quantization: Creation of Nothing from the Void (with I. Volovich), UNC-Chapel Hill Report IFP-347-UNC. March 1989. 15 pages.

167. The Giovanelli-Haynes Protogalaxy and Dark Matter Candidates (with T. Piran), UNC-Chapel Hill Report IFP-363-UNC. Hebrew University Report HEBREW-10. September 1989. 5 pages.

168. SU(15) Grand Unification (with B. H. Lee), Phys. Rev. Lett. 64, 619-621 (1990)

169. Non-Archimedean Geometry and Applications to Particle Theory, in Differential Geometric Methods in Physics, Editors : L. L. Chau and W. Nahm, Plenum Press (1990) pages 399-408.

170. Status of P-Adic Strings, in Superstrings and Particle Theory, Editors: L.Clavelli and B.Harms, World Scientific (1990) pages 238-250

171. Stability Analysis of P-Adic String Solitons (with H. Nishino), Phys. Lett. 242B, 354-356 (1990)

172. Cosmogenesis and Primary Quantization. (with I. Volovich), Mod. Phys. Lett. A5, 1825-1832 (1990)

173. Families in Confining Theory of Quarks, Sarks and Leptons (with Y. J. Ng), Phys. Rev. D42, 3242-3245 (1990)

174. Classical P-Adic Spacetime (with B. Dragovic and B. V. Urosevic), Mod. Phys. Lett. A5, 1521-1528 (1990)

175. Three Families in Confining Theory of Quarks, Sarks and Leptons, in PASCOS-90, Editors: P. Nath and S. Reucroft, World Scientific (1991) pages 327-332.

176. Higgs Sector and Proton Decay in SU(15) Grand Unification (with T. W. Kephart), Phys. Rev. D42, 3892-3894 (1990)

177. Natural Strong CP Conservation in Flipped Physics (with T. W. Kephart), Phys. Rev. Lett. 65, 820-822 (1990)

178. Strong-Coupled Sark Model, in 1990 International Workshop on Strong Coupling Gauge Theories and Beyond, Editors: T. Muta and K. Yamawaki, World Scientific (1991) pages 118-125.

179. Sarks as Additional Fermions (with J. Agrawal, Y. J. Ng, H. Nishino, and O. Yasuda), Nucl. Phys. B351, 161-182 (1991)

180. Beyond Planck Energy to Non-Archimedean Geometry (with I. Aref'eva), Mod. Phys. Lett. A6, 313-316 (1991)

181. The Wave Function of the Universe and P-Adic Gravity (with I. Aref'eva, B. Dragovic, and I. Volovich), Int. J. Mod. Phys. A6, 4341-4358 (1991)

182. Theta-Parameter Evolution and Topological-Charge Matching Conditions in Chiral-Color Theory (with T.W. Kephart), Phys.Rev.D43, 624-625 (1991)

183. Axigluon Mass Bound from e+e- Annihilation (with F. Cuypers and A. F. Falk), Phys. Lett. 259B, 173-174 (1991)

184. Method for Constructing Models with Strong CP Invariance (with T. W. Kephart), Phys. Rev. Lett. 66 1666-1668 (1991)

185. Strong and Weak CP in a Model with a New Gauged U(1) Symmetry (with D. Ng), Phys. Rev. D43, 3034-3039 (1991)

186. Heavy Axion on Sark Bridge Between Higgs and Technicolor Models (with T. Yanagida), Mod. Phys. Lett. A6, 1589-1596 (1991)

187. Simplified Symmetric Quark Mass Matrices and Flavor Mixing (with Y. Okada), Mod. Phys. Lett. A6, 2169-2172 (1991)

188. Model Building and New Particles, in Particle Phenomenology in the 90's, Editors: A. Datta, P. Ghose, and A. Raychaudhuri, World Scientific (1992) pages 193-204

189. Dynamical Electroweak Symmetry Breaking in a Sark Model (with O. Yasuda), Phys. Rev. D44, 3709-3710 (1991)

190. Strong CP Conservation in Aspon Model, in PASCOS-91, Editors:P. Nath and S. Reucroft, World Scientific (1992) pages 255-262.

191. Light Leptoquarks as Possible Signature for Strong-Electroweak Unification, Mod. Phys. Lett. A7, 559-562 (1992)

192. Baryon Conservation and SU(15) GUT, in Particle Phenomenology in the 90's, Editors: A. Datta, P. Ghose, and A. Raychaudhuri, World Scientific (1992) pages 255-262

193. Phenomenology of the Aspon Model of CP Violation (with D. Ng, T. W. Kephart and T. J. Weiler), Phys. Rev. Lett. 68, 2129-2132 (1992)

194. SU(15) and Difermions, in Physics and Experiments with Linear Colliders, Editors: R.Orava, P.Eerola and M.Nordberg, World Scientific (1992) Volume II, pages 543-546

195. Consistency Checks of Grand Unification (with U. Amaldi, W. de Boer, H. Furstenau, and J. T. Liu), Phys. Lett. 281B, 374-382 (1992)

196. Bottom Quark Mass Predictions in Non-Supersymmetric SU(5) Unification (with J. T. Liu and M. Yamaguchi), Phys. Lett. B277, 130-136 (1992)

197. Dilepton Gauge Bosons: Present Status and Future Prospects (with D. Ng), Phys. Rev. D45, 4240-4245 (1992)

198. Are 100GeV Electron-Electron Collisions Interesting? Mod. Phys. Lett. A7, 2017-2019 (1992)

199. Lower Bound on Dilepton Mass from Polarized Muon Decay (with E. D. Carlson), Phys. Lett. 283B, 123-124 (1992)

200.Possible Solution of Strong CP Problem in Generalized Unimodular Gravity (with Y. J. Ng and H. Van Dam), J. Math. Phys. 33, 3881-3882 (1992)

201. Dilepton Production in e-p and e-e+ Colliders (with J. Agrawal and D. Ng), Nucl. Phys. B386,267-279 (1992)

202. Chiral Dilepton Model and the Flavor Question, Phys. Rev. Lett. 69, 2889-2891 (1992)

203. Lorentz Invariance and No Ghost Theorem for Effective String (with M. Yamaguchi), UNC-Chapel Hill Report IFP-420-UNC. January 1992. 8 pages.

204. Examples of Building Testable Models for the SSC, in Topics on Quantum Gravity and Beyond, Festschrift for L.Witten, Editors: F.Mansouri and J.J.Scanio, World Scientific(1993) pages 377-381

205. Chiral Aspon Model: An Alternative Fully Gauged Model of CP Violation (with T. W. Kephart), Phys.Rev. D47,3655 - 3657 (1993)

206. Searching for Dileptons in Z Decay (with T. W. Kephart, D. Ng and T. C. Yuan), Phys. Lett. B317, 369-370 (1993)

207. Search for Heavy Leptons at Hadron Colliders (with D. Ng, M. Sher and Y. Yuan), Phys. Rev. D48, 3128-3135 (1993)

208. Heterotic Parafermionic Superstring (with J. T. Liu), Phys. Rev. Lett. 70, 130-133 (1993)

209. On the Properties of Dileptons, in The Fermilab Meeting, DPF '92, Editors: C.H.Albright, P.H.Casper, R.Raja and J.Yoh, World Scientific (1993) pages 1259-1261

210. Dileptons in e-e- Collisions, in Workshop on Physics and Experiments with Linear e+e- Colliders, Editors: F.A.Harris, S.Olsen, S.Pakvasa and X.Tata, World Scientific(1993) Vol. II, pages 546-550.

211. Clarification of 331 Model, in Workshop on on Physics and Experiments with Linear e+e- Colliders, Editors:F.A.Harris,S.Olsen,S.Pakvasa and X.Tata, World Scientific(1993) Vol. II, pages 911-916.

212. Leptoquarks in e-e- Colliders (with J. Blümlein), in Workshop on Physics and Experiments with Linear e+e- Colliders, Editors: F.A.Harris, S.Olsen, S. Pakvasa and X. Tata. World Scientific(1993) Vol. II, pages 926-928.

213. Testing Extensions of the Standard Model, in Salamfestschrift, Editors: A.Ali, J.Ellis, and S.Randjbar-Daemi, World Scientific (1994) pages 173-185.

214. Cubic Mass Relation in the SU(3)xU(1) Electroweak Theory (with P. Krastev and J.T.Liu), Mod. Phys. Lett. 9A, 761-770 (1994)

215. Influence of Additional Fermions and Gauge Bosons on Epsilon Primed Over Epsilon (with J. Agrawal), Nucl. Phys. B419, 254-278 (1994)

216. Motivated and Falsifiable Models Beyond the Standard One, in New Physics with New Experiments, Editors: Z.Ajduk, S Pokorski and A. K. Wroblewski, World Scientific (1994) pages 274 - 286.

217. Phenomenology of an SU(3) X SU(3) X U(1) Model of Flavor (with J.T.Liu, D.Ng and B.C.Rasco), Mod. Phys. Lett. 9A,1975 - 1984 (1994).

218. Beyond the Standard Model on the Supercollider, in Walifest - MRST15, New Directions in the Application of Symmetry Principles to Elementary Particle Physics, Editor: J. Schechter, World Scientific (1994) pages 62 - 74.

219. Outcome from Spontaneous CP Violation for B Decays (with A. W. Ackley, B. Kayser and C. N. Leung), Phys. Rev. D50, 3560 - 3563 (1994)

220. Partial Derivation of the Transformation Properties of Quarks and Leptons (with R. N. Mohapatra), Phys. Rev. D50, 3569 -3571 (1994).

221. New Gauge Bosons for Multi-TeV pp Scattering, in Proceedings of the Cornelius Lanczos International Centennial Conference, Editors: J.D.Brown, M.T.Chu, D.C.Ellison and R.J.Plemmons, Siam publishers (1994) pages 623-625.

222. Minimal Family Symmetry (with T. W. Kephart), Phys. Rev. D51, 1 - 4 (1995).

223. Retrospective on P-Adic String Theory, in Proceedings of the Steklov Mathematical Institute of the Academy of Science of Russia, 203, Moscow, Nauka (1994) pages 287-291.

224. Spontaneous CP Violation and Dedicated Studies of B Mesons, in Yukawa Couplings and the Origins of Mass. Editor: P. Ramond. International Press, Boston (1996) pages 133-135.

225. Spontaneous CP violation and Neutral B Meson Decays, in Proceedings of WHEPP-3, Madras. Pramana (India) 45 (Supplement) 109-112 (1995).

226. Family Symmetry, in Proceedings of WHEPP-3, Madras. Pramana (India) 45 (Supplement) 113-116 (1995).

227. The Third Family is Different, in PASCOS 94, Proceedings of the Fourth International Symposium on Particles, Strings and Cosmology. Editor: K. C. Wali. World Scientific (1995) pages 63 - 80.

228. Quantum Anomalies and Three Families, in Quantum Coherence and Reality, Editors: J. S. Ananden and S.L. Safko, World Scientific (1994) pages 355-357.

229. How the Top Family Differs, in Beyond the Standard Model IV, Editors: J. Gunion, T. Han and J. Ohnemus, World Scientific (1995) pages 463-465.

230. Simple Non-Abelian Finite Flavor Groups and Fermion Masses (with T. W. Kephart), Int. J. Mod. Phys. 10A , 4689-4704 (1995).

231. Massive Superstrings are Black Holes (with T. W. Kephart), Mod. Phys. Lett. A10 , 2571-2578 (1995).

232. The Decay b -> s + photon in the 3-3-1 Model (with J. Agrawal and J. T. Liu), Int. J. Mod. Phys. A11, 2263 - 2280 (1996).

233. Quark Mass Textures within a Finite Non-Abelian Dicyclic Group (with O. C. W. Kong), Phys. Rev. Lett. 75, 781-784 (1995).

234. Dilepton Production in High Luminosity Multi-GeV Electron Scattering (with B. C. Rasco), Phys. Rev. Lett. 75, 1899-1902 (1995).

235. Identification as Black Holes of All Massive Superstring States, in Proceeding of Unified Symmetry: In the Small and in the Large II, Editors: B. N. Kursunoglu, S. Mintz and A. Perlmutter, Plenum Press (1995). pages 197 - 202.

236. Constraining $\alpha_S(M(Z))$ from the Hidden Sector (with B. Keszthelyi, B.D. Wright, and T.W. Kephart), UNC-Chapel Hill Report IFP-714-UNC.

Vanderbilt University Report VAND-TH-95-5. November **1995**. 12 pages.

237. Effective Lagrangians and Parity-Conserving Time-Reversal Violation at Low Energies (with J. Engel and R. Springer), Phys. Rev. D53, 5112 - 5114 (1996).

238. Treating Top Differently from Charm and Up, in Particle Theory and Phenomenology, Editors: K.E. Lassila, J. Qiu, A. Sommerer, G. Valencia, K. Whisnant and B.L. Young. World Scientific (1996) pages 229-245. hep-ph/9507351.

239. Influence of Addional Fermions and Gauge Bosons on Epsilon Primed Over Epsilon, in Proceedings of the International Europhysics Conference on High Energy Physics, Editors: J. Lemonne, C. Vander Velde and F. Verbeure, World Scientific. (1996) pages 164 - 166.

240. Flavors and Horizontal Symmetries, in From the Standard Model to Grand Unified Theories, Proceedings of the Sixth Yukawa International Seminar, Editors: M. Bando, K. Inoue and T. Kubo. Prog. Theor. Phys. Suppl. 123, 327 - 335 (1996).

241. Dicyclic Horizontal Symmetry and Supersymmetric Grand Unification (with O. C. W. Kong), Phys. Rev. D53, 2293-2295 (1996).

242. Horizontal Symmetry for Quark and Squark Masses in Supersymmetric SU(5) (with O. C. W. Kong), Phys. Rev. Lett. 77, 1699 - 1702 (1996).

243. Narrow Resonance in e-e- Scattering due to Additional Gauge Boson, in Proceedings of the Electron-Electron Linear Collider Workshop, Editor: C. A. Heusch. Int. J. Mod. Phys. 11A, 1621 - 1626 (1996).

244. Constraining the QCD Coupling from the Superstring Hidden Sector, in Neutrino Mass, Dark Matter, Gravitational Waves, Monopole Condensation, and Light Cone Quantization, Editors: B.N. Kursunoglu, S.L. Mintz, and A. Perlmutter, Plenum Press (1996) pages 323-330.

245. An Elusive Z' Coupled to Beauty, in Effective Theories and Fundamental Interactions. Editor: A. Zichichi. World Scientific (1997) pages 433-448. hep-ph/9610320.

246. Elusive Z' Coupled to Quarks of Third Generation (with M. B. Wise and B. D. Wright), Phys. Rev. D54, 5820 - 5823 (1996)

247. How to Define the Minimal Supersymmetric Aspon Model (with O.C.W. Kong), UNC-Chapel Hill Report IFP-726-UNC. December 1996. 14 pages.

248. Quarks, Squarks and Textures (with O. C. W. Kong), Phys. Rev. D55, 5501-5507 (1997). hep-ph/9603372.

249. Z_8' Orbifold Models, Supersymmetry Breaking, and Cosmology (with B. Keszthelyi, T. W. Kephart and B. D. Wright), Mod. Phys. Lett. A12, 439-446 (1997). hep-ph/9611445.

250. Production of Two Non-Conjugate Leptoquarks in e-e- Collision (with F. Cuypers and R. Rueckl), Phys. Lett. B390, 221-226 (1997) hep-ph/9610464.

251. Cabibbo Mixing and the Search for CP Violation in the Aspon Model (with S. L. Glashow), Phys. Rev. D55, 1691-1692 (1997). hep-ph/9609496.

252. Strong CP and Low-Energy Supersymmetry (with O.C.W. Kong), Phys. Lett. B402,297-302 (1997). hep-ph/9612452.

253. Orthogonal Mixing and CP Violation, in High Energy Physics and Cosmology. Editors: B. N. Kursunoglu, S.L.Miatz and A. Perlmutter. Plenum Publishing Corporation (1997) pages 67-78.

254. CP Violation in the Aspon Model, in Proceedings of the Second International Conference on B Physics and CP Violation. Editors: T.E. Browder, F.A. Harris and S. Pakvasa. World Scientific. (1998) p. 382-386. hep-ph/9705461.

255. Leptoquarks, in Beyond the Standard Model V, Editors: G. Eigen, P. Osland and B. Stugu. AIP Conference Proceedings No. 415, American Institute of Physics (1997) pages 281-287. hep-ph/9706220.

256. S and T for Leptoquarks and Bileptons (with M. Harada), UNC-Chapel Hill Report IFP-748-UNC. November 1997. 4 pages.

257. Model Building, in Fundamental Particles and Interactions. Frontiers in Contemporary Physics. An International Lecture and Workshop Series at Vanderbilt University. Editors: R.S. Panvini and T.J. Weiler. AIP Press.(1998) pages 34-38. hep-ph/9706237.

258. Models for New Physics, in Proceedings of the Blois 97 Workshop, Recent Advances in Hadron Physics. Editors: K. Kang, S.K. Kim and C. Lee. World Scientific Publishing Co.(1998) p. 304-313.

259. Longevity and Highest-Energy Cosmic Rays (with B. Keszthelyi and Y.J. Ng), Int. J. Mod. Phys. D8, 1-6 (1999). astro-ph/9709080.

260. Long-Lived Quarks? (with P.Q. Hung) UNC-Chapel Hill Report IFP-745-UNC. Phys. Rev. D58, 057704-1 to 057704-4(1998). hep-ph/9711218.

261. Constraints from Precision Electroweak Data on Leptoquarks and Bileptons (with M. Harada), UNC-Chapel Hill Report IFP-748-UNC. Phys. Rev. D58. 095013-1 to 095013-10 (1998). hep-ph/9711448.

262. Bileptons - Status and Prospects, in Second e-e- Workshop at Santa Cruz. Int. J. Mod. Phys. A13, 2345-2349 (1998). hep-ph/9711281.

263. Leptoquarks Revisited, in Physics of Mass. Editors: B.N. Kursunoglu, S.L.Mintz and A. Perlmutter. Kluwer Academic/ Plenum Publishers. (1998) pages 75-80.

264. Kaon Spontaneous CP Violation Reevaluated (with M. Harada), Phys. Rev. D59, 017901-1 to 017901-3 (1999). hep-ph/9803416.

265. Alternative Solution of Strong CP, in Proceedings of Fifth IFP Workshop: Axions. Gainesville, FL. Editor: P. Sikivie. Nucl. Phys. (Proc. Supp.) 72, 36-39 (1999). hep-ph/9804423.

266. Highest-Energy Cosmic Rays, in Proceedings of PASCOS98, Editor: P. Nath. World Scientific (1999). pages 60-65. astro-ph/9804219.

267. Spontaneous CP Violation, in Particle Physics and Cosmology, First Tropical Workshop. High Energy Physics, Second Latin American Symposium.

San Juan, Puerto Rico 1998. Editor: J. P. Nieves. AIP Conference Proceedings (1998). pages 160-164. hep-ph/9804395.

268. Bileptons from Muon Collider Backward Scattering (with X. Guan), Mod. Phys. Lett. A13, 2621-2625 (1998). hep-ph/9805402.

269. Neutron Electric Dipole Moment and Spontaneous CP Breaking (with M. Harada), Phys. Lett. B436, 369-371 (1998). hep-ph/9805489.

270. Cosmic Background Radiation Temperature Anisotropy: Position of First Doppler Peak (with Y.J. Ng and R.N. Rohm), Mod. Phys. Lett. A13, 2541 - 2549 (1998). astro-ph/9806118.

271. Quarks and Leptons beyond the Third Generation (with P.Q. Hung and M. Sher), Physics Reports 330, 263-348 (2000). hep-ph/9903387.

272. Aspects of Soft and Spontaneous CP Violation (with M. Harada), Phys. Rev D59, 036004-1 to 036004-9 (1999). hep-ph/9809402.

273. Neutrino Masses in R Parity Violating Supersymmetry (with L. Clavelli), UNC-Chapel Hill Report IFP-764-UNC. hep-ph/9811326.

274. ADS/CFT String Duality and Conformal Gauge Theories, Phys. Rev. D60, 041901-1 to 041901-3 (1999). hep-th/9812117.

275. Position of First Doppler Peak, in COSMO-98, Second International Workshop on Particle Physics and the Early Universe. Editor: D.O. Caldwell. AIP Publishing (1999) pages 172-175. astro-ph/9901012.

276. Is the Cosmological Constant Non-Zero? In Confluence of Cosmology, Massive Neutrinos, Elementary Particles, and Gravitation. Editors: B.N. Kursunoglu, S.L. Mintz and A. Perlmutter. Plenum Publishing (1999) pages 85-94. astro-ph/9901013.

277. Conformal N=0 d=4 Gauge Theories from AdS/CFT Superstring Duality? (with W.F. Shively) Phys. Lett. B454, 49-52 (1999). hep-th/9902168.

278. Unification with Enlarged Kaluza-Klein Dimensions (with A. Rasin), Phys. Lett. 460B, 313-324 (1999). hep-ph/9903479.

279. Conformal Approach to Particle Phenomenology (with C. Vafa), Harvard University Report HUTP-99/A018. hep-th/9903226.

280. Mersenne Primes. Polygonal Anomalies and String Theory Classification (with T.W. Kephart), Phys. Rev. D60, 087901-1 to 087901-4 (1999). hep-th/9904212.

281. Conformality and Gauge Coupling Unification, Phys. Rev. D60, 085004-1 to 085004-3 (1999). hep-th/9905042.

282. Can the Zee Ansatz for Neutrino Masses be Correct? (with S.L. Glashow) Phys. Lett. B461, 95-98 (1999). hep-ph/9906375.

283. Conformality from Field-String Duality on Abelian Orbifolds, Phys. Rev. D60, 121901-1 to 121901-3 (1999). hep-th/9907051.

284. Application of Conformal Gauge Theories Derived from Field-String Duality, in Proceedings of Beyond the Desert '99. Editors: H.V. Klapdor-Kleingrothaus

and I.V. Krivosheina. Institute of Physics Publishing (2000) pages 69-86. hep-th/9908167.

285. Conformality and Brodsky, in Proceedings from JLab/University of Georgia Workshop on "The Transition from Low to High Q Form Factors." Organizing Committee: G. Strobel and D. Mack. Thomas Jefferson National Accelerator Facility (1999) pages 113-126. hep-ph/9909447.

286. Nonabelian Discrete Symmetries, Fermion Mass Textures and Large Neutrino Mixing (with A. Rasin), Phys. Lett. 478B, 424-433 (2000). hep-ph/9910522.

287. Chiral Fermions and AdS/CFT Duality for a Nonabelian Manifold (with T.W. Kephart), Phys. Lett. B 485, 403-407 (2000). hep-th/9912028.

288. Bilepton Resonance in Electron-Electron Scattering, in Proceedings of the Third International Conference on Electron-Electron Scattering at TeV Energies. University of California at Santa Cruz, December 10-12, 1999. Int. J. Mod. Phys. A15, 2455-2460 (2000). hep-ph/0002017.

289. Relic Neutrinos and Z-Resonance Mechanism for Highest-Energy Cosmic Rays (with J.L. Crooks and J.O. Dunn), Ap.J. Lett.546, L1-L3 (2001). astro-ph/0002089.

290. Conformality, Particle Phenomenology and the Cosmological Constant, in Quantum Gravity, Generalized Theory of Gravitation, and Superstring Theory-Based Unification. Editors: B.N. Kursunoglu, S.L. Mintz and A. Perlmutter. Kluwer Academic / Plenum Publishers. (2000) pages 13 - 23. hep-th/0002053.

291. Seeking Gauge Bileptons in Linear Colliders (with A. Rasin), Phys. Lett. B 482, 129-132 (2000). hep-ph/0002135.

292. Quintessence and CMB (with J. L. Crooks, J.O. Dunn and Y.J. Ng), UNC-Chapel Hill Report No. IFP-784-UNC. astro-ph/0005406.

293. Minimal Model for Neutrino Masses and Mixings, in Proceedings of Second Tropical Conference on Particle Physics and Cosmology, Editor: J. Nieves. AIP Conference Proceedings (2000). pages 18-23. hep-ph/0005206.

294. Economic Model for Neutrino Masses and Mixings, in Proceedings of DPF2000 (August 2000) Columbus, Ohio. Editors: K.K. Gan and R. Kass. Int. J. Mod. Phys. A16, Suppl. 1C, 701-703 (2001). hep-ph/0008278.

295. AdS/CFT Duality and Conformality for Non-Abelian Orbifold, in Proceedings of DPF2000 (August 2000) Columbus, Ohio. Editors: K.K. Gan and R. Kass. Int. J. Mod. Phys. A16, Suppl. 1C, 975-977 (2001). hep-th/0008202.

296. Quintessence Model and Cosmic Microwave Background, in Proceedings of DPF2000 (August 2000) Columbus, Ohio. Editors: K.K. Gan and R. Kass. Int. J. Mod. Phys. A16, Suppl. 1C, 1050-1052 (2001). astro-ph/0008412.

297. CMB with Quintessence: Analytic Approach and CMBFAST (with James L. Crooks, James O. Dunn, Y.Jack Ng and Ryan Rohm), Mod. Phys. Lett. A16, 63-73 (2001). astro-ph/0010404.

298. Finite N AdS/CFT Correspondence for Abelian and Nonabelian Orbifolds, and Gauge Coupling Unification, J. Math. Phys. 42, 2915-2928 (2001). hep-th/0011165.

299. Classification of Conformality Models Based on Nonabelian Orbifolds (with T.W. Kephart), Phys. Rev. D64, 086007-1 to 086007-33 (2001). hep-th/0011186.

300. Quintessence and Cosmic Microwave Background, in the Role of Neutrinos, Strings, Gravity, and Variable Cosmological Constant in Elementary Particle Physics. Editors: B.N. Kursunoglu, S.L. Mintz and A. Perlmutter. Kluwer Academic/Plenum Publishers. (2001). pages 15-31. astro-ph/0102344.

301. Model of Soft CP Violation Using Scalars with Quarks Number Two (with S.L. Glashow and T. Yoshikawa), Phys. Rev. Lett. 87, 011801 (2001). hep-ph/0103022.

302. Phenomenology of Anti-de-Sitter-Conformal Field Theory Duality, in Proceedings of Cairo International Conference on High Energy Physics. Editors: S. Khalil, Q. Shafi and H. Tallat. Rinton Press (2001) pages 126-134. hep-ph/0104149.

303. Prediction of Sine Squared Theta in the Conformal Approach to Coupling Unification (with R.N. Mohapatra and S. Suh), Phys. Lett. 520B, 331-336 (2001). hep-ph/0104211.

304. Gauge Field Theories, Second Edition (BOOK) John Wiley & Sons, Inc. (2000) pp i-vii and 1-343.

305. Spontaneous CP Violation: An Alternative to the Standard Model, Int. J. Mod. Phys. A17 S1, 47-57 (2002). 2001: A Spacetime Odyssey. Editors M.J. Duff and J.T. Liu. World Scientific (2002). pages 47-57. hep-ph/0107240.

306. Finite-N Conformality and Gauge Coupling Unification, in String Theory, 10th Tohwa University International Symposium on String Theory. Editors: H. Aoki and T. Tada. AIP Conference Proceedings #607. (2002). pages 88-96. hep-th/0108011.

307. PASCOS 2001 (BOOK) Eighth International Symposium on Particles, Strings and Cosmology. Editor with Y. J. Ng. Rinton Press (2001). Pages i-xiv and 1 to 427.

308. Modified Dispersion Relations from Closed Strings in Toroidal Cosmology (with M. Bastero-Gil and L. Mersini), Phys. Rev. D65, 106002-1 to 106002-12 (2002). hep-th/0110167.

309. Zee Model Confronts SNO Data (with M. Oh and T. Yoshikawa), Phys. Rev. D65, 073014-1 to 073014-6 (2002). hep-ph/0110300.

310. Non-Associative Loops for Holger Bech Nielsen (with S.L. Glashow, T.W. Kephart and R.M. Rohm) Festschrift dedicated to the 60th Birthday of Holger Bech Nielsen. Editors: N.M. Borstnik, C.D. Froggatt and D. Lukman. DMFA-Zaloznistvo. Ljubljana. (2001) pages 56-63. hep-th/0111292.

311. Zeroes of the Neutrino Mass Matrix (with S.L. Glashow and D. Marfatia), Phys. Lett. 536B, 79-82 (2002). hep-th/0201008.

312. Dark Energy from Strings, in Coral Gables Conference 2001 on Cosmology and Elementary Particle Physics. Editors: B. Kursunoglu, S.L. Mintz and A. Perlmutter. AIP Conference Proceedings Volume 624 (2002). Pages 59-68. hep-th/0202063.

313. Sufficiently Small $\bar{\theta}$ in $SU(3)^3 \times S_3$ Unification Model (with K. Chalut, H. Cheng, K. Stowe and T. Yoshikawa), Mod. Phys. Lett. A17, 1513-1524 (2002). hep-ph/0204074.

314. Majorana Mass Zeroes from Triplet VEV without Majoron Problem (with M.C. Oh and T. Yoshikawa), Phys. Rev. D66, 033006-1 to 033006-8 (2002). hep-ph/0204273.

315. Perturbative Inaccessibility of Conformal Fixed Points in Nonsupersymmetric Quiver Theories (with P. Minkowski), CERN preprint CERN-TH/2002-180. hep-th/0208024.

316. Strong-Electroweak Unification at About 4 TeV, Mod. Phys. Lett. A18, 1377-1382 (2003). hep-ph/0208044.

317. Cosmological Sign of Neutrino CP Violation (with S.L. Glashow and T. Yanagida), Phys. Lett. B548, 119-121 (2002). hep-ph/0208157.

318. How to Test Stringy Dark Energy, Phys. Lett. B555, 139-143 (2003). astro-ph/0209037.

319. Conformality and Unification of Gauge Couplings. Proceedings of the First International Conference on String Phenomenology. Editors: S.A. Abel, A.E. Faraggi, A. Ibarra and M. Plumacher. World Scientific (2003). pages 124-129. hep-th/0209121.

320. Connecting Link Between Leptogenesis and Oscillations. Third Tropical Workshop on Particles and Cosmology: Neutrinos, Branes and Cosmology. Editors: J.F. Nieves and C.N. Leung. AIP Conference Proceedings (2003). Pages 128-132. hep-ph/0209273.

321. The Fate of Dark Energy (with T. Takahashi), Phys. Lett. B557, 135-138 (2003). astro-ph/0211544.

322. Stability Issues for Dark Energy, Mod. Phys. Lett. A19, 801-806 (2004). hep-th/0302007.

323. Quiver Gauge Theory and Unification at About 4 TeV. Proceedings of SCGT02 in Nagoya, Japan. December 2002. Editors: M. Harada, Y. Kikukawa and K. Yamawaki. World Scientific Publishing.(2003). pages 250-265. hep-th/0302057.

324. Robustness and Predictivity of 4 TeV Unification (with R.M. Rohm and T. Takahashi), Phys. Lett. B570, 67 (2003). hep-ph/0302074.

325. AdS/CFT Correspondence and QCD with Quarks in Fundamental Representations, Phys. Lett. B567, 265-267 (2003). hep-th/0305160.

326. Cosmic Degeneracy with Dark Energy Equation of State (with J.L. Crooks, J.O. Dunn, H.R. Norton and T. Takahashi), Astropart. Phys. 20, 361-367 (2003) astro-ph/0305495.

327. Grand Unification at About 4 TeV. Proceedings of SUGRA20. Editor: P. Nath. Rinton Press (2004). pages 407-417. hep-th/0306029.

328. MSSM from AdS/CFT (with T.W. Kephart), Phys. Lett. B585, 24-28 (2004). hep-ph/0306053.

329. Consistency Conditions for AdS/CFT Embeddings (with T.W. Kephart), Int. J. Mod Phys. A19, 593-598 (2004). hep-th/0306207.

330. Dark Energy Present and Future, in Neutrinos, Flavor Physics and Precision Cosmology. Proceedings of Fourth Tropical Workshop in Cairns, Australia. Editors: J.F. Nieves and R.R. Volkas. AIP Conference Proceedings. Volume 689.(2003) pages 197-205. astro-ph/0307071.

331. AdS/CFT-Inspired Unification at About 4 TeV, Proceeding of the 2nd International Conference on String Phenomenology, Editors: V. Sanz, S. Abel, J. Santiago and A. Faraggi, World Scientific (2004). pages 160 - 166. hep-ph/0309275.

332. Status of the Standard Model of Particle Phenomenology, Proceedings of The Euroconference on Symmetries Beyond the Standard Model, Editors: N.M. Borstnik, H.B. Nielsen, C.D. Froggatt and D. Lukman, DMFA-ZALOZNISTVO, Ljublyana. (2003), pages 1-5. hep-ph/0310217.

333. AdS/CFT Correspondence and Unification at About 4 TeV, Proceedings of The Euroconference on Symmetries Beyond the Standard Model, Editors: N.M. Borstnik, H.B. Nielsen, C.D. Froggatt and D. Lukman. DMFA-ZALOZNISTVO. Ljublyana. (2003). pages 16-22. hep-th/0310171.

334. On Deviations from Bimaximal Neutrino Mixing (with S. Petcov and W. Rodejohann), Nucl. Phys. B287, 31-54 (2004). hep-ph/0401206.

335. Candidates for Inflaton in Quiver Gauge Theory (with T. Takahashi), Phys. Rev. D, (in press). hep-ph/0402119.

336. Aspects of Neutrino Mass Matrices, talk at Coral Gables 2003 (this proceedings). hep-ph/0403164.

337. Bigger Rip with No Dark Energy (with T. Takahashi), Astropart. Phys. (in press). astro-ph/0405333.

338. AdS/CFT Inspired String Phenomenology, talk at Pheno 2004 Symposium, Madison, WI. hep-th/0405226.

339. Possible Gauge Theoretic Origin for Quark-Lepton Complementarity (with R.N. Mohapatra). hep-ph/0407139.

340. String Approach to QCD Quarks in Fundamental Representations, talk at 8th Workshop on Nonperturbative QCD, IAP, Paris. hep-ph/0407142.

341. Enhanced Big Rip without Conventional Dark Energy, talk at 20th IAP Colloquium on CMB, IAP, Paris. astro-ph/0407353.

342. String Theory and Unification, talk at PASCOS 2004 / Nath Fest, World Scientific (in press). hep-th/0409010.

343. Dark Energy - A Pedagogic Review, keynote talk at joint session of 5th Rencontres du Vietnam, Hanoi, August 2004 (in press). astro-ph/0409166.

2 Astrophysics and Cosmology

GEOMETRIC EXPLANATION FOR AN ACCELERATING UNIVERSE

EDWARD J. BACINICH*

Alpha Omega Research Foundation Inc
1048 S. Ocean Blvd. Palm Beach Fl 33480

Abstract. The appearance and interpretation of an accelerating universe may be an observed distortion resulting from a universe defined by spherical geometry. The annihilation of Planck and anti-Planck mass is paramount in explaining the Big-Bang [1]. In a model similar to the standard model of a Riemannian-Friedmann-Lemaitre hypersphere, the primordial energy of the Big-Bang is released in the form of electromagnetic-like radiation that expands radiantly in every 4D direction from time zero in the structure of hyper-waves carried by Planck and anti-Planck bosons. The resulting geometry shows that time is synonymous with the lightspeed expansion of our universe. In this model we find that time is not parallel but radiant. This implies that time is a vector - where every place we observe has a unique time direction (angle) with a magnitude (age) and a light cone. The result of this condition is the illusion that the further back we look from our position in spacetime, space appears to be contracted and time appears to run slower both exponentially and logarithmically. Simply stated, we can not rely on observations unless we understand the geometric distortions inherent in curved photon paths.

1. Introduction

The model attributes spacetime expansion of our universe to a hyperspherical electromagnetic-like wave in a quantized Riemannian geometry [1-4]. Here the Big-Bang (BB) acts as an emitter of super energetic radiation which is carried by Planck and anti-Plank bosons locked in step in a wave train of concentric (parallel) universes in which a single wave happens to be our quantum vacuum and universe.

Our universe is a 4D-quantum wave expanding in all 4D-time directions at lightspeed (time vectors). The very motion of this hyperwave establishes time and space where time is our empirical measurement of the lightspeed expansion of space (the quantum vacuum). On this hyper-wave surface, ordinary electromagnetic radiation follows a four-dimensionally curved trajectory at the speed of light [3,4]. Geometrically, the trajectory must follow the natural log to the base e in a great nautilus-like spiral motion on the lightspeed expanding hyperwave surface.

2. Red-Shift z

In a constant lightspeed expanding hyperspherical wave universe we discover that red-shift z and the angle in ø radians have a direct one to one correlation. We find that z = ø radians [3] and emits red shifted light from the source which has nothing to do with the stretching of either space or wavelengths due to the expansion of the universe. Instead they are the result of virtually infinite 4D-time vectors.

Current models assume that time is parallel as opposed to radiant. These models lead to the assumption that the universe is accelerating where time is going the same rate and direction for distant objects as in our local position. This is only true in a global perspective where all real objects are simultaneously in the present with the same radial magnitude (thus age) from the BB but expanding in different time directions (4D angles). However we can not observe this. We can only observe the past and not the present. If contemporary astrophysical models assume that all time observations for past objects are constant with ours, then they are not accounting in their interpretation of data for dramatically smaller magnitudes of time vectors in the extreme past, (Fig.2.). Since every observed place has its own unique space-time vector radian from the BB origin, any position in the universe can only observe all other positions moving slower in time relative to themselves. We observe photons on a curved light-cone where the 45° light cone angle to the local time direction changes progressively and exponentially in relation to our position. This changing orientation of light cones correctly describes the illusion that we observe. But, it should be restated that in reality time is constant for all locations; a conclusion that can be understood from the geometric diagram in the bisection of a 4D universe down its central axis (figure 1).

Consider an asteroid on a collision course with the earth. When it is first discovered, it is one light year away in spacetime. Being one year behind us in time, how could it ever hit us and why should we worry? It must by some means make up the one-year time differential to enter into our same space and time frame. Since we move in time at lightspeed in a 4D-direction only, the asteroid being back in time, must appear to move faster than our rate of time if it is to collide with us.

The Doppler effect is nothing more than a consequence of time. When we see a blue shifted object, it means that we see the object approaching us and moving uniformly through time at an increased rate. When we see a red shifted object, it means that it is moving uniformly away from us and time is going slower at a decreased rate. In other words, in relation to the observer, a blue shifted object is catching up with our rate of time and a red shifted object is receding from us and progressively moving slower than our rate of time. This is what we observe, but this is not reality. Right now we can only observe images in various positions of the past where no real object can ever be observed by us in our present leading time frame.

All observations are merely images of the past. Notice that because of our relativistic time expansion, the z component of x, y, z in 3 space is contracted flat. Thus there is no real inside or outside to our lightspeed-expanding universe.

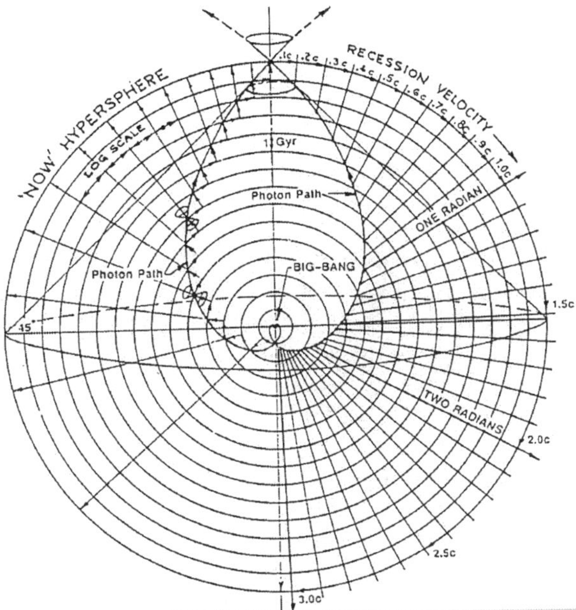

Figure 1. Bisection of a 4D Universe cut down its central axis

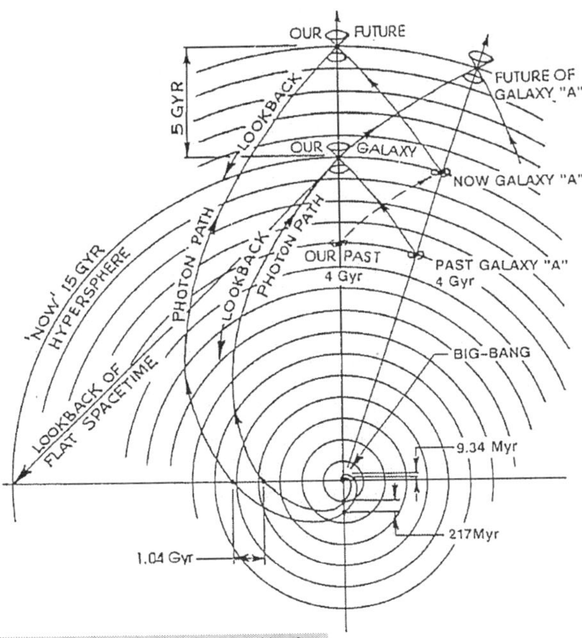

Figure 2. Geometric time dilation

3. *Space, distance, location and nothing*

The quantum vacuum is composed of hexagonal wave-packets containing three Planck and three anti-Planck bosons. Space can be defined as where the bosons are and exist. Between the bosons is a quantum state of nothingness in increments of the V_o quanta [1]. 'Distance' can be defined as the number of Planck wave-packets on an arc between time line radians. At the BB with a zero radius magnitude of $r_i/V_o \sim 4.33 \times 10^{25}$ $(cm)^o$ the finite number of Planck bosons given to us is $\sim 6 \times 10^{189}$. The number of wave-packets between an arc of one centimeter is $\sim 10^{99}$ packets. The total energy density is $\sim 10^{120}$ GeV/cm^3 x $10^{189} = 10^{309}$ GeV/cm^3 for the finite total energy density of our universe. As the universe expands 'nothingness' increases but the number of wave-packets in one centimeter is the same. This 'nothingness' is paramount in describing the expansion of the universe.

As the universe expands at lightspeed (which is the speed of time) the quantum state of nothingness increases between time-line vectors of any two wave-packets. There is no increase of 3D space because the number of boson packets between time-line-radials remain constant through time, except for gravitational considerations. The process of gravity actually decreases (only marginally) the amount of Planck bosons between time vector angles. Thus as the universe expands, there is a gradual decreasing of space rather than the creation of more space as popular theories suggest.

Our hyperspherical model shows that time must be radiant and not parallel [1]. At red shift $ø = \pi$, the time direction is opposite to ours. Looking in all directions from our position, where all photons focus on the observer at focal-point zero (FP_o), we will see the same point at red-shift $z = \pi$ spread all over the sky where all photons focus from a previous epoch at focal point $FP_1(\pi)$. The same is true of FP_2 (2π), FP_3 (3π), FP_4 (4π), etc., all spread over the sky and would appear to us as the CMB. We know that no object from our position occupies the focal positions of F_1, F_2 otherwise we could not clearly see beyond π ($z = 3.14159$ radians) or 2π ($z = 6.28318$ radians).

Time runs slower (dilated) by e to the power of ø radians or to the power of red shift z by,

$$t_{dil} = e^{ø}, \tag{0.1}$$

(see ref 3, Table II, P515).

Observed recession velocity is the product of the speed of light c and ø radians. However, because of time dilation, the observed recession velocity is,

$$v_{rec} = cø/e^{ø}, \tag{0.2}$$

where c is the speed of light, red shift z = ø radians and e is the natural log to the base 2.7182818. The age of the universe at red shift z, is given by whatever we say the universe's age is now (say 15 Gyr) divided by the e to the power of z = ø radians.

$$\text{Age then} = \text{Age now}/e^{ø} \qquad (0.3)$$

The half-life of any age universe always has a red-shift that is z = 0.693 radians because $e^{0.693} = 2$. This shows that time (i.e. the frequency of events) appears to run slower by one half. At this and at every radian, photons are emitted red shifted at the source because of time dilation of angle ø which produces the observed red shift.

Although there is a 4D-lightspeed expansion (time), it should be restated that there is no 3D spatial expansion. What appears to expand in our 4D diagram is the circumference which represent x, y and z of 3-space where z is contracted flat because of the relativistic lightspeed time expansion. It is the quantum state of nothingness that increases as the universe expands in time. Thus any circumference on the hyperspherical expanding wave surface has the same number of Planck bosons originally given to us that comprises our quantum vacuum and is spacetime.

Past spacetime is distorted when viewed from our own *present unique and leading* position. *Unique*: because all other objects move successively in different time directions the further in the past we look. *Leading*: because along any line of sight, we only see images that are in the past and are not the real objects in our present spacetime. Therefore the definition of length between two points can be described as the 4D time angles between them. The greater the angle between time vectors emanating radiantly from the BB, the further the objects appear to be in the past. It should also be noted by inspection of cited papers [2-4] that other than peculiar velocities and streaming, a galaxy at red shift z = 1 for example was, is, and always will be, one radian away. The question that arises is why doesn't the galaxy move farther away and increase its recession velocity along with its time angle (z = 1 radiant) as the universe ages? We find that the only motion the galaxy shows as the universe ages is to move further back in time from our position (thereby increasing its line of sight distance) without any increase in recession velocity or time angle, see fig.2.and ref. 3, fig. 5, P 512. This geometrical fact by itself could be interpreted as global accelerated expansion.

In our model, what we experience as *time* is simply the radiant 'c' expansion of our wave-like hyperspherical universe moving away from the BB. Every time direction has an equal and opposite time direction. Symmetric and asymmetrical time coexists in this model without paradox.

Notice that the net effect of gravity at any age along the circumference of a wave-surface universe adds up to zero, c expansion rate is not affected.

The Hubble Constant can be calculated in the following method which shows that space itself does not expand. As an example we use a galaxy 'A' at red shift z = ø = 1.1 radians, giving us a hypersphere of 4,993 Myrs in a 15,000 Myr universe - using equation (0.3). Here we use the arc length expressed in Mpc from galaxy 'A' to our spacetime then (a), instead of the photon look-back trajectory from galaxy 'A' to our spacetime position now (b), which we believe is illusionary. The great spiral trajectory of photons from time past is influenced not only by the lightspeed motion of time but also by crossing different time vector directions. The arc length is given by,

$$\text{arc length} = \text{ø(age then)}, \tag{0.4}$$

where age then is given by equation (0.3). The Hubble constant is given by,

$$H_0 = \text{recession velocity / arc length}, \tag{0.5}$$

where recession velocity is given by equation (0.2).

Now, $H_0 = cø/e^{ø}/ø(now/e^{ø})$ which reduces to the lightspeed expansion velocity c. Thus,

$$H_0 = c \,/\, \text{age now}, \tag{0.6}$$

where c is the 4D lightspeed and age now ~ radius now in Mpc.

For example; in a 13,700 Myr universe ~ radius now of 4,200.39 Mpc we have, $H_0 = 71.37$ km /s / Mpc.

Conclusion

Popular theories in cosmology are based on the assumption of Euclidean geometry. Our model is based on a spherical universe that is in essence non-Euclidean. The dynamics of special and general relativity are a reality of a radiantly moving relativistic geometry, which gives us spacetime. Looking along the photon path from our present position, time appears to run progressively and exponentially slower. Thus the luminous flux of Ia type supernovas appear dimmer than their corresponding red-shifts would indicate. The reason for this is that the frequency of events 'υ' appear to run slower which means that the wavelengths increase giving us emitted red-shifted photons at the source. Consequently the number of photons we see as measured by our time is reduced progressively and exponentially the further back we look. as all light of the 'observed' universe focuses upon the observer's unique and leading time position. See ref. 5 and 6 for general background.

References

1. E.J. Bacinich, "Super Unification of All Forces", AIP Proceedings 672, 2003, p.197, Short Distance Behavior of Fundamental Interactions, edited by B.N. Kursunogly et al.
2. E.J. Bacinich, "Einstein's Elevator", AIP Conference Proceedings, 624. 2002, p.51, Coral Gables Conference on Cosmological and Elementary Particle Physics, edited by B.N. Kursunogly et al.
3. E.J. Bacinich & T.A. Kriz, "Photon Trajectory Attributes of an Expanding Hypersphere", Physics Essays Volume 8 number 4, 1995.
4. E.J. Bacinich & T.A. Kriz, "The Arrow of Time in an Expanding 3-Sphere" Physics Essays Volume 12 number 1, March 1999.
5. P.J.E. Peebles, Principles of Physical Cosmology (Princeton University Press, 1993).
6. H.C. Ohanian and R. Ruffini, Gravitation and Spacetime (second edition) (Norton &Co, NY and London 1994).

Edward J. Bacinich* Director, Alpha Omega Research Foundation, a 501c - 3 Corporation.
Website: http:///www.alphaomegafoundation.org
Email: research@alphaomegafoundation.org

A THEORETICAL VALUE FOR THE NEWTON GRAVITATION CONSTANT FROM THE GEM THEORY OF FIELD UNIFICATION AND THE KURSUNOGLU-BRANDENBURG HYPOTHESIS OF MASSIVE GAMMA-RAY BURSTERS

J. E. BRANDENBURG

Florida Space Institute-University of Central Florida
Kennedy Space Center, Florida, 32899, USA

Abstract

Theoretical progress on an approach to the Einstein-Schrodinger-Kursunoglu problem of EM – Gravity unification: the GEM (Gravity-Electro-Magnetism) theory, a combination of the Einstein-Sahkarov and Kaluza-Klein approaches, is summarized as well a collaborative hypothesis with the late Behram Kursunoglu concerning the Gamma Ray Bursters as " field unification events." A derivation of the value of the Newton-Gravitation Constant, G, based on the splitting of both EM from gravity and electrons from protons with the appearance of the Kaluza-Klein fifth dimension in a U(1) symmetry, is presented. This results in the formula: $G = e^2/(4\pi\varepsilon_o m_p m_e)\ \alpha \exp(-2(m_p/m_e)^{1/2}) = 6.668 \times 10^{-11}$ newton-m^2-kg^{-2} where m_p and m_e are the proton and electron masses respectively and α is the fine structure constant. This value is within experimental uncertainty (1.5 parts per thousand) of the presently accepted value 6.673×10^{-11} newton-m^2-kg^{-2}. In the Newtonian limit the portion of the Kaluza-Klein action that is quadratic in first derivatives of the metric and in Poynting Flux appears in the form of a "Vacuum Bernoulli Equation" showing Gravitational energy density to be equated to an EM dynamic pressure that is quadratic in the local Poynting Flux: $g^2/(2\pi G) + S^2/(c^2 L) =$ Constant, where g and S are the local gravity and Poynting vector magnitudes, respectively, and where L is the Lagrangian density of the EM field.

I Introduction: The GEM theory and Its Implications

The GEM theory[1,2] is an attempt to create a geometric theory to solve the Einstein-Schrodinger-Kursunoglu problem of unification of Gravity and Electromagnetism. The theory limits itself to protons and electrons. The theory is still in an early state of development, being described as a "Bohr Model" of unification by analogy with the earliest quantum mechanical model of the hydrogen atom, and is based on an extension of the work of Einstein, Kaluza and Klein [3], and Sahkarov [4] and Zeldovich [5]. The theory is based on two postulates: 1. Gravity consists of a an organized pattern of ExB drift or Poynting cells. 2.

The separate appearance of gravity from electromagnetism and protons from electrons comes from the Planck scale and occurs with the appearance and development of the Kaluza-Klein fifth dimension

The GEM theory begins with the Hilbert action principle, which allows the derivation of the fundamental equations for vacuum fields from the extremization of the action integrals:

$$W = (16\pi G)^{-1} \int (R - 2\Lambda)\sqrt{-g}\,dx^4 \qquad (1a)$$

$$L = \int g^{\mu\upsilon} \partial_\mu \phi \partial_\upsilon \phi^* \sqrt{-g}\,dx^4 \qquad (1b)$$

where R is the Ricci curvature scalar of General relativity , Λ is the cosmological constant and G is the Newton Gravitation constant, and where the action integral L for a massless quantum field is also present. In the concept of Sahkarov , the Λ term represented a zeroth order ground state EM field or ZPF(Zero Point Fluctuation) whereas R term represented first order perturbation on this field caused by space-time curvature. This integral was used by Sahkarov to relate the force of gravity to a radiation pressure , or Poynting flux, due to the ZPF of the massless field , which he identified with the EM ground state. In his article, Sahkarov derived the value of G, as the "metric elasticity of space" by assuming that the Planck Length $r_P = (G\hbar/c^3)^{1/2}$, where $\hbar$ is the rationalized Planck's constant $\hbar = h/2\pi$, the short wavelength cutoff for the ZPF spectrum, and that space-time curvature formed a perturbation on this spectrum. The value of the integral is determined by the cutoff frequency near the Planck frequency $\omega_p = c/r_p$ and this theory has been extended by Puthoff[6] Thus , in the Sahkarov model of gravity , g, the gravity acceleration is related to S the Poynting vector.

The physical nature of this ZPF gravity model can be seen simply by noting that just as hot spheres in a cold enclosure will repel each other due to mutual radiation pressure, so cold spheres in a hot enclosure will , from local Poynting vector inflow, attract on another due to mutual shadowing (see Figure 1).

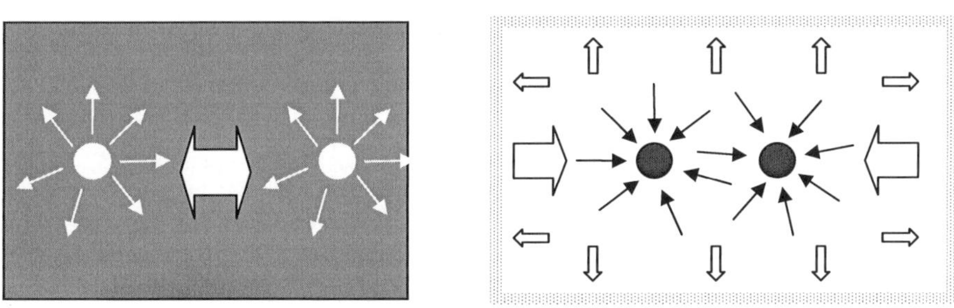

A. B.

Figure 1. A. Two hot spheres in a cold box, small arrows mark Poynting vector and the block arrow shows mutual repulsion. B. Two cold spheres in a cold box

small arrows mark Poynting flow and large block arrows show mutual attraction. Smaller block arrows show outward radiation pressure on the walls.
Thus Sahkarov showed that a self-consistent model of gravity as a EM radiation pressure from ZPF could be formulated.

In The GEM theory the ZPF is proposed to create a set of "ExB drift" cells that produces a gravity acceleration. In the GEM theory the EM fields are postulated to be arrays of ExB drift cells.[1] The ExB drift is well a known phenomena in plasma physics and affects all charged particles , regardless of charge or mass .[7]

$$V_D = \frac{(ExB)c}{B^2} = \frac{4\pi S}{B^2} \tag{2}$$

The GEM expression for gravity satisfies the equivalence principle because E , and thus S vanishes in the frame of the moving particle. This is because E $=E_o$-vxB in this frame, and even in a case where E increases with time, $E_{,t} = E_{o,t} - v_{,t}$ xB, so that the particles will experience no local electric field in their own frame.

II Kaluza-Klein and the GEM theory

Kaluza and Klein extended the Hilbert action principle by adding an additional dimension that is "compactified" to small fixed length "ε"so that we have for an action principle for a universe full of hydrogen and gravity and EM fields

$$W' = (16\pi G)^{-1} \int K\sqrt{-g}\,dx^5 \tag{3}$$

where the energy density K can be written , in the limit of the new dimension being compactified

$$\gamma_{ab} = \begin{bmatrix} g_{\tau\upsilon} + \xi\kappa^2 A_\tau A_\upsilon & \xi\kappa A_\tau \\ \xi\kappa A_\upsilon & \xi \end{bmatrix} \tag{4}$$

$$K = R - 2\Lambda + \kappa^2\left[\xi\left[\frac{E^2 - B^2}{8\pi}\right] + \gamma_{ab}\Theta^{ab}\right] \tag{5}$$

where E and B are the electric, and magnetic field strengths, and $\kappa^2 = 16\pi G$.
Where we have defined the metric tensor over dimensions 1-5 as a or b and dimensions 1-4 of conventional space-time as the Greek letters τ or υ. Since the

new dimension is limited in extent to length ε, we can write all quantum fields as of the form of a linear expansion

$$\phi = \sum_{n=0}^{\infty} C_n \phi_n (x_\mu) \exp(\frac{inx_5}{2\pi\varepsilon}) \qquad (6a)$$

This gives a new quantum Lagrangian

$$L' = \int \sum_{n=o}^{\infty} [\ ((\partial_\mu + \xi \frac{i\kappa n}{2\pi\varepsilon} A_\mu)\phi_n)^2 - \frac{\xi n^2}{(2\pi\varepsilon)^2} |\phi_n|^2 \ ]\sqrt{-g}\, dx^4 \varepsilon \qquad (6b)$$

so that element $\gamma_{55} = \xi$ is the element controlling the length of the line element in the fifth dimension with itself, and where we have defined the five dimensional particle stress tensor $\Theta^{a\ b}$ as the sum per unit volume V . Note that in the equation for the quantum Lagrangian the form has gone from that of a massless quantum field to that of particles with both charge and mass.

For the action K, coordinates of protons and electrons respectively for the stress tensor are defined conventionally, from which proper times along their trajectories, τ_p and τ_e respectively are determined , and λ is a trajectory parameter. We now define from the Kaluza-Klein formalism

$$\sigma = \frac{d\tau_p}{d\lambda} \bigg/ \frac{d\tau_e}{d\lambda} = \left(\frac{m_p}{m_e}\right)^{1/2} = 42.8503... \qquad (7)$$

Which can be thought of as the controlling parameter differentiating protons and electrons in their geodesic trajectories. Kaluza-Klein theory thus allowed, by a small increase in dimensionality, the explicit appearance of EM fields, and particles with both charge and mass, which they chose to be protons and electrons, along with gravity, from one variational principle. This constitutes a very basic description of the Cosmos as a whole, whose major known particle constituents are protons and electrons. If one conceives that the action principle assumes a massless EM field then the appearance of the fifth dimension allows trapping or scattering of massless quanta to create rest mass, charge, and hence particles. This is similar to other concepts of mass creation form massless quantum fields [8,9].

If one concentrates for a moment on the Hilbert-Kaluza-Klein action and writes this in near Newtonian limit one can write the field terms of the action as where t and H are the traces of the Landau-Lipshitz pseudo tensor for gravity stress [11] and a linearized form of the Einstein operator. On expansion of the EM Lagrangian and selection of those parts which vanish together in the free falling frame in the GEM model, we obtain

$$(16\pi G)^{-1} K_{field} = \frac{R}{16\pi G} + \frac{E^2 - B^2}{8\pi} = \frac{-g^2}{2\pi G} + \frac{S^2}{u_o c^2} = 0 \qquad (8)$$

where we choose S being associated with the gravity fields in the GEM model. This last expression is called "Vacuum Bernoulli Equation" because it relates a static gravity energy density to a dynamic EM momentum flux. This expression satisfies the equivalence principle because both terms vanish in a free falling frame.

III The Value of the Gravitation Constant

In the GEM theory the birth and "full deployment" of the Kaluza-Klein compact fifth dimension gives rise to the separate appearance of both the force field pair gravity and EM , but simultaneously gives rise to the particle field pair proton and electron. It does this by allowing a new degree of freedom to appear in turbulence at the Planck scale.

The appearance of the fifth dimension, in addition to the four dimensions of space-time, allows then the coupled appearance of both proton-electron matter fields (lepton and baryon number) and the EM fields. This is similar to a concept first proposed by Sakharov [12]

In a geometric interpretation, the signature of the coupled appearance of particles and EM fields in the GEM theory is the relation between curvature radius of the Planck Length $r_P = (G\hbar/c^3)^{1/2}$ and what is termed the "mesoscale" radius , typical of classical particle radii

$$r_o = \frac{e^2}{m_o c^2} \qquad (9)$$

where $m_o = (m_p m_e)^{1/2}$ .This is based on the postulate that baryon and lepton number disappear at the Planck scale coincidentally with the separate identity of Gravity and EM fields. The appearance of the Kaluza-Klien fifth dimension is thus the appearance of particles- electrons and protons with their classical radii. In the GEM model both protons and electrons begin in a U(1) symmetric field , the simplest possible field symmetry consistent with QED, for a mesoscale "union" particle with mass m_o. This union field exists at the Planck length

$$m = m_o \cos(\phi) + i m_o \sin(\phi) \qquad (10)$$

The angle ϕ, we will consider in this model corresponds to charge state and is thus quantized as a canceling pair $\pm\phi_o$, even in the Planck Scale. The appearance of the compact fifth dimension allows this angle to become an imaginary rotation angle to give two real particle masses corresponding to an "up" quantum state and "down" quantum state from the U(1) symmetry

$$m = m_o \exp(\pm\phi_o(\xi)) \qquad (11)$$

So that for This means that even though mass symmetry is broken in terms of the new 5 space we experience, it is preserved in terms of a geometry involving the imaginary angles in the original U(1) symmetry. That is, the new particle dimension looks symmetric in the space of imaginary angle.

To obtain a smooth transition to the union field as curvature collapses to the Planck scale the angle ϕ_o must be dependent on curvature near the Planck length but very insensitive to it at larger curvatures, where the new fifth dimension is fully developed. Based on the lack of observation of proton decay, lepton and baryon numbers are obviously strongly conserved. The simplest way to obtain this mixture of scale sensitivity with curvature r_c is for the rotation angle to have the dependence

$$\phi_o(\eta) = \ln(\ln(r_c(\xi)/r_P)) \qquad (12)$$

where we choose as simple expression for r_c

$$r_c(\xi) = \xi r_o \qquad (13)$$

Therefore, in the GEM model, the separate appearance of protons and electrons is, like the separate appearance of EM and gravity forces, linked to the appearance and full development of the Kaluza-Klein fifth dimension. The physical description of this dimension is thus as a particle classical radius r_o.
The expansion of the effective curvature to r_o , the mesoscale radius of classical particle radii then yields

$$\left(\frac{m_p}{m_e}\right)^{1/2} = \ln\left({}^{r_o}\!/_{r_P}\right) \qquad (14)$$

This formula can be inverted to obtain an expression for G (in esu units, $e^2 = e^2/4\pi\varepsilon_o$ MKS)

$$G = \left({}^{e^2}\!/_{m_p m_e}\right)\alpha \exp\left(-2\left({}^{m_p}\!/_{m_e}\right)^{1/2}\right) = 6.668x10^{-8}\, dyne - cm^2 gm^2 \qquad (15)$$

where α is the fine structure constant.

This formula gives the value of the Newton gravitation constant to a within 0.7 part per thousand of its accepted value of 6.673×10^{-8} dyne-cm^2 g^2, its presently accepted value, with an experimental accuracy of 1.5 parts per thousand. (CODATA). The constant is found to be linear in the fine structure constant, which is consistent with the Sahkarov model of Gravity being a higher order EM effect in the limit of weak fields.

IV The Kursunoglu Brandenburg Hypothesis of Gamma Ray Bursters

It was my great honor to discuss the great problems of physics with Behram Kursunoglu in his final days. We both agreed that to ponder and enquire into the deep secrets of the universe was one of the greatest and most sublime pleasures a human can experience.

There are many paths to the truth, like different paths to the summit of a mountain, and some are more difficult and dangerous than others. Thus, I think, it is with theories of unification. Behrum and I had a rare opportunity to compare the differing sights we saw on our respective paths. He chose to follow in the same formal path of Einstein and Schrodinger , men whose brilliance no one can question. He did this by inserting the F tensor into the metric, invoking a new length to be introduced for unification. I, on the other hand , was drawn by the elegance of making EM and gravity appear through the magic of Kaluza Kline formalism of a hidden fifth dimension, but I noted also that this also requires a new length to appear, in this case to keep the magical fifth dimension hidden. In discussing our models for EM Gravity unification thus we both agreed that the heart of successful unification theory in physics was understanding the appearance of new scale lengths in nature, which in his theory was the unification length near the Planck length, and in my theory is both the Planck and the mesoscale lengths.

In our wide ranging discussions, we also found that we both agreed that the most massive of GRBs (Gamma Ray Bursters) might be evidence that nature , in extreme conditions, would produce catastrophic Gravity-EM Unification (GEMU) events, where vast amounts of gravity field energy might be converted directly into EM energy[12]. We both concentrated on the processes near in-spiraling black hole pairs where spacetime was highly curved and dynamic. We arrived at a physical model for this unification event when I pointed out that Hawking evaporation of micro-black holes was an already recognized mechanism for gravity-EM energy conversion. That is, a micro-black hole is a pure gravity field, and after it evaporates it is pure EM radiation field. What was needed then was a model for moving dynamic gravity energy to very short wavelengths, where it could evaporate into EM radiation via the Hawking mechanism. Thus the model we developed was of the in-spiral of two black holes, whose final motion would become relativistic since their frequency of rotation would increase catastrophically because their orbits would decay due to gravitational radiation, and this radiation would itself develop synchrotron radiation at many harmonics to the rotation frequency of the in-spiraling black holes.

$$\omega_{peak} \approx \gamma^2 \omega_{rotation} \qquad (16)$$

This synchrotron effect can in turn, hypothetically, lead to a gravitational turbulent cascade to shorter wavelengths leading to "shredding" of the event horizons into min-black holes who then evaporate by the Hawking process into gama rays, with a luminosity for the micro-black holes of,

$$L = A\,(n_{eff}\,/\,2)\,\sigma_{Stefan-Boltzman}\,T^4 \qquad (17)$$

where n_{eff} is the number of particles whose rest mass is exceeded by kT and kT is defined by the gravity field at the surface of the micro-black holes.

$$kT = \frac{\hbar g}{2\pi c} \qquad g = \frac{GM}{r_s} \qquad (18)$$

We felt happy with this hypothetical model, and believed it would become more rigorous and be supported by observations in the future. Thus, some of Behram's last musing's were on the relation of EM-Gravity unification to the most brilliant cataclysms observed in the nature.

References

1. Brandenburg, J. E. (1992) "Unification of Gravity and Electromagnetism in the Plasma Universe", IEEE Transaction in Plasma Science Vol 20, 6, p 944.
2. Brandenburg, J. E. (1995) "A Model Cosmology Based on Gravity-Electromagnetism Unification", Astrophysics and Space Science, 227, p 133 also in
3. Klein, Oskar (1926) "Quantum Theory and Five-Dimensional Relativity", Zeitschrift fur Physik, 37, 895.
4. Sakharov A.D. (1968) "Vacuum quantum fluctuations in curved space and the theory of gravitation" Sov. Phys. Doklady 12, 1040-1041.
5. Zel'dovich, Ya. B. (1967), Cosmological constant and elementary particles, Sov. Phys.--JETP Lett. 6, 316-317.
6. Puthoff, H.E., (1989) "Gravity as a zero-point fluctuation force", Phys. Rev. A 39, 2333.
7. J.D. Jackson (1975) *Classical Electro-Dynamics*, Wiley and Sons, NY p 205
8. Haisch, B. , Rueda, A., and Puthoff, H.E., "Advances in the Proposed Electromagentic Zero-Point Field Theory of Inertia", AAIA 98-3143.
9. P.R. Holland (1992) "Mass As A Quantum Effect", Phys. Rev. A, Rapid Com. 46, 7, 1Oct.
10. Misner C.W. , Thorne Kip S. , and Wheeler J.A.,(1973) *Gravitation*, W. H. Freeman and Company, San Francisco, p465
11. A.D Sakharov, (1967) JETP Lett., **5**, 24
12. Behram N. Kursunoglu, (2003) "The Layered Structure of the Universe".

ECCENTRIC INFLATION AND WMAP*

ROMAN V. BUNIY
Department of Physics and Astronomy
Vanderbilt University
Nashville TN, 37235
USA
E-mail: roman.buniy@vanderbilt.edu

For uniform arrangements of magnetic fields, strings, or domain walls (together with the cosmological constant and non-relativistic matter), exact solutions to the Einstein equations are shown to lead to a universe with ellipsoidal expansion. We argue the results can be used to explain some features in the WMAP data. The magnetic field case is the easiest to motivate and has the highest possibility of yielding reliable constraints on observational cosmology.

1. Introduction

The first year WMAP results [3, 4, 5] contain interesting large-scale features which warrant further attention [6, 7]. One glaring feature is the suppression of power at large angular scales ($\theta \gtrsim 60^\circ$), which is reflected most distinctly in the reduction of the quadrupole C_2 and octopole C_3. This effect was also seen in the COBE results [8, 9]. While one can argue that such an effect could just be statistical, it does not seem unreasonable to try to model such behavior by altering the cosmological model from the standard big bang plus inflation scenario. More intriguingly, the quadrupole C_2 and octupole C_3 are found to be aligned. In particular, the $l = 2$ and 3 powers are found to be concentrated in a plane P inclined about 30° to the Galactic plane. In a coordinate system in which the equator is in the plane P, the $l = 2$ and 3 powers are primarily in the $m = \pm l$ modes. The axis of this system defines a unique ray and supports an idea of power on the axis is suppressed relative to the power in the orthogonal plane. These effects

*The calculations presented in this talk appeared in short form in Ref. [1] and a full exposition is to appear in Ref. [2].

seem to suggest one (longitudinal) direction may have expanded differently from the other two (transverse) directions in P.

We approach the issue of global anisotropy of the Universe by a simple modification of the conventional Friedman-Robertson-Walker (FRW) model. To achieve this, one has to consider an energy-momentum tensor which either is spatially non-spherical or spontaneously becomes non-spherical. Such a situation could occur when magnetic fields [10] or cosmic defects [11] are present. Moreover, it is known that large scale magnetic fields exist in the universe, perhaps up to cosmological scales [10, 12].

As a modest step toward understanding the form, significance and implications of an asymmetric universe, we will modify the standard spherically symmetric FRW cosmology to a form with only planar symmetry [13]. Our choice of energy-momentum will result in non-spherical expansion from an initially spherical symmetric configuration: an initial co-moving sphere will evolve into an ellipsoid that can be either prolate or oblate depending on the choice of matter content.

For the sake of clarity, we first give general equations for cosmologies with planar symmetry, (The universe looks the same from all points but they all have a preferred direction.) and then investigate one case in detail—a universe filled with dust, uniform magnetic fields and cosmological constant. This is perhaps the most easily motivated, exactly solvable case to consider and it will give us a context in which to couch the discussion of other examples with planar symmetry and cases where planar symmetry is broken.

To set the stage, consider an early epoch in the universe at the onset of cosmic inflation, where strong magnetic fields have been produced in a phase transition [14, 15, 16, 17, 18]. Assuming the magnitude of the magnetic field and vacuum energy (Λ) densities initially are about the same, we will find that eventually Λ dominates. It was estimated [16] that the initial magnetic field energy produced in the electroweak phase transition was within an order of magnitude of the critical density. Other phase transitions may have even higher initial field values [17, 18], or high densities of cosmic defects. Hence, it is not unphysical to consider a universe with B-fields and Λ of comparable magnitudes. If the B-fields are aligned in domains, then some degree of inflation is sufficient to push all but one domain outside the horizon.

2. Planar geometry

To make the simplest directionally anisotropic universe, we modify FRW spherical symmetry of space-time into planar symmetry; the most general form for the metric then is [13] $(g_{\mu\nu}) = \mathrm{diag}(1, -e^{2a}, -e^{2a}, -e^{2b})$, where a and b are functions of t and z; the xy-plane is the plane of symmetry. We also impose translational symmetry along the z-axis; the functions a and b now depend only on t.

(Examples of planar-symmetric spaces include space uniformly filled with either uniform magnetic fields, static alligned strings, or static stacked walls; i.e., the defects are at rest with respect to the cosmic backgound frame. This situation with defects is artificial or at best contrived, but could perhaps arise in brane world physics where static walls could be static or walls beyond the horizon could be connected by static strings. We will not pursue these details here. Of course, to these any spherically-symmetric contributions can be added: vacuum energy, matter, radiation.)

To support the symmetry of space-time, the energy-momentum tensor for the matter has to have the same symmetry, $(T^{\mu}{}_{\nu}) = (8\pi G)^{-1}\mathrm{diag}(\xi, \eta, \eta, \zeta)$. Here the energy density ξ, transversal η and longitudinal ζ tension densitites are functions of time only. The corresponding Einstein equations are

$$\dot{a}^2 + 2\dot{a}\dot{b} = \xi, \tag{1}$$

$$\ddot{a} + \ddot{b} + \dot{a}^2 + \dot{a}\dot{b} + \dot{b}^2 = \eta, \tag{2}$$

$$2\ddot{a} + 3\dot{a}^2 = \zeta. \tag{3}$$

We also need the equation expressing covariant conservation of the energy-momentum [a direct consequence of Eqs. (1)–(3)]:

$$\dot{\xi} + 2\dot{a}(\xi - \eta) + \dot{b}(\xi - \zeta) = 0. \tag{4}$$

For an adiabatic process, the entropy in a comoving volume is conserved. For the matter with the equation of state $p = w\rho$, this leads to[a] $\rho = \rho_{\mathrm{i}}\, e^{-(1+w)(2a+b)}$ and, as a result, the anisotropic part of energy-momentum is conserved as well:

$$\dot{\tilde{\xi}} + 2\dot{a}(\tilde{\xi} - \tilde{\eta}) + \dot{b}(\tilde{\xi} - \tilde{\zeta}) = 0. \tag{5}$$

For the cases of magnetic field, strings and walls, the set of quantities $(\tilde{\xi}, \tilde{\eta}, \tilde{\zeta})$ take, respectively, the values $(\epsilon, -\epsilon, \epsilon)$, $(\epsilon, 0, \epsilon)$ and $(\epsilon, \epsilon, 0)$.

[a]The subsript "i" refers to the moment of transition from isotropic to anisotropic dynamics.

Various relations and inequalities based on energy conditions, many of which involve the eccentricity of the expansion can be derived for general ξ, η and ζ; for details, see Ref. [2].

3. Magnetic field

Extending the analysis of Ref. [1], we now present the exact solution for the case of cosmological constant, dust ($w = 0$) and uniform B-field [2].

Conservation of the anisotropic part of the energy-momentum, Eq. (5), gives $a = \frac{1}{4}\ln(\epsilon_{\rm i}/\epsilon)$. Using this result in Eq. (3), we find that $\epsilon(t)$ is given implicitly by

$$t - t_{\rm i} = \tfrac{1}{4}\int_{\epsilon}^{\epsilon_{\rm i}} {\rm d}\epsilon \left[\tfrac{1}{3}\lambda\epsilon^2 + \tfrac{1}{3}\epsilon_{\rm i}^{-\frac{3}{4}}(4\epsilon_{\rm i} + \rho_{\rm i})\epsilon^{\frac{11}{4}} - \epsilon^3\right]^{-\frac{1}{2}}. \tag{6}$$

Eq. (1) and $\rho = \rho_{\rm i}\, e^{-(2a+b)}$ give

$$b = \tfrac{1}{2}\ln\frac{\lambda\epsilon^{-\frac{1}{2}} + (4\epsilon_{\rm i} + \rho_{\rm i})\epsilon_{\rm i}^{-\frac{3}{4}}\epsilon^{\frac{1}{4}} - 3\epsilon^{\frac{1}{2}}}{(\lambda + \rho_{\rm i})\epsilon_{\rm i}^{-\frac{1}{2}} + \epsilon_{\rm i}^{\frac{1}{2}}} + \ln\left[1 + F_{\rm m}(\epsilon)\right], \tag{7}$$

where

$$F_{\rm m}(\epsilon) = \frac{3\rho_{\rm i}}{8\epsilon_{\rm i}}\left(1 + \frac{\lambda + \rho_{\rm i}}{\epsilon_{\rm i}}\right)^{\frac{1}{2}}\int_{\epsilon/\epsilon_{\rm i}}^{1} {\rm d}x\, x^{-\frac{1}{4}}\left[\frac{\lambda}{\epsilon_{\rm i}} + \left(4 + \frac{\rho_{\rm i}}{\epsilon_{\rm i}}\right)x^{\frac{3}{4}} - 3x\right]^{-\frac{3}{2}}. \tag{8}$$

A careful inspection of the above solution shows that space is oblate, $e^{a-b} \geq 1$, and its oblateness monotonically increases from its initial value (unity) to its asymptotic value. More magnetic field increases the anisotropy; more matter reduces anisotropy, but neither can change an oblate ellipsoid into a prolate one.

The asymptotics for large t are as follows:

$$\epsilon \sim \epsilon_{\rm i}\, e^{-4(\lambda/3)^{\frac{1}{2}}(t - t_{\rm i} + \tau)}, \tag{9}$$

$$a \sim (\lambda/3)^{\frac{1}{2}}(t - t_{\rm i} + \tau), \tag{10}$$

$$b \sim (\lambda/3)^{\frac{1}{2}}(t - t_{\rm i} + \tau) - \tfrac{1}{2}\ln\left[1 + (\epsilon_{\rm i} + \rho_{\rm i})/\lambda\right] + \ln\left[1 + F_{\rm m}(0)\right], \tag{11}$$

$$\rho \sim \rho_{\rm i}[1 + (\epsilon_{\rm i} + \rho_{\rm i})/\lambda]^{\frac{1}{2}}\,[1 + F_{\rm m}(0)]^{-1} e^{-(3\lambda)^{\frac{1}{2}}(t - t_{\rm i} + \tau)}, \tag{12}$$

where

$$\tau = \tfrac{1}{4}\int_0^{\epsilon_{\rm i}} {\rm d}\epsilon \left\{\left(\tfrac{1}{3}\lambda\epsilon^2\right)^{-\frac{1}{2}} - \left[\tfrac{1}{3}\lambda\epsilon^2 + \tfrac{1}{3}\epsilon_{\rm i}^{-\frac{3}{4}}(4\epsilon_{\rm i} + \rho_{\rm i})\epsilon^{\frac{11}{4}} - \epsilon^3\right]^{-\frac{1}{2}}\right\}. \tag{13}$$

4. Strings

From the conservation equation there follows $a = \frac{1}{2}\ln(\epsilon_i/\epsilon)$, and then Eq. (3) gives $\epsilon(t)$ implicitly via

$$t - t_i = \tfrac{1}{2}\int_{\epsilon}^{\epsilon_i} d\epsilon \left[\tfrac{1}{3}\lambda\epsilon^2 + \epsilon^3 + \tfrac{1}{3}\epsilon_i^{-\frac{3}{2}}(\rho_i - 2\epsilon_i)\epsilon^{\frac{7}{2}}\right]^{-\frac{1}{2}}. \tag{14}$$

Similarly to the case of the magnetic field, we solve Eq. (1) and find

$$b = \tfrac{1}{2}\ln\frac{3 + \lambda\epsilon^{-1} + (\rho_i - 2\epsilon_i)\epsilon_i^{-\frac{3}{2}}\epsilon^{\frac{1}{2}}}{1 + (\lambda + \rho_i)\epsilon_i^{-1}} + \ln\left[1 + F_s(\epsilon)\right], \tag{15}$$

where

$$F_s(\epsilon) = \frac{3\rho_i}{4\epsilon_i}\left(1 + \frac{\lambda + \rho_i}{\epsilon_i}\right)^{\frac{1}{2}} \int_{\epsilon/\epsilon_i}^{1} dx\, x^{\frac{1}{2}} \left[\frac{\lambda}{\epsilon_i} + 3x + \left(\frac{\rho_i}{\epsilon_i} - 2\right)x^{\frac{3}{2}}\right]^{-\frac{3}{2}}. \tag{16}$$

A careful inspection of the above solution shows that space is oblate, $e^{a-b} \geq 1$, and its oblateness monotonically increases from its initial value (unity) to its asymptotic value. More string density increases the anisotropy; more matter reduces anisotropy, but neither can change an oblate ellipsoid into a prolate one.

The asymptotics for large t are as follows:

$$\epsilon \sim \epsilon_i\, e^{-2(\lambda/3)^{\frac{1}{2}}(t - t_i + \tau)}, \tag{17}$$

$$a \sim (\lambda/3)^{\frac{1}{2}}(t - t_i + \tau), \tag{18}$$

$$b \sim (\lambda/3)^{\frac{1}{2}}(t - t_i + \tau) - \tfrac{1}{2}\ln\left[1 + (\rho_i + \epsilon_i)/\lambda\right] + \ln\left[1 + F_s(0)\right], \tag{19}$$

$$\rho \sim \rho_i[1 + (\epsilon_i + \rho_i)/\lambda]^{\frac{1}{2}}[1 + F_s(0)]^{-1} e^{-(3\lambda)^{\frac{1}{2}}(t - t_i + \tau)}, \tag{20}$$

where

$$\tau = \tfrac{1}{2}\int_0^{\epsilon_i} d\epsilon \left\{\left(\tfrac{1}{3}\lambda\epsilon^2\right)^{-\frac{1}{2}} - \left[\tfrac{1}{3}\lambda\epsilon^2 + \tfrac{1}{3}\epsilon_i^{-\frac{3}{2}}(\rho_i - 2\epsilon_i)\epsilon^{\frac{7}{2}} + \epsilon^3\right]^{-\frac{1}{2}}\right\}. \tag{21}$$

5. Walls

Energy-momentum conservation gives $b = \ln(\epsilon_i/\epsilon)$. Eq. (3) is easily solved to give

$$a = \tfrac{2}{3}\ln\frac{e^{(3\lambda)^{\frac{1}{2}}(t - t_i)} - \sigma}{1 - \sigma} - (\lambda/3)^{\frac{1}{2}}(t - t_i), \tag{22}$$

where

$$\sigma = \frac{[1 + (\rho_i + \epsilon_i)/\lambda]^{\frac{1}{2}} - 1}{[1 + (\rho_i + \epsilon_i)/\lambda]^{\frac{1}{2}} + 1}. \tag{23}$$

Using Eqs. (1) and (22) together with $\rho = \rho_i e^{-(2a+b)}$, we find

$$\epsilon = \epsilon_i \gamma \frac{(\gamma^3 - \sigma)^{\frac{1}{3}}}{\gamma^3 + \sigma} F_w(\gamma), \tag{24}$$

where

$$F_w(\gamma) = \left[\frac{(1-\sigma)^{\frac{1}{3}}}{1+\sigma} + \frac{\rho_i}{2\lambda} \frac{(1-\sigma)^{\frac{4}{3}}}{1+\sigma} \frac{\gamma^3 - 1}{\gamma^3 + 1} + \frac{3\epsilon_i}{2\lambda} \int_1^\gamma \mathrm{d}x \, \frac{(x^3 - \sigma)^{\frac{4}{3}}}{(x^3 + \sigma)^2} \right]^{-1} \tag{25}$$

and $\gamma = e^{(\lambda/3)^{\frac{1}{2}}(t - t_i)}$.

Inspecting the above solution shows that space is prolate, $e^{a-b} \leq 1$, and its prolateness monotonically increases from its initial value (unity) to its asymptotic value. More wall density increases the anisotropy; more matter reduces anisotropy, but neither can change an prolate ellipsoid into a oblate one.

The asymptotics for large t are as follows:

$$\epsilon \sim \epsilon_i \, e^{-(\lambda/3)^{\frac{1}{2}}(t-t_i)} F_w(\infty), \tag{26}$$

$$a \sim (\lambda/3)^{\frac{1}{2}}(t - t_i) - \tfrac{2}{3} \ln (1 - \sigma), \tag{27}$$

$$b \sim (\lambda/3)^{\frac{1}{2}}(t - t_i) - \ln F_w(\infty), \tag{28}$$

$$\rho \sim \rho_i (1 - \sigma)^{\frac{4}{3}} \, e^{-\mu(t-t_i)} F_w(\infty). \tag{29}$$

6. Conclusions

To apply the results of this paper, it will be necessary to consider how the spectrum of density perturbations are effected by asymmetric expansion. Since perturbations get laid down by quantum fluctuations and then asymmetrically expanded in our model, any initial spherical perturbation becomes ellipsoidal. After a while the expansion becomes spherically symmetric, but as long as perturbations remain outside the horizon they stay ellipsoidal. Only after they enter our horizon will they be able to adjust (they will probabily start to oscillate between prolate and oblate with frequency that depends on size and overdensity). So if the perturbations are just entering at last scattering they should be ellipsoidal. The smaller they are at last scattering, the more they have oscillated and if damped, the closer to spherical they should be. Hence, the large perturbations (corresponding to small l) will have a better memory of our eccentric phase. This would give what seems to be hinted at in the WMAP observations–more distortion of the low l modes.

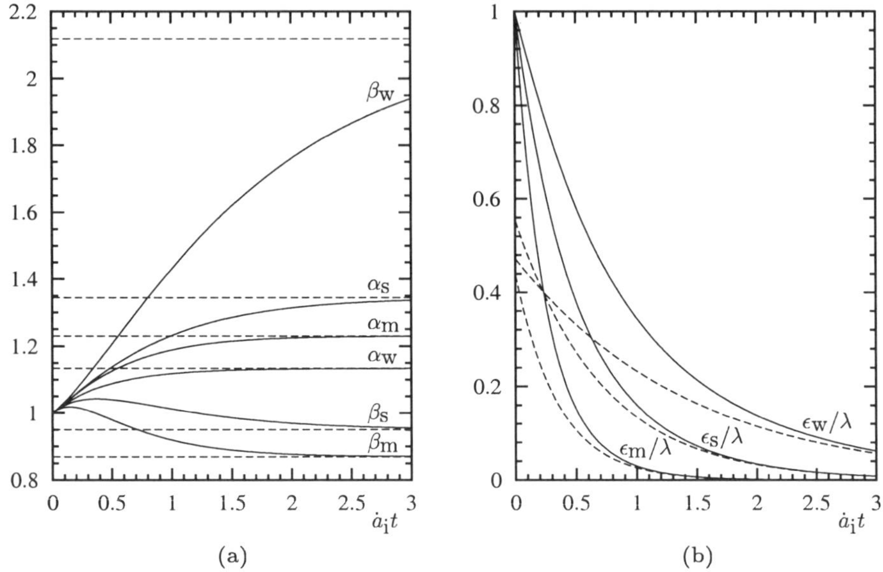

Figure 1. (a) The ratios of the asymmetric scale factors to the symmetric scale factors, $\alpha = e^{a-c}$ and $\beta = e^{b-c}$, where $c = (\frac{1}{3}\lambda)^{\frac{1}{2}}(t - t_i)$. In all three cases the same initial conditions are used. Solid lines are exact solutions and dashed lines are asymptotics for large t. (Matter is not included.) (b) The ratio ϵ/λ for the same three cases and initial conditions as in (a).

To summarize, what we need are modes that expanded eccentrically to be entering the horizon at the time of last scattering and then to feed into the Sachs-Wolfe calculation. This is a most interesting and challenging calculation, since it requires a full reanalysis of the density perturbations in eccentric geometry [19].

Acknowledgments

This work is supported in part by U.S. DoE grant number DE-FG05-85ER40226.

References

1. A. Berera, R. V. Buniy and T. W. Kephart, arXiv:hep-ph/0311233.
2. A. Berera, R. V. Buniy and T. W. Kephart, "Asymmetric inflation: exact solutions," in preparation.
3. C. L. Bennett *et al.*, Astrophys. J. Suppl. **148**, 1 (2003).
4. D. N. Spergel *et al.*, Astrophys. J. Suppl. **148**, 175 (2003).

5. G. Hinshaw *et al.*, Astrophys. J. Suppl. **148**, 135 (2003).
6. M. Tegmark, A. de Oliveira-Costa and A. Hamilton, arXiv:astro-ph/0302496.
7. A. de Oliveira-Costa, M. Tegmark, M. Zaldarriaga and A. Hamilton, arXiv:astro-ph/0307282.
8. G. F. Smoot *et al.*, Astrophys. J. **396**, L1 (1992).
9. A. Kogut *et. al*, arXiv:astro-ph/9601060.
10. P. P. Kronberg, Rept. Prog. Phys. **57**, 325 (1994).
11. M. B. Hindmarsh and T. W. Kibble, Rept. Prog. Phys. **58**, 477 (1995).
12. S. D. Wick, T. W. Kephart, T. J. Weiler and P. L. Biermann, Astropart. Phys. **18**, 663 (2003).
13. A. H. Taub, Annals Math. **53**, 472 (1951).
14. G. K. Savvidy, Phys. Lett. B **71**, 133 (1977).
15. S. G. Matinian and G. K. Savvidy, Nucl. Phys. B **134**, 539 (1978).
16. T. Vachaspati, Phys. Lett. B **265** (1991) 258.
17. K. Enqvist and P. Olesen, Phys. Lett. B **329**, 195 (1994).
18. A. Berera, T. W. Kephart and S. D. Wick, Phys. Rev. D **59**, 043510 (1999).
19. A. Berera, R. V. Buniy and T. W. Kephart, "Asymmetric inflation: density perturbation," in preparation.

DILATON SIGNATURES IN THE ELECTROMAGNETIC SPECTRA OF STARS

ROBERTO CASADIO*, SERGIO FABI† AND BENJAMIN HARMS†

*Dipartimento di Fisica, Università di Bologna and I.N.F.N., Sezione di Bologna, via Irnerio 46, 40126 Bologna, Italy
†Department of Physics and Astronomy, The University of Alabama, Box 870324, Tuscaloosa, AL 35487-0324, USA

Abstract. We study the propagation of classical electromagnetic waves on the simplest four-dimensional spherically symmetric metric with a dilaton background field. Solutions to the relevant equations are obtained perturbatively in a parameter which measures the strength of the dilaton field (hence parameterizes the departure from Schwarzschild geometry). The loss of energy from outgoing modes is estimated as a back-scattering process against the dilaton background, which would affect the luminosity of stars with a dilaton field. The radiation emitted by a freely falling point-like source on such a background is also studied by numerical methods.

INTRODUCTION

The occurrence of violations of the Equivalence Principle is a subject of continuing investigation in the community of general relativists, both theoretically and experimentally (for a review see [1] and References therein). A possible source for such effects would be the existence of a scalar component of gravity, which naturally emerges in string theories and is usually referred to as the *dilaton*. Solar system tests rule out such a possibility to a very high precision in the weak field regime [1]. However, it cannot be excluded that the picture is different in regions of strong gravitational fields, such as those that can be found around very dense compact sources. It is therefore interesting to study solutions of the Einstein field equations with a scalar field.

This talk is based on a study of the propagation of electro-magnetic (EM) waves on the background of the Janis-Newman-Winicour (JNW) solution [2]. The corresponding metric is spherically symmetric and asymptotically flat, but the static dilaton field is not trivial. Although the JNW space-time is known to contain a naked singularity, it is possible to make it physically useful by suitably cutting the space outside the singular point and then sewing a (spherically symmetric) regular inner patch to complete the space-time manifold. Provided this procedure is carried out consistently, one can then describe compact astrophysical sources such as stars with a dilaton field.

A more complicated problem is to obtain the spectrum of the radiation emitted by a freely falling, electrically charged, point-like source in the JNW background. The wave equations now contain the energy-momentum density of the falling particle and must be solved numerically for various values of the perturbing parameter in order to compare the emitted power to the analogous case in Schwarzschild. A preliminary analytical analysis will show that some effect is already expected at large distance from the center, however most radiation would be emitted near the singularity where tidal forces on the EM field of the falling particle become stronger. Since naked singularities seem to be possible intermediate stages in the gravitational collapse of regular matter (see, e.g. [3] and References therein), one can view the falling source as a component of the forming collapsed object. Our results may then contribute to the understanding of the (in)stability of naked singularities.

JNW SOLUTION AND SCHWARZSCHILD LIMIT

The JNW metric is a static, spherically symmetric and electrically neutral solution of the field equations obtained by varying the action

$$S = \frac{1}{16}\int d^4x\sqrt{-g}\left[R-\frac{1}{2}(\nabla\Phi)^2\right]-\int d^4x\sqrt{-g}\,e^{-q\Phi}F^2 ,$$

in which R is the scalar curvature of the metric $g_{\mu\nu}$, whose determinant is denoted by g, F is the EM field strength and note that the dilaton Φ is included inside the gravitational Lagrangian. Further, the coupling constant $q > 0$ between the EM field and the dilaton has also been exponentiated.

For the purpose of studying linear perturbations in the NP formalism [4], it is convenient to write the JNW metric [2] as

$$\mathrm{d}s^2 = \frac{\Delta}{\rho^2}\,\mathrm{d}t^2 - \frac{\rho^2}{\Delta}\,\mathrm{d}r^2 - \rho^2\,\mathrm{d}\Omega^2 , \tag{1}$$

where $\mathrm{d}\Omega^2 = \mathrm{d}\theta^2 + \sin^2\theta\,\mathrm{d}\phi^2$, $\rho^2 = \left(r-r_-\right)^{1-a}\left(r-r_+\right)^{1+a}$, $\Delta = \left(r-r_-\right)\left(r-r_+\right)$, and $a \in (-1,0)$. The dilaton field is then

$$\Phi = \sqrt{1-a^2}\,\ln\left(\frac{r-r_-}{r-r_+}\right) , \tag{2}$$

and $r = r_+ > r_-$ represents a (point-like, since $\rho(r_+) = 0$) naked singularity.

We now introduce a *quasi-Schwarzschild regime* [corresponding to a "small" deviation away from the Schwarzschild metric] for the region outside the "horizon" ($0 < r_+ < r$). It is defined by $r_- = 0$, $a = -1+2x^2$ with $0 \le x \ll 1$. In particular, for small x, one finds that the metric (1) reduces to Schwarzschild,

$$\mathrm{d}s^2 \simeq \left[1-(1-2x^2)\frac{r_h}{\rho}\right]\mathrm{d}t^2 - \left[1+(1-2x^2)\frac{r_h}{\rho}\right]\mathrm{d}\rho^2 - \rho^2\,\mathrm{d}\Omega^2 ,$$

where $r_h \equiv r_+$, provided the ADM mass is given by the relation

$$M = \left(1-2x^2\right)\frac{r_h}{2} .$$

The relevant PPN parameters [1] are thus $\alpha = \gamma = \beta = 1$.

In the approximation $r \gg r_h$, the derivative of the dilaton to leading order in x becomes

$$\partial_r \Phi \simeq -\frac{2x}{r-r_h}\frac{r_h}{r} \sim -2x\frac{r_h}{r^2} ,$$

from which one can see that x is a measure of the dilaton strength for large r (note that $\partial_r \Phi$ diverges for $r \to r_h$ for all values of $x > 0$). However, the values of the PPN parameters imply that deviations from the Schwarzschild solution in the weak field zone ($r \gg r_h$) are negligible and no further bound for x can be read off from solar test measurements. Since no direct measurement in the strong field regime is available, the strength of the dilaton field also cannot be bounded for $r \sim r_h$. This naturally leads to the consideration of EM waves that reach infinity after propagating in the strong field zone.

DILATONIC STARS

Maxwell waves in Newman-Penrose formalism

We first choose the contravariant components of the NP tetrad $l^i = \left[\frac{\rho^2}{\Delta}, 1, 0, 0\right]$, $n^i = \left[\frac{1}{2}, -\frac{\Delta}{2\rho^2}, 0, 0\right]$ and $m^i = \left[0, 0, \frac{1}{\sqrt{2\rho^2}}, \frac{i \operatorname{cosec}\theta}{\sqrt{2\rho^2}}\right]$. The Maxwell field is now completely described by the three complex scalars

$$\begin{aligned}
\phi_0 &\equiv F_{ij} l^i m^j = \frac{\rho^3}{\sqrt{2}\Delta}\left[E^\theta + B^\varphi + i\sin\theta\left(E^\varphi - B^\theta\right)\right] \\
\phi_1 &\equiv \frac{1}{2}F_{ij}\left(l^i n^j + m^{i*} m^j\right) = \frac{\rho^4 E^r}{2\Delta^2}\left(1 + \frac{\Delta^2}{4\rho^4}\right) \qquad (3)\\
\phi_2 &\equiv F_{ij} m^{i*} n^j = -\frac{\rho}{2\sqrt{2}}\left[E^\theta - B^\varphi + i\sin\theta\left(E^\varphi + B^\theta\right)\right] ,
\end{aligned}$$

where E^i and B^i are the usual electric and magnetic components. Further, the time and φ dependence of the EM waves is taken to be $\phi_i \sim \exp(+i\omega t - im\varphi)$, where m is an integer. The ϕ_i must then satisfy the equations [5]

$$(\hat{D} - 2\tilde{\rho})\phi_1 - (\hat{\delta}^* + \tilde{\pi} - 2\tilde{\alpha})\phi_0 + \tilde{\kappa}\phi_2 = J_1 \qquad (4)$$

$$(\hat{\delta} - 2\tilde{\tau})\phi_1 - (\hat{\Delta} + \tilde{\mu} - 2\tilde{\gamma})\phi_0 + \tilde{\sigma}\phi_2 = J_3 \qquad (5)$$

$$(\hat{D} - \tilde{\rho} + 2\tilde{\varepsilon})\phi_2 - (\hat{\delta}^* + 2\tilde{\pi})\phi_1 + \tilde{\lambda}\phi_0 = -J_4 \qquad (6)$$

$$(\hat{\delta} - \tilde{\tau} + 2\tilde{\beta})\phi_2 - (\hat{\Delta} + 2\tilde{\mu})\phi_1 + \tilde{\nu}\phi_0 = -J_2 , \qquad (7)$$

with the dilatonic currents

$$\begin{aligned}
J_1 &= \frac{q}{2}\left[(\phi_1 + \phi_1^*)\hat{D} - \phi_0\hat{\delta}^* - \phi_0^*\hat{\delta}\right]\Phi \\
J_2 &= \frac{q}{2}\left[(\phi_1 + \phi_1^*)\hat{\Delta} - \phi_2\hat{\delta} - \phi_2^*\hat{\delta}^*\right]\Phi
\end{aligned}$$

$$J_3 = \frac{q}{2}\left[(\phi_1 - \phi_1^*)\,\hat{\delta} - \phi_0\,\hat{\Delta} + \phi_2^*\,\hat{D}\right]\Phi$$
$$J_4 = \frac{q}{2}\left[(\phi_1 - \phi_1^*)\,\hat{\delta}^* + \phi_0^*\,\hat{\Delta} - \phi_2\,\hat{D}\right]\Phi\,.$$

In the above, hatted quantities represent the differential operators $\hat{D} = \partial_r + i\frac{\rho^2}{\Delta}\,\omega$, $\hat{\Delta} = -\frac{\Delta}{2\rho^2}\left(\partial_r - i\frac{\rho^2}{\Delta}\,\omega\right)$, $\hat{\delta} = \frac{1}{\sqrt{2\rho^2}}\left(\partial_\theta - \frac{m}{\sin\theta}\right)$ and $\hat{\delta}^* = \frac{1}{\sqrt{2\rho^2}}\left(\partial_\theta + \frac{m}{\sin\theta}\right)$. Tilded quantities are spin coefficients, those of which that do not vanish being given by

$$\tilde{\rho} = -\frac{2r + r_+(a-1) - r_-(a+1)}{2(r-r_-)(r-r_+)}\,, \quad \tilde{\mu} = -\frac{2r + r_+(a-1) - r_-(a+1)}{4(r-r_-)^{1-a}(r-r_+)^{1+a}}$$
$$\tilde{\alpha} = -\tilde{\beta} = -\frac{\sqrt{2}\,\mathrm{cotan}\,\theta}{4(r-r_-)^{\frac{1-a}{2}}(r-r_+)^{\frac{1+a}{2}}}\,, \quad \tilde{\gamma} = \frac{a(r_- - r_+)}{4(r-r_-)^{1-a}(r-r_+)^{1+a}}\,.$$

Quasi-Schwarzschild solutions

We are interested in deviations from Schwarzschild induced by a background dilaton (2) which is experimentally unobservable in the weak field regime. We will therefore look for solutions to Eqs. (4)-(7) in the quasi-Schwarzschild regime as previously defined, bearing in mind that the parameter x is constrained to be small by solar system tests as we mentioned before.

One can then discard all terms of order x^2 (and higher) and finally obtains

$$\left(\partial_r + \frac{2}{r} + \frac{i\omega r}{r-r_h}\right)\phi_1 - \frac{\sqrt{2}}{2r}\left(\partial_\theta + \frac{m+\cos\theta}{\sin\theta}\right)\phi_0 = -\tilde{x}\frac{r_h\,(\phi_1+\phi_1^*)}{r(r-r_h)} \tag{8}$$
$$\left(\partial_r + \frac{1-i\omega r}{r-r_h}\right)\phi_0 + \frac{\sqrt{2}}{r-r_h}\left(\partial_\theta - \frac{m}{\sin\theta}\right)\phi_1 = -\frac{\tilde{x}r_h}{r-r_h}\left(\frac{\phi_0}{r} + \frac{2\phi_2^*}{r-r_h}\right) \tag{9}$$
$$\left(\partial_r + \frac{1}{r} + \frac{i\omega r}{r-r_h}\right)\phi_2 - \frac{\sqrt{2}}{2r}\left(\partial_\theta + \frac{m}{\sin\theta}\right)\phi_1 = -\tilde{x}\frac{r_h}{r}\left(\frac{\phi_0^*}{2r} + \frac{\phi_2}{r-r_h}\right) \tag{10}$$
$$\left(\partial_r + \frac{2}{r} - \frac{i\omega r}{r-r_h}\right)\phi_1 + \frac{\sqrt{2}}{r-r_h}\left(\partial_\theta - \frac{m-\cos\theta}{\sin\theta}\right)\phi_2 = -\tilde{x}\frac{r_h\,(\phi_1+\phi_1^*)}{r(r-r_h)}\,, \tag{11}$$

where the left hand sides equated to zero are just the wave equations for the EM waves in Schwarzschild, and $\tilde{x} \equiv qx > 0$ is an effective coupling constant between the EM field and the dilaton. We can now proceed to consider particular cases.

General case

For $\tilde{x} = 0$, we know that the set of first order Eqs. (8)-(11) are separable and can be transformed into a (separable) second order equation for ϕ_1 which plays the role of a (spin 0) potential for the physical modes ϕ_0 and ϕ_2. The angular parts of $\phi_0 \equiv R_0(r)\,S_0(\theta)$

and $\phi_2 \equiv R_2(r)\,S_2(\theta)$ are then given by spin-weighted spherical harmonics ${}_sP_l^m$ of spin $s=\pm 1$ respectively. For $\tilde{x}\neq 0$ the same procedure still works but becomes more involved because of the currents appearing in Eqs. (8)-(11), as we shall shortly show.

Since spherical symmetry is preserved, we can assume that the angular parts are still given by suitable superpositions of ${}_sP_l^m$'s,

$$\phi_i = \sum_{l\geq |m|,1} R_l^{(1)}(r)\,{}_{1-i}P_l^m(\theta)\,, \quad i=0,1,2$$

where ${}_0P_l^m(\theta)$ is the ordinary Legendre polynomial and the radial functions must be determined. Upon substituting the above into Eqs. (8) and (11), the angular dependence factors out and one is left with two radial equations for each mode l of ϕ_0 and ϕ_2,

$$R_l^{(0)} = -\frac{\sqrt{2}r}{l\,(l+1)}\left[\left(\partial_r+\frac{2}{r}+\frac{i\,\omega\, r}{r-r_h}\right)R_l^{(1)}+\frac{\tilde{x}\,r_h\,\tilde{R}_l^{(1)}}{r\,(r-r_h)}\right]$$
$$R_l^{(2)} = \frac{r-r_h}{\sqrt{2}\,l\,(l+1)}\left[\left(\partial_r+\frac{2}{r}-\frac{i\,\omega\, r}{r-r_h}\right)R_l^{(1)}+\frac{\tilde{x}\,r_h\,\tilde{R}_l^{(1)}}{r\,(r-r_h)}\right],$$

where $\tilde{R}_l^{(1)} \equiv R_l^{(1)}+(-1)^m R_l^{(1)*}$ (note that the presence of the complex conjugation is responsible for the mixing of outgoing with ingoing modes). These equations completely determine $R_l^{(0)}$ and $R_l^{(2)}$ in terms of $R_l^{(1)}$. The latter must satisfy Eqs. (9) and (10), which, after some algebra, are seen to be equivalent to

$$\left[\partial_r^2+\frac{4r-3r_h}{r-r_h}\,\partial_r+\frac{2-l\,(l+1)}{r\,(r-r_h)}+\frac{\omega^2\,r^2}{(r-r_h)^2}\right]R_l^{(1)} = -\frac{2\,\tilde{x}\,r_h}{r\,(r-r_h)}\left(\partial_r+\frac{1}{r}\right)\tilde{R}_l^{(1)}\,,$$

hence showing the consistency of the set of four Eqs. (8)-(11).

Zero radial oscillations

>From Eqs. (3), one finds that if one imposes $E^r=0$ so that there is no radial oscillation in the EM field, the complex scalar $\phi_1=0$. Let us then consider this case and also define $W_0\equiv\phi_0$ and $W_2\equiv 2\,\phi_2^*$, which must then solve the equations

$$\left[\partial_\theta+\frac{m+\cos(\theta)}{\sin(\theta)}\right]W_0=\left[\partial_\theta-\frac{m-\cos(\theta)}{\sin(\theta)}\right]W_2^*=0 \tag{12}$$

$$\left(\partial_r+\frac{1-i\,\omega\, r}{r-r_h}\right)W_0=\frac{r}{r-r_h}\left(\partial_r+\frac{1}{r}-\frac{i\,\omega\, r}{r-r_h}\right)W_2=-\frac{\tilde{x}\,r_h}{r-r_h}\left(\frac{W_2}{r-r_h}+\frac{W_0}{r}\right). \tag{13}$$

Note that the angular parts of $W_0\equiv S_0(\theta)\,R_0(r)$ and $W_2\equiv S_2(\theta)\,R_2(r)$ satisfy Eqs. (12). They are thus given by $S_0=\frac{\tan^m(\theta/2)}{\sin(\theta)}$ and $S_2=\frac{\cot^m(\theta/2)}{\sin(\theta)}$. For the particular case $m=0$ the angular parts therefore coincide and, consequently, the remaining Eqs. (13) become purely radial equations for $R_0(r) = e^{i\sigma(r)}X(r)$ and $R_2(r) = e^{-i\sigma(r)}Y(r)$. The real phase σ is the same for both fields and satisfies the purely Schwarzschild equation

$$\frac{d\sigma}{dr} = \frac{\omega r \sigma}{r - r_h} , \tag{14}$$

and the functions X and Y must solve

$$\left(\partial_r + \frac{1}{r - r_h}\right) X = \frac{r}{r - r_h}\left(\partial_r + \frac{1}{r}\right) Y = -\frac{\tilde{x} r_h}{r - r_h}\left(\frac{Y}{r - r_h} + \frac{X}{r}\right) . \tag{15}$$

Hence, $\phi_0 \sim R_0$ represents the ingoing modes while $\phi_2 \sim R_2^*$ stands for the outgoing modes. The presence of terms (proportional to $\tilde{x}$) which mix ingoing with outgoing modes immediately implies that outgoing radiation emitted by the central source would partly back-scatter into ingoing waves. Thus the expected dilaton effect is to make the source less bright (see the next Subsection for more quantitative estimates).

Since $E^2 \sim B^2 \sim 1/r^2$ for large r, admissible solutions to (15) should be of order $1/r$ or higher at large r, as happens in the case of the ϕ_i's. In fact, the general solutions are given by

$$X = \frac{C}{r - r_h}\left[\left(1 - \frac{r_h}{r}\right)^{2\tilde{x}} + D\right] , \quad Y = \frac{C}{r}\left[\left(1 - \frac{r_h}{r}\right)^{2\tilde{x}} - D\right] , \tag{16}$$

where C and D are constants to be determined from the initial conditions. Note also that the limit $\tilde{x} \to 0$ yields solutions of the homogeneous (purely Schwarzschild) equations of the form

$$\phi_0 = \frac{A}{\sin(\theta)}\frac{e^{+i\sigma(r)}}{r - r_h} , \quad \phi_2 = \frac{B}{\sin(\theta)}\frac{e^{-i\sigma(r)}}{r} ,$$

with $A = C(1 + D)$ and $B = C(1 - D)$.

The exchange of energy between ingoing and outgoing waves becomes more apparent by redefining the constants so as to write the solution as

$$\begin{aligned} \phi_1 &\simeq \frac{A(1+B)}{\sin(\theta)}\frac{e^{+i\sigma}}{r - r_h}\left[1 + \frac{2\tilde{x}}{1+B}\ln\left(1 - \frac{r_h}{r}\right)\right] \\ \phi_2 &\simeq \frac{A(B-1)}{\sin(\theta)}\frac{e^{-i\sigma}}{r}\left[1 - \frac{2\tilde{x}}{1-B}\ln\left(1 - \frac{r_h}{r}\right)\right] \end{aligned} \tag{17}$$

where the constants $A = A_{lm}(\omega)$ and $B = B_{lm}(\omega)$ must be determined from the initial conditions. There is obviously no initial condition (choice of A and B) which can produce just ingoing (ϕ_0) or just outgoing (ϕ_2) waves for all values of $r > r_h$.

Darkened stars

Let us consider the particular solution found in Eq. (17). We denote by $r_e > r_h$ the outer radius of the central star and consider the case $B = -1$, corresponding to a purely outgoing mode ϕ_2 in Schwarschild ($\tilde{x}=0$). The ratio between the energy of such an EM wave ($\sim \phi_2^2$) at the point of observation $r=r_o$ and at the point of emission $r=r_e$ would be $\Gamma_{\text{Schw}} \simeq \left[\frac{\phi_2^{(0)}(r_o)}{\phi_2^{(0)}(r_e)}\right]^2 \sim \left(\frac{r_e}{r_o}\right)^2$. The same quantity in JNW (evaluated to leading order in $\tilde{x}$) is obtained from the solution in Eqs. (17),

$$\Gamma_{\text{JNW}} \sim \left(\frac{r_e}{r_o}\right)^2 \left[\frac{1-\tilde{x}\ln\left(1-\frac{r_h}{r_o}\right)}{1-\tilde{x}\ln\left(1-\frac{r_h}{r_e}\right)}\right]^2 .$$

On assuming that $r_h \ll r_e \ll r_o$ (a safe assumption to protect the star from the central singularity), one finds

$$\Gamma \simeq \Gamma_{\text{Schw}} \left[1 - 2\tilde{x}\frac{r_h}{r_e}\right] ,$$

so that the star appears dimmer than in the Schwarzschild case.

FREELY FALLING SOURCE

A possible source of electromagnetic radiation from a dilatonic black hole (or naked singularity) is the radiation emitted by a charged particle falling freely towards the singularity. In fact, the curved background distorts the EM field of the falling charge and causes the latter to release energy as though it were being accelerated in a Lorentz frame.

For practical purposes, it is more convenient to regard the falling charge Q as distorting the EM field $F_{\mu\nu}^{(0)}$ of the given background, and expand the exact $F_{\mu\nu}$ to linear order in Q as

$$F_{\mu\nu} = F_{\mu\nu}^{(0)} + Q F_{\mu\nu}^{(1)} .$$

Since in our case $F_{\mu\nu}^{(0)} = 0$, the perturbed EM tensor elements $F_{\mu\nu}^{(1)}$ are determined by the Maxwell-dilaton equation

$$\partial_\nu \left(\sqrt{-g}\, e^{-q\Phi} F^{(1)\mu\nu}\right) = 4\pi \sqrt{-g}\, j^\mu , \tag{18}$$

where $Q\, j^\mu$ is the current associated with the falling charge. The metric tensor elements are given in Eq. (2) and, in order to solve the set of Eqs. (18), the (antisymmetric) perturbations $F_{\mu\nu}^{(1)}$ are conveniently expanded in tensor harmonics [6]

$$F_{\mu\nu}^{(1)} = \begin{bmatrix} 0 & f_{01} Y_l^m & f_{02}\,\partial_\theta Y_l^m & -f_{02}\sin(\theta)\,\partial_\phi Y_l^m \\ * & 0 & f_{12}\,\partial_\theta Y_l^m & f_{12}\,\partial_\phi Y_l^m \\ * & * & 0 & 0 \\ * & * & 0 & 0 \end{bmatrix} , \tag{19}$$

where the f_{ij}'s depend only on r and t (and carry integer indices l and m which we omit), the Y_l^m's are again the normalized spherical harmonics, and the asterisks $(*)$ denote the negative of the corresponding transposed tensor element. The field tensor given in Eq. (19) is for electric multipoles. Further, using the field equations all of the tensor elements $F_{\mu\nu}^{(1)}$ can be put in terms of a single element, say f_{12}.

The function f_{12} is assumed to be of the form $f_{12}(r,t) = \hat{f}_{12}(r)\exp(i\omega t)$. Upon substitution of this expression into the equation for f_{12} and expanding to first order in $\tilde{x} \equiv qx$, the function $\hat{f}_{12}$ satisfies the equation

$$\frac{\mathrm{d}^2\hat{f}_{12}}{\mathrm{d}r^2} + \frac{r_h(3+2\tilde{x})}{r(r-r_h)}\frac{\mathrm{d}\hat{f}_{12}}{\mathrm{d}r} + \left[\frac{\omega^2 r^2}{(r-r_h)^2} - \frac{r_h(2r-3r_h)}{r^2(r-r_h)^2}(1+2\tilde{x}) - \frac{l(l+1)}{r(r-r_h)}\right]\hat{f}_{12}$$
$$= -\frac{e^{i\omega T(r)}}{(r-r_h)^2}\left[1 - 2\tilde{x}\ln\left(1-\frac{r_h}{r}\right)\right], \tag{20}$$

where $T(r) = r_h\left[2\frac{r^{1/2}}{r_h^{1/2}} - \frac{2r^{3/2}}{3r_h^{3/2}} + \ln\frac{\sqrt{r/r_h}-1}{\sqrt{r/r_h}+1}\right]$ is the time (to order $\tilde{x}^2$) for the particle to fall from ∞ to the point r.

Let us note the presence of terms of order $\tilde{x}$ both in the l.h.s. and r.h.s. of Eq. (20). In particular, the latter makes the source S stronger with respect to the Schwarzschild case (for $r > r_h$ the log is negative), and one expects a higher emission of energy from the falling charge.

Numerical results

The amplitude $\hat{f}_{12}$ defined by Eq. (20) is purely oscillatory as a function of r for $r \to \infty$. It is the Fourier transform of the retarded product of the radial position R of the observer times the component of the magnetic field transverse to the direction of the propagation of the radiation. The product of $\hat{f}_{12}$ times its complex conjugate is the energy per unit frequency per solid angle [7]. Once the function $\hat{f}_{12}$ has been determined, the energy distribution corresponding to the angular mode l and averaged over the complete solid angle is given by

$$\left\langle\frac{\mathrm{d}E}{\mathrm{d}\omega}\right\rangle = \frac{l(l+1)}{2\pi}\hat{f}_{12}\hat{f}_{12}^* \,. \tag{21}$$

The energy distribution (21) is plotted in Fig. 1 for unit charge Q and singularity mass M $(= r_h/2)$ for the angular mode $l = 1$. The total energy of the emitted radiation is

$$\Delta E \simeq C_{(4)}\frac{Q^2}{M} ,$$

where $C_{(4)}$ is the area under the curve. The frequency distribution curve is greater for the case when a dilaton is present ($\tilde{x} > 0$, higher curve in Fig. 1) than for the Schwarzschild case ($\tilde{x} = 0$, lower curve in Fig. 1). A charged particle being accelerated near a dilatonic

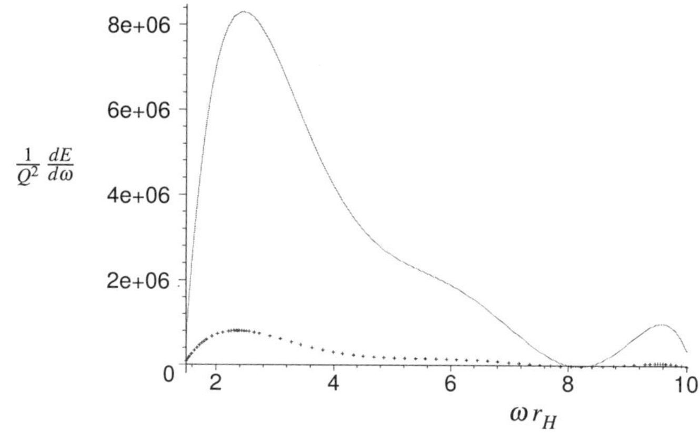

FIGURE 1. Energy distribution emitted in the $l = 1$ mode by a particle of unit charge falling in a JNW background of mass $r_h/2$ for $\tilde{x} > 0$ (higher curve) and $\tilde{x} = 0$ (lower curve).

singularity feels the combined effect of the scalar and tensor gravitational fields and hence is accelerated to a higher speed in a given time than an identical particle in the same tensor field alone. Both curves show a secondary peak, but the one for the dilaton curve is much more pronounced. We argue that the secondary peak is due to the curvature of space-time, because it disappears only for $r_h = 0$. For the dilaton the secondary peak at $\omega r_h \approx 9.6$ is an appreciable fraction of the higher peak at $\omega r_h \approx 2.7$. Such a peak should be relatively easy to detect in the frequency distribution of radiation from a star.

CONCLUSIONS

We have applied the resulting expansions to two cases of physical interest. The first case is that of a star with a dilatonic field which we have analyzed by means of the NP formalism. We find that its luminosity is reduced by the dilaton background. Such a reduction would, for example, affect the luminosity-to-distance relations that are used to determine the distance of astrophysical objects.

In the second case we have described the emission from a point-like charged particle freely falling in the chosen background. For nonzero $\tilde{x}$ the frequency distribution curve $dE/d\omega$ differs significantly from that obtained for the Schwarzschild metric. Models of the radiation from compact astrophysical sources incorporating this effect would provide a test of the existence of a scalar component of gravity. Possible extensions of the present work would be to repeat the above analysis for gravitational [8] and scalar waves [9].

ACKNOWLEDGMENTS

This work was supported in part by the U.S. Department of Energy under Grant no. DE-FG02-96ER40967.

REFERENCES

1. C.M. Will, *Theory and experiment in gravitational physics*, 2nd ed. (Cambridge University Press, 1993); Living Rev. Rel. **4** (2001) 4.
2. A.I. Janis, E.T. Newman and J. Winicour, Phys. Rev. Lett. **20** (1968) 878.
3. T. Harada, H. Iguchi and K. Nakao, Prog. Theor. Phys. **107**, 449 (2002).
4. S. Chandrasekhar, *The Mathematical Theory of Black Holes* (Oxford University Press, Oxford, 1983).
5. R. Casadio, B. Harms, Y. Leblanc and P.H. Cox, Phys. Rev. D **56**, 4948 (1997).
6. F. Zerilli, Phys. Rev. D **9**, 860 (1974); M. Johnston, R. Ruffini, F. Zerilli, Phys. Lett. **B49**, 185 (1974).
7. See, for example, J.D. Jackson, *Classical Electrodynamics*, 3rd Edition, John Wiley & Sons, New York, 1998.
8. V. Cardoso and J. P. S. Lemos, Phys.Rev. D **67** (2003) 084005.
9. V. Cardoso, J. P. S. Lemos, Phys.Rev. D **66** (2002) 064006.

BIG CRUNCH/BIG BANG STRING COSMOLOGIES

BEN CRAPS* AND BURT OVRUT†

*Enrico Fermi Institute, University of Chicago

†Department of Physics, University of Pennsylvania

Abstract

We study Big Crunch/Big Bang cosmologies that correspond to exact worldsheet superconformal field theories of type II strings. Generically, the Big Crunch fluctuation spectrum is altered when passing through the bounce singularity. The change in the spectrum is characterized by a function which is momentum and time-dependent. We compute this function explicitly and show that it indicates "entanglement entropy" with spacelike separated regions in the geometry. We demonstrate that in the Milne vacuum limit the fluctuation spectrum is unaltered as it passes through the singularity.

References

[1] Ben Craps and Burt A. Ovrut, "Global Fluctuation Spectra in Big Crunch/Big Bang String Vacua," Phys. Rev. **D69** (2004) 066001. arXiv: hep-th/0308057

DIRECT DETERMINATIONS OF THE REDSHIFT BEHAVIOR OF THE PRESSURE, ENERGY DENSITY, AND EQUATION OF STATE OF THE DARK ENERGY AND THE ACCELERATION OF THE UNIVERSE

RUTH A. DALY

Department of Physics
Berks-Lehigh Valley College
Penn State University
Reading, PA 19601-6009
E-mail: rdaly@psu.edu

S. G. DJORGOVSKI

Division of Physics, Mathematics, and Astronomy
California Institute of Technology
MS 105-24, Pasadena, CA 91125
E-mail: george@astro.caltech.edu

One of the goals of current cosmological studies is the determination of the expansion and acceleration rates of the universe as functions of redshift, and the determination of the properties of the dark energy that can explain these observations. Here the expansion and acceleration rates are determined directly from the data, without the need for the specification of a theory of gravity, and without adopting an a priori parameterization of the form or redshift evolution of the dark energy. We use the latest set of distances to SN standard candles from Riess et al. (2004), supplemented by data on radio galaxy standard ruler sizes, as described by Daly & Djorgovski (2003, 2004). We find that the universe transitions from acceleration to deceleration at a redshift of $z_T \approx 0.4$, with the present value of $q_0 = -0.35 \pm 0.15$. The standard "concordance model" with $\Omega_0 = 0.3$ and $\Lambda = 0.7$ provides a reasonably good fit to the dimensionless expansion rate as a function of redshift, though it fits the dimensionless acceleration rate as a function of redshift less well. The expansion and acceleration rates are then combined with a theory of gravity to determine the pressure, energy density, and equation of state of the dark energy as functions of redshift. Adopting General Relativity as the correct theory of gravity, the redshift trends for the pressure, energy density, and equation of state of the dark energy out to $z \sim 1$ are determined, and are found to be generally consistent with the concordance model; they have zero redshift values of $p_0 = -0.6 \pm 0.15$, $f_0 = 0.62 \pm 0.05$, and $w_0 = -0.9 \pm 0.1$.

1. Introduction

One way to determine the expansion and acceleration rates of the universe as functions of redshift is through studies of the coordinate distance to sources at different redshift. This can be accomplished with a variety of techniques including the use of supernovae and radio galaxies (e.g. Riess et al. 2004; Perlmutter et al. 1999; Riess et al. 1998; Daly 1994; Guerra & Daly 1998; Guerra, Daly, & Wan 2000). The techniques fall into two broad categories: the integral and the differential approaches.

The former, traditional approach involves the integration of a theoretically predicted expansion rate over redshift to obtain predicted coordinate distances to different redshifts; the difference between these predicted coordinate distances and the observed coordinate distances is then minimized to obtain the best fit model parameters. This approach usually requires the specification of a theory of gravity (generally taken to be General Relativity; GR) and a parameterization of the redshift evolution of the dark energy. Maor, Brustein, & Steinhardt (2001) and Barger & Marfatia (2001) discuss how difficult it is to extract the redshift behavior of the dark energy using this method. Some techniques have been developed to extract the redshift behavior of the dark energy using the integral method (e.g. Starobinsky 1998; Huterer & Turner 1999, 2001; Saini et al. 2000; Chiba & Nakamura 2000; Maor, Brustein, & Steinhardt 2001; Golaith et al. 2001; Wang & Garnavich 2001; Astier 2001; Gerke & Efstathiou 2002; Weller & Albrecht 2002; Padmanabhan & Choudhury 2002; Tegmark 2002; Huterer & Starkman 2003; Sahni et al. 2003; Alam et al. 2003; Wang & Freese 2004; Wang et al. 2004; Wang & Tegmark; Nessier & Perivolaropoulos 2004; Gong 2004; Zhu, Fujimoto, & He 2004; Elgaroy & Multamaki 2004; Huterer & Cooray 2004; Alam, Sahni, & Starobinsky 2004).

2. The Methodology

The differential approach has been investigated by Daly & Djorgovski (2003, 2004), and we offer a brief summary here. It is well known (e.g. Weinberg 1972; Peebles 1993; Peebles & Ratra 2003) that the dimensionless expansion rate $E(z)$ can be written as the derivative of the dimensionless coordinate distances $y(z)$; the expression is particularly simple when the space curvature term is equal to zero. In this case,

$$\left(\frac{\dot{a}}{a}\right) H_0^{-1} \equiv E(z) = (dy/dz)^{-1} , \tag{1}$$

where a is the cosmic scale factor, and Hubble's constant is $H_0 = (\dot{a}/a)$ evaluated at zero redshift. This representation follows directly from the Friedman-Robertson-Walker line element, and does not require the use of a theory of gravity. Similarly, in a spatially flat universe (as convincingly demonstrated by CMBR measurements, Spergel et al. 2003), it is shown in Daly & Djorgovski (2003) that the dimensionless deceleration parameter

$$-\left(\frac{\ddot{a}a}{\dot{a}^2}\right) \equiv q(z) = -[1 + (1+z)(dy/dz)^{-1}\ d^2y/dz^2] \tag{2}$$

also follows directly from the FRW line element, and is independent of any assumptions regarding the dark energy or a theory of gravity. Thus, measurements of the dimensionless coordinate distance to sources at different redshifts can be used to determine dy/dz and d^2y/dz^2, which can then be used to determine $E(z)$ and $q(z)$, and these direct determinations are completely model-independent, as discussed by Daly & Djorgovski (2003).

In addition, if a theory of gravity is specified, the measurements of dy/dz and d^2y/dz^2 can be used to determine the pressure, energy density, and equation of state of the dark energy as functions of redshift (Daly & Djorgovski 2004); we assume the standard GR for this study. *These determinations are completely independent of any assumptions regarding the form or properties of the dark energy or its redshift evolution.* Thus, we can use the data to determine these functions directly, which provides an approach that is complementary to the standard one of assuming a physical model, and then fitting the parameters of the chosen function.

In a spatially flat, homogeneous, isotropic universe with non-relativistic matter and dark energy Einstein's equations are $(\ddot{a}/a) = -(4\pi G/3)\ (\rho_m + \rho_{DE} + 3P_{DE})$ and $(\dot{a}/a)^2 = (8\pi G/3)\ (\rho_m + \rho_{DE})$, where ρ_m is the mean mass-energy density of non-relativistic matter, ρ_{DE} is the mean mass-energy density of the dark energy, and P_{DE} is the pressure of the dark energy. Combining these equations, we find $(\ddot{a}/a) = -0.5[(\dot{a}/a)^2 + (8\pi G)\ P_{DE}]$.

Defining the critical density at the present epoch in the usual way, $\rho_{oc} = 3H_0^2/(8\pi G)$, it is easy to show that $p(z) \equiv (P_{DE}(z)/\rho_{oc}) = (E^2(z)/3)\ [2q(z) - 1]$. Combining this expression with eqs. (1) and (2) we obtain the pressure of the dark energy as a function of redshift in terms of first and second derivatives of the dimensionless coordinate distance y (Daly & Djorgovski 2004)

$$p(z) = -(dy/dz)^{-2}[1 + (2/3)\ (1+z)\ (dy/dz)^{-1}\ (d^2y/dz^2)]\ . \tag{3}$$

Thus, the pressure of the dark energy can be determined directly from measurements of the coordinate distance. In addition, this provides a direct measure of the cosmological constant for Friedmann-Lemaitre models since in these models $p = -\Omega_\Lambda$. If more than one new component is present, this pressure is the sum of the pressures of the new components.

Similarly, the energy density of the dark energy can be obtained directly from the data

$$f(z) \equiv \left(\frac{\rho_{DE}(z)}{\rho_{oc}}\right) = (dy/dz)^{-2} - \Omega_0(1+z)^3 \ , \tag{4}$$

where $\Omega_0 = \rho_{om}/\rho_{oc}$ is the fractional contribution of non-relativistic matter to the total critical density at zero redshift, and it is assumed that this non-relativistic matter evolves as $(1+z)^3$. If more than one new component is present, then f includes the sum of the mean mass-energy densities of the new components.

The equation of state $w(z)$ is defined to be the ratio of the pressure of the dark energy to it's energy-density $w(z) \equiv P_{DE}(z)/\rho_{DE}(z)$. As shown by Daly & Djorgovski (2004), the equation of state is

$$w(z) = -\frac{[1 + (2/3)\ (1+z)\ (dy/dz)^{-1}\ (d^2y/dz^2)]}{[1 - (dy/dz)^2\ \Omega_0\ (1+z)^3]} \ . \tag{5}$$

Here, w is the equation of state of the dark energy; if more than one new component contributes to the dark energy, w is the ratio of the sum of the total pressures of the new components to their total mean mass-energy densities.

3. Results and Conclusions

The results presented here follow those presented by Daly & Djorgovski (2003, 2004), where more details can be found. The data used here includes 20 radio galaxies (RG) compiled by Guerra, Daly, & Wan (2000) and the "gold" supernova (SN) sample compiled by Riess et al. (2004). We note that in the redshift interval where the two sets of coordinate distances (RG and SN) overlap, the agreement is excellent, suggesting that neither one is affected by some significant bias, and allowing us to combine them for this study.

Measurements of luminosity distances and angular size distances are easily converted to coordinate distances, $y(z)$. Using some robust numerical differentiation method, these can be used to determine the first and second derivatives as functions of redshift, which can be combined to determine the

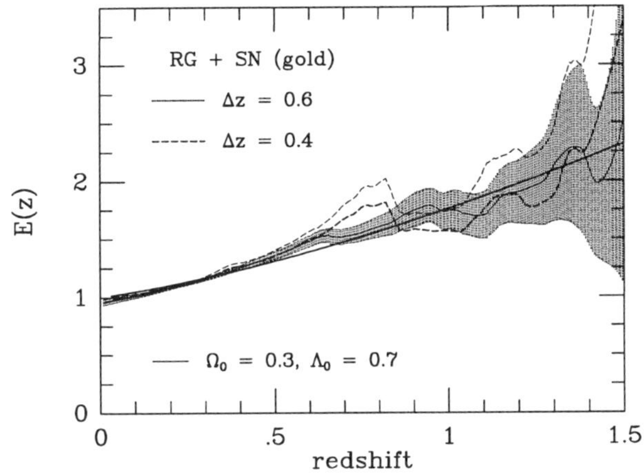

Figure 1. The derived values of the dimensionless expansion rate $E(z) \equiv (\dot{a}/a)H_0^{-1} = (dy/dz)^{-1}$ obtained with window functions of width $\Delta z = 0.4$ and their 1 σ error bars (dashed lines) and 0.6 (dotted line and hatched error range). At zero redshift, the value of E is $E_0 = 0.97 \pm 0.03$. The value of $E(z)$ predicted in a spatially flat universe with a cosmological constant $\Lambda = 0.7$ and mean mass density $\Omega_0 = 0.3$ is also shown.

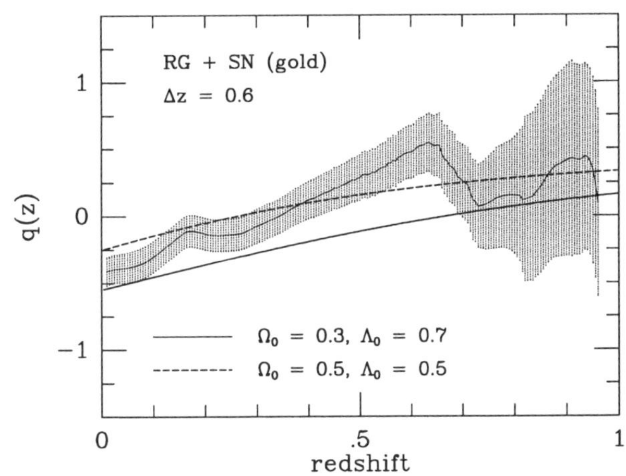

Figure 2. The derived values of deceleration parameter $q(z)$ and their 1 σ error bars obtained with window function of width $\Delta z = 0.6$ applied to the radio galaxy and gold supernovae samples. The universe transitions from acceleration to deceleration at a redshift $z_T \approx 0.4$, with an uncertainty difficult to quantify due to large fluctuations at $z > 0.5$, caused by the sparseness of the data at higher redshifts. The present value is $q_0 = -0.35 \pm 0.15$. Solid and dashed lines show the expected dependence in the standard Friedmann-Lemaitre models with zero curvature, for two pairs of values of Ω_0 and Λ_0.

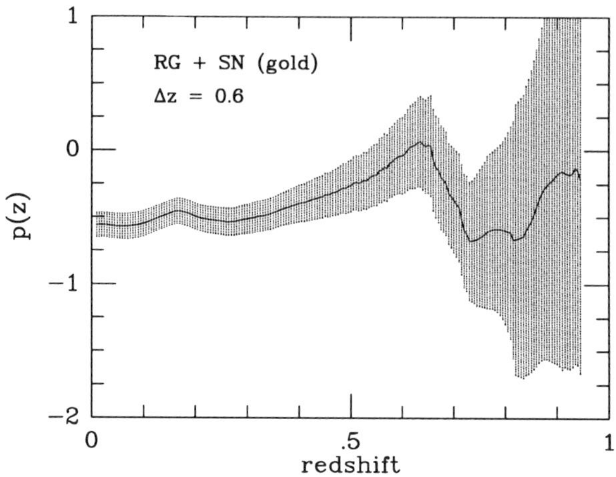

Figure 3. The derived values of dark energy pressure $p(z)$, obtained with window function of width $\Delta z = 0.6$. This derivation of $p(z)$ requires a choice of theory of gravity, and GR has been adopted here. The present value is $p_0 = -0.6 \pm 0.15$. Note that in the standard Friedmann-Lemaitre models, $p_0 = -\Lambda_0$, and thus we have a direct measurement of the value of the cosmological constant, which is also fully consistent with other modern measurements.

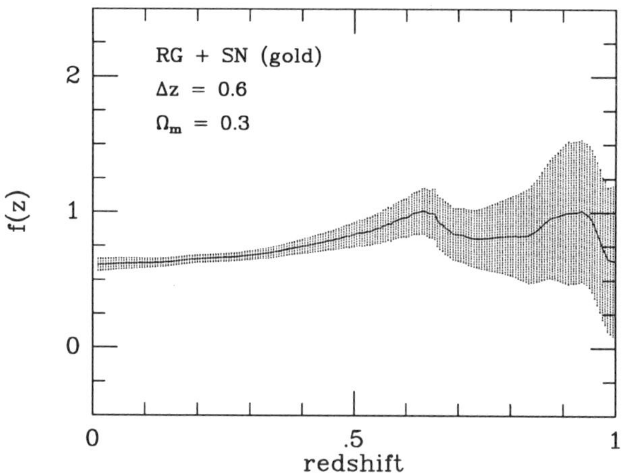

Figure 4. The derived values of the dark energy density fraction $f(z)$, obtained with window function of width $\Delta z = 0.6$. This derivation of $f(z)$ requires of theory of gravity and the value of Ω_0 for the nonrelativistic matter; GR has been adopted here, and $\Omega_0 = 0.3$ is assumed. The present value is 0.62 ± 0.05, and the trend is consistent with $f(z) = const.$ out to $z \approx 1$.

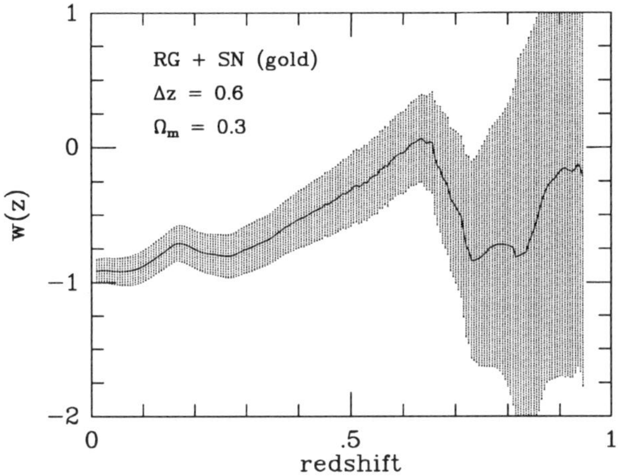

Figure 5. The derived values of the dark energy equation of state parameter $w(z)$, obtained with window function of width $\Delta z = 0.6$. This derivation of $w(z)$ requires of theory of gravity and the value of Ω_0; GR has been adopted here, and $\Omega_0 = 0.3$ is assumed. The present value is $w_0 = -0.9 \pm 0.1$, consistent with cosmological constant models.

dimensionless expansion and acceleration rates of the universe as functions of redshift, and the pressure, energy density, and equation of state of the dark energy as functions of redshift, as described above.

We see that the universe transitions from acceleration to deceleration at a redshift of about 0.4 (consistent with determinations by Daly & Djorgovski 2003, Riess et al. 2004, and Alam, Sahni, & Starobinsky 2004); and our determination only depends upon the assumption that the universe is homogeneous, isotropic, and spatially flat. Assuming GR, we solve for the pressure, energy density, and equation of state of the dark energy. Each is generally consistent with remaining constant to a redshift of about 0.5 and possibly beyond, but determining their behavior at higher redshifts is severely limited by the available data.

As more and better data become available, this methodology can be used to determine the evolution of the dark energy properties and the observed kinematics of the universe with an increasing precision and confidence.

Acknowledgments

This work was supported in part by the U. S. National Science Foundation under grants AST-0206002, and Penn State University (RAD), and by the Ajax Foundation (SGD). Finally, we acknowledge the great work and efforts of many observers who obtained the valuable data used in this study.

References

1. Alam, U., Sahni, V., Saini, T. D., & Starobinsky, A. A. 2003, MNRAS, 344, 1057
2. Alam, U., Sahni, V., & Starobinsky, A. 2004, astro-ph/0403687
3. Astier, P. 2001, Phys. Lett. B, 500, 8
4. Barger, V., & Marfatia, D. 2001, Phys. Lett. B, 498,
5. Bennett, C., et al. (the WMAP team) 2003, ApJ, 583, 1
6. Chiba, T., & Nakamura, T. 2000, Phys. Rev. D, 62, 121301
7. Daly, R. A. 1994, ApJ, 426, 38
8. Daly, R. A., & Djorgovski, S. 2003, ApJ, 597, 9
9. Daly, R. A., & Djorgovski, S. 2004, ApJ, in press (astro-ph/0403664)
10. Elgaroy, O., & Multamaki, T. 2004, astro-ph/0404402
11. Gerke, B. F., & Efstathiou 2002, MNRAS, 335, 33
12. Goliath, M., Amanullah, T., Astier, P. Goobar, A., & Pain, R. 2001, A&A, 380, 6
13. Gong, Y. 2004, astro-ph/0401207
14. Guerra, E. J., & Daly, R. A. 1998, ApJ, 493, 536
15. Guerra, E. J., Daly, R. A., & Wan, L. 2000, ApJ, 544, 659
16. Huterer, D., & Cooray, A. 2004, astro-ph/0404062
17. Huterer, D., & Starkman, G. 2003, Phys. Rev. Lett. 90, 031301
18. Huterer, D., & Turner, M. S. 1999, Phys. Rev. D, 60, 081301
19. Huterer, D., & Turner, M. S. 2001, Phys. Rev. D, 64, 123527
20. Maor, I., Brustein, R., & Steinhardt, P. 2001, Phys. Rev. Lett., 86, 6
21. Nessier, S., & Perivolaropoulos, L. 2004, astro-ph/0401556
22. Padmanabhan, T., Choudhury, T. R. 2002, MNRAS, 344, 823
23. Peebles, P. J. E. 1993, Principles of Physical Cosmology, Princeton University Press
24. Peebles, P. J. E., & Ratra, B. 2003, Rev. Mod. Phys., 75, 559
25. Perlmutter et al. 1999, ApJ, 517, 565
26. Riess, R. G., Strolger, L., Tonry, J., Casertano, S., Ferguson, H. G., Mobasher, B., Challis, P., Filippenko, A. V., Jha, S., Li, W., Chornock, R., Kirshner, R. P., Leibundgut, B., Dickinson, M., Livio, M., Giavalisco, M., Steidel, C. C., Benitez, N., & Txvetanov, Z. 2004, ApJ, in press
27. Riess, A. G. et al. 1998, AJ, 116, 1009
28. Sahni, V., Saini, T. D., Starobinsky, A. A., & Alam, U. 2003, J. Exp. Theor. Phys. Lett., 77, 201
29. Saini, T., Raychaudhury, S., Sahni, V., & Starobinsky, A. A. 2000, Phys. Rev. Lett., 85, 1162
30. Spergel, D., et al. (the WMAP team) 2003, ApJS, 148, 175
31. Tegmark, M. 2002, Phys. Rev. D66, 103507
32. Starobinsky, A., 1988, JETP Lett., 68, 757
33. Wang, Y., & Garnavich, P. 2001, ApJ, 552, 445
34. Wang, Y., & Freese, K. 2004, astro-ph/0402208
35. Wang, Y., Kostov, V., Freese, K., Frieman, J. A., & Gondolo, P. 2004, astro-ph/0402080
36. Wang, Y., & Tegmark, M 2004, astro-ph/0403292
37. Weinberg, S. 1972, Gravitation and Cosmology, John Wiley & Sons
38. Weller, J., & Albrecht, A. 2002, Phys. Rev. D, 65, 103512
39. Zhu, Z., Fujimoto, & He, X. 2004, astro-ph/0403228

COSMOLOGY WITH CLUSTERS OF GALAXIES

M. DONAHUE
Michigan State University
Dept. of Physics and Astronomy
East Lansing, MI 48824-2320 USA
E-mail: donahue@pa.msu.edu

Cluster observations provide unique and useful constraints on cosmological parameters. The contents of clusters and the rate of their formation are very sensitive to the mean matter density (Ω_M and the normalization and shape of the spectrum of initial density perturbations near the size scale of $\sim 8h^{-1}$ Mpc. Future and on-going cluster studies constrain Ω_Λ (acceleration) and the equation of state of the "dark energy," particularly in conjunction with either constraints from the cosmic microwave background or Type Ia supernovae of white dwarfs.

1. The Cluster-Cosmology Connection

Clusters of galaxies are the largest gravitationally relaxed structures in the universe. Because of their sheer size, clusters of galaxies are likely to be "fair samples" of the universe. The matter in a cluster of galaxies is from a volume defined by a very large radius $r \sim 10 - 25$ Mpc(hΩ/0.3)$^{-1/3}$. Here, h is the Hubble constant $H_0/100$ km s^{-1}, Ω is a normalized density such that $\Omega = \rho/\rho_c$ where ρ_c is the the critical density. The critical density is the maximum density a non-accelerating universe could have and not eventually re-collapse, $\rho_c = 3H_0^2/(8\pi G) = 1.9 \times 10^{-29} h^2$ g cm^{-3}. In this paper, I use a Hubble constant of 70 km/s/Mpc throughout unless parameterized by h.

Clusters occupy a special place in the hierarchy of structures in the universe. They sit on the top of the pyramid, as the rarest and most massive systems ($10^{14} - 10^{15}$ $M_\odot$, where $M_\odot$ is a solar mass or 2×10^{30} kg.) In a cold dark matter dominated universe, as the most massive systems, clusters were the latest to decouple from the expansion and collapse.

In this meeting we heard some discussion about cold dark matter (CDM). I can say that the hypothesis of cold dark matter is alive and well. In the context of the current theory of gravity, CDM is a viable explanation for a wide range of astronomical observables, from flat rotation curves of galaxies, to the growth of perturbations and galaxy formation, to multiple constraints on the baryon density. I recommend the book Cosmological Physics by John A. Peacock (available from Cambridge University Press).

1.1. *Why bother with clusters?*

One could argue that we already get some cosmological constraints from Type Ia supernovae and cosmic microwave background experiments. In response, University of Chicago professor Mike Turner likes to quote Carl Sagan: "Extraordinary claims require extraordinary proof." The constraints resulting from these experiments are nothing short of extraordinary: not only is the density of matter less than what we might have anticipated (based on inflation predictions or even a sense of aesthetics), but the universe appears to be accelerating its expansion, a result akin to tossing a ball up in the air and seeing it fly away with ever-increasing speed. Such baffling results demand independent confirmation however it can be done.

Clusters provide a third check on these claims. Cluster contents and cluster evolution are very sensitive to Ω_m and the normalization of the initial power spectrum, σ_8. With sufficient statistics, observations of $z = 0.5 - 1.5$ clusters also allow constraints on Ω_Λ (the acceleration parameter, ~ 0.7 in the current picture, $\Omega_\Lambda + \Omega_M = 1$ if the geometry is flat.)

As I will discuss, the constraints from clusters are complementary to those of the other cosmological estimators. The constraints from clusters are bounded by ellipsoidal contours. These uncertainty contours elongate in different directions when compared to the parameter space bounds from other techniques. On the $\Omega_M - \Omega_\Lambda$ plane, the CMB constraints align along a line $\Omega_M + \Omega_\Lambda = 1$ (e.g. Spergel et al. 2003). The supernovae constraints tilt the opposite direction, roughly along a line $\Omega_\Lambda - \Omega_M = 0.5$ (e.g. Knop et al. 2004). The cluster constraints from both the baryon fraction and the cluster evolution observations lie approximately along a vertical line where $\Omega_M = 0.3$. All three of these lines intersect at the "concordance model" of $\Omega_\Lambda \sim 0.7$, $\Omega_M \sim 0.3$.

2. Multi-wavelength Views of Clusters of Galaxies

2.1. *Optical*

In the optical bandpass, clusters of galaxies look like what you might expect: clusters of up to thousands of galaxies. The average redshift[a] of these galaxies in combination with the assumed geometry of spacetime gives the distance to the cluster[b], and the dispersion of redshifts is proportional to the dispersion of line-of-sight orbital velocities of the cluster members. The velocity dispersion of a cluster provides an estimate of the cluster binding mass, via the classic virial theorem.

2.2. *X-ray Views*

From orbiting X-ray satellites such as the Chandra X-ray Observatory and the X-ray Multi Mirror (XMM) Newton Observatory, we get a very different view of clusters of galaxies. With X-ray vision, clusters appear not as groups of galaxies, but as giant, optically-thin, glowing gas clouds. Clusters are among the most luminous X-ray sources known. The intergalactic gas in a cluster of galaxies is confined by the same dark matter potential that confines the galaxies. The gas and the galaxies are distributed rather similarly, concentrated towards the center. The gas is hot, at the virial temperature of the clusters. Typical temperatures of clusters range between $kT \sim 2-12$ keV, with poor clusters like Virgo at the cool end and the most massive clusters known, such as MS0451.6-0306, at the hotter end. The main source of heat for the gas is gravitational, so to at least zeroth order its temperature directly accounts for the binding energy of the cluster gravitational potential. The X-ray surface brightness of the gas is proportional to the integral of $n_{elec}^2 T^{1/2}$ along the line of sight. X-ray detectors simultaneously acquire images and spatially-resolved spectra of clusters. The spectra provide sky maps of kT. Together with the surface brightness of the cluster and the assumption of symmetry, such data allow the derivation of cluster electron density, binding mass as a function of radius, and gas mass as a

[a] A redshift $z = (\lambda_{observed} - \lambda_{rest})/\lambda_{rest}$ of the wavelengths (λ) of the spectral features in a galaxy spectrum is a direct measure of the ratio ($(1+z)^{-1}$) of the scale factor at the time the photon was emitted to the time it was observed in our telescopes.

[b] One measure of distance is the lookback time. The estimate of the lookback time depends on cosmological parameters, which is why astronomers prefer to speak about redshifts rather than the derived distances or lookback times. For the concordance model, $z = 0.5$ corresponds to a lookback time of 5 billion years; $z = 1$ corresponds to 7.7 billion years ago, $z = 3$ corresponds to 11.5 billion years ago.

function of radius. The density of the gas falls off like r^{-3} or so at large radii. In the most massive clusters, the galaxies contribute little or nothing to the total X-ray emission[c]

3. Cluster Mass Budget

From all these observables, we can assemble a picture of the typical mass budget of a cluster of galaxies. A massive cluster is typically about 10^{15} $M_\odot$. Hot gas outweighs the stars in the cluster's galaxies by a factor of 6-10, a startling fact which implies that most of the baryons reside outside of galaxies. Dark matter outweighs the hot gas by a factor of about 8. Since each mass estimate has different dependencies on cosmological parameters such as the Hubble constant, the ratios also have residual dependencies on these parameters. Therefore we can constrain the cosmological parameters by measuring these ratios in combination with information from other experiments.

3.1. *Baryonic Mass Fraction and Mass to Light Ratios*

The baryonic mass fraction (M_b/M_{grav}) in clusters of galaxies is thought to be representative of the baryonic mass fraction of the entire universe, because clusters form by accreting matter from an enormous volume of space. The density of baryons in the universe, Ω_b, can be measured in a couple of ways. One is to measure deuterium in high-redshift, distant gas clouds, and use models of primoridial nucleosynthesis to infer a baryon density. Deuterium is created in the first 3 minutes of the universe, and only destroyed in stars. If the baryon density is higher, hydrogen burning is more efficient, and the resulting primordial deuterium abundance is lower. The current best estimate of Ω_b is by Burles & Tytler (1998), $\Omega_b = 0.019 \pm 0.001h^{-2}$. Mohr et al. (1999) and others have measured the gas mass fraction in clusters and estimated $\Omega_b/\Omega_m = 0.08 \pm 0.01h^{-3/2}$. Combining these estimates results in $\Omega_m \sim 0.3 \pm 0.05$. (Recently Allen et al. (2004) have taken a similar strategy, in combination with WMAP cosmological background constraints, to place constraints on other parameters such as w.)

Mass-to-light ratios tend to imply somewhat lower Ω_m. The mass-to-optical light ratio required to "close" the universe is approximately 1200

[c]I also discussed Sunyaev-Zel'dovich maps of the hot gas in clusters, as seen in relief against the cosmic microwave background; see Ref. 7 for a recent review. I discussed gravitational lensing, but lack space here; see Ref 1 for discussion.

in solar units (solar mass/solar luminosity). Most galaxies have a mass-to-light ratio of approximately 5-10, while clusters tend to have mass-to-light ratios between 100-400 (e.g. Bahcall et al. 2000). Carlberg, Yee & Ellingson (1997) studied a well-defined sample of X-ray luminous clusters to estimate mass-to-light ratios and infer that $\Omega_m \sim 0.2$. I've glossed over the details regarding specific bandpasses and sensitivity to star formation histories here. Such concerns have led people to look to infrared light, which is less affected by the presence or absence of blue, bright but rare main sequence stars. Lin, Mohr, and Stanford (2003) use near-infrared observations to infer $\Omega_m \sim 0.2 \pm 0.03$. The differences between the estimates from baryon fractions and from mass to light are now statistically significant, and may suggest that massive clusters are NOT fair samples of light. This conclusion would follow if star formation were somehow less efficient in clusters of galaxies.

4. Constraints from Cluster Evolution

The rate large scale structure forms via gravitational processes depends sensitively on the average density of mass. Clusters form quickly in high density regions; in low density regions they form slowly or not at all (e.g. Borgani & Guzzo 2001). In particular, in a universe with a $\Omega_m = 1$, large scale structures grow rapidly at all scales at all times. In low density universe with $\Omega_m = 0.3$, those same structures initially grow at the same rate, but at $z \sim 1 - 2$, the matter density of the universe departs from the critical density and structure formation slows down dramatically. Therefore, by comparing the universe now to the universe as it was at $z = 0.5$ or at $z = 1.0$, we can determine how fast structures formed and by inference, the average density of matter.

5. Cluster evolution and dark energy

Right now we don't know much about dark energy, aside that it's not matter (hence the term "energy") and that it probably has something to do with the energy content of the vacuum. The mass-energy density of the dark energy is Ω_Λ. But until we know the nature of the dark energy - that is, how it behaves - we cannot predict, for example, the fate of the universe.

In this section, I will describe how the cluster evolution can be used to constrain the equation of state of dark energy. The parameter w is $P = w\rho$. If $w = -1$, the dark energy can be described by a "cosmological constant", and the universe accelerates forever. (w may even vary with time.) In other theoretical scenarios, Ω_Λ can diminish or even change sign. In another intriguing scenario, Ω_Λ increases until the "Big Rip" occurs, tearing apart space time in everything from galaxies to atoms.

5.1. *The experiment*

Two factors determine the number density of clusters present at a given mass and a given redshift: the evolution of clusters and the cosmological volume element (see Voit 2004).

The size of the cosmological volume element depends somewhat on w. Since clusters do not evolve very rapidly between $z = 0.5 - 0.0$, the number of clusters at a given redshift < 0.5 depends mainly on the volume element at that redshift, rather than any evolutionary effect. (Figure 1.)

In the linear regime of gravitational growth, the growth of a perturbation is decoupled from the size of the perturbation. All density perturbations grow linearly at the same rate, which can be described by a growth function dependent only on redshift. Figure 2 shows the normalized value of the growth function as a function of redshift and for the same set of cosmological parameters as in Figure 1. In universes with lower matter densities, growth slows down at low redshift, while in an Einstein-de Sitter universe, growth continues inversely proportional to the scale factor at all times.

With the growth function and the volume element, all that is required to compute the expected number density of clusters of a given mass and given redshift is the primordial perturbation spectrum, suitably normalized and filtered by processes encountered by density perturbations as the universe expands. The spectrum of gaussian density perturbations starts out scale-free with a slope of $n = -1$. The weakly-interacting matter fluctuations are decoupled from the radiation field (unlike the baryons, which are coupled with radiation until the recombination epoch about 300,000 years after the Big Bang). The perturbations grow steadily with time, and as each perturbation scale enters the horizon, the perturbation may oscillate.

Because the radiation density dominates the mass-energy budget at early times, the perturbation spectrum does not remain scale-free. Silk damping(Ref 16) by radiation erases the response of baryons to small-scale density perturbations and stalls the growth of the dark matter fluctuations until the radiation density is no longer important. This process suppresses some of the power at smaller angular scales and creates a feature in the perturbation spectrum. Given this modified power spectrum and its nor-

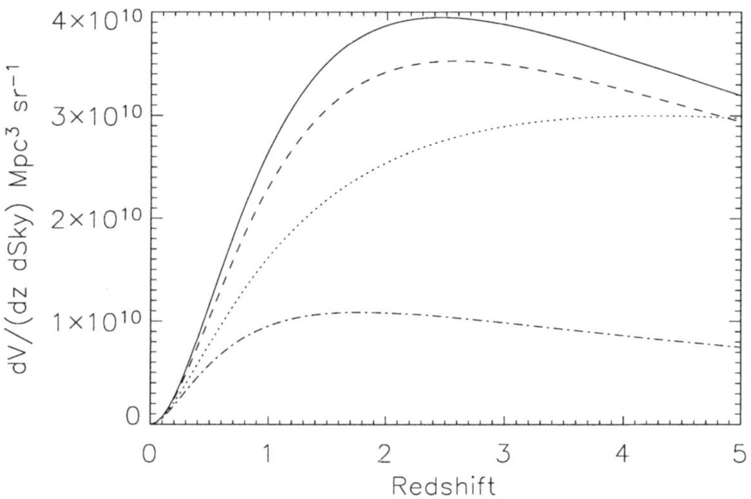

Figure 1. Differential volume element $dV/dz(z)$, plotted in units of cubic Megaparsecs (Mpc^3) per steradian on the sky. For the figures in this paper, the lines are coded to the same convention. The concordance model $\Omega_M = 0.3$, $\Omega_\Lambda = 0.7$, Hubble's constant $H_0 = 70$ km/s/Mpc, and $w = -1$ is plotted with a solid line. The same model, but with $w = -0.8$ is plotted with a dashed line. An open model where $\Omega_\Lambda = 0$ and $\Omega_M = 0.3$ is plotted with a dotted line, and the Einstein-de Sitter model where $\Omega_M = 1$ and $\Omega_\Lambda = 0$ is plotted with a dash-dot line pattern. Note that the volume element grows out to redshift 2, but then starts to decrease in the Λ-dominated cosmologies, but continues a mild increase in open cosmologies. The volume elements for open and Λ-dominated cosmologies are much larger than the volume element for a Einstein-de Sitter universe. The observable universe is a bigger and older place for open and Λ-dominated cosmologies.

malization (σ_8), a mass function of dark matter halos for any given redshift (or epoch) can be computed. The cluster mass function changes with redshift and is sensitive to cosmological parameters. (Figure 3.)

By integrating a mass function from a minimum mass to infinity, one can also calculate the number of clusters above a minimum mass as a function of redshift (Figure 4ab). The most massive clusters evolve the most rapidly. Current cluster samples easily distinguish between the $\Omega_M = 1.0$ and the $\Omega_M = 0.3$ models, which differ by factors of 100-10,000 depending on the mass threshold. Distinguishing between $w = -1$ and $w = -0.8$ is more subtle - a 10% difference is expected, starting around $z > 0.5$. We do not have to observe clusters at incredibly high redshifts to accomplish this,

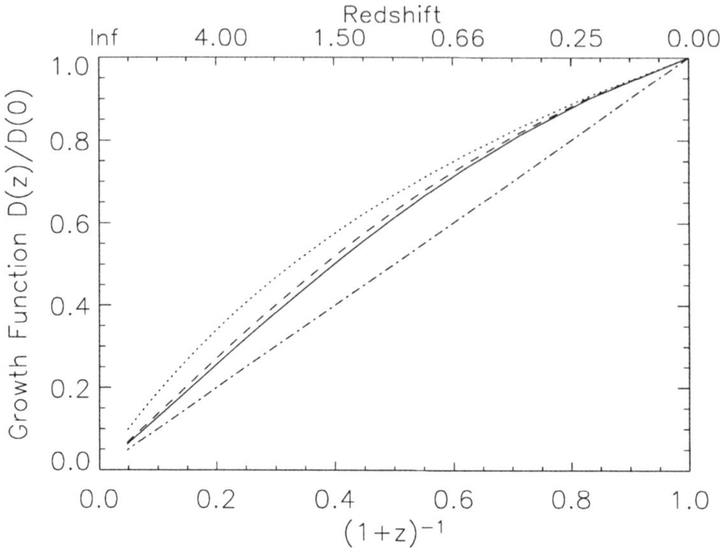

Figure 2. Growth function as a function of redshift and cosmology (key is the same as in Figure 1. All lines are normalized to 1.0 at the present time $z = 0$. The lower x-axis is $1/(1+z)$. In an Einstein deSitter universe, perturbations grow $\propto (1+z)^{-1}$, or the scale factor of the universe's expansion, as this plot shows. The upper x-axis is redshift (z), for comparison to other plots and the discussion in the text.

however. The $z \sim 0.5 - 1.5$ clusters may be the optimal sample for this experiment, since at $z < 0.5$, $w = -0.8$ models predict fewer cluster than $w = -1$ models while for $z > 0.7$, the expectation is reversed.

5.2. *Systematics*

It is crucial to understand and to be able to quantify the selection of any sample of clusters used for this experiment. X-ray selection has proved to be superior for selecting clusters because of its relative immunity to contamination by foregrounds and backgrounds, and because of clear correlations between global X-ray properties such as X-ray luminosity or X-ray temperature and mass (e.g. Voit et al. 2003).

However, X-ray cluster studies, like the Type Ia supernovae studies, are vulnerable to unanticipated evolution. We need to both (1) understand the physics of those correlations well enough to predict the evolution and (2) calibrate the correlation empirically and sufficiently well to minimize the effects of on the assessment of cosmological parameters.

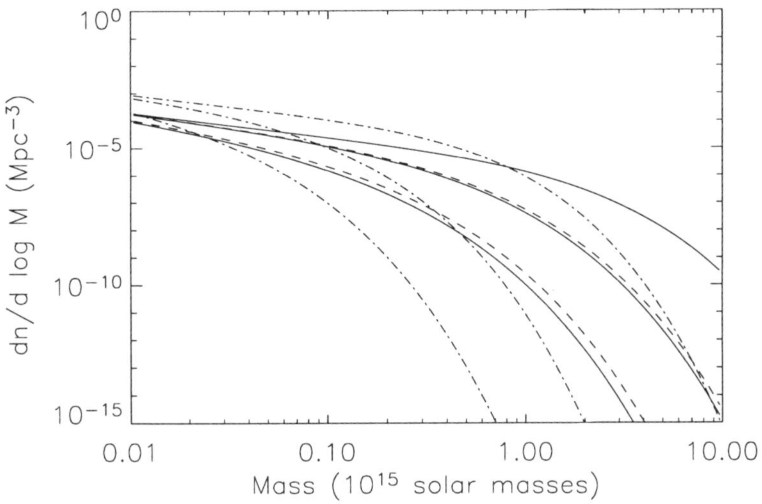

Figure 3. The cluster mass function for 3 different epochs ($z = 0, 1, 2$) are plotted for 3 different cosmologies. Cluster number density always increases, so each curve triplet is $z = 0, 1, 2$ from left to right. The normalization (σ_8) for each cosmology is chosen such that the same number of clusters (integrated over a range of masses) is predicted for $z = 0$.Three notes: (1) that even at low redshift, the slopes of the mass function depend on cosmological parameters. So it is possible to use the slope of the observed cluster mass function to constrain Ω_M (Reiprich & Böhringer 2002). (2) The cluster mass function evolves much more strongly when Ω_M is high. (3) The slope and normalization of the mass function of clusters at the present epoch is not affected at all by changes in the w parameter; differences (between the solid line and the dashed line) increases at higher redshift, but remain relatively moderate.

Cluster surveys will produce catalogs of thousands of clusters, with properties such as their X-ray luminosities. The basic problem is relating observable properties to the cluster masses, particularly at high redshift where the low-redshift calibrations may no longer apply. Majumdar & Mohr (2003) realized this issue and calculated that the detailed calibration of 100 clusters uniformly distributed between $z = 0.1 - 1.2$ improves the constraints of larger surveys. Such surveys include the Sunyaev-Zeldovich surveys of the South Pole Telescope (http://astro.uchicago.edu/spt), and the ESA Planck Telescope (expected launch date 2007), and the DUO X-ray survey, proposed to NASA's small missions program, which will survey 10,000 square degrees for clusters of galaxies.

Improved physical models are currently in progress (e.g. da Silva et al.

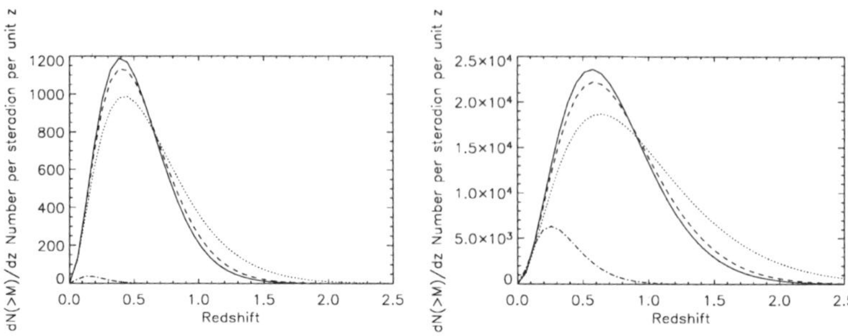

Figure 4. The number of clusters per steraidian as a function of redshift and above a mass threshold for 4 different cosmologies. The line keys and the cosmologies are the same as in Figure 1. The left panel is for very massive, rare clusters ($M > 10^{15}$ $M_{\odot}$). The right panel is for somewhat less massive clusters that are far more common ($M > 3 \times 10^{14}$ $M_{\odot}$). With current samples, it is easy to differentiate between $\Omega_M = 0.3$ and $\Omega_M = 1$ cosmologies with ~ 50 high mass clusters at a range of redshifts $z = 0 - 1$). More clusters will be needed to differentiate predictions from cosmologies with different w parameters. However, the clusters do not need to be even higher redshift – more clusters at the redshifts being probed now and at only somewhat higher redshifts are required ($z = 0 - 1.5$).

2004; Motl et al 2004). Hydrodynamic simulations of the gas properties are now starting to incorporate feedback processes from star formation and active galaxies. The shocks and compression of the baryons accompanying the merging or tidal interactions of dark matter halos are now being taken into account when predicting a cluster's luminosity not only after the cluster has relaxed, but while the cluster is in the process of relaxing into a near-equilibrium state. Recent spectroscopic observations of cluster cores by Peterson et al. (2003) have revealed that we don't understand how gas cools in a realistic gravitational potential well. Processes of radiative cooling and star formation occur on scales below that sampled by the simulations. Processes of energy transport such as conduction, convection, and turbulence are only beginning to be explored (Sparks et al. 2004).

Theoretical and empirical investigations should continue in parallel. Observers should acquire detailed observations of 100 or more clusters with current X-ray telescopes. Those calibrations will pay off when the large cluster surveys produce their catalogs of thousands of clusters. Chandra, XMM, and the Hubble Space Telescope (HST) have finite lives in orbit. Chandra has spatial resolution capabilties that will not be matched by any mission currently in NASA's queue. The questions are critical; the resources are limited and relatively short-lived; the time to calibrate is now.

Acknowledgements

This work was supported by Chandra grant, SAO GO3-4159X, and Hubble Space Telescope grant HST-GO-6668.

References

1. Allen, S. W., Schmidt, R. W., Fabian, A. C., Ebeling, H. MNRAS, 342, 287 (2003).
2. Allen, S. W., Schmidt, R. W., Ebeling, H., Fabian, A. C., van Speybroeck, L. MNRAS, in press. (astro-ph/0405340).
3. Bahcall, N. A., Cen, R., Daveé, R., Ostriker, J. P., & Yu, Q. ApJ, 541, 1 (2000).
4. Borgani, S. & Guzzo, L., Nature, 409, 39 (2001).
5. Burles, S. & Tytler, D., ApJ, 507, 732 (1998).
6. Carlberg, R. G., Yee, H. K. C., Ellingson, E. ApJ, 478, 462 (1997).
7. Carlstrom, J. E., Holder, G. P., Reese, E. D., ARA&A, 40, 643 (2002).
8. da Silva, A. & Clef Collaboration, SF2A-2004: Semaine de l'Astrophysique Francaiese, EdP-Sciences Conference Series, p. 147.
9. Knop, R. A. et al. ApJ, 598, 102 (2004).
10. Lin, Y-T., Mohr, J. J., Stanford, S. A. ApJ, 591, L749 (2003).
11. Majumdar, S. & Mohr, J. J. ApJ, 585, 603 (2003).
12. Mohr, J. J., Mathiesen, B., Evrard, A. E. ApJ, 517, 627 (1999).
13. Motl, P. M. et al. ApJ, 606, 635, (2004).
14. Peterson, J. R. et al. ApJ, 590, 207 (2003).
15. Reiprich, T. H. & Böhringer, H. ApJ, 567, 716 (2002).
16. Silk, J. ApJ, 151, 459 (1968).
17. Sparks, W. B., Donahue, M., Jordan, A., Ferrarese, L, Cote, P. ApJ, 607, 294, (2004).
18. Spergel, D. N., et al. ApJS, 148, 175 (2003).
19. Voit, G. M., Balogh, M. L., Bower, R. G., Lacey, D. G., Bryan, G. L. ApJ, 593, 272 (2003).
20. Voit, G. M. Rev. Mod. Physics, in press (2004).

TEV ANTINEUTRINOS FROM CYGNUS OB2*

HAIM GOLDBERG

Department of Physics, Northeastern University, Boston, MA 02115

High energy cosmic ray experiments have identified an excess from the region of the Galactic Plane in a limited energy range around 10^{18} eV (EeV). This is very suggestive of neutrons as candidate primaries, because the directional signal requires relatively-stable neutral primaries, and time-dilated neutrons can reach Earth from typical Galactic distances when the neutron energy exceeds an EeV. We here point out that if the Galactic messengers are neutrons, then those with energies below an EeV will decay in flight, providing a flux of cosmic antineutrinos above a TeV which is *observable* at a kilometer-scale neutrino observatory. The expected event rate per year above 1 TeV in a detector such as IceCube, for example, is 20 antineutrino showers (all flavors) and a 1° directional signal of 4 $\bar{\nu}_\mu$ events. A measurement of this flux can serve to identify the first extraterrestrial point source of TeV antineutrinos.

An intriguing anisotropy in the cosmic ray spectrum has emerged in the energy range near an EeV. The Akeno Giant Air Shower Array (AGASA) has revealed a correlation of the arrival direction of the cosmic rays to the Galactic Plane (GP) at the 4σ level [1]. The GP excess, which is roughly 4% of the diffuse flux, is mostly concentrated in the direction of the Cygnus region, with a second spot towards the Galactic Center (GC) [2]. Evidence at the 3.2σ level for GP enhancement in a similar energy range has also been reported by the Fly's Eye Collaboration [3]. The existence of a point-like excess in the direction of the GC has been confirmed via independent analysis [4] of data collected with the Sydney University Giant Airshower Recorder (SUGAR). This is a remarkable level of agreement among experiment using a variety of techniques.

Independent evidence may be emerging for a cosmic accelerator in the Cygnus spiral arm. The HEGRA experiment has detected an extended TeV γ-ray source in the Cygnus region with no clear counterpart and a spectrum not easily accommodated with synchrotron radiation by electrons [5]. The

*This work is supported in part by the US National Science Foundation

difficulty to accommodate the spectrum by conventional electromagnetic mechanisms has been exacerbated by the failure of CHANDRA and VLA to detect X-rays or radiowaves signaling acceleration of any electrons [6]. The model proposed is that of a proton beam, accelerated by a nearby mini-quasar or possibly Cygnus X-3, interacting with a molecular cloud to produce pions that are the source of the gamma rays. Especially intriguing is the possible association of this source with Cygnus-OB2, a cluster of more than 2700 (identified) young, hot stars with a total mass of $\sim 10^4$ solar masses [7]. Proton acceleration to explain the TeV photon signal requires only 0.1% efficiency for the conversion of the energy in the stellar wind into cosmic ray acceleration. Also, the stars in Cygnus-OB2 could be the origin of time-correlated, clustered supernova remnants forming a source of cosmic rays. By cooperative acceleration their energies may even exceed the ~1 PeV cutoff of individual remnants and accommodate cosmic rays up to the ankle, where the steeply falling ($\propto E^{-3.16\pm0.08}$) cosmic ray spectrum flattens to $E^{-2.8\pm0.3}$ [8].

All evidence points to a transition from galactic to extragalactic sources above several EeV of primary energy. The steepness of the falloff between the knee (about 3 PeV) and the ankle (about 10 EeV) is expected from supernova shock models, and may indicate that we are witnessing the high energy end of the galactic flux. The extension of the nominal PeV cutoff beyond the ankle can be understood as a collective effect of stellar winds originating in the region of multiple supernova explosions. These provide a second acceleration to the particles and boost their energies far beyond the values expected from their single-shock encounter. Strong additional support for this picture emerges from accumulating evidence (in a bi-modal proton-iron model) for a dominant Fe component in the flux [9]. The importance of the heavy component is apparent all the way down to the region of several PeV [10], with the spectral index hardening slightly to 3.02 ± 0.03 below 500 PeV. *An immediate consequence of this nucleus-dominance picture is the creation of free neutrons via nuclei photodisintegration on background photon fields.* These liberated neutrons are presumably responsible for the observed directional signals. *This implies that it may not be a coincidence that the signal appears first at energies where the neutron lifetime allows propagation distances of galactic scales, i.e., 10 kpc.*

For every surviving neutron at $\sim$ EeV, there are many neutrons at lower energy that decay via $n \to p + e^- + \overline{\nu}_e$. The decay mfp of a neutron is $c\,\Gamma_n\,\overline{\tau}_n = 10\,(E_n/\text{EeV})$ kpc, the lifetime being boosted from its rest-frame value $\overline{\tau}_n = 886$ seconds to its lab value via $\Gamma_n = E_n/m_n$. The proton is bent by the Galactic magnetic field, the electron quickly loses energy via synchrotron radiation, and the $\overline{\nu}_e$ travels along the initial neutron direction, producing a directed TeV energy beam whose flux is calculable. *We show in this talk that the expected $\bar{\nu}$ flux from the direction of the Cygnus region is measurable in IceCube* [11]. Furthermore, by the same logic, a km–scale Mediterranean detector, if designed with sufficiently low threshold, can see the $\bar{\nu}$ flux pointing toward the GC source as well. The GZK neutrinos [12] and the "essentially guaranteed" $\bar{\nu}$ flux calculated here probably constitute the best motivated cosmic neutrino fluxes. Of these two neutrino fluxes, the expected event rate for the galactic beam is higher: 4 $\bar{\nu}_\mu$ events per year and 16 in $\bar{\nu}_e + \bar{\nu}_\tau$ showers.

We turn to the calculation. The basic formula that relates the neutron flux at the source (dF_n/dE_n) to the antineutrino flux observed at Earth $(dF_{\overline{\nu}}/dE_{\overline{\nu}})$ is [13]:

$$\frac{dF_{\overline{\nu}}}{dE_{\overline{\nu}}}(E_{\overline{\nu}}) = \int dE_n \, \frac{dF_n}{dE_n}(E_n) \left(1 - e^{-\frac{D\, m_n}{E_n\, \overline{\tau}_n}}\right) \int_0^Q d\epsilon_{\overline{\nu}} \, \frac{dP}{d\epsilon_{\overline{\nu}}}(\epsilon_{\overline{\nu}}) \times \int_{-1}^{1} \frac{d\cos\overline{\theta}_{\overline{\nu}}}{2} \, \delta\left[E_{\overline{\nu}} - E_n \, \epsilon_{\overline{\nu}} \, (1+\cos\overline{\theta}_{\overline{\nu}})/m_n\right] . \qquad (1)$$

The variables appearing in Eq. (1) are the antineutrino and neutron energies in the lab ($E_{\overline{\nu}}$ and E_n), the antineutrino angle with respect to the direction of the neutron momentum, in the neutron rest-frame ($\overline{\theta}_{\overline{\nu}}$), and the antineutrino energy in the neutron rest-frame ($\epsilon_{\overline{\nu}}$). The parameters appearing in Eq. (1) are the neutron mass and rest-frame lifetime (m_n and $\overline{\tau}_n$), and the distance to the neutron source (D). dF_n/dE_n is the neutron flux at the source, or equivalently, the neutron flux that would be observed from the source region in the absence of neutron decay. Finally, $\frac{dP}{d\epsilon_{\overline{\nu}}}(\epsilon_{\overline{\nu}})$ is the normalized probability that the decaying neutron in its rest-frame produces a $\overline{\nu}_e$ with energy $\epsilon_{\overline{\nu}}$; $\int_0^Q d\epsilon_{\overline{\nu}} \, \frac{dP(\epsilon_{\overline{\nu}})}{d\epsilon_{\overline{\nu}}} = 1$ defines the normalization, where the maximum neutrino energy in the neutron rest frame is $Q \equiv m_n - m_p - m_e = 0.79$ MeV, and the minimum neutrino energy is zero in the massless limit. For the decay of unpolarized neutrons, there is no angular dependence in $\frac{dP}{d\epsilon_{\overline{\nu}}}$.

The E_n integration in Eq. (1) is effectively cut off at $\sim$ EeV, the energy beyond which a neutron is stable over a 10 kpc path-length, and it is truly cut off by the $E_{\rm max}$ of the neutron spectrum. The expression in parentheses in Eq. (1) is the decay probability for a neutron with lab energy E_n, traveling a distance D. We will take D to be the 1.7 kpc distance from Earth to Cygnus OB2.

The Galactic anisotropy observed by the various collaborations spans the energy range 0.8 to 2.0 EeV. The lower cutoff specifies that only neutrons with EeV energies and above have a boosted $c\tau_n$ sufficiently large to serve as Galactic messengers. The upper cutoff reflects an important feature of photodisintegration at the source: heavy nuclei with energies in the vicinity of the ankle will fragment to neutrons with energies about an order of magnitude smaller. To account for the largest neutron energies, it may be necessary to populate the heavier nucleus spectrum in the region above the ankle. This is not a problem – one fully expects the emerging harder extragalactic spectrum to overtake and hide the steeply falling galactic population. It is not therefore surprising that in order to fit the spectrum in the anisotropy region and maintain continuity to the ankle region without introducing a cutoff, the AGASA Collaboration required a spectrum $\propto E^{-3}$ or steeper [1].

A detailed scenario for ultrahigh energy nuclei (parents of the anisotropy neutrons) originating in a pulsar close to the Cygnus OB2 region has been recently described [14]. The sequence begins with the one-shot acceleration in the spinning neutron star [15], resulting in an E^{-1} spectrum. Softening of the spectrum to E^{-2} ensues through gravitational wave losses during spindown [16]. Following this scheme, we assume that some of the nuclei are captured in the dense region of the source, attaining sufficient diffusion in milliGauss magnetic fields. The resulting time delay of several thousand years[17] produces a further steepening of the injection power law spectrum. Note that once diffusion has been established, additional Rayleigh steps in the Galactic magnetic field do not change the spectral index significantly. In their random traversal of the OB association, the nuclei undergo photodisintegration on far infrared thermal photons populating molecular clouds with temperatures of 15-100 K [18]. Taking an average photodisintegration cross section of 40 mb, we find an interaction time between 4 and 1300 yr, allowing sufficient neutron production to explain the anisotropy.

To incorporate the preceding discussion in our work, we take in what follows a single power law neutron spectrum with a spectral index of 3.1, representing an average over the PeV-EeV energy region. Specifically, $dF_n/dE_n = C\, E_n^{-3.1}$, with the normalization constant fixed near an EeV to the observed excess. The constant C is determined by integrating dF_n/dE_n over a bin (E_1, E_2) with the result $F_n \approx 9$ km^{-2} yr^{-1} [2].

The typical energy for the antineutrino in the lab is that of the decaying neutron times $Q/m_n \sim 10^{-3}$. Thus, the stability of neutrons at $\gtrsim$ EeV implies a PeV upper limit for the produced antineutrinos. The increasing abundance of neutrons below an EeV in turn implies an increasing neutrino flux as energies move below a PeV.

Nuclei with Lorentz factor $\sim 10^6$ are synthesized in all supernovae. Hadronic interactions with the HII population (density 30 cm^{-3} [6]) and photodisintegration from ultraviolet photons emitted from OB stars results in a flux of PeV neutrons. From the measured Lyman emission of the O stars [7], the O÷B temperature, and luminosity characteristics of the B stars [19], we obtain a photon number density in the ultraviolet of ~ 230 cm^{-3} for the core region ~ 10 pc. This implies a collision time of about 0.1 Myr, comparable to the time scale for hadronic interactions. However, in contrast to hadronic interactions, significant photodisintegration occurs in the outer 20 pc of the source. The photon density is reduced to > 25 cm^{-3}, which lengthens the reaction time to ~ 1 Myr. The diffusion time (~ 1.2 Myr) is a bit smaller than the age of the cluster ~ 2.5 Myr [20], and somewhat higher than the reaction time. This is sufficient to permit over 90% efficiency for photodisintegration over the lifetime of the source. The effective volume for photodisintegration is then about a factor of 27 larger than for hadronic interactions, and the net result of all these considerations is that the PeV neutron population is about an order of magnitude greater than TeV charged pions resulting from hadronic collisions [21].

Eq. (1) may be sufficiently approximated by inserting a δ function averaging $\frac{dP}{d\epsilon_{\overline{\nu}}}(\epsilon_{\overline{\nu}}) = \delta(\epsilon_{\overline{\nu}} - \epsilon_0)$, with ϵ_0 taken as the average $\epsilon_{\overline{\nu}}$ in the decay. Substituted into Eq. (1), with an additional step function cutoff on the exponential one obtains

$$\frac{dF_{\overline{\nu}}}{dE_{\overline{\nu}}}(E_{\overline{\nu}}) = \frac{m_n}{2\,\epsilon_0} \int_{\frac{m_n\, E_{\overline{\nu}}}{2\,\epsilon_0}}^{E_n^{\max}} \frac{dE_n}{E_n}\, \frac{dF_n}{dE_n}(E_n)\,. \tag{2}$$

A direct $\overline{\nu}_e$ event in IceCube will make a showering event, which, even if seen, provides little angular resolution. In the energy region below 1 PeV, IceCube will resolve directionality only for ν_μ and $\bar{\nu}_\mu$. Fortunately, neutrino oscillations rescue the signal. Since the distance to the Cygnus region greatly exceeds the $\overline{\nu}_e$ oscillation length $\lambda_{\rm osc} \sim 10^{-2} \frac{E_{\bar{\nu}}}{\rm PeV}$ parsecs (taking the solar oscillation scale $\delta m^2 \sim 10^{-5} {\rm eV}^2$), the antineutrinos decohere in transit. The arriving antineutrinos are distributed over flavors, with the muon antineutrino flux $F_{\bar{\nu}_\mu}$ given by the factor $\frac{1}{4}\sin^2(2\,\theta_\odot) \simeq 0.20$ times the original $F_{\bar{\nu}_e}$ flux. The $\bar{\nu}_\tau$ flux is the same, and the $\bar{\nu}_e$ flux is 0.6 times the original flux. Here we have utilized for the solar mixing angle the most recent SNO result $\theta_\odot \simeq 32.5^\circ$ [22], along with maximal mixing for atmospheric ν_μ-ν_τ neutrinos and a negligible ν_e component in the third neutrino eigenstate.

The integral neutrino flux $F_{\overline{\nu}}(> E_{\overline{\nu}}) \equiv \int_{E_{\overline{\nu}}} dE_{\overline{\nu}} \frac{dF_{\overline{\nu}}}{dE_{\overline{\nu}}}$ is particularly useful for experiments having a neutrino detection-efficiency that is independent of neutrino energy, or nearly so. IceCube is an example of such an experiment. Our calculated integral flux is shown in Fig. 1. As mentioned above, the nuclear photodisintegration threshold leads to an infrared cutoff on the primary neutron energy at the source, which in turn leads to a low energy cutoff $\sim$ TeV on the integral flux.

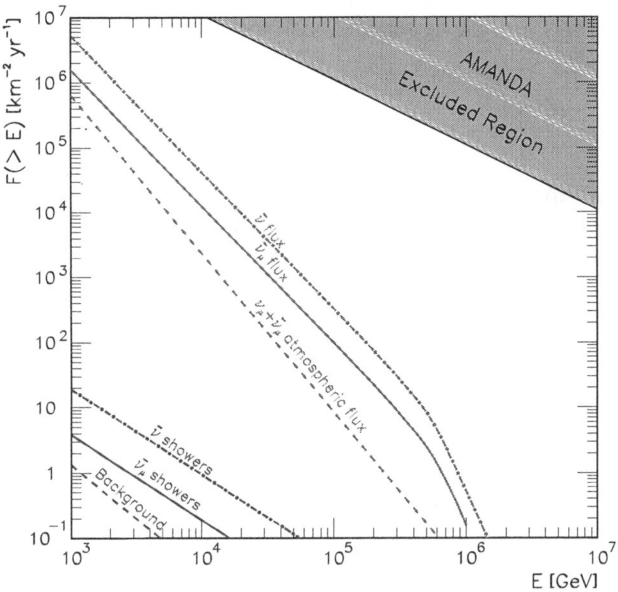

Figure 1. Integrated flux of $\bar{\nu}_\mu$ (solid) and $\bar{\nu}_\mu + \bar{\nu}_e + \bar{\nu}_\tau$ (dashed-dotted) predicted to arrive at Earth from the direction of the Cygnus region. Also shown is the integrated $\nu_\mu + \bar{\nu}_\mu$ atmospheric flux for an angular bin of $1^\circ \times 1^\circ$. The shaded band indicates the region excluded by the AMANDA experiment [23]. The expected number of showers $\bar{\nu}_\mu$ (solid) and $\bar{\nu}_\mu + \bar{\nu}_e + \bar{\nu}_\tau$ (dashed-dotted) to be detected (say in IceCube) are plotted on the bottom-left. The expected background for the same angular bin is indicated by the dashed line.

We now estimate the signal-to-noise ratio at IceCube. The angular resolution of the experiment $\approx 0.7^\circ$ allows a search window of $1^\circ \times 1^\circ$ [24]. We begin with the "noise". The event rate of the atmospheric ν-background that will be detected in the search bin ($\Delta\Omega_{1^\circ\times 1^\circ} \approx 3 \times 10^{-4}$ sr) is given by

$$\left.\frac{dN}{dt}\right|_{\rm background} = A_{\rm eff} \int dE\, J_{\nu+\bar\nu}(E)\; p(E)\; \Delta\Omega_{1^\circ\times 1^\circ}\,, \qquad (3)$$

where $A_{\rm eff}$ is the effective area of the detector, $J_{\nu+\bar\nu}(E)$ is the $\nu_\mu + \bar\nu_\mu$ atmospheric flux in the direction of the Cygnus region (about 40° below the horizon) [25], and $p(E) \approx 1.3 \times 10^{-6}\,(E/{\rm TeV})^{0.8}$ denotes the probability (generic to ice/water detectors) that a ν (or $\bar\nu$) with energy E on a trajectory through the detector produces a signal [26]. For a year of running at IceCube and $E_{\bar\nu}^{\rm min} = 1$ TeV, from Eq. (3) one obtains a background of 1.5 events. Poisson statistics then imply that a signal $\geq$ 3.5 events is significant at the 95% CL [27]. The number of $\bar\nu$ showers in the signal, for energies above $E_{\bar\nu}^{\rm min}$, is given by

$$\left.\frac{dN}{dt}\right|_{\rm signal} = A_{\rm eff} \int_{E_{\bar\nu}^{\rm min}} dE_{\bar\nu}\, \frac{dF_{\bar\nu}}{dE_{\bar\nu}}(E_{\bar\nu})\; p(E_{\bar\nu})\,. \qquad (4)$$

For a year of running at IceCube, one expects 20 neutrino showers (all flavors) with energies $\geq$ 1 TeV, of which 4 $\bar\nu_\mu$ events will cluster within 1° of the source direction, comfortably above the stated CL.

IceCube is not sensitive to TeV neutrinos from the GC, as these are above the IceCube horizon, where atmospheric muons will dominate over any signal. However, other kilometer-scale neutrino detectors, such as those planned for the Mediterranean Sea, may see the GC flux. Additionally, Southern Auger should see the cosmic ray excess in the direction of the GC [28] and Northern Auger should be sensitive to the Cygnus region.

We conclude that in a few years of observation, IceCube will attain 5σ sensitivity for discovery of the Fe$\to n \to \bar\nu_e \to \bar\nu_\mu$ cosmic beam, providing the "smoking ice" for the GP neutron hypothesis.

Acknowledgments

I would like to thank my collaborators Luis Anchordoqui, Francis Halzen and Tom Weiler. I am also grateful to John Beacom, Jim Cronin, Concha Gonzalez-Garcia, Todor Stanev, and Alan Watson for valuable discussions.

References

1. N. Hayashida *et al.* [AGASA Collaboration], Astropart. Phys. **10**, 303 (1999) [arXiv:astro-ph/9807045].
2. M. Teshima *et al.*, Proc. 27th International Cosmic Ray Conference, (Copernicus Gesellschaft, 2001) p.341.

3. D. J. Bird *et al.* [HiRes Collaboration], Astrophys. J. **511**, 739 (1999).
4. J. A. Bellido, R. W. Clay, B. R. Dawson, and M. Johnston-Hollitt Astropart. Phys. **15**, 167 (2001) [arXiv:astro-ph/0009039].
5. F. A. Aharonian *et al.* [HEGRA Collaboration], Astron. Astrophys. **393**, L37 (2002) [arXiv:astro-ph/0207528].
6. Y. Butt *et al.*, Astrophys. J. **597**, 494 (2003) [arXiv:astro-ph/0302342].
7. J. Knodlseder, astro-ph/0007442.
8. M. Nagano *et al.*, J. Phys. G **18**, 423 (1992).
9. See *e.g.,* M. T. Dova, M. E. Mancenido, A. G. Mariazzi, T. P. Mc Cauley, and A. A. Watson, astro-ph/0305351.
10. S. P. Swordy *et al.*, Astropart. Phys. **18**, 129 (2002).
11. IceCube is the neutrino telescope under construction at the South Pole. `http://icecube.wisc.edu/`
12. The GZK or "cosmogenic" neutrinos are created as secondaries from the decay of charged pions produced in collisions between ultrahigh energy nucleons and the cosmic microwave background. K. Greisen, Phys. Rev. Lett. **16**, 748 (1966), G. T. Zatsepin and V. A. Kuzmin, JETP Lett. **4**, 78 (1966) [Pisma Zh. Eksp. Teor. Fiz. **4**, 114 (1966)]. For an update on GZK neutrinos, see *e.g.,* R. Engel, D. Seckel and T. Stanev, Phys. Rev. D **64**, 093010 (2001); Z. Fodor, S. D. Katz, A. Ringwald and H. Tu, JCAP **0311**, 015 (2003).
13. L. A. Anchordoqui, H. Goldberg, F. Halzen and T. J. Weiler, Phys. Lett. B (to be published) [arXiv:astro-ph/0311002].
14. W. Bednarek, Mon. Not. Roy. Astron. Soc. **345**, 847 (2003).
15. P. Blasi, R. I. Epstein and A. V. Olinto, Astrophys. J. **533**, L123 (2000) [arXiv:astro-ph/9912240].
16. J. Arons, Astrophys. J. **589**, 871 (2003) [arXiv:astro-ph/0208444].
17. W. Bednarek and R. J. Protheroe, Astropart. Phys. **16**, 397 (2002).
18. C. D. Wilson, C. E. Walker and M. D. Thornley, arXiv:astro-ph/9701245.
19. M. M. Hanson, Astrophys. J. **597**, 957 (2003) [arXiv:astro-ph/0307540].
20. J. Knodlseder, M. Cervino, D. Schaerer, P. von Ballmoos and G. Meynet, arXiv:astro-ph/0104074.
21. This estimate makes use of secondary populations in high energy hadronic interactions as given in [J. Knapp, D. Heck and G. Schatz, Nucl. Phys. Proc. Suppl. **52B**, 136 (1997)], with a cut on $\pi^{\pm}$ with energies > 2 TeV.
22. S. N. Ahmed *et al.* [SNO Collaboration], nucl-ex/0309004.
23. J. Ahrens [AMANDA Collaboration], Phys. Rev. Lett. **92**, 071102 (2004).
24. J. Ahrens *et al.* [IceCube Collaboration], Astropart. Phys. **20**, 507 (2004).
25. P. Lipari, Astropart. Phys. **1**, 195 (1993).
26. T. K. Gaisser, F. Halzen and T. Stanev, Phys. Rept. **258**, 173 (1995) [Erratum-ibid. **271**, 355 (1996)] [arXiv:hep-ph/9410384].
27. G. J. Feldman and R. D. Cousins, Phys. Rev. D **57**, 3873 (1998) [arXiv:physics/9711021]; N. Gehrels, Astrophys. J. **303**, 336 (1986).
28. R. W. Clay *et al.*, astro-ph/0308494.

NEW PHYSICS AT LOW ENERGIES AND DARK MATTER-DARK ENERGY TRANSMUTATION

E. I. GUENDELMAN AND A. B. KAGANOVICH

Physics Department, Ben Gurion University of the Negev, Beer Sheva 84105, Israel

E-mail: guendel@bgumail.bgu.ac.il ; alexk@bgumail.bgu.ac.il

A field theory is proposed where the regular fermionic matter and the dark fermionic matter can be different states of the same "primordial" fermion fields. In regime of the fermion densities typical for normal particle physics, the primordial fermions split into three families identified with regular fermions. When fermion energy density becomes comparable with dark energy density, the theory allows transition to new type of states. The possibility of such Cosmo-Low Energy Physics (CLEP) states is demonstrated by means of solutions of the field theory equations describing FRW universe filled with homogeneous scalar field and uniformly distributed nonrelativistic neutrinos. Neutrinos in CLEP state are drawn into cosmological expansion by means of dynamically changing their own parameters. One of the features of the fermions in CLEP state is that in the late time universe their masses increase as $a^{3/2}$ ($a = a(t)$ is the scale factor). The energy density of the cold dark matter consisting of neutrinos in CLEP state scales as a sort of dark energy; this cold dark matter possesses negative pressure and for the late time universe its equation of state approaches that of the cosmological constant. The total energy density of such universe is less than it would be in the universe free of fermionic matter at all.

1. Main ideas of the Two Measures Theory and the scale invariant model

The Two Measures Theory (TMT) is a generally coordinate invariant theory where the action has the form

$$S = \int L_1 \Phi d^4x + \int L_2 \sqrt{-g} d^4x \tag{1}$$

including two Lagrangians L_1 and L_2 and two measures of the volume elements (Φd^4x and $\sqrt{-g}d^4x$ respectively). One is the usual measure of integration $\sqrt{-g}$ in the 4-dimensional space-time manifold equipped by the metric $g_{\mu\nu}$. Another is also a scalar density built of four scalar fields φ_a ($a = 1, 2, 3, 4$), $\Phi = \varepsilon^{\mu\nu\alpha\beta}\varepsilon_{abcd}\partial_\mu\varphi_a\partial_\nu\varphi_b\partial_\alpha\varphi_c\partial_\beta\varphi_d$. It is assumed that the Lagrangians L_1 and L_2 are functions of the matter fields, the dilaton field, the metric, the connection (or spin-connection) but not of the "measure fields" φ_a. Varying φ_a, we get $B^\mu_a \partial_\mu L_1 = 0$ where $B^\mu_a = \varepsilon^{\mu\nu\alpha\beta}\varepsilon_{abcd}\partial_\nu\varphi_b\partial_\alpha\varphi_c\partial_\beta\varphi_d$. Since $Det(B^\mu_a) = \frac{4^{-4}}{4!}\Phi^3$ it follows that if $\Phi \neq 0$, $L_1 = sM^4 = const$ where $s = \pm 1$ and M is a constant of integration with the dimension of mass. Important feature of TMT that is responsible for many interesting and desirable results of the field theory models studied so far[1]-[10] (see also [5]) consists of the assumption that all fields, including also metric, connection (or vierbein and spin-connection) and the measure fields φ_a are independent dynamical variables.

As it has been shown earlier, in TMT there is no a need to postulate the existence of three species for each type of fermions (like three neutrinos, three charged leptons, etc.) but rather this is achived as a dynamical effect of TMT in normal particle physics conditions. The matter content of our model includes the dilaton scalar field ϕ, two so-called primordial fermion fields (the neutrino primordial field ν and the electron primordial field E) and electromagnetic field A_μ. Generalization to the non-Abelian gauge models including Higgs fields and quarks is straightforward[7]. To simplify the presentation of the ideas we ignore also the chiral properties of neutrino; this can be done straightforward and does not affect the main results.

Keeping the general structure (1), it is convenient to represent the action in the following form:

$$
\begin{aligned}
S = & \int d^4x e^{\alpha\phi/M_p}(\Phi + b\sqrt{-g})\left[-\frac{1}{\kappa}R(\omega, e) + \frac{1}{2}g^{\mu\nu}\phi_{,\mu}\phi_{,\nu}\right] \\
& - \int d^4x e^{2\alpha\phi/M_p}[\Phi V_1 + \sqrt{-g}V_2] - \int d^4x\sqrt{-g}\frac{1}{4}g^{\alpha\beta}g^{\mu\nu}F_{\alpha\mu}F_{\beta\nu} \\
& + \int d^4x e^{\alpha\phi/M_p}(\Phi + k\sqrt{-g})\frac{i}{2}\sum_i \overline{\Psi}_i\left(\gamma^a e^\mu_a \overrightarrow{\nabla}^{(i)}_\mu - \overleftarrow{\nabla}^{(i)}_\mu \gamma^a e^\mu_a\right)\Psi_i \\
& - \int d^4x e^{\frac{3}{2}\alpha\phi/M_p}\left[(\Phi + h_\nu\sqrt{-g})\mu_\nu\overline{\nu}\nu + (\Phi + h_E\sqrt{-g})\mu_E\overline{E}E\right] \qquad (2)
\end{aligned}
$$

where where Ψ_i $(i = \nu, E)$ is the general notation for the primordial fermion fields ν and E, V_1 and V_2 are constants, $F_{\alpha\beta} = \partial_\alpha A_\beta - \partial_\beta A_\alpha$, μ_ν and μ_E are the mass parameters, $\overrightarrow{\nabla}^{(\nu)}_\mu = \overrightarrow{\partial}_\mu + \frac{1}{2}\omega^{cd}_\mu\sigma_{cd}$, $\overrightarrow{\nabla}^{(E)}_\mu = \overrightarrow{\partial}_\mu + \frac{1}{2}\omega^{cd}_\mu\sigma_{cd} + ieA_\mu$; $R(\omega, e) = e^{a\mu}e^{b\nu}R_{\mu\nu ab}(\omega)$ is the scalar curvature, e^μ_a and ω^{ab}_μ are the vierbein and spin-connection; $g^{\mu\nu} = e^\mu_a e^\nu_b \eta^{ab}$ and $R_{\mu\nu ab}(\omega) = \partial_\mu\omega_{\nu ab} + \omega^c_{\mu a}\omega_{\nu cb} - (\mu \leftrightarrow \nu)$; constants b, k, h_i are dimensionless parameters.

The action (2) is invariant under the global scale transformations

$$e^a_\mu \to e^{\theta/2} e^a_\mu, \quad \omega^\mu_{ab} \to \omega^\mu_{ab}, \quad \varphi_a \to \lambda_a \varphi_a \quad where \quad \Pi\lambda_a = e^{2\theta}$$

$$A_\alpha \to A_\alpha, \quad \phi \to \phi - \frac{M_p}{\alpha}\theta, \quad \Psi_i \to e^{-\theta/4}\Psi_i, \quad \overline{\Psi}_i \to e^{-\theta/4}\overline{\Psi}_i. \tag{3}$$

One can show that except for a few special choices providing positivity of the energy and the right chiral structure in the Einstein frame, Eq.(2) describes *the most general TMT action satisfying the formulated above symmetries.*

2. Constraint and equations of motion in the Einstein frame

Variation of the measure fields φ_a with the condition $\Phi \neq 0$ leads, as we have already seen in Sec.2, to the equation $L_1 = sM^4$ where L_1 is now defined, according to Eq. (1), as the part of the integrand of the action (2) coupled to the measure Φ. The appearance of a nonzero integration constant sM^4 spontaneously breaks the scale invariance (3). One can see that the measure Φ degrees of freedom appear in all the equations of motion only through dependence on the scalar field $\zeta \equiv \Phi/\sqrt{-g}$. In particular, the gravitational and all matter fields equations of motion include noncanonical terms proportional to $\partial_\mu\zeta$. It turns out that with the set of the new variables (ϕ and A_μ remain the same)

$$\tilde{e}_{a\mu} = e^{\frac{1}{2}\alpha\phi/M_p}(\zeta + b)^{1/2} e_{a\mu}, \quad \Psi'_i = e^{-\frac{1}{4}\alpha\phi/M_p}\frac{(\zeta + k)^{1/2}}{(\zeta + b)^{3/4}}\Psi_i \tag{4}$$

which we call the Einstein frame, the spin-connections become those of the Einstein-Cartan space-time and the noncanonical terms proportional to $\partial_\mu\zeta$ disappear from all equations of motion. Since $\tilde{e}_{a\mu}$, ν' and E' are invariant under the scale transformations (3), spontaneous breaking of the scale symmetry (3) (by means of equation $L_1 = sM^4$) is reduced in the new variables to the *spontaneous breaking of the shift symmetry* $\phi \to \phi + const.$

The gravitational equations in the Einstein frame take the standard GR form $G_{\mu\nu}(\tilde{g}_{\alpha\beta}) = \frac{\kappa}{2}T^{eff}_{\mu\nu}$ where

$$T^{eff}_{\mu\nu} = K_{\mu\nu} + \tilde{g}_{\mu\nu}V_{eff}(\phi;\zeta) + T^{(em)}_{\mu\nu} + T^{(f,can)}_{\mu\nu} + T^{(f,noncan)}_{\mu\nu} \tag{5}$$

Here $K_{\mu\nu} = \phi_{,\mu}\phi_{,\nu} - \frac{1}{2}\tilde{g}_{\mu\nu}\tilde{g}^{\alpha\beta}\phi_{,\alpha}\phi_{,\beta}$, $G_{\mu\nu}(\tilde{g}_{\alpha\beta})$ is the Einstein tensor in the Riemannian space-time with the metric $\tilde{g}_{\mu\nu}$; the function $V_{eff}(\phi;\zeta)$ has the form

$$V_{eff}(\phi;\zeta) = \frac{b\left(sM^4 e^{-2\alpha\phi/M_p} + V_1\right) - V_2}{(\zeta + b)^2}; \tag{6}$$

$T_{\mu\nu}^{(em)}$ is the canonical energy momentum tensor for the electromagnetic field; $T_{\mu\nu}^{(f,can)}$ is the canonical energy momentum tensor for (primordial) fermions ν' and E' in curved space-time including also standard electromagnetic interaction of E'. $T_{\mu\nu}^{(f,noncan)}$ is the *noncanonical* contribution of the fermions into the energy momentum tensor

$$T_{\mu\nu}^{(f,noncan)} = -\tilde{g}_{\mu\nu}\sum_i F_i(\zeta)\overline{\Psi'}_i\Psi'_i \equiv \tilde{g}_{\mu\nu}\Lambda_{dyn}^{(ferm)} \tag{7}$$

where $i = \nu', E'$ and

$$F_i(\zeta) \equiv \frac{\mu_i}{2(\zeta+k)^2(\zeta+b)^{1/2}}[\zeta^2 + (3h_i - k)\zeta + 2b(h_i - k) + kh_i]. \tag{8}$$

The structure of $T_{\mu\nu}^{(f,noncan)}$ shows that it is originated by fermions but behaves as a sort of variable cosmological constant. This is why we will refer to it as *dynamical fermionic* Λ *term* $\Lambda_{dyn}^{(ferm)}$. One has to emphasize the substantial difference of the way $\Lambda_{dyn}^{(ferm)}$ emerges here as compared to the models of the condensate cosmology (see for example Ref.[12]). As we will see in the next sections, Λ_{dyn}^{ferm} becomes negligible in gravitational experiments with observable matter. However it may be very important for some astrophysics and cosmology problems.

The dilaton ϕ field equation in the new variables reads

$$\Box\phi - \frac{\alpha}{M_p(\zeta+b)}\left[sM^4e^{-2\alpha\phi/M_p} - \frac{(\zeta-b)V_1 + 2V_2}{\zeta+b}\right] = -\frac{\alpha}{M_p}\sum_i F_i\overline{\Psi'_i}\Psi'_i \tag{9}$$

where $\Box\phi = (-\tilde{g})^{-1/2}\partial_\mu(\sqrt{-\tilde{g}}\tilde{g}^{\mu\nu}\partial_\nu\phi)$.

Equations for the primordial leptons in terms of the variables (4) take the standard form of fermionic equations in the Einstein-Cartan space-time where the standard electromagnetic interaction presents also. All the novelty consists of the form of the ζ depending "masses" $m_i(\zeta)$ of the primordial fermions ν', E':

$$m_i(\zeta) = \frac{\mu_i(\zeta + h_i)}{(\zeta+k)(\zeta+b)^{1/2}} \qquad i = \nu', E'. \tag{10}$$

It should be noticed that change of variables we have performed by means of Eq.(4) provide also a conventional form of the covariant conservation law of fermionic current $j^\mu = \overline{\Psi}'\gamma^a\tilde{e}_a^\mu\Psi'$.

The scalar field ζ in the above equations is defined by the constraint which is the consistency condition of equations of motion, see for example[5].

In the Einstein frame (4) the constraint takes the form

$$-\frac{1}{(\zeta+b)^2}\left\{(\zeta-b)\left[sM^4e^{-2\alpha\phi/M_p}+V_1\right]+2V_2\right\}=\sum_i F_i\overline{\Psi}_i'\Psi_i'. \qquad (11)$$

Generically the constraint (11) determines ζ as a very complicated function of ϕ, $\overline{\nu'}\nu'$ and $\overline{E'}E'$. However, there are a few very important particular situations where the theory allows exact solutions of great interest[6,7].

3. Some limiting cases of physical interest

In a typical particle physics situation, say detection of a single fermion, the measurement implies a localization of the fermion which is expressed in developing a very large value of $|\overline{\Psi}'\Psi'|$. According to the constraint (11) this is possible if $F_i(\zeta)\approx 0$, $i=\nu', E'$ (which gives two constant solutions for ζ) or $\zeta\approx -b$. These solutions allow to describe the effect of splitting of the primordial fermions into three generations of the regular fermions (for details see [6,7],citeproceed-2001). It is interesting also that for the first two generations (which we associate with the solutions where $F_i(\zeta)\approx 0$) their coupling to the dilaton ϕ is automatically strongly suppressed, as it follows from Eq.(9), which provides a solution of the fifth force problem (about the role of the symmetry $\phi\to\phi+const$ see Ref.[11])

In the case of the complete absence of massive fermions the constraint determines ζ as the function of ϕ: $\zeta=b-2V_2/(V_1+sM^4e^{-2\alpha\phi/M_p})$. The effective potential of the scalar field ϕ results then from Eq.(6)

$$V^{(0)}_{eff}(\phi)\equiv V_{eff}(\phi;\zeta)|_{\overline{\psi'}\psi'=0}=\frac{(V_1+sM^4e^{-2\alpha\phi/M_p})^2}{4[b(V_1+sM^4e^{-2\alpha\phi/M_p})-V_2]}. \qquad (12)$$

Assuming $bV_1>V_2$ and $s=+1$ we see that the asymptotic (as $\phi\to\infty$) value of $V^{(0)}_{eff}$ is the positive cosmological constant $\Lambda^{(0)}=\frac{V_1^2}{4(bV_1-V_2)}$. If $2V_2>bV_1>V_2$ then $V^{(0)}_{eff}$ has the absolute minimum $V^{(0)}_{eff,min}=V_2/b^2$ at $\phi=\phi_{min}=(M_p/2\alpha)\ln[bM^4/(2V_2-bV_1)]$.

4. Cosmo-Low Energy Physics states

It turns out that besides the normal fermion vacuum where the fermion contribution to the constraint is totally negligible, TMT predicts possibility of so far unknown states which can be realized, for example, in astrophysics and cosmology. Let us study a toy model[10] where in addition to the homogeneous scalar field ϕ, the spatially flat universe is filled also with uniformly distributed nonrelativistic neutrinos as a model of dark matter. Spreading of the neutrino wave packets during their free motion lasting a long time yields extremely small values of $\overline{\Psi}'\Psi' = u^{\dagger}u$ (u is the large component of the Dirac spinor Ψ'). There is a solution where the decaying fermion contribution $u^{\dagger}u \sim \frac{const}{a^3}$ to the constraint is compensated by approaching $\zeta \to -k$. Then solving (11) for ζ we have to take into account both sides of the constraint. After averaging over typical cosmological scales (resulting in the Hubble low), the constraint (11) reads

$$-(k+b)\left(sM^4e^{-2\alpha\phi/M_p}+V_1\right)+2V_2+(b-k)^2\frac{n_0^{(\nu)}}{a^3}F_\nu(\zeta)|_{\zeta\approx -k}=0 \quad (13)$$

where $F_\nu(\zeta)|_{\zeta\approx -k}=\mu_\nu(h_\nu-k)(b-k)^{1/2}(\zeta+k)^{-2}+O((\zeta+k)^{-1})$ and $n_0^{(\nu)}$ is a constant determined by the total number of the cold neutrinos.

Cosmological equations are then as following

$$\left(\frac{\dot{a}}{a}\right)^2=\frac{1}{3M_p^2}\left[\rho_\phi+\rho_{clep}\right] \quad (14)$$

$$\ddot{\phi}+3\frac{\dot{a}}{a}\dot{\phi}+\frac{2\alpha k}{(b-k)^2M_p}M^4e^{-2\alpha\phi/M_p}+O\left((\zeta+k)e^{-2\alpha\phi/M_p}\right)=0 \quad (15)$$

where

$$\rho_\phi=\frac{1}{2}\dot{\phi}^2+\frac{bV_1-V_2}{(b-k)^2}+\frac{b}{(b-k)^2}sM^4e^{-2\alpha\phi/M_p}+O(\zeta+k), \quad (16)$$

$$\rho_{clep}=\left[\frac{\mu_\nu(h_\nu-k)}{(\zeta+k)(b-k)^{1/2}}-F_\nu(\zeta)|_{\zeta\approx -k}\right]\frac{n_0^{(\nu)}}{a^3} \quad (17)$$

The corrections $O(\zeta+k)$ behave as $O(\zeta+k)\propto a^{-3/2}$ and it turns out that they are negligible if no fine tuning of the parameters of the theory will be done. We choose $s=+1$ and assume here that $V_1>0,\ V_2>0$ *and* $b>0,\ k<0,\ h_\nu<0,\ h_\nu-k<0,\ b+k<0$. In the case we work with a positive fermion energy solution, in order to provide positivity of the effective neutrino mass (see Eq.(10)) in the CLEP state, we should consider the regime where $\zeta=-k-\varepsilon$, $\varepsilon>0$. Then $F_\nu(\zeta)|_{\zeta\approx -k}<0$ and both of the terms in Eq.(17) are positive. The first term in Eq.(17) results from the canonical part $T_{00}^{(\nu,can)}$ of the neutrino energy-momentum tensor after making use of the equations for neutrino field, neglecting the terms proportional to 3-momenta of the neutrinos and averaging. The second term in Eq.(17) comes from the dynamical fermionic $\Lambda_{dyn}^{(ferm)}$ term.

Neglecting in (17) terms of the order of $(\zeta+k)^{-1}$ as compared to the terms of the order of $(\zeta+k)^{-2}$ and using again the constraint (13) we obtain for the pressure and density of the uniformly distributed neutrino in the CLEP state

$$P_{clep} = -\rho_{clep} = \frac{2V_2 + |b+k|V_1}{(b-k)^2} + \frac{|b+k|}{(b-k)^2} M^4 e^{-2\alpha\phi/M_p} \tag{18}$$

which is typical for the dark energy sector including both a cosmological constant and an exponential ϕ-potential (compare (18) with (16)). The accuracy of this approximation grows as $a(t) \to \infty$.

The total energy density and the total pressure (including both the scalar field ϕ and neutrinos in CLEP state) in the framework of the explained above approximations can be represented in the form

$$\rho_{dark}^{(total)} \equiv \rho_\phi + \rho_{clep} = \frac{1}{2}\dot{\phi}^2 + U_{dark}^{(total)}(\phi) \tag{19}$$

$$P_{dark}^{(total)} \equiv P_\phi + P_{clep} = \frac{1}{2}\dot{\phi}^2 - U_{dark}^{(total)}(\phi), \tag{20}$$

where the effective potential $U_{dark}^{(total)}(\phi) = \Lambda + V_{quint}(\phi)$, where

$$\Lambda = \frac{V_2 + |k|V_1}{(b-k)^2}, \qquad V_{quint}(\phi) = \frac{|k|}{(b-k)^2} M^4 e^{-2\alpha\phi/M_p}. \tag{21}$$

This means that the evolution of the late time universe in the state with $\zeta \approx -k$ proceeds as it would be in the standard field theory model (non-TMT) including *both the cosmological constant and the quintessence-like field ϕ with the exponential potential.* Note that to provide the observable energy densities (for example, $\Lambda \sim \rho_{crit}$, where ρ_{crit} is the present day critical energy density) there is no need of fine tuning of the dimensionfull parameters V_1 and V_2 but instead one can assume that the dimensionless parameters b, k are very large.

It is very interesting to compare the effective potential $V_{eff}^{(0)}(\phi)$, Eq.(12), predicted for the universe filled only with the homogeneous scalar field (for short, a state "absent of fermions"), on the one hand, with the effective dark sector potential $U_{dark}^{(total)}(\phi)$ for the universe filled both with the homogeneous scalar field and with the uniformly distributed nonrelativistic neutrinos (for short, "CLEP state"), on the other hand. The *remarkable result* consists in the fact that if $bV_1 > V_2$, which is needed for positivity of $\Lambda^{(0)}$, then

$$V_{eff}^{(0)}(\phi) - U_{dark}^{(total)}(\phi) \equiv \frac{\left[\frac{b+k}{2}\left(V_1 + M^4 e^{-2\alpha\phi/M_p}\right) - V_2\right]^2}{4(b-k)^2\left[b\left(V_1 + M^4 e^{-2\alpha\phi/M_p}\right) - V_2\right]} > 0. \tag{22}$$

This means that (for the same value of $\dot{\phi}^2$) *the universe in "the CLEP state" has a lower energy density than the one in the "absent of fermions" state.* One should emphasize that this result does not imply at all that ρ_{clep} is negative.

For illustration of what kind of solutions one can expect, let us take the *particular value* for the parameter α, namely $\alpha = \sqrt{3/8}$. Then in the framework of the explained above approximations, the cosmological equations allow the following analytic solution:

$$\phi(t) = \frac{M_p}{2\alpha}\varphi_0 + \frac{M_p}{2\alpha}\ln(M_p t), \qquad a(t) \propto t^{1/3}e^{\lambda t}, \tag{23}$$

where

$$\lambda = \frac{1}{M_p}\sqrt{\frac{\Lambda}{3}}, \qquad e^{-\varphi_0} = \frac{2(b-k)^2M_p^2}{\sqrt{3}|k|M^4}\sqrt{\Lambda}. \tag{24}$$

The mass of the neutrino in such CLEP state increases exponentially in time and its ϕ dependence is double-exponential:

$$m_\nu|_{CLEP} \sim a^{3/2}(t) \sim t^{1/2}e^{\frac{3}{2}\lambda t} \sim \exp\left[\frac{3\lambda e^{-\varphi_0}}{2M_p}\exp\left(\frac{2\alpha}{M_p}\phi\right)\right]. \tag{25}$$

References

1. E.I. Guendelman and A.B. Kaganovich, *Phys. Rev.* **D53**, 7020 (1996); *ibid.* **D55**, 5970 (1997); *ibid.* **D56**, 3548 (1997); *ibid.* **D57**, 7200 (1998); *Mod. Phys. Lett.* **A12**, 2421 (1997); *ibid.* **A13**, 1583 (1998).
2. E.I. Guendelman, *Phys. Lett.* B**412**, 42 (1997); E.I. Guendelman, gr-qc/0303048; E.I. Guendelman and E. Spallucci, hep-th/0311102.
3. E.I. Guendelman and A.B. Kaganovich, *Phys. Rev.* **D60**, 065004 (1999).
4. E.I. Guendelman, *Mod. Phys. Lett.* **A14**, 1043 (1999); *Class. Quant. Grav.* **17**, 361 (2000); gr-qc/0004011; *Mod. Phys. Lett.* **A14**, 1397 (1999); gr-qc/9901067; hep-th/0106085; *Found. Phys.* **31**, 1019 (2001).
5. E.I. Guendelman and A.B. Kaganovich, Coral Gables Conference 2001: *AIP Conf. Proc.* **624**, 220-229 (2002).
6. E.I. Guendelman and A.B. Kaganovich, *Int. J. Mod. Phys.* A**17**, 417 (2002).
7. E.I. Guendelman and A.B. Kaganovich, *Mod. Phys. Lett.* A**17**, 1227 (2002)
8. A.B. Kaganovich, *Phys. Rev.* **D63**, 025022 (2001).
9. E.I. Guendelman and O. Katz, *Class. Quant. Grav.* **20**, 1715 (2003).
10. E.I. Guendelman and A.B. Kaganovich, gr-qc/0312006.
11. S.M. Carroll, *Phys. Rev. Lett.* **81**, 3067 (1998).
12. R.D. Ball and A.M. Matheson, *Phys. Rev.* D **45**, 2647 (1992); L. Parker and Y. Zhang, ibid., **47**, 416 (1993); R.H. Brandenberger and A.R. Zhitnitsky, ibid., **55**, 4640 (1997); I. Dymnikova and M. Khlopov, *Mod. Phys. Lett.* A **15**, 2305 (2000); D.G. Caldi and A. Chodos, hep-ph/9903416; M. Nishiyama, M. Morita and M. Morikawa, astro-ph/0403571.

PULSAR KICKS AND DARK MATTER FROM A STERILE NEUTRINO

ALEXANDER KUSENKO*

Department of Physics and Astronomy, UCLA, Los Angeles, CA 90095-1547
RIKEN BNL Research Center, Brookhaven National Laboratory, Upton, NY 11973

The observed velocities of radio pulsars, which range in the hundreds kilometers per second, and many of which exceed 1000 km/s, are not explained by the standard physics of the supernova explosion. However, if a sterile neutrino with mass in the 1–20 keV range exists, it would be emitted asymmetrically from a cooling neutron star, which could give it a sufficient recoil to explain the pulsar motions. The same particle can be the cosmological dark mater. Future observations of X-ray telescopes and gravitational wave detectors can confirm or rule out this explanation.

1. Introduction

Pulsar velocities range from 100 to 1600 km/s[1,2]. Their distribution leans toward the high-velocity end, with about 15% of all pulsars having speeds over (1000 km/s)[2]. The origin of these velocities is a long-standing puzzle. Pulsars are born in supernova explosions, so it would be natural to look for an explanation in the dynamics of the supernova[3]. However, state-of-the-art 3-dimensional numerical calculations[4] show that even the most extreme asymmetric explosions do not produce pulsar velocities greater than 200 km/s. Earlier 2-dimensional calculations[5] claimed a maximal pulsar velocity up to 500 km/s to be possible. Of course, even this size of the kick was way too small to explain the large population of pulsars with speeds above 1000 km/s.

Given the absence of a "standard" explanation, one is compelled to consider alternatives, possibly involving new physics. One of the reasons why the standard explanation fails is because most of the energy is carried

*Work partially supported by the DOE grant DE-FG03-91ER40662 and the NASA ATP grant NAG5-13399.

away by neutrinos, which escape isotropically. The remaining energy must be distributed with a substantial asymmetry to account for the large pulsar velocities. In contrast, only a few per cent anisotropy in the distribution of neutrinos would give the pulsar a kick of required magnitude.

Neutrinos are *produced* anisotropically, but they *escape* isotropically. The asymmetry in production comes from the asymmetry in the basic weak interactions in the presence of a strong magnetic field. Indeed, if the electrons and other fermions are polarized by the magnetic field, the cross section of the urca processes, such as $n+e^+ \rightleftharpoons p+\bar{\nu}_e$ and $p+e^- \rightleftharpoons n+\nu_e$, depends on the orientation of the neutrino momentum. Depending on the fraction of the electrons in the lowest Landau level, this asymmetry can be as large as 30%, much more than one needs to explain the pulsar kicks[6]. However, this asymmetry is completely washed out by scattering of neutrinos on their way out of the star[12].

If, however, the same interactions produced a particle which had even weaker interactions with nuclear matter than neutrinos, such a particle could escape the star with an asymmetry equal its production asymmetry.

It is intriguing that the same particle can the dark mater.

The simplest realization of this scenario is a model that adds only one singlet fermion to the Standard Model. The SU(2)×U(1) singlet, a sterile neutrino, mixes with the usual neutrinos, for example, with the electron neutrino.

For a sufficiently small mixing angle between ν_e and ν_s, only one of the two mass eigenstates, ν_1, is trapped. The orthogonal state,

$$|\nu_2\rangle = \cos\theta_m|\nu_s\rangle + \sin\theta_m|\nu_e\rangle, \tag{1}$$

escapes from the star freely. This state is produced in the same basic urca reactions ($\nu_e+n \rightleftharpoons p+e^-$ and $\bar{\nu}_e+p \rightleftharpoons n+e^+$) with the effective Lagrangian coupling equal the weak coupling times $\sin\theta_m$.

We will consider two ranges of parameters, for which the $\nu_e \to \nu_s$ oscillations occur on or off resonance. First, let us suppose that a resonant oscillation occurs somewhere in the core of the neutron star. Then the asymmetry in the neutrino emission comes from shift in the resonance point depending on the magnetic field[7]. Second, we will consider the off-resonance case, in which the asymmetry comes directly from the weak processes, as described above[8].

2. Resonant, Mikheev-Smirnov-Wolfenstein oscillations

Neutrino oscillations in a magnetized medium are described by an effective potential[9]

$$V(\nu_s) = 0 \tag{2}$$

$$V(\nu_e) = -V(\bar{\nu}_e) = V_0\,(3\,Y_e - 1 + 4\,Y_{\nu_e}) \tag{3}$$

$$V(\nu_{\mu,\tau}) = -V(\bar{\nu}_{\mu,\tau}) = V_0\,(Y_e - 1 + 2\,Y_{\nu_e}) \;+ \frac{eG_F}{\sqrt{2}}\left(\frac{3N_e}{\pi^4}\right)^{1/3}\frac{\vec{k}\cdot\vec{B}}{|\vec{k}|} \tag{4}$$

where Y_e (Y_{ν_e}) is the ratio of the number density of electrons (neutrinos) to that of neutrons, $\vec{B}$ is the magnetic field, $\vec{k}$ is the neutrino momentum, $V_0 = 10\,\mathrm{eV}\,(\rho/10^{14}\mathrm{g\,cm^{-3}})$. The magnetic field dependent term in equation (4) arises from polarization of electrons and *not* from a neutrino magnetic moment, which is small and which we will neglect.

The condition for resonant MSW oscillation $\nu_i \leftrightarrow \nu_j$ is

$$\frac{m_i^2}{2k}\cos 2\theta_{ij} + V(\nu_i) = \frac{m_j^2}{2k}\cos 2\theta_{ij} + V(\nu_j) \tag{5}$$

where $\nu_{i,j}$ can be either a neutrino or an anti-neutrino.

In the presence of the magnetic field, the condition (5) is satisfied at different distances r from the center, depending on the value of the $(\vec{k}\cdot\vec{B})$ term in (5). The average momentum carried away by the neutrinos depends on the temperature of the region from which they escape. The deeper inside the star, the higher is the temperature during the neutrino cooling phase. Therefore, neutrinos coming out in different directions carry momenta which depend on the relative orientation of $\vec{k}$ and $\vec{B}$. This causes the asymmetry in the momentum distribution.

The surface of the resonance points is

$$r(\phi) = r_0 + \delta\,cos\,\phi, \tag{6}$$

where $cos\,\phi = (\vec{k}\cdot\vec{B})/k$ and δ is determined by the equation $(dN_n(r)/dr)\delta \approx e\left(3N_e/\pi^4\right)^{1/3}B$. This yields[7]

$$\delta = \frac{e\mu_e}{\pi^2}\,B \Bigg/ \frac{dN_n(r)}{dr}\,, \tag{7}$$

where $\mu_e \approx (3\pi^2 N_e)^{1/3}$ is the chemical potential of the degenerate (relativistic) electron gas.

Assuming a black-body radiation luminosity $\propto T^4$, the asymmetry in momentum distribution is[7]

$$\frac{\Delta k}{k} = \frac{4e}{3\pi^2} \left(\frac{\mu_e}{T} \frac{dT}{dN_n} \right) B. \tag{8}$$

To calculate the derivative in (8), we use the relation between the density and the temperature of a non-relativistic Fermi gas. Finally,

$$\frac{\Delta k}{k} = \frac{4e\sqrt{2}}{\pi^2} \frac{\mu_e \mu_n^{1/2}}{m_n^{3/2} T^2} B = 0.01 \left(\frac{B}{3 \times 10^{15} \mathrm{G}} \right) \tag{9}$$

if the neutrino oscillations take place in the core of the neutron star, at density of order $10^{14}\,\mathrm{g\,cm^{-3}}$. The neutrino oscillations take place at such a high density if one of the neutrinos has mass in the keV range, while the other one is much lighter. The magnetic field of the order of $10^{15} - 10^{16}$ G is quite possible inside a neutron star, where it is expected to be higher than on the surface. (In fact, some neutron stars, dubbed magnetars, appear to have surface magnetic fields of this magnitude.)

Some comments are in order. First, a similar kick mechanism, based entirely on active neutrino oscillations (and no steriles) could also work if the resonant oscillations took place between the electron and tau neutrinospheres[13]. This, however, would require the mass difference between two neutrinos to be of the order of 100 eV, which is ruled out. Second, the neutrino kick mechanism was criticized incorrectly by Janka and Raffelt[14]. It was subsequently shown by several authors[15,16] that Janka and Raffelt made several mistakes, which is why their estimates differ from eq. (9).

3. Off-resonant oscillations

For somewhat lighter masses, the resonant condition is not satisfied anywhere inside the star. In this case, however, the off-resonant production of sterile neutrinos in the core can occur through ordinary urca processes. A weak-eigenstate neutrino has a $\sin^2\theta$ admixture of a heavy mass eigenstate ν_2. Hence, these heavy neutrinos can be produced in weak processes with a cross section suppressed by $\sin^2\theta$.

Of course, the mixing angle in matter θ_m is not the same as it is in vacuum, and initially $\sin^2\theta_m \ll \sin^2\theta$. However, as Abazajian, Fuller, and Patel[10] have pointed out, in the presence of sterile neutrinos the mixing angle in matter quickly evolves toward its vacuum value. When $\sin^2\theta_m \approx$

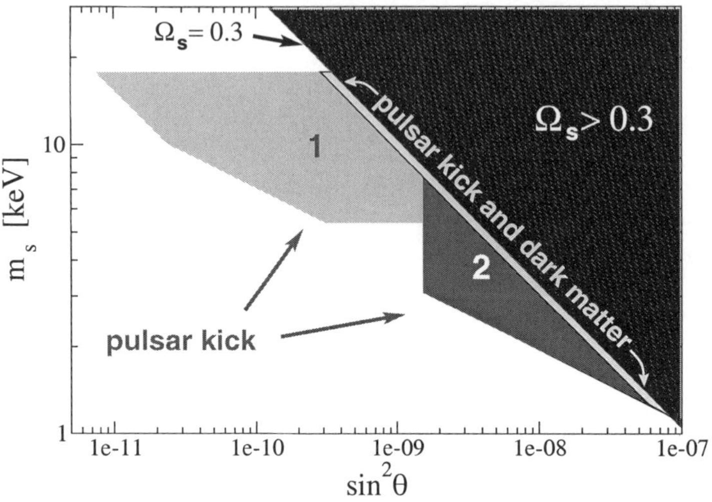

Figure 1. The range of parameters for the sterile neutrino mass and mixing. Regions 1 and 2 correspond to parameters consistent with the pulsar kicks for (1) resonant and (2) off-resonant transitions, respectively. Both regions overlap with a band in which the sterile neutrino is dark matter.

$\sin^2\theta$, the production of sterile neutrinos is no longer suppressed, and they can take a fraction of energy out of a neutron star.

Sterile neutrinos escape with a sizable asymmetry due to weak interactions of fermions polarized by the magnetic field. (Once again, we neglect a neutrino magnetic moment and consider only the matter fermions.) The resulting asymmetry can explain the pulsar kicks if the mass and mixing angle fall inside region 2 in Fig. 1.

4. Sterile neutrinos as dark matter; observational consequences

The parameter space allowed for the pulsar kicks[8] overlaps nicely with that of dark-matter sterile neutrinos[10,11]. Sterile neutrinos in this range may soon be discovered[17]. Relic sterile neutrinos with mass in the 1-20 keV range can decay into a lighter neutrino and a photon. The X-ray photons should be detectable by the X-ray telescopes. Chandra and XMM-Newton can exclude part of the parameter space[17]. The future Constellation-X can probably explore the entire allowed range of parameters.

In the event of a nearby supernova, the neutrino kick can produce gravity waves that could be detected by LIGO and LISA[18,19].

Active-to-sterile neutrino oscillations can give a neutron star a kick. However, if a black hole is born in a supernova, it would not receive a kick, unless it starts out as a neutron star and becomes a black hole later, because of accretion. (The latter may be what happened in SN1987A, which produced a burst of neutrinos, but no radio pulsar.) If the central engines of the gamma-ray bursts are compact stars, the kick mechanism acting selectively on neutron stars and not black holes could probably explain the short bursts as interrupted long bursts[20].

Since one does not expect a significant correlation between the magnetic field inside a hot neutron star (while this field is, presumably, growing via the dynamo effect) and the eventual exterior field of a radio pulsar, the neutrino kick mechanism does not predict any $B - v$ correlation.

To summarize, the nature of cosmological dark matter is still unknown. We know that at least one particle beyond the Standard Model must exist to account for dark matter. This particle may come as part of a "package" if supersymmetry is right. However, it may be that the dark matter particle is simply an SU(2)×U(1) singlet fermion, which has a small mixing with neutrinos. In the latter case, the same dark-matter particle would be emitted anisotropically from a supernova with an asymmetry sufficient to explain the pulsar kick velocities.

References

1. See, *e. g.*, A. G. Lyne and D. R. Lorimer, Nature 369 (1994) 127; J. M. Cordes and D. F. Chernoff, Astrophys. J. **505**, 315 (1998); B. M. S. Hansen and E. S. Phinney, Mon. Not. R. Astron. Soc. **291**, 569 (1997); C. Fryer, A. Burrows, and W. Benz, Astrophys. J. **496**, 333 (1998).
2. Z. Arzoumanian, D. F. Chernoff and J. M. Cordes, Astrophys. J. **568**, 289 (2002).
3. I. S. Shklovskii, Sov. Astr., **13**, 562 (1970) [Astr. Zh. **46**, 715 (1970)].
4. C. L. Fryer, Astrophys. J. **601**, L175 (2004)
5. A. Burrows and J. Hayes, Phys. Rev. Lett. **76**, 352 (1996); L. Scheck *et al.*, arXiv:astro-ph/0307352.
6. O. F. Dorofeev, V. N. Rodionov and I. M. Ternov, Sov. Astron. Lett. **11**, 123 (1985).
7. A. Kusenko and G. Segrè, Phys. Lett. B **396**, 197 (1997)
8. G. M. Fuller, A. Kusenko, I. Mocioiu, and S. Pascoli, Phys. Rev. D **68**, 103002 (2003).
9. J. F. Nieves and P. B. Pal, Phys. Rev. **D40** 1693 (1989); J. C. D'Olivo,

J. F. Nieves and P. B. Pal, *ibid.*, 3679 (1989); S. Esposito and G. Capone, Z. Phys. **C70** (1996) 55; J. C. D'Olivo, J. F. Nieves and P. B. Pal, Phys. Rev. Lett., **64**, 1088 (1990); H. Nunokawa, V. B. Semikoz, A. Yu. Smirnov, and J. W. F. Valle, Nucl. Phys. **B 501**, 17 (1997).
10. K. Abazajian, G. M. Fuller and M. Patel, Phys. Rev. D **64**, 023501 (2001)
11. S. Dodelson and L. M. Widrow, Phys. Rev. Lett. **72**, 17 (1994); X. d. Shi and G. M. Fuller, Phys. Rev. Lett. **82**, 2832 (1999) A. D. Dolgov and S. H. Hansen, Astropart. Phys. **16**, 339 (2002).
12. A. Vilenkin, Astrophys. J. **451**, 700 (1995); A. Kusenko, G. Segrè, and A. Vilenkin, Phys. Lett. B 437, 359 (1998); P. Arras and D. Lai, astro-ph/9806285.
13. A. Kusenko and G. Segrè, Phys. Rev. Lett. **77**, 4872 (1996)
14. H. T. Janka and G. G. Raffelt, Phys. Rev. D **59**, 023005 (1999).
15. A. Kusenko and G. Segrè, Phys. Rev. **D59**, 061302 (1999).
16. M. Barkovich, J. C. D'Olivo, R. Montemayor and J. F. Zanella, Phys. Rev. D **66**, 123005 (2002).
17. K. Abazajian, G. M. Fuller and W. H. Tucker, Astrophys. J. **562**, 593 (2001)
18. L. C. Loveridge, Phys. Rev. D **69**, 024008 (2004) [arXiv:astro-ph/0309362].
19. H. J. Mosquera Cuesta, Astrophys. J. **544**, L61 (2000); Phys. Rev. D **65**, 061503 (2002); H. J. Mosquera Cuesta and K. Fiuza, arXiv:astro-ph/0403529.
20. A. Kusenko and D. V. Semikoz, arXiv:astro-ph/0312399.

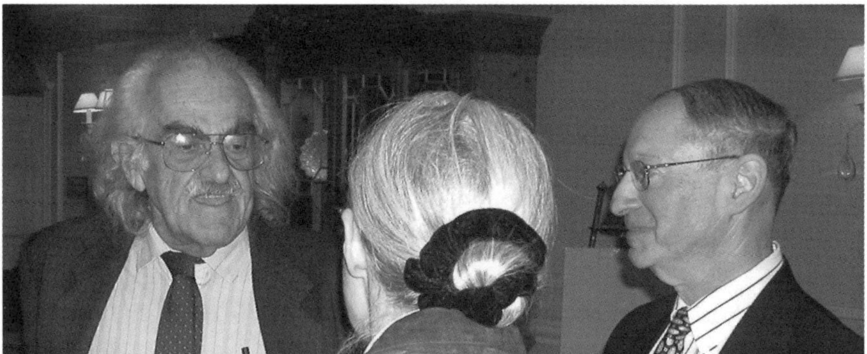

LIGO

SYDNEY MESHKOV
Caltech

This talk is dedicated to the memory of my close and loyal friend, Behram Kursunoglu. Behram and I go back a long way, back to a motel room at Argonne National Laboratory where the idea for the Coral Gables conferences was hatched in 1963, the era of symmetries. Behram always insisted that I should get involved in Gravity, and he was right – it just took a while. In 1994 I started working on LIGO, a wonderful device for Gravitational Wave detection. Behram was very happy when this happened.

The main contents of my talk are contained in five recent publications.

1. First Upper Limits from LIGO on Gravitational Wave Bursts.
   By LIGO Collaboration (B. Abbott et al.), May 2004
   Phys. Rev. D69:102001, 2004

2. Analysis of First LIGO Science Data for Stochastic Gravitational Waves.
   By LIGO Scientific Collaboration (B. Abbott et al), Dec. 2003
   e-print Archive: gr-qc/0312088

3. Analysis of LIGO Data for Gravitational Waves from Binary Neutron Stars
   By LIGO Scientific Collaboration (B. Abbott et al.), Aug. 2003
   e-print Archive: gr-qc/0308069

4. Setting Upper Limits on the Strength of Periodic Gravitational Waves Using the First Science Data from the GEO 600 and LIGO Detectors
   By LIGO Collaboration (B. Abbott et al.), Aug. 2003
   e-print Archive: gr-qc/0308050
   Phys. Rev. D69:082004, 2004

5. Detector Description and Performance for the First Coincidence Observations Between LIGO and GEO
   By LIGO Collaboration (B. Abbott et al.), Aug. 2003
   e-print Archive: gr-qc/0308043
   Nucl. Instrum. Meth.A517:154-179, 2004

ON THE MASS OF A UNIVERSE WITH A QUANTUM STRING-LIKE BEGINNING

A. J. MEYER, II

International Scientific Projects, Inc.
PO Box 3477
Westport Connecticut, 06880 USA
Email: ajmeyer@InternationalScientificProjects.org

Abstract

In this paper the mass of a universe, which began as a gravitationally closed, maximally spinning, Planck density quantum string is derived.

A universe beginning in such an initial state has been called a Super Spin Model universe. The total mass, M, of the Super Spin Model universe is shown to be a function of only four fundamental parameters ħ, c, G, e, such that: $M = 4\pi^2(\hbar c/G)^{1/2} exp(\hbar c/e^2)$.

The present paper primarily consists of a brief derivation of the above equation and a short synopsis of the Super Spin Model.

Keywords: Planck, primaton, geon, Kerr, string, cosmology, super spin model, unification, gravitation, rotating universe, non-stationary, axi-symmetric, space-time, event horizon.

Introduction

This is a brief review of the Super Spin Model, wherein the relationship:

$$M = 4\pi^2(\hbar c/G)^{1/2} exp(\hbar c/e^2)$$

was established for a non-stationary axi-symmetric space-time, where Mc^2 is the space-time's total conserved energy.[1]

The present paper primarily consists of a brief derivation of the above equation and a short synopsis of the Super Spin Model (SSM).

At some point near the beginning, the universe apparently was in a super dense energy state of about 10^{92} gm·c²/cm³.

Other than introducing the somewhat ad-hoc mechanism of inflation, how could the nascent universe expand and overcome the intense gravitational fields that normally would reduce it to a state of permanent and complete gravitational collapse?

According to the SSM, the reason the universe expands is due to its initial rotating toroidal string-like structure - its topology. If the energy were distributed in the form of a maximally rotating Planck density string, then

---

[1] Three additional independent equations for the mass 'M' of such non-stationary axi-symmetric space-times were also derived that led to linkages amongst: Newton's universal gravitational constant 'G', the fundamental quantum of electric charge 'e', the fundamental quantum of angular momentum – Planck's constant 'ħ', the speed of light 'c' in vacuo and the electron and proton masses, m_e and m_p. This set of connections, in turn, led to linkages among the four fundamental forces: the Gravitational, Electromagnetic, Strong, and Weak. These linkages are discussed in Meyer (1995).

because of the initial balance of the centrifugal and gravitational forces, and due to its small surface area, the string would begin to evaporate into its "ergosphere" and in spite of its huge density would naturally begin to expand into its other, initially "compressed," dimensions. Such a string would already have been "pre-inflated" in one spatial dimension, while "compressed" in the others.

To reiterate, the SSM universe is postulated to have begun as a gravitationally closed, maximally rotating Planck density geon, i.e. a four dimensional closed string with three of its four dimensions "curled up" or compacted within contiguous Planck areas of their locally mutually orthogonal surfaces. Such a geon,[2] with mass M and angular momentum magnitude J, has certain aspects that resemble the ring singularity of an extreme Kerr black hole. One may visualize this beginning state as a circle of light– a *geonic* pseudo ring singularity – a closed string of maximum energy density light consisting of Planck mass photons – "*primatons.*" See [Wheeler (1955)] and [Meyer (1980)].

The SSM is partially inspired by the Kerr family [Kerr (1963)] of stationary solutions for the empty space, axi-symmetric, Einstein-Maxwell, source free equations. However, an expanded Bekenstein-Hawking thermodynamics is used which incorporates both the inner and outer event horizon areas as measures of entropy, thereby yielding complete consistency with the Third Law of Thermodynamics.[3]

By employing a continuous variation of parameters methodology along with a parsimonious set of initial boundary constraints, the stationary metric is extended to a non-stationary axi-symmetric space-time possessing non-stationary "evolving" event horizons.

Due to their parametric simplicity, the uncharged Kerr family provides a rich laboratory in which to perform "gedanken versuchen." This uncharged family is determined by just two parameters: E and **J**, where $E=Mc^2$ is the total energy of the space-time and **J** is its total angular momentum. This family is represented by the following metric:

$$ds^2 = (2R_g r - \sigma^2)du^2/\sigma^2 - 2aR_z^2 sin^2\theta d\phi du/\sigma^2 + R_\phi^4 sin^2\theta d\phi^2/\sigma^2$$
$$+ 2dudr - 2a sin^2\theta dr d\phi + \sigma^2 d\theta^2.$$

Where:

$R_g \equiv GM/c^2 \equiv$ Gravitational radius of space-time,

$a \equiv$ Space-time's specific angular momentum radius = J/Mc,

$\sigma^2 \equiv r^2 + a^2 cos^2\theta \equiv$ Space-time's rotation radial coordinate offset,

$R_z^2 \equiv 2R_g r$,

---

[2] Geons are gravitational electromagnetic entities. They are regions consisting of electromagnetic radiation, which have sufficient energy density to become gravitationally confined. That is, light itself forms a sort of black hole. Wheeler developed this concept in 1954.

[3] It can be readily shown that the Third Law of Thermodynamics is satisfied if the area of the inner event horizon of a Kerr black hole is treated as a measure of the hole's negative entropy and, as usual, the outer event horizon area is treated as a measure of its positive entropy.

$\gamma^2 \equiv R_z{}^2 - r^2 - a^2$,
$R_\phi{}^4 \equiv (r^2 + a^2)^2 + \gamma^2 a^2 sin^2\theta$,
Mc^2 = Conserved total energy of space-time,

And:

$J \equiv |\mathbf{J}| \equiv aMc$ ≡ Scalar value of space-time's angular momentum,

such that:

$-\infty \leq r \leq +\infty$,
$-\infty \leq u \leq +\infty$,
$0 \leq \phi \leq 2\pi$,
$0 \leq \theta \leq \pi$.

Note, there exist two non-negative event horizon radii, r_+ and r_-, which are the conjugate solutions of the equation: $\gamma^2(r) = 0$, which are: $r_\pm = R_g \pm [R_g{}^2 - a^2]^{1/2}$. They can be re-written as:

$$r_\pm = R_g[1 + sin(\pm\,\Phi)] \Rightarrow$$
$$a^2 = r_+r_- = R_g{}^2 cos^2(\Phi) \text{ and } r_+ + r_- = 2R_g.$$

Hence, the space-time's specific angular momentum radius, 'a' is parameterized in terms of M and $\pm\Phi$, so that:

$$a = a(\pm\Phi) = R_g\, cos(\Phi) = GMcos(\Phi)/c^2.$$

The angle 'Φ' is therefore a measure of the magnitude of the space-time's angular momentum (and/or expansion state). That is:

$$J(M, \pm\,\Phi) = a(\pm\Phi)Mc = R_g Mccos(\Phi) = GM^2cos(\Phi)/c = |\,\mathbf{J}(M, \pm\,\Phi)|.$$

As is easily seen, $\Phi = \pm 0$ implies a maximally rotating Kerr space-time and $\Phi = \pm\,\pi/2$ implies a static Schwarzschild space-time.

The Description of the SSM Universe (U) as a Rotating *Non-Stationary* Axi-Symmetric Space-Time

One way to produce a model of a non-stationary axi-symmetric space-time, is to characterize the entire family of stationary uncharged Kerr solutions as the set of ordered pairs: $\{< M, \mathbf{J} >\} = \{< M, \mathbf{J}(\pm\,\Phi) > \ni [-\pi/2 \leq \Phi \leq \pi/2]\}$.

Such a parameterization gives the entire family of Kerr solutions in terms of only two parameters 'M' and 'Φ' and only two universal constants 'G' and 'c.' That is any member of the Kerr metric family can be represented as the ordered pair: $< M, \mathbf{J}(\Phi) > = < M, GM^2cos(\Phi)/c >$.

By treating $\pm\Phi$ as a dual-time dependent variable, it is possible to transform the Kerr metric for the *stationary family*, wherein each member contains a pair of *completed* inner and outer event horizons, into a *single dual-valued time-varying metric*. In other words, the *family* of metrics is transformed into a *single* metric function of Φ, which governs the evolution of the *non-stationary incomplete* inner and outer event horizons, $H(r_\pm(\Phi),\theta_\pm(\Phi))$, that emanate from the "creation ring" and *grow* in both $\pm\Phi$ directions, starting from $\Phi = 0$, where:

$r_\pm(\pm 0) = R_g$ and $\theta_\pm(\pm 0) = \pi/2$. The expansion trajectories of the event horizons' altitudinal "rims" are governed by: $\theta_\pm(\Phi) = \pi/2 \pm |\Phi|$.

For the stationary metric, when $\gamma^2(r) = 0 \Rightarrow r_\pm = R_g \pm [R_g^2 - a^2]^{1/2}$, then $ds_\pm = 0$. However, this is not generally true when the metric coefficients are written as a function of the parameter Φ. That is, $ds_\pm/d\Phi$ is generally $\neq 0$, for the hypersurfaces, $\gamma^2(r_\pm) = 0$, when $\pm\Phi$ is a dual-time dependent variable.

Rewriting the metric in terms of its past directed and future directed non-stationary event horizons,[4] we get:

$$ds_\pm^2 = \sigma_\pm^2 d\theta^2 - 2a\sin^2\theta dr_\pm d\phi + 2dr_\pm du_\pm + \sin^2\theta[R_{z\pm}^2 d\phi - a\sin\theta du_\pm]^2/\sigma_\pm^2.$$

Where the boundary conditions of 'U' are:

$$0 \leq r_- \leq R_g \leq r_+ \leq 2R_g,$$
$$0 \leq \theta_- \equiv \pi/2 - |\Phi| \leq \theta \leq \pi/2 + |\Phi| \equiv \theta_+,$$
$$0 \leq \phi \leq 2\pi,$$
$$-\infty \leq u \leq +\infty.$$

And: $r_\pm = R_g[1 + \sin(\pm\Phi)]$ are the *event horizon radii,*

$\sigma_\pm^2 = r_\pm^2 + a^2\cos^2\theta$ are the *rotation-offset radii,*

$R_{z\pm}(\Phi) = [2R_g r_\pm]^{1/2} = [r_\pm^2 + a^2]^{1/2} = R_g[2(1 \pm \sin\Phi)]^{1/2}$ are the *radii of gyration* of the event horizons, and

$\Omega_\pm(\Phi) = c\, d\phi/du_\pm = -g_{u\phi}(\Phi)c/g_{\phi\phi}(\Phi) = ac/(R_{z\pm})^2 = c\cos\Phi/[2R_g(1 \pm \sin\Phi)]$ are the *dual event horizon rotation rates.*

For a constant M, one can see that the *moments of inertia* are: $I_\pm = M R_{z\pm}^2$, since: $I_\pm\Omega_\pm = J(\pm\Phi) = a(\pm\Phi)Mc = R_g Mc \cos(\Phi) = GM^2\cos(\Phi)/c = |\mathbf{J}(\pm\Phi)|$.

The above formulations are applicable to the subset **U** of the axi-symmetric space-time manifold,[5] $\mathfrak{M}$, such that **U** is isomorphic to the intrinsically non-stationary region, which is partially bounded by or swept out by the two "incomplete" event horizons with radii: $r_\pm = r(\Phi(t_\pm))$. Notice at $\Phi = \pm\pi/2$, the **U** becomes identical to a Schwarzschild black hole with complete and stationary event horizons. It is also critical to notice, that for constant Φ (the stationary case), the $r_\pm$ are the loci of null hyper-surfaces, but for variable Φ (the non-stationary case) the $r_\pm$ loci are null only where:

$$\cos^2\theta = \sin^2\Phi \Rightarrow \theta = \theta_\pm \equiv \pi/2 \pm \Phi.$$

The inner and outer event horizon areas are calculated by:

$$A(\pm\Phi) = \int_{(\pi/2-\Phi)}^{(\pi/2+\Phi)} \int_0^{2\pi} \sqrt{(g_{\phi\phi}\, g_{\theta\theta})}\, d\phi d\theta = 4\pi R_{z\pm}^2 \sin(\pm\Phi) =$$

$$8\pi R_g^2 \sin(\pm\Phi)[1 + \sin(\pm\Phi)].$$

---

[4] See Meyer (1995)

[5] There might also exist an independent **anti** $\mathfrak{M} \equiv -\mathfrak{M}$, simultaneously created with equal energy and angular momentum, but with opposite parity.

Therefore, the total *net* horizon area[6] for the SSM Universe is reckoned as:

$$A_U^{(2)}(\Phi) = A(+\Phi) + A(-\Phi) = 16\pi(R_g\, sin\Phi)^2.$$

Moreover, the spatial volume[7] or hyper-surface area of the SSM U is calculated as:

$$A_U^{(3)}(\Phi) = 4\pi R_g^3 sin\Phi(5\Phi - 3 sin\Phi cos\Phi) \equiv V_U(\Phi).$$

We also find that the space-time 4-volume, (ignoring the $\sqrt{-1}$ coefficient) is:

$$A_U^{(4)}(\Phi) = (8\pi c/3)\, t_s(\Phi)R_g^3\, sin^2\Phi\, (3 + 2\, sin^2\Phi - sin^4\Phi).$$

Of course, the circumference $C_{\gamma\pm}$ is the one-dimensional "area":

$$A_U^{(1)}(\Phi) = 4\pi R_g.$$

A net time interval,[8] $t_s(\Phi)$, defined as the difference between the future directed, $u_+(\Phi)/c$, and past-directed, $u_-(\Phi)/c$, geodesic time displacements, is reckoned by:

$$ct_s(\Phi) \equiv [u_+(\Phi) - u_-(\Phi)] = R_g[-ln(1 - sin|\Phi|) - ln(1 + sin|\Phi|)] = -R_g\, ln(1 - sin^2\Phi).$$

In general: $ct_s(\pm\Phi) = -R_g\, ln(cos^2\Phi)$. Note that at maximum expansion, when $\Phi=\pm\pi/2$, then:

$$ct_s(\pm\pi/2) = -R_g ln[1 - sin^2(\pm\pi/2)] = +\infty.$$

Observe that the past-directed[9] time displacement: $t_-(\Phi) = R_g\, ln(1 + sin|\Phi|)/c$, is always finite.

The expansion parameter 'Φ' grows in both directions away from an origin $\Phi \sim \pm 0$. That is, the space-time expands both ways in time.[10] In the beginning, the space-time or "creation ring" is a closed cosmic string with an *inflated* circumference. Since **J**, the angular momentum vector, is a cosine function of Φ, then by increasing $|\Phi|$, the angular momentum magnitude is decreased and the space expands. As Φ approaches $\pm\pi/2$, the outer event horizon approaches the Schwarzchild radius and the inner event horizon radius approaches zero.

See **Figs. 1**, **2** and **3**, which are schematics that attempt to depict how the SSM universe evolves.

---

[6] Note that at $\Phi = 0$, the net event horizon area is zero, which of course means that the Bekenstein-Hawking entropy is also zero.

[7] This spatial volume calculation is worked out in Meyer (1995).

[8] Meyer (1995), (2000)

[9] Whereas, the maximum past time is always finite, the maximum future time is always infinite.

[10] Notice that both directions and senses of the temporal dimension, i.e. time and anti-time began (were created) at $\Phi = 0$, along with the other spatial dimensions. It is simple but interesting to observe that anti-time trajectories cannot reach the "terra incognita" beyond or before the universe was created.

The Initial Conditions

Let $\Phi_\chi \approx \pm 0$ be the values of Φ when the density of the toroidal geonic string universe is equal to the Planck photon density.[11] This occurs when:

$$\rho \equiv M / V_U(\Phi_\chi) = M / 4\pi R_g^3 \Phi_\chi^2 = \rho_\varpi = c^5 / 8\pi^2 G^2 \hbar$$
$$\Rightarrow \Phi_\chi = \pm\sqrt{\pi} / N_\varpi.$$

Therefore, for the small angle at creation $\Phi = \Phi_\chi$ one gets:

$$ct_s(\Phi_\chi) = -R_g \, ln(1 - sin^2 \Phi_\chi) \approx R_g \Phi_\chi^2 = \pi R_g / N_\varpi^2 = \pi\hbar/Mc \equiv ct_\chi,$$

which is *one half*[12] the Compton wavelength of the SSM universe. Note for a universe of mass $M \sim 10^{56}$gm, the Compton time, or creation interval t_χ is about 10^{-104}sec.

Nevertheless, immediately after the instant of creation at $t_\chi \sim 10^{-104}$ sec, the **U**'s *minimum* spatial dimension is already on the order of the Planck length. It therefore follows that the lowest bound of any future temporal increment must be on the order of the Planck time.

If the universe started as a non-rotating bubble membrane fluctuation, then because of the extreme "thinness" of the membrane, there is no apparent reason that such a non-rotating bubble would be stable. It should "flash back" into the vacuum from whence it sprang within its Compton time,[13] $t_c = 2\pi\hbar/Mc^2$.

However, in the SSM, the angular momentum of the maximally spinning string induces a coarser granularity by a factor of N_ϖ upon both the spatial and temporal quanta of the fluctuation's active region. That is, the previously indivisible quanta become Planck sized.

Thus, the angular momentum of the creative process produces a real space time, "inflating," by a factor of N_ϖ, the universe's original indivisible spatial and temporal quanta, i.e., its Compton length and time, to the Planck length and time, (r_ϖ and t_ϖ), which become its new indivisible spatial and temporal quanta.

The rotating string topology thereby produces a "one-way valve" which "closes all the hatches" and stops the massive virtual fluctuation from "sinking" back into the vacuum within $\sim 10^{-104}$ sec. In other words, the Creator "writes a check" to the vacuum for about 10^{56} grams of energy and then "changes the *banking rules* before it can clear."

---

[11] In this paper, the Greek letter varpi 'ϖ' denotes a Planckian parameter, so that the Planck mass is: $m_\varpi = (\hbar c/G)^{1/2}$, the Planck momentum magnitude is: $p_\varpi = m_\varpi c$, the Planck radius is: $r_\varpi = \hbar/p_\varpi$, the Planck wavelength is: $\lambda_\varpi = 2\pi r_\varpi$, the Planck time is: $t_\varpi = \lambda_\varpi/c$ and so forth.

[12] This might mean there were two universes created as a quantum fluctuation of the vacuum. That is, there might be another **U** created with opposite parity, so that the net angular momentum would be zero. In this paper, we shall just be concerned with one of them.

[13] See discussion in Meyer (1995)

At creation, when $\Phi = \Phi_\chi = \pm\sqrt{\pi}/N_\varpi$:
$\exists$ a gravitationally closed, Planck density string of *length*[14] $\equiv C_{\gamma\pm} = 4\pi R_g$.
$\exists$ maximum electromagnetic energy density[15] $= \rho_\varpi c^2 = c^7/8\pi^2 G^2\hbar$.
$\exists$ maximum action for the gravitationally closed space= $|\mathbf{J}(0)| = GM^2/c$.
$\exists$ maximum string tension = $\mathcal{T}_\varpi = Mc^2/C_{\gamma\pm} = c^4/4\pi G$.
$\exists$ maximum vacuum polarization[16] energy = $\varepsilon_{evac\varpi} = e^2(c^3/\hbar G)^{1/2} = kT_{evac\varpi}$.
$\exists$ maximum photon energy= $\varepsilon_{\gamma\varpi} = (\hbar c^5/G)^{1/2} \equiv m_\varpi c^2$.

Derivation of the String's Mass

Along the string's ergosphere, the Unruh temperature [Unruh (1976)] is reckoned as:

$T_{evac\varpi} = \hbar\kappa_{evac\varpi}/2\pi ck$, where: $\kappa_{evac\varpi} = 2\pi\partial^2\mathbf{r}/\partial t^2|_\varpi = e^2c/r_\varpi\hbar$ and $r_\varpi = \lambda_\varpi/2\pi$.

And when $|\Phi| \le |\Phi_\chi|$, then: $r_\pm(\Phi) = R_g \pm \delta_r \approx R_g$ and $\theta_\pm(\Phi) = \pi/2 \pm \delta_\theta \approx \pi/2$.

In other words, a main postulate is that the initial fluctuation is a non-singular coherent toroidal electromagnetic disturbance of Planck density, ρ_ϖ. This fluctuation is formally identical to a toroidal geon, or a closed circular string of mass N_ϖ in Planck units, with total initial angular momentum magnitude:

$$J(0) = [\mathbf{J}(0)\bullet\mathbf{J}(0)]^{1/2} = \hbar[N_\varpi^4 + 2N_\varpi^2 + 3]^{1/2} \approx \hbar N_\varpi^2 = R_g Mc,$$
where: $N_\varpi \equiv M/m_\varpi$.

After a little algebra, one finds that the initial volume of the six dimensional phase space occupied by the string-like universe at Φ_χ is:

$$V_0^{(6)}(\Phi_\chi) = \int dV^{(3)} \int dp^{(3)} = 8\pi^2 R_g r_\varpi^2 p_\varpi^3 = 8\pi^2 N_\varpi \hbar^3.$$

However, the total phase space volume at $\pm\Phi_\chi$ is the sum of the volumes of all cubic action compartments $\hbar^3$. This volume is calculated as:

$$V_S^{(6)}(\Phi_\chi) = [(\mathbf{J}_x(\Phi_\chi) \pm \hbar) \times (\mathbf{J}_y(\Phi_\chi) \pm \hbar)]\bullet(\mathbf{J}_z(\Phi_\chi) \pm \hbar).$$

Now, since the metric is axi-symmetric, we can choose: $\mathbf{J}_x(\Phi_\chi) = \mathbf{J}_y(\Phi_\chi) = \mathbf{0}$ and $|\mathbf{J}_z(\Phi_\chi)| = R_g Mc$. It then follows that: $V_S^{(6)}(\Phi_\chi) = (R_g Mc \pm \hbar)\hbar^2 = N_\varpi^2\hbar^3 \pm O(\hbar^3)$. Moreover, the ratio of the string's initial sub-phase space volume, $V_0^{(6)}(\Phi_\chi)$, (*with all photons in the Planck state*) to the total phase space volume $V_S^{(6)}(\Phi_\chi)$, should be equal to the Bose-Einstein expectation number, $\langle n_\varpi \rangle$.

---

[14] $\forall\Phi \ni \gamma(r_\pm(\Phi)) = 0$, $C_{\gamma\pm} = \int_0^{2\pi}[g_{\phi\phi(\gamma=0)}]^{1/2}d\phi = 4\pi R_g$ = circumference of *both* inner and outer event horizons at the "equator" (the space orthogonal to the axis of symmetry, $\mathbf{J}(0)$, i.e. $\theta = \pi/2$.) Note, *the circumference is constant for all* Φ and hence, *is independent* of the event horizon radii.

[15] See the discussions about maximum energy density electromagnetic quanta in Meyer (1980), (1995).

[16] See Meyer (1995).

That is: $\langle n_\varpi \rangle = V_0^{(6)}(\Phi_\chi)/V_S^{(6)}(\Phi_\chi) = 8\pi^2 N_\varpi \hbar^3/N_\varpi^{\,2} \hbar^3 = 8\pi^2/N_\varpi =$
$2/[exp(\varepsilon_{\gamma\varpi}/kT_{evac\varpi}) - 1] = 2/[exp(\varepsilon_{\gamma\varpi}/kT_{evac\varpi}) - 1)] = 2/[exp(\hbar c/e^2) - 1)]$.

Since, $\hbar c/e^2 = \alpha^{-1} \approx 137$, we obviously can neglect the '-1' term and therefore get:

$$N_\varpi = 4\pi^2 exp(\hbar c/e^2) = 4\pi^2 \, exp(1/\alpha) = M/m_\varpi.$$

Therefore, the mass, M, of this rotating string universe is reckoned as:

$$M = 4\pi^2(\hbar c/G)^{1/2} exp(\hbar c/e^2) = 2.8062 \times 10^{56} \text{ gm}.$$

Entropy Formulation

It has been found that the thermodynamics of stationary black holes becomes consistent with classical thermodynamics by introducing an extended interpretation of the "Bekenstein-Hawking"[17, 18] entropy formula, by augmenting it to include the inner event horizon area as a measure of "negentropy."[19]

By also extending this augmentation to the incomplete and evolving event horizons of the non-stationary SSM U, the net entropy of **U** as a function of Φ then becomes:

$$S(\pm\Phi) = 4\pi k(N_\varpi sin\Phi)^2 \geq 0, \forall\Phi.$$

It is easy to see that entropy will increase along both positive and negative temporal directions as both temporal displacements increase away from the origin in both positive and negative time.

There is a low initial entropy value upon the creation of the Planck density string[20] at $\Phi = \Phi_\chi$. This value is:

$$S(\Phi_\chi) = 4\pi k(N_\varpi sin\,\Phi_\chi)^2 = 4\pi^2 k,$$

where $\Phi_\chi = \pm\sqrt{\pi}/N_\varpi$ and 'k' is Boltzman's constant.

Since $ct_s(\pm\Phi) = -R_g ln(1 - sin^2\Phi_\chi)$, then as $\Phi \rightarrow \pm\pi/2$, and $t_s \rightarrow \infty$, the **U** approaches its Schwarzchild limit, and the net event horizon area and net entropy approach their maxima.[21] From the above equation for $S(\pm\Phi)$ the maximum entropy is therefore:

$$S(\pm\pi/2) = 4\pi k N_\varpi^{\,2} = 64\pi^5 k \, exp(2/\alpha) = 1.51287 \times 10^{139} \text{ (erg/°K)}.$$

---

[17] Bekenstein (1973)
[18] Hawking (1974)
[19] Since the dual gravitational "*strengths*" or "*accelerations,*" $\kappa_\pm$, for stationary Kerr black holes are reckoned by: $\kappa_\pm = \pm c^2(R_g^2 - (J/Mc)^2)^{1/2}/2R_g\,[R_g \pm (R_g^2 - (J/Mc)^2)^{1/2}] = 0$. When $R_g = J/Mc$, both event horizon temperatures are calculated as: $T_\pm = \hbar\kappa_\pm/2\pi kc = 0$. However, the un-augmented Bekenstein-Hawking entropy is: $S_{BH} = 2\pi kGM^2/\hbar c >> 0$, which leads to a violation of the Third Law.
[20] In the SSM, since the space-time is non-stationary and its string-like initial state is far from equilibrium, the initial net entropy is near zero, whereas its initial temperature is very large.
[21] The excess angular momentum energy, $\Delta E_J = Mc^2[1-1/\sqrt{2}]$, that is not taken up by the expansion, see Christodoulu (1970), is conjectured to be taken up by the generation of "flywheels" surmised to be a species of neutrinos, see Meyer (1995). Some of the excess may also be stored in the rotational energy of galaxies, stars and other spinning and rotating objects.

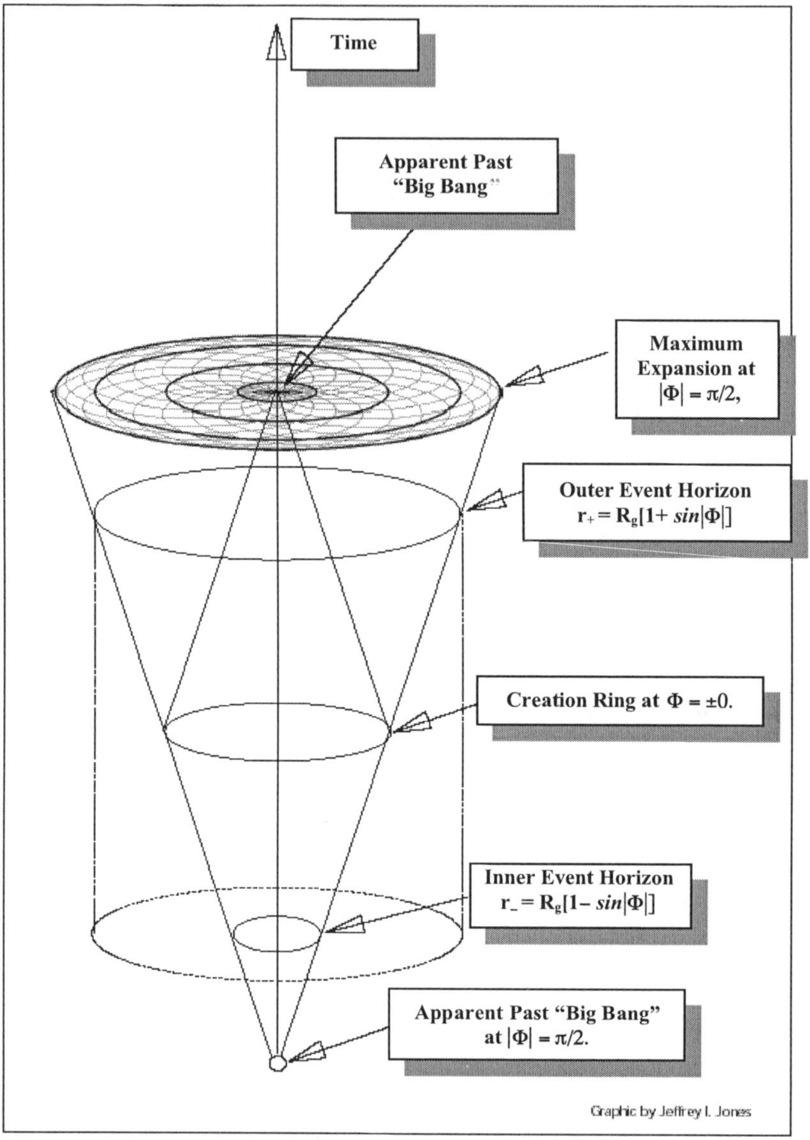

Figure 1. Portrait of evolution of SSM universe

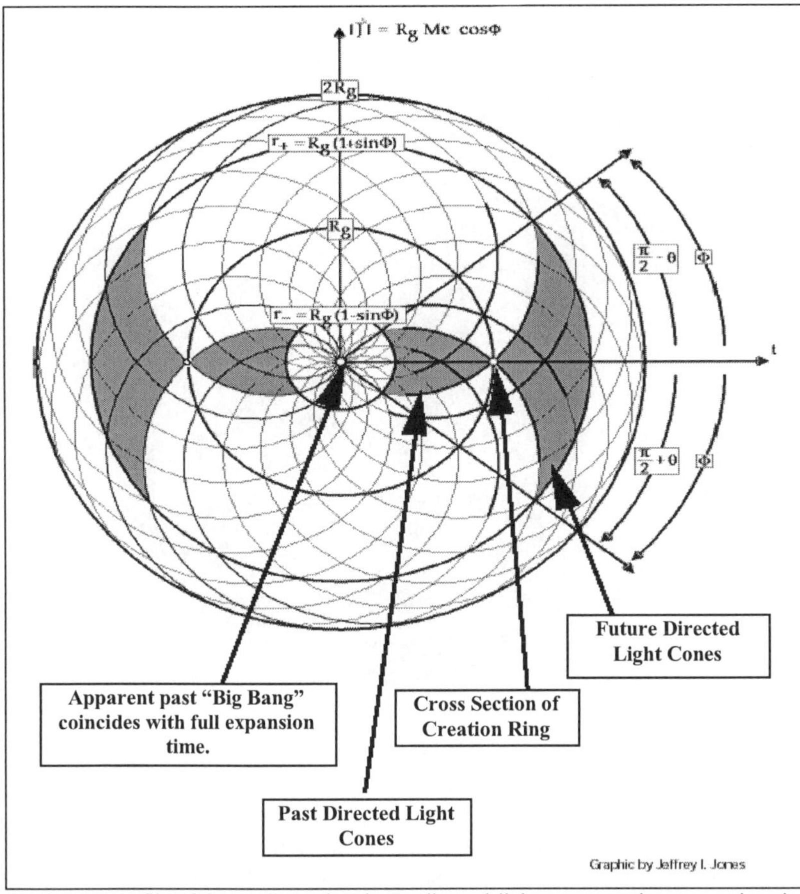

Figure 2. Profile of future directed and past directed light cones starting at creation ring as projected onto the following space: $\{< \pm t, \theta, \phi > \ni (\pi/2 - \Phi \leq \theta \leq \pi/2 + \Phi) \wedge (\phi = \text{constant})\}$.

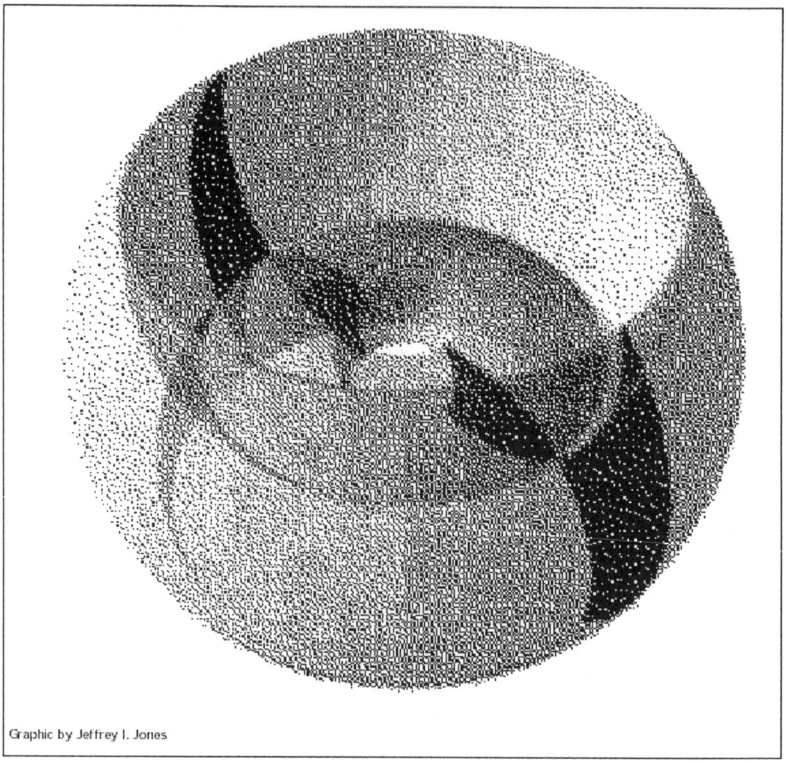

Figure 3 (Solid View of Figure's 1 & 2)

This illustrates the loci of future and past-directed light cones, starting at creation ring as projected in the following space: $\{< \pm t, \theta, \phi > \ni (\pi/2 - \Phi \leq \theta \leq \pi/2 + \Phi) \wedge (0 \leq \phi \leq 2\pi)\}$. The darkest region is a two-dimensional $\{< \pm t, \theta >\}$ profile of light cones as shown in Figure 2. Only two-dimensional projections of three-space are shown in the above solid.

Acknowledgments

I am very grateful for the friendship, inspiration, and thoughtful encouragement that Professor Behram Kursunoglu gave me over many years. I miss him greatly. I also wish to thank Professor Arnold Perlmutter for much helpful advice, Professor Jeffrey I. Jones, of Queensland Technological University for graphics and technical expertise, and to my wife Connie, and daughters Michelle and Eva, for their prayers and support in getting this paper and presentation together. Above all, I wish to acknowledge and thank the eternal *"God (Who) used beautiful mathematics in Creating the World!"* as P.A.M. Dirac proclaimed.

References

Christodoulu, D. 1970. Physical Review Letters, 25, 1596-1597.
Bekenstein, J.D. 1973. Physical Review, D7, 2333-2346.
Hawking, S.W. 1974 Nature 248, 30.
Kerr, R. 1963. Physical Review Letters, 11, 237.
Meyer, A.J. 1980. "Primatons, Maximum Energy Density Quanta" in Recent Developments in High Energy Physics (editors A. Perlmutter and L.F. Scott), Plenum Press, New York
Meyer, A.J. 1995. Physics Essays 8 (4), pp. 524-586
Meyer, A.J. 2000. "The Unification of G and e" in Quantum Gravity, Generalized Theory of Gravitation and Superstring Theory-Based Unification (editors Behram N. Kursunoglu, Stephen L. Mintz and Arnold Perlmutter), Kluwer Academic / Plenum Publishers, New York
Unruh, W.G. 1976. Physical Review, D14, 870-892.
Wheeler, J.A. 1955. Physical Review, 97, No. 2, 511-536.13.

VARIABLE SPEED OF LIGHT COSMOLOGY AND BIMETRIC GRAVITY: AN ALTERNATIVE TO STANDARD INFLATION

J. W. MOFFAT

Perimeter Institute for Theoretical Physics, Waterloo, Ontario, N2J2W9, Canada
and
Department of Physics, University of Toronto, Toronto, Ontario, M5S1A7, Canada
e-mail: jmoffat@perimeterinstitute.ca

Abstract. A scalar-tensor bimetric gravity model of early universe cosmology is reviewed. The metric frame with a variable speed of light (VSL) and a constant speed of gravitational waves is used to describe a Friedmann-Robertson-Walker universe. The Friedmann equations are solved for a radiation dominated equation of state and the power spectrum is predicted to be scale invariant with a scalar mode spectral index $n_s = 0.97$. The scalar modes are born in a ground state superhorizon and the fluctuation modes are causally connected by the VSL mechanism. The cosmological constant is equated to zero and there is no significant dependence on the scalar field potential energy. A possible way of distinguishing the bimetric gravity model from standard inflationary models is discussed.

1. Bimetric Gravity Theory

In the scalar-tensor bimetric gravity theory, the metric of spacetime is given by [1,2,3,4]:[a]

$$\hat{g}_{\mu\nu} = g_{\mu\nu} + B\partial_\mu\phi\partial_\nu\phi, \tag{1}$$

where $B = [\text{length}]^2$ and ϕ is a scalar field (biscalar field). The metric $\hat{g}_{\mu\nu}$ is called the 'matter' metric and $g_{\mu\nu}$ is the gravitational metric. The biscalar field ϕ is a component of the gravitational field.

The model consists of a self-gravitating scalar field coupled to matter through the matter metric $\hat{g}_{\mu\nu}$ with the action

$$S = S_{\text{grav}} + S_\phi + \hat{S}_{\text{M}}, \tag{2}$$

[a]Bekenstein [5] considered bimetric (disformal) gravity theories in the context of lensing effects in generalized gravity theories, not as variable speed of light cosmological theories

where

$$S_{\text{grav}} = -\frac{1}{\kappa}\int d\mu(R[g] + 2\Lambda). \tag{3}$$

Here, $\kappa = 16\pi G/c^4$, Λ is the cosmological constant, $d\mu = \sqrt{-g}d^4x$, $\mu = \sqrt{-g}$, $\hat{\mu} = \sqrt{-\hat{g}}$ and c denotes the currently measured speed of light. We have the relation $\mu = \sqrt{K}\hat{\mu}$ where $K = 1 - B\hat{g}^{\mu\nu}\partial_\mu\phi\partial_\nu\phi$. We use the metric signature $\eta_{\mu\nu} = \text{diag}(+1,-1,-1,-1)$ where $\eta_{\mu\nu}$ is the flat Minkowski metric.

The minimally-coupled scalar field action is

$$S_\phi = \frac{1}{\kappa}\int d\mu\left[\frac{1}{2}g^{\mu\nu}\partial_\mu\phi\partial_\nu\phi - V(\phi)\right]. \tag{4}$$

From this action we can derive

$$\bar{g}^{\mu\nu}\hat{\nabla}_\mu\hat{\nabla}_\nu\phi + KV'[\phi] = 0, \tag{5}$$

where

$$\bar{g}^{\mu\nu} = \hat{g}^{\mu\nu} + \frac{B}{K}\hat{\nabla}^\mu\phi\hat{\nabla}^\nu\phi - \kappa B\sqrt{K}\hat{T}^{\mu\nu}, \tag{6}$$

and $\hat{\nabla}^\mu$ denotes the covariant derivative with repsect to $\hat{g}_{\mu\nu}$.

2. Bimetric Gravity Cosmology

We can choose either $\hat{g}_{\mu\nu}$ or $g_{\mu\nu}$ to be comoving metric frames in an FRW universe. If we choose $\hat{g}_{\mu\nu}$ as the comoving metric, then

$$d\hat{s}^2 \equiv \hat{g}_{\mu\nu}dx^\mu dx^\nu$$

$$= c^2dt^2 - R^2(t)\left(\frac{dr^2}{1-kr^2} + r^2(d\theta^2 + \sin^2\theta d\phi^2)\right), \tag{7}$$

and

$$ds^2 \equiv g_{\mu\nu}dx^\mu dx^\nu$$

$$= c^2(1 - \frac{B}{c^2}\dot{\phi}^2)dt^2 - R^2(t)\left(\frac{dr^2}{1-kr^2} + r^2(d\theta^2 + \sin^2\theta d\phi^2)\right). \tag{8}$$

From (8) the speed of gravitational waves (gravitons) is given by $c_g(t) = c(1 - \frac{B}{c^2}\dot{\phi}^2)^{1/2}$, while the speed of light is constant. Alternatively, if we choose $g_{\mu\nu}$ as the comoving metric, we have

$$ds^2 \equiv g_{\mu\nu} dx^\mu dx^\nu$$

$$= c_g^2 dt^2 - R^2(t)\left(\frac{dr^2}{1-kr^2} + r^2(d\theta^2 + \sin^2\theta d\phi^2\right) \tag{9}$$

and

$$d\hat{s}^2 \equiv \hat{g}_{\mu\nu} dx^\mu dx^\nu$$

$$= c^2(1 + \frac{B}{c^2}\dot{\phi}^2)dt^2 - R^2(t)\left(\frac{dr^2}{1-kr^2} + r^2(d\theta^2 + \sin^2\theta d\phi^2)\right). \tag{10}$$

Now, the speed of light is given by the time dependent quantity $c_\gamma(t) = c(1 + \frac{B}{c^2}\dot{\phi}^2)^{1/2}$, while the speed of gravitational waves $c_g = c$ is constant.

In general relativity (GR), one can always perform a diffeomorphism transformation which removes the time dependence of the speed of light. In the bimetric gravity theory there are two speeds: one associated with the speed of light (photons) c_γ in the VSL metric frame, and another with the speed of gravitational waves (gravitons) c_g in the gravitational metric frame with the dimensionless ratio $\gamma = c_g/c_\gamma$. In contrast to GR, we cannot simultaneously remove the time dependence in both c_γ and c_g by a diffeomorphism transformation. Thus, by regauging clocks, we cannot make both c_γ and c_g simultaneously constant. This makes the time dependence of either c_γ or c_g a non-trivial feature of the theory.

We will first consider the metric frame in which the matter metric $\hat{g}_{\mu\nu}$ is comoving. Then, the Friedmann equation is of the form

$$H^2 + \frac{c_g^2 k}{R^2} = \frac{8\pi\bar{G}}{3}\rho_M + \frac{1}{3}c_g^2\bar{\Lambda}. \tag{11}$$

Here, the effective gravitational constant $\bar{G}(t) = G(1 - \frac{B}{c^2}\dot{\phi}^2)^{3/2}$ and the effective cosmological constant is

$$\bar{\Lambda} = \Lambda + \frac{1}{2}\left(\frac{1}{2c_g^2}\dot{\phi}^2 + V[\phi]\right). \tag{12}$$

When the speed of gravitons $c_g(t) \to 0$ in the early universe, then $\bar{G}(t) \to 0$ and the Friedmann equation (11) becomes

$$H^2 = \frac{1}{12}\dot{\phi}^2, \tag{13}$$

where we have set $\Lambda = V[\phi] = 0$. We can show by solving the wave equation for ϕ for a radiation universe with the equation of state, $p = c^2\rho/3$, that $\dot{\phi}^2 \to$ constant as $t \to 0$, so that we obtain an initially inflationary solution

$R \sim \exp(At)$ where $A = \dot{\phi}/\sqrt{12}$ = constant. Note that the derivation of initial inflation *does not depend significantly* on a choice of potential $V[\phi]$. The initial inflationary phase is produced by the assumption that the speed of gravitons becomes very small in the early universe.

We shall now choose the gravitational metric $g_{\mu\nu}$ as the comoving frame. In the $\hat{g}_{\mu\nu}$ matter metric frame (we call it the VSL metric frame) we have: $K(t) = 1/(1 + \frac{B}{c^2}\dot{\phi}^2)$, so that $c_\gamma(t) = cK^{-1/2}(t)$ and the speed of photons $c_\gamma(t) \to \infty$ as $K(t) \to 0$.

The matter stress-energy tensor (using $\hat{u}^0 = c$) is

$$\hat{T}^{00} = K\rho, \quad \hat{T}^{ij} = \frac{p}{R^2}\gamma^{ij}, \tag{14}$$

and the conservation laws give

$$\dot{\rho} + 3H\Big(\rho + \frac{p}{c^2}\Big) = 0. \tag{15}$$

We adopt the radiation equation of state $p = \frac{1}{3}c^2\rho$. It is useful at this point to introduce the following quantities derived from the constant B: $H_B^2 = \frac{c^2}{12B}$ and $\rho_B = 1/(2\kappa c^2 B)$, where the latter comes from $H_B^2 = (1/6)\kappa c^4\rho_B$

The Friedmann equation is given by

$$H^2 + \frac{kc^2}{R^2} = \frac{1}{3}c^2\Lambda + \frac{1}{6}\left(\frac{1}{2}\dot{\phi}^2 + c^2V[\phi]\right) + \frac{1}{6}\kappa c^4\sqrt{K}\rho. \tag{16}$$

The scalar field equation is

$$(1 - \kappa c^2 BK^{3/2}\rho)\ddot{\phi} + 3H(1 + \kappa B\sqrt{K}p)\dot{\phi} + c^2V'[\phi] = 0. \tag{17}$$

We postulate that in the very early universe K is very small, so that $B\dot{\phi}^2 \gg c^2$ and $c_\gamma(t) \to \infty$. We will set $V[\phi] = 0$ and $\Lambda = 0$. This gives as $K \to 0$ the Friedmann equation

$$H^2 + \frac{kc^2}{R^2} = \frac{1}{12}\dot{\phi}^2 \tag{18}$$

and

$$\ddot{\phi} + 3H\dot{\phi} = 0. \tag{19}$$

Eq.(19) has the solution $\dot{\phi} = \sqrt{12}H_B\left(\frac{R_{pt}}{R}\right)^3$, where we have chosen the arbitrary constant of integration R_{pt} to parameterize the time at which $K = 1/2$. This gives $K = 1/[1 + (R_{pt}/R)^6]$, so that it indicates the time at which the gravitational and matter metrics are close to coinciding. The subscript *pt* indicates that this is the 'phase transition', where standard local Lorentz symmetry with a single light cone will be restored–it is also the end of the period that will appear as inflation in the comoving gravitational metric frame.

The Friedmann equation is then

$$H^2 = H_B^2 \left(\frac{R_{pt}}{R}\right)^6 \tag{20}$$

and the biscalar field ϕ dominates at early times yielding $R \sim t^{1/3}$ and $H \sim 1/3t$.

This is not an inflationary solution, and so would seem to not solve the horizon and flatness problems. We are working in the comoving gravitational metric frame and the speed of light propagation is much larger than the speed of gravitational waves $c_g = c$. The coordinate distance that light would travel since the initial singularity is

$$d_H = c \int \frac{dt}{\sqrt{K}R} \propto \frac{1}{t^{1/3}}, \tag{21}$$

where we used $R \sim t^{1/3}$. This diverges as $t \to 0$, showing that there is no matter particle horizon in spacetime. This solves the horizon problem and correlates fluctuations 'born' superhorizon. The flatness problem can also be solved, because $\Omega_\phi \to 1$ as $t \to 0$ (see ref. [6]) where $\Omega_\phi = \dot{\phi}^2/12H^2$.

3. Derivation of Primordial Scalar Power Spectrum

In the comoving gravitational metric $g_{\mu\nu}$, the significant decrease in the radiation density ρ_r due to $K \to 0$ as $t \to 0$ gives a minimally-coupled Einstein-Klein-Gordon field, and so scalar mode fluctuations $\phi = \phi_0(t) + \delta\phi(t, \vec{x})$ about the background cosmological solution can be determined. Ignoring gravitational back-reaction, the equation for the perturbation of the biscalar field is

$$\frac{d^2\delta\phi_{\vec{k}}}{dt^2} + 3H\frac{d\delta\phi_{\vec{k}}}{dt} + \frac{c^2\vec{k}^2}{R^2}\delta\phi_{\vec{k}} = 0, \tag{22}$$

where $\delta\phi(\vec{x}, t) = (2\pi)^{-2/3} \int d^3k \exp[-i\vec{k}\cdot\vec{x}]\delta\phi_{\vec{k}}$. The solution for our spacetime is given by Bessel functions

$$\delta\phi_{\vec{k}} = A(\vec{k})J_0(\xi) + B(\vec{k})Y_0(\xi), \quad \xi = \frac{ck}{R_{pt}H_B}\left(\frac{R}{R_{pt}}\right)^2. \tag{23}$$

Since $\xi > 1$ corresponds to the modes passing back inside the horizon at the present time, we see that in the early universe modes of interest satisfy $\xi \ll 1$. Thus, we cannot assume a scenario where the modes are 'born' in the quantum vacuum in the early universe. *Instead, we assume that the fluctuation modes are born 'superhorizon' in a ground state.*

Hollands and Wald [7] assume that modes are 'born' or 'emerge' from a fundamental description of spacetime, rather than prior to some time at scales smaller than a length scale ℓ_0. When the modes pass through the constant sound horizon: $c_\gamma(t)/H(t) \sim \sqrt{B}\dot{\phi}/H(t) \sim$ const. ($c_\gamma(t) \sim 1/t$ and $H(t) \sim 1/t$). The modes are described by a normalized plane wave of the flat spacetime wave operator

$$\delta\phi_{\vec{k}}(t_k) = \sqrt{\frac{\kappa\hbar c^2}{(2\pi R_k)^3 2\omega_k}}\cos(\omega_k t_k - \vec{k}\cdot\vec{x} + \delta), \quad \omega_k = \frac{ck}{R_k}, \tag{24}$$

where $R_k = R(t_k)$ is the scale at which a mode is born.

The fluctuations born superhorizon *are causally correlated by the large speed of light or superluminal VSL mechanism in the early universe.*

If one assumes that one should use this as initial data, then we match not only the initial perturbation but also its time derivative. Doing so and keeping only the dominant contribution as $\xi \to 0$ gives

$$\delta\phi_{\vec{k}} \approx \sqrt{\frac{9\kappa\hbar c^2}{(2\pi R_k)^3 32\omega_k}}\cos(\omega_k t_k - \vec{k}\cdot\vec{x} + \delta)\ln(\xi_k)J_0(\xi), \tag{25}$$

where ξ_k represents ξ evaluated at $R = R_k$. The new contribution here results from the fact that when matching onto the initial state and its derivative the Bessel function $Y_0(z)$ is logarithmically divergent when $\xi \to 0$. The Hollands and Wald wave function did not have this additional logarithmic term, which will lead to slight deviations from a scale invariant spectrum.

We obtain the spectrum of scalar field fluctuations to be $\mathcal{P}_{\delta\phi} = \frac{9}{2(2\pi)^3}$ $\left(\frac{\ell_P}{\ell_0}\right)^2 \ln^2(\xi_k)$, and the curvature power spectrum is then found in the usual way (recall that $\dot{\phi}^2 = 12H^2$ in this spacetime):

$$\mathcal{P}_{\mathcal{R}} = \frac{H^2}{\dot{\phi}^2}\mathcal{P}_{\delta\phi} = \frac{3}{8(2\pi)^3}\left(\frac{\ell_P}{\ell_0}\right)^2 \ln^2(\xi_k), \tag{26}$$

where we have used the Hollands-Wald condition $R_k = k\ell_0$ and $\xi_k = \sqrt{12B}$ $\ell_0^2 k^3/2R_{pt}^3$.

From this the spectral index is calculated to be $n_s = 1 + \frac{d\ln\mathcal{P}_{\mathcal{R}}}{d\ln k} = 1 + \frac{6}{\ln(\xi_k)}$ and the running of the spectral index is calculated from: $\alpha_s = dn_s/d\ln k$, from which we find the relation $\alpha_s = -\frac{1}{6}(1-n_s)^2$.

In the large scale limit, we therefore have

$$\delta_H = \frac{2}{5}\sqrt{\mathcal{P}_{\mathcal{R}}} \approx \frac{2}{5}\sqrt{\frac{3}{8(2\pi)^3}}\left(\frac{\ell_P}{\ell_0}\right)|\ln(\xi_k)|. \tag{27}$$

4. Comparison with the Data

Assuming that $\ell_0 \approx \sqrt{12B}$ and evaluating the logarithm at the pivot point, $ck \sim 7R_0H_0$, this simplifies to $\delta_H \approx \ell_P/\ell_0 \approx 10^{-5}$, and so we can match the amplitude of the CMB fluctuations by assuming that $\ell_0 \sim \sqrt{12B} \approx 10^5\,\ell_P$. This choice predicts [1] $n_s \approx 0.97, \quad \alpha_s \approx -4.5 \times 10^{-4}$. These results agree well with the WMAP data $n_s = 0.99 \pm 0.04$ [8].

We can predict the acoustical waves power spectrum and fit it to the corresponding WMAP data that can be detected at the surface of last scattering. By adopting the ΛCDM data from WMAP and our prediction $n_s = 0.97$, we obtain a fit to the data as good as the fit obtained from an inflationary model using an appropriate slow-roll potential for the inflaton.

An important piece of data obtained from the WMAP observations is the cross polarization E-T temperature power. Since our primordial fluctuations are born 'superhorizon' and are causally correlated by the superluminal VSL mechanism early in the universe, we obtain a good fit to the cross correlated polarization data. Again the fit to the data is competetive with inflationary model predictions. This data confirms that a superluminal 'causal' mechanism, such as that provided by a generic inflationary model or the VSL bimetric gravity model, is necessary to fit the E-T polarization data obtained by WMAP.

5. Conclusions

We have shown that the VSL mechanism in the bimetric gravity model, choosing the gravitational metric as comoving, predicts a scale invariant power spectrum with a spectral index $n_s = 0.97$, which agrees well with the WMAP data. The mechanism for producing this result is quite different from standard inflationary models, in which a 'slow rolling' potential is chosen from a large possible number of potential models [9].

The basic postulate is that $c_\gamma(t)$ becomes very large in the early universe, solving the flatness and horizon problems. It is possible that the VSL mechanism can avoid the initial fine-tuning problems of generic inflationary models.

A possible observational test to distinguish the bimetric gravity model from standard slow-roll inflationary models is to observe the tensor mode (B-mode) and the ratio $r = T/S$ in the power spectrum. The predictions r in the VSL bimetric gravity theory and generic inflation models may be sufficiently different in magnitude to decide which model agrees better with observations. However, the pure gravitational tensor modes have not been detected by WMAP.

In the VSL metric frame in the bimetric gravity model, we are not required to have a large vacuum energy in the initial phase of the universe, so that no fine tuning is needed to obtain the tiny cosmological constant that is favored by the WMAP and supernovae data. This is in contrast to inflationary models which require a very large vacuum energy in the early universe to generate enough e-folds of inflation.

Acknowledgments

This research was supported by the Natural Sciences and Engineering Research Council of Canada.

References

1. M. A. Clayton and J. W. Moffat, Phys. Let. **B477**, 269 (2000), gr-qc/9910112.
2. M. A. Clayton and J. W. Moffat, Phys. Let. **B506**, 177 (2001), gr-qc/0101126.
3. M. A. Clayton and J. W. Moffat, Int. J. Mod. Phys. **D12**, 697 (2003), astro-ph/0203164.
4. M. A. Clayton and J. W. Moffat, JCAP 0307 (2003) 004, gr-qc/0304058.
5. J. D. Bekenstein, Phys. Rev. **D48**, 3641 (1993), gr-qc/9211017; J. D. Bekenstein and R. H. Sanders, Astrophys. J. **429**, 480 (1994), astro-ph/9311062.
6. J. W. Moffat, Int. J. Mod. Phys. **D2**, 351 (1993), gr-qc/9211020; J. Magueijo, Rept. Prog. Phys. **66**, 2025 (2003), astro-ph/0305457.
7. S. Hollands and R. M. Wald, Gen. Rel. Grav. **34**, 2043 (2002), gr-qc/0205058; hep-th/0210001.
8. D. N. Spergel, et. al., Astrophys. J. Suppl. **148**, 175 (2003), astro-ph/0302209.
9. A. D. Linde, *Particle Physics and Inflationary Cosmology*, Contemporary Concepts in Physics. Harwood Academic Publishers, Chur, Switzerland, 1990.

COSMIC NEUTRINO PHYSICS AND ASTROPHYSICS*

THOMAS J. WEILER
Department of Physics & Astronomy, Vanderbilt University,
Nashville, TN 37235†
E-mail: tom.weiler@vanderbilt.edu

Over the next decade or two, neutrino telescopes will map out the neutrino sky, analogous to the way the electromagnetic sky has been mapped for centuries. Like light and unlike cosmic-rays, the neutrinos will point back to their sources. Unlike light, the neutrinos are not attenuated at high energies and so will allow us to see farther into space, and deeper into sources. We illustrate with specific examples the promise which neutrino astronomy at energies from a TeV to a ZeV holds to study astrophysics and particle physics.

1. Introduction to Cosmic-Rays and Cosmic Neutrinos

The highest energy ever recorded for a cosmic-ray (CR) event is 3×10^{20} eV. This record was attained in 1991 by the Fly's Eye collaboration. The energy is truly macroscopic, as 3×10^{20} eV equals 50 Joules, about the energy that one man-power can achieve with a macroscopic object such as a baseball, golf ball, or a bullet. So then, what makes Nature's acceleration of CRs unusual? It is the fact that whereas a human partitions the energy over an Avogadro number of nuclei, Nature accelerates even a single proton to this energy. Nature is $\gtrsim 10^{24}$ times more efficient.

At Fermilab, the highest-energy accelerator in the world at present boosts protons to a TeV. A few years from now, the LHC at CERN will accelerate protons to 7 TeV. Nature apparently has "Zevatrons" which accelerate particles eight orders of magnitude beyond what our terrestrial accelerators provide, to order (ZeV=10^{21} eV).

In EECR physics, size certainly matters. The observed CR spectrum falls as roughly the third power of energy, so that at 10^{20}eV the incident flux

*This work is supported by DOE, NASA, CFCP, KIPAC, and Vanderbilt University's Discovery and sabbatical programs.
†present address: Theory Division, CERN,CH-1211, Geneva 23, Switzerland

at Earth is just one particle per square kilometer per century. As AGASA and HiRes are larger than Fly's Eye, so is Auger larger than them. An even larger experiment is proposed. The Extreme Universe Space Observatory (EUSO) proposes to put a 1.5m Fresnel lens on the International Space Station (ISS), to image the fluorescent signal from showers 400 km below in the Earth's atmosphere. Within a 60° viewing angle, the EUSO lens watches a "cloud chamber" with a truly enormous area, $\sim \pi\,(400/\sqrt{3}\,\mathrm{km})^2 \sim 10^5\,\mathrm{km}^2$. Multiplying this field-of-view (FOV) by the vertical column density of the atmosphere, which is 10 meters water equivalent (we), one finds a target volume of $10^3\,\mathrm{km}^3$ we. At a ton per m^3, the EUSO target volume is a teraton! EUSO's launch would be in 2009, with the three-year mission ending in 2012, exactly 100 years after Hess' discovery of the cosmic origin of CRs.

So what has this got to do with neutrinos? First of all, a large detector like Auger or EUSO can view neutrino-initiated events in addition to nucleon-initiated events. Even at EE, the νN cross-section is roughly six orders of magnitude smaller than the hadronic cross-section, so clearly, huge-area detectors are required. To spin it more positively, if the differential (in energy) neutrino flux were the same as the CR flux with its E^{-3} spectrum, then the smaller probability for neutrino interaction is compensated by moving two orders of magnitude lower in energy. The small neutrino cross-section leads to the neutrino signature: the atmosphere is nearly transparent to neutrinos, and so incident neutrinos will interact low in the atmosphere where the target density is greatest. Hadronic primaries with their large cross-section interact high in the atmosphere.

The Zevatrons which accelerate protons to 10^{20}eV or higher, whatever they may be are probably also producing pions at a similar energy. Pions result easily from $p\gamma$ and pp interactions. They typically carry 20% of the parent energy. The charged pions decay to three neutrinos and an $e^{\pm}$. Each final quantum carries away a quarter of the pion energy. Thus, neutrinos with about 5% of the parent energy are expected. E_{max} for the proton spectrum is unknown (and perhaps unknowable, given the "GZK absorption" on the CMB at EE). A measurement of $E_\nu^{\mathrm{max}} \sim 0.05 \times E_{\mathrm{max}}$ for neutrinos could help determine this. Of course, there may also be cosmic neutrino sources beyond the $p \to \pi \to \nu$ chain.

There are some advantages which neutrino astrophysics holds over CR astrophysics, and the small neutrino cross-section is actually one of them. For example, the cosmic ambiance in which the neutrinos are created is probably transparent to neutrinos. Thus, whereas nucleons may well suffer scatterings and energy losses as they emerge from their cosmic engines, neutrinos are expected to emerge with their full energy intact, and to carry with them unique information about the central engines (accreting black holes being the popular choice). Also, a small neutrino cross-section means that cosmic neutrinos are not absorbed by the intervening molasses, in contrast to cosmic nucleons and photons. Above $E_{\rm GZK}$ is a background-free playground where only neutrino primaries are allowed to play.

Another advantage derives from the zero charge of the neutrino. CRs, being charged protons and nuclei, are bent by intervening magnetic fields including our Galactic field. Thus they do not point back to their sources. On the other hand, neutrinos, being neutral, are unbent and so point to their sources. Neutrino hot-spots should be found on the sky. There is some evidence for this already in the AGASA data set, but its significance must await the truly staggering cluster-statistics projected for the Auger and especially EUSO data sets.

In addition to the large-area CR detectors which may "moonlight" as neutrino observatories, there are large-area dedicated neutrino detectors operating (Baikal, AMANDA, RICE), and even larger ones under construction or proposed (ANTARES, NESTOR, IceCube, ANITA, ASHRA, NEMO, SALSA). Each of these experiments exploits a sub-surface medium of water, ice, or salt (domes), transparent to light or radio, to detect a Cerenkov signal. A penetrating neutrino interacts to produce a hadronic jet and/or a muon, which then emits the Cerenkov signal. The signal from a hadronic jet is different from the signal from a muon. All neutrino-nucleon interactions produce a hadronic jet. Only the ν_μ charged current (CC) reaction produces a muon. Thus, an experiment which can independently measure jets and muons can provide information on neutrino flavor physics, via the $\nu_\mu/\nu_{\rm any}$ ratio. IceCube seems capable of measuring both signals.

2. Cosmic Neutrino Flavor Physics

At production in the cosmic sources, let the ratios of the neutrino flavors e, μ, and τ, be written as $w_e : w_\mu : w_\tau$, with $\sum w_\beta = 1$. The relative fluxes of each mass eigenstate ($j = 1, 2, 3$) are then given by $w_j = \sum_\alpha w_\alpha |U_{\alpha j}|^2$. The $U_{\alpha j}$ are elements of the neutrino mass-to-flavor mixing matrix, defined by $|\nu_\alpha\rangle = \sum_j U^*_{\alpha j} |\nu_j\rangle$, fundamental to particle physics. The propagating mass eigenstates acquire relative phases, giving rise to flavor oscillations. However, the phases, $\sim \delta m^2\, L/2\,E$ with δm^2 being the neutrino mass-squared difference, will be erased by the uncertainties in distance L and energy E, which lead to $\delta m^2\, L/E \gg 1$. Thus the neutrinos arriving at

Earth are an incoherent mixture of mass eigenstates with the proportions given above. Consequently, a measurement at Earth of the cosmic neutrino flavor will return

$$P_{\nu_\alpha \text{ detected}} = \sum_j |U_{\alpha j}|^2 \sum_\beta w_\beta \, |U_{\beta j}|^2 \,. \tag{1}$$

For three neutrino species, as we assume in this talk, there is now strong evidence from atmospheric and reactor data suggesting that ν_μ and ν_τ are maximally mixed and U_{e3} is nearly zero. This twin happenstance, $\theta_{\rm atm} = 45°$ and $\Re(U_{e3}) = 0$, is sufficient to generate "ν_μ-ν_τ interchange symmetry." One consequence of this remarkable symmetry is that in the process of decohereing, any initial flavor ratio having $w_e = \frac{1}{3}$ will arrive at Earth with democratic ratios of $1:1:1$. Since neutrinos from astrophysical sources are expected to arise dominantly from the decays of charged pions (and kaons) and their muon daughters, their initial flavor ratios of nearly $1:2:0$ should arrive at Earth democratically distributed. So there is a fairly robust prediction of $1:1:1$ flavor ratios for measurements of astrophysical neutrinos.

The ν_μ-ν_τ symmetry ensures that ν_μ and ν_τ arrive at Earth with equal probability. Thus there is a but a single independent flavor-ratio to be determined, which we may take to be $\phi_e : \phi_\mu$. This ratio can be inferred from the ratio of measured rates of shower and muon-track events, identifiable and distinguishable in IceCube. It has been shown that for a "typical" neutrino spectrum of $E^2_{\nu_\mu} dN_{\nu_\mu}/dE_{\nu_\mu} = 10^{-7}$ GeV cm^{-2} s^{-1} for one year, IceCube could determine the ratio of the muon events to shower events with uncertainties on the order of 20% [1]. Fortunately, in some interesting theoretical scenarios, the predictions for deviations from a 1:1:1 flavor ratio are so extreme as to not require greater precision.

In Fig. 1, we show the relationship between the muon-to-shower ratio and ν_e fraction, assuming ν_μ-ν_τ symmetry and an E^{-2} power law spectrum. Note that the ν_e fraction of 1/3, expected from known particle physics, can be measured to a range of 0.26-0.37 or 0.18-0.44 for the two fluxes shown in Fig. 1, sufficient to test interesting speculations, to which we now turn.

Could the flavor ratios at Earth differ significantly from $1:1:1$? The astrophysics could be different than is outlined here. For example, if the charged pion decays promptly but the daughter muon loses energy before decaying, then at high energy the flux may approximate $\sim 0:1:0$. This leads to mass-eigenstate ratios of $\sim 1:2:3$, and Earthly ratios of $\sim 1:2:2$. Another possibility, mentioned in the introduction, is that there

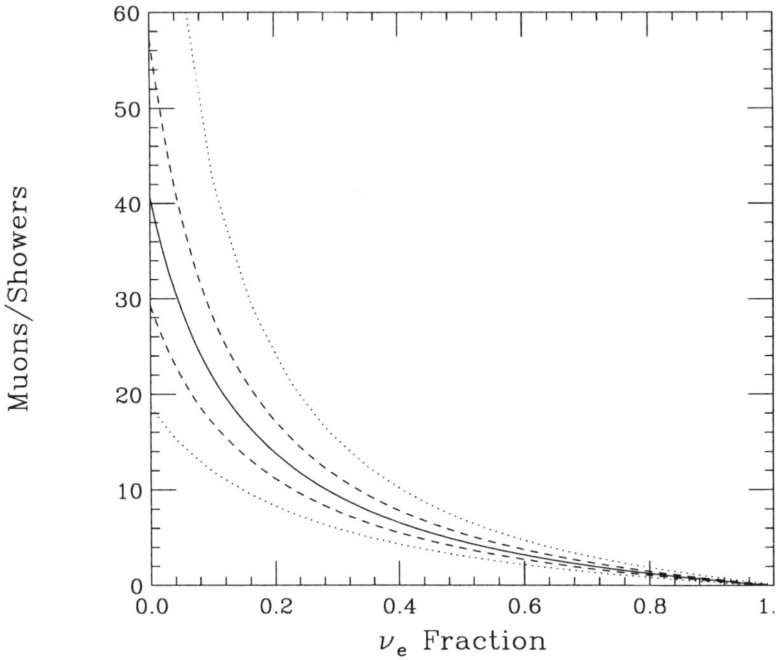

Figure 1. The relationship between the muon-to-shower ratio and the ν_e fraction, assuming ν_μ-ν_τ symmetry and an E^{-2} power law spectrum. The muon energy threshold is 100 GeV, and the shower threshold is 1 TeV. Horizontal sources are assumed. The solid line is the central predicted value. The dashed (dotted) lines represent the 68% confidence interval for a spectrum of $E^2_{\nu_\mu} dN_{\nu_\mu}/dE_{\nu_\mu} = 10^{-7}$ GeV cm^{-2} s^{-1} for one year (for an exposure five times smaller).

may be an astrophysical flux of energetic cosmic neutrons, which decay to pure $\bar{\nu}_e$, giving an initial ratio of $1:0:0$. This leads to mass-eigenstate ratios of $\sim 7:3:0$, and Earthly ratios of $\sim 3:1:1$.

Alternatively, the particle physics could be different than assumed here (three stable neutrinos). We are aware of three mechanisms by which flavor ratios for neutrinos from pion decay could differ substantially from the 1:1:1 expected in the Standard Model (SM). They are (i) neutrino decay; (ii) new oscillations of cosmic-wavelength – especially motivated are mass matrices leading to pseudo-Dirac neutrinos [2], or exhibiting mirror symmetry; and (iii) new flavor-resonances, as might result e.g., from a preferred basis other than the flavor- or mass-bases (spontaneously-broken Lorentz Invariance provides such a basis), or from time- or density-varying neutrino masses [3]. Here we will discuss only the simplest possibility, neutrino decay [4].

2.1. *Neutrino Decay Signatures*

In the SM of particle physics, all fermions (spin $\frac{1}{2}$ particles) come in triplicate, labeled by a "generation" index. The heavier two generations decay down to the lightest one, except possibly for the neutrinos. Radiative two-body decay modes of the neutrino are constrained by photon appearance searches to have very long lifetimes, and three-body decays are strongly constrained by bounds on anomalous $Z\nu\bar{\nu}$ couplings. In contrast, the limits (set by solar neutrino data) on the decay modes $\nu_i \to \nu_j + X$ and $\nu_i \to \overline{\nu}_j + X$, where X denotes a very light or massless particle, e.g. a Majoron, are too weak to eliminate the possibility of astrophysical neutrino decay by a factor of about $10^7 \times (L/100\text{ Mpc}) \times (10\text{ TeV}/E)$. The dependence of the inferred lifetime on L/E is a consequence of Special Relativity: the lab-frame survival-probability of the neutrino is $\exp(-L/\tau_{\text{lab}}) = \exp(-Lm_j/E\tau_j)$, where τ_j is the rest-frame lifetime of ν_j. The challenge of using cosmic baselines to measuring neutrino lifetimes [5] is worth the effort, for a discovery would mean physics much more exotic than neutrino mass and mixing alone.

The most natural possibility with unstable neutrinos is that the heaviest two mass eigenstates both completely decay to the lightest (plus a Majoron). In this case,[6] the resulting relative flavor flux at Earth is just $|U_{\alpha 1}|^2$, the flavor projection of the sole surviving mass-eigenstate. This simple result is independent of neutrino energy, dynamics at the source, or decay-chain details. Recently, it has become clear that the particular values and small uncertainties on the neutrino mixing parameters lead for the first time to very distinctive predictions of flavor-ratio signatures for neutrino decay [4]. A consequence of the ν_μ-ν_τ symmetry is that each mass eigenstate contains an equal fraction of ν_μ and ν_τ. Thus, unitarity and $|U_{\mu 1}|^2 = |U_{\tau 1}|^2$ lead to Earthly ratios of $2|U_{e1}|^2 : (1 - |U_{e1}|^2) : (1 - |U_{e1}|^2)$. Note how different this result is from that for three stable neutrinos. Inputting the measured value of $|U_{e1}|^2$, one finds $\phi_e : \phi_\mu : \phi_\tau = 6 : 1 : 1$.

So far, we have assumed a "Normal Hierarchy," where the two mass-states contributing to solar oscillations are lighter than ν_3. Another possibility is the "Inverted Hierarchy,' where the two solar states are heavier than ν_3. In this case, the surviving neutrino is ν_3, and the flavor ratio at Earth is $2\,|U_{e3}|^2 : (1 - |U_{e3}|^2) : (1 - |U_{e3}|^2) \approx 0 : 1 : 1$.

Interestingly, both the normal and inverted spectra have extreme $\phi_{\nu_e} : \phi_{\nu_\mu}$ ratios, $6 : 1$ and $0 : 1$ respectively. Either provides a very useful diagnostic. Assuming no new physics besides decay, a ratio greater than 1 suggests the normal hierarchy, while a ratio smaller than 1 suggests an inverted hierarchy. Moreover, these ratios are so dramatic that decay cannot be mimicked by different neutrino flavor ratios at the source or by other non-standard neutrino interactions [4]. These results, and a comparison with ratios from the possible but unexpected $0:1:0$ initial flavor ratio, are summarized in Table (1).

Table 1. Summary of the muons/showers ratios expected for selected scenarios [4]

ratios at source $\phi_e:\phi_\mu:\phi_\tau$	decays	ratios at Earth $\phi_e:\phi_\mu:\phi_\tau$	muons/showers from Fig. 1 central value
$1:2:0$	none	$1:1:1$	9
	Normal	$6:1:1$	1.5
	Inverted	$0:1:1$	40
$0:1:0$	none	$1:2:2$	14

It is unlikely that the ν_μ-ν_τ symmetry is exact. However, variation of the mixing angles θ_{23} and θ_{13} about the symmetry limit 45° and 0, respectively, leads to only small ($\lesssim 20\%$) deviations, *when the CP-violating phase δ is ignored.* In contrast, we find that the flavor ratios are very sensitive to the CP phase δ, when it is included and allowed to vary [7]. This is shown in Fig. (2), for the normal hierarchy decay to ν_1. As the value expected in the no-decay case is 1, the effects of decay are always pronounced. On the other hand, for the case of an inverted hierarchy (or no decay), varying θ_{13} has little effect. Determination of the phase δ is of crucial importance in terrestrial long-baseline oscillation experiments, since CP- and T-violating observables in neutrino oscillations are proportional to $\sin\delta$. Since measurement of δ in terrestrial experiments will be an extremely challenging task, it is intriguing to find an example in astrophysics where the effect of varying δ is huge.

3. "Almost Guaranteed" Neutrino Fluxes

The predicted cosmogenic or GZK neutrino flux is often termed "almost guaranteed." It peaks at and above an EeV, and extends to $\sim$ ZeV. The reaction of origin is $N + \gamma_{\rm CMB} \to N' + \pi + \cdots$, where the intermediate state is dominated by the Δ resonance. For nucleon sources at $\gtrsim 100$ Mpc, essentially all of the nucleon energy above $E_{\rm GZK}$ goes into pion production, via repeated GZK-scattering. Two-thirds of the pions are charged. Each charged pion decays to three neutrinos and an $e^\pm$. The neutrinos carry

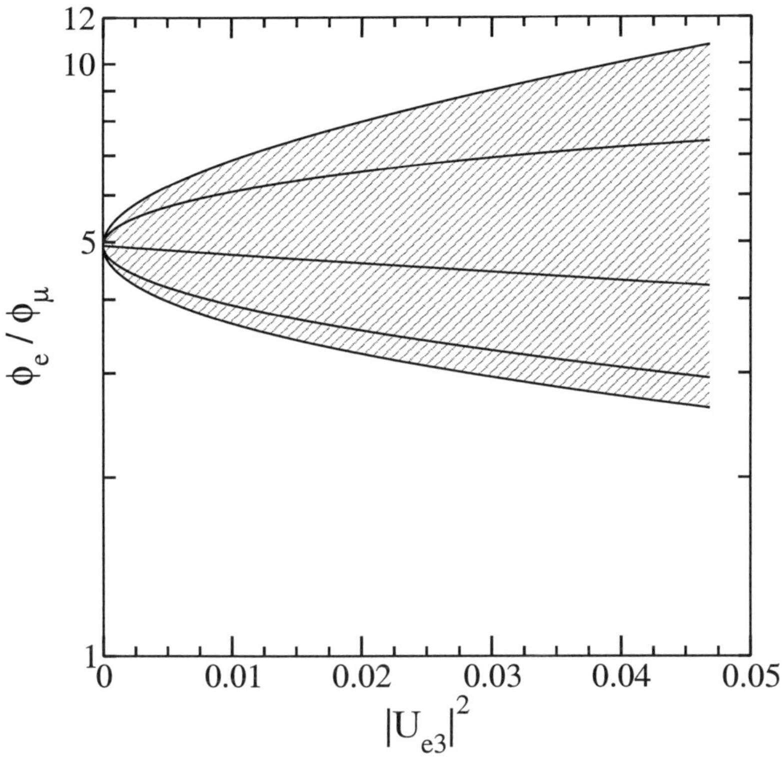

Figure 2. Variation of the ratio ϕ_e/ϕ_μ with $|U_{e3}|^2 = \sin^2\theta_{13}$ in the allowed range. From bottom to top, the solid curves correspond to $\delta = (0, \pi/4, \pi/2, 3\pi/4, \pi)$, with the hatched region showing the full allowed range. The atmospheric angle is $\theta_{23} = 45°$ and the solar angle is $\theta_{12} = 32.5°$.

away 3/4 of the $\pi^{\pm}$ energy. The normalization and spectrum of the GZK neutrinos depends on two features of the original EE nucleon spectrum above the observed end-point: the spectral index and the maximum energy $E_{\rm max}$. The almost guaranteed cosmogenic neutrino flux is shown in Fig. (3) twice. One representation is the minimal one, with a spectrum peaking at an EeV. It is perhaps accessible to Auger, probably to ANITA, and certainly to SALSA. The other representation is the most optimistic one, peaking at 10^{20}eV. Besides the mentioned experiments, it is easily accessible to EUSO.

Also shown in Fig. (3) are present and projected neutrino flux limits at EE, and a sample of predicted EE neutrino fluxes. I don't have the space to explain either here, so I refer the reader to Eberle et al. [8] for details. What is obvious, however, is that in a few years time, the EE neutrino observatories will either discover an EE neutrino flux, or rule out many models. Already the GLUE and FORTE experiments are impacting the viability of the Z-burst mechanism [9] of EE events.

Here I present a new claim [10] for a second "almost guaranteed" cosmic neutrino flux. This one is of galactic origin, and peaks near a PeV.

The AGASA experiment has revealed a 4% excess of CR events in the Galactic Plane (GP) [11], in the directions of the Cygnus region (along our spiral arm) and toward the Galactic Center (GC) [12]. The significance is 4σ. An intriguing feature of the excess is that it exists only within a decade of energy centered about an EeV. Fly's Eye and SUGAR data support this discovery. The data is remarkable in (at least) three ways:
(i) somehow, even in our own relatively quiescent Galaxy, acceleration of CRs to $\gtrsim$ EeV is happening;
(ii) the primaries are likely not protons, because charged particles are bent by Galactic magnetism;
(iii) the excess is observed only in the energy decade centered on an EeV, not below and not above.

The HEGRA experiment has detected an extended TeV γ-ray source in the Cygnus region. Especially intriguing is the possible association of this source with Cygnus-OB2, a cluster of more than 2700 identified, young, hot OB-stars with a total mass of $\sim 10^4$ solar masses. The high-density of young massive stars in Cygnus-OB2 could provide nearly-simultaneous supernovae, leading to overlapping stellar winds or SNR shocks which co-operatively accelerate CRs beyond $E_{\rm max} \sim 1\,$PeV of individual remnants to the EeV energy of the observed excess.

Another clue as to origin is evidence for a dominant Fe component in the flux [13]. A consequence is that one may expect a flux of free neutrons via nuclei photo-disintegration on background photon fields at the source. Liberated neutrons would naturally explain the observed directional signals. First of all, neutrons are not bent by Galactic magnetic fields. Secondly, the decay mfp of a neutron is $c\,\Gamma_n\,\overline{\tau}_n = 10\,(E_{\rm n}/{\rm EeV})$ kpc, the lifetime being boosted from its rest-frame value $\overline{\tau}_n = 886$ seconds to its lab value via $\Gamma_n = E_n/m_n$. Thus, neutrons with energy much below an EeV would decay before reaching Earth from a distance $D \sim$ several kpc, of order of the distance to Galactic objects and the Galactic Center. Neutrons with energy much above an EeV are presumably scarce due to the falling source spectrum.

For every surviving neutron at $\sim$ EeV, there are many neutrons at lower energy that decay to $p + e + \bar{\nu}_e$. Collectively, these decays produce

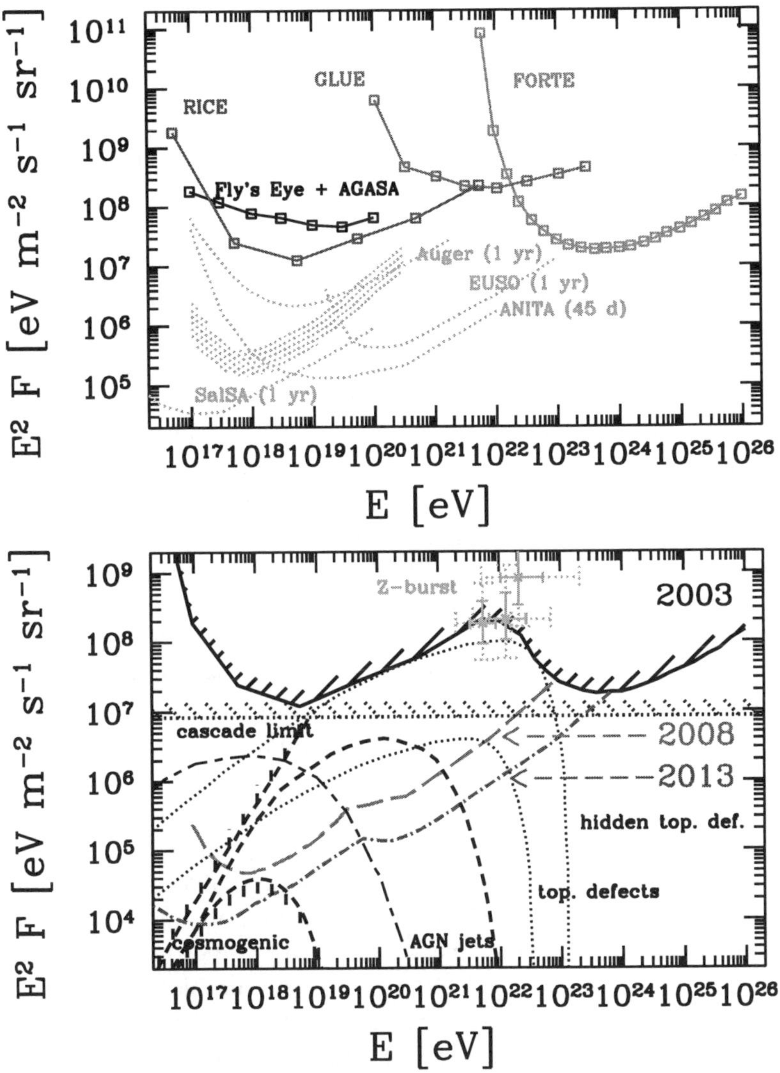

Figure 3. Current status and next decade prospects for EE ν physics, expressed in terms of diffuse neutrino fluxes per flavor, from Eberle et al. [8]

Top: Upper limits from various experiments, and projected sensitivities in the near future corresponding to one event per energy decade and indicated duration.

Bottom: Road-map for improvement by the years 2008 and 2013, under reasonable assumptions. Also shown is a wide sample of predictions for EE ν fluxes.

a directed beam of "almost guaranteed" $\bar{\nu}$'s pointing back to the source. The typical energy for the $\bar{\nu}$ in the lab is that of the parent neutron times $Q/m_n \sim 10^{-3}$ (the Q-value for β-decay is $m_n - m_p - m_e = 0.78$ MeV). Thus, the stability of neutrons at $\gtrsim$ EeV implies a PeV upper limit for the produced antineutrinos. The increasing abundance of neutrons below an EeV in turn implies an increasing neutrino flux as energies move below a PeV. The calculation of this neutrino flux is very straightforward, and normalized to the observed CR excess.

One interesting feature is that, even though the initial neutrinos are pure $\bar{\nu}_e$, the arriving antineutrinos are distributed over flavors. This results form neutrino oscillation and decoherence over cosmic distance, as discussed earlier. The muon antineutrino flux $F_{\bar{\nu}_\mu}$ is given by the factor $\frac{1}{4}\sin^2(2\,\theta_\odot) \simeq 0.20$ times the original $F_{\bar{\nu}_e}$ flux. The $\bar{\nu}_\tau$ flux is the same, and the $\bar{\nu}_e$ flux is 0.60 times the original flux. Thus, no matter what flavor a neutrino detector may be sensitive to, the detector is presented with a benchmark signal from our own Galaxy. Moreover, these predicted flavor ratios differ considerably from the $1:1:1$ expected from the pion chain for neutrino production, and therein offer another tst of the model.

The GZK neutrinos and this β-decay Galactic $\bar{\nu}$ flux probably constitute the best motivated cosmic neutrino fluxes. Of these two neutrino fluxes, the expected event rate for the Galactic beam is higher at IceCube: 4 $\bar{\nu}_\mu$ events per year and 16 in $\bar{\nu}_e + \bar{\nu}_\tau$ showers. The background from the atmosphere is only 1.5 events per year. The predicted signal is comfortably above 95% CL after just one year. After a few years, the 5σ discovery criterion is reached, providing the "smoking" ice for the GP neutron hypothesis. Additional confirmation of the neutron origin may soon be available directly from Auger: Auger South should see the cosmic ray excess in the direction of the GC and Auger North should be sensitive to the Cygnus region. I addition, the neutrino flux from the GC source is measurable in a km–scale Mediterranean detector, if designed with sufficiently low threshold.

The explanation of the Galactic excess presented here depends on dilation of the neutron lifetime, due to standard special relativity (SR). Consequently, a byproduct of confirmation of the neutron origin of the excess would be validation of SR to an unprecedented boost factor of $\Gamma \approx 10^9$, many orders of magnitude beyond present limits. Additional confirmation of the neutron origin may soon be available directly from Auger: Auger South should see the cosmic ray excess in the direction of the GC and Auger North should be sensitive to the Cygnus region.

4. Conclusions

EECR physics strongly suggests the existence of extragalactic cosmic neutrinos. The GZK photopion mechanism provides the likely link. In addition, our own Galactic Plane is a surprising accelerator of primaries most likely including neutrons, to EeV energies. The β-decay anti-neutrinos from these neutrons should be observable in IceCube. With the growth of neutrino observatories, even flavor identification of cosmic neutrinos on a statistical basis becomes possible, opening up a window for discoveries in particle physics not otherwise accessible to experiment. The example presented here is a vastly improved search for neutrino decay. The future of astronomy looks bright in neutrinos.

Acknowledgments

I thank my collaborators on the materials presented here. They are L. Anchordoqui, J. Beacom, N. Bell, B. Eberle, H. Goldberg, F. Halzen, D. Hooper, J. Learned, S. Pakvasa, A. Ringwald, and L. Song.

References

1. J. Beacom, N. Bell, D. Hooper, S. Pakvasa, and T. Weiler, Phys. Rev. D68 (2003) 093005 [arXiv:hep-ph/0307025].
2. J. Beacom, N. Bell, D. Hooper, J. Learned, S. Pakvasa, and T. Weiler, Phys. Rev. Lett. 92 (2004) 011101 [arXiv:hep-ph/0307151].
3. P. Q. Hung and H. Päs, astro-ph/0311131.
4. J. Beacom, N. Bell, D. Hooper, S. Pakvasa, and T. Weiler, Phys. Rev. Lett. 90 (2003) 181301 [arXiv:hep-ph/0211305]; J. Beacom and N. Bell, Phys. Rev. D65 (2002) 113009.
5. T. Weiler, J, Learned, W. Simmons, and S. Pakvasa, arXiv:hep-ph/9411432.
6. Y. Farzan and A. Y. Smirnov, Phys. Rev. D65 (2002) 113001.
7. J. Beacom, N. Bell, D. Hooper, S. Pakvasa, and T. Weiler, Phys. Rev. D69 (2004) 017303 .
8. B, Eberle, A. Ringwald, L. Song, and T. Weiler, arXiv:hep-ph/0401203.
9. T. J. Weiler, Astropart. Phys. 11 (1999) 303-316 [arXiv: hep-ph/9710431]; D.Fargion, B.Mele, and A.Salis, Astrophys. J. 517 (1999) 725.
10. L. Anchordoqui, H. Goldberg, F. Halzen, and T. J. Weiler, arXiv:astro-ph/0311002.
11. N. Hayashida *et al.* [AGASA Collaboration], Astropart. Phys. **10**, 303 (1999).
12. M. Teshima *et al.*, Proc. 27th International Cosmic Ray Conference, (Copernicus Gesellschaft, 2001) p.341.
13. See *e.g.,* M. T. Dova, M. E. Mancenido, A. G. Mariazzi, T. P. Mc Cauley, and A. A. Watson, astro-ph/0305351.

3 Lepton Experiments

THE STATUS OF MINOS

T. BERGFELD, FOR THE MINOS COLLABORATION
University of South Carolina
Columbia, SC 29208

This paper gives an introduction to the MINOS experiment, discusses the ongoing construction of the NUMI beam line, and MINOS detectors and gives an update on their current status. The physics goals and expected sensitives are given for a nominal running period.

1. Introduction

In the past few years, the evidence for neutrino oscillations continues to mount in both the atmospheric and solar mass ranges[1,2,3]. The MINOS experiment is designed to provide detailed measurements regarding neutrinos oscillations significantly improving our understanding of this phenomenon. These measurements will be done to high precision using a well-controlled neutrino beam from the NUMI Beamline at Fermilab. The data will come from two nearly identical detectors located to intercept the neutrino beam. The near detector, located 1km downstream from the hadron target, is used to control the systematics and understanding the properties of the beam. The far detector is located 735km downstream in a new cavern in the Soudan Mine in northern Minnesota and provides the oscillation measurement.

2. The NUMI Beamline

The NUMI Beamline transports 120 GeV/c protons from the Fermilab Main Injector and focuses them on a 1m long thick carbon target. The beam is designed to have an intensity of 4×10^{13} protons every 1.9 seconds[4], although the initial intensity is expected to be less. These protons interact with the carbon target producing secondary particles, mainly pions, which are focused with two magnetic horns into a 675m long 2m diameter decay pipe where they decay into neutrinos. At the end of the decay pipe, the resulting hadrons and muons are monitored to ensure an accurate understanding of the beamline before they are absorbed by rock leaving only the neutrinos continuing on in the beam.

One advantage of the current two horn design is the ability to adjust the energy spectrum of the resulting neutrinos. The two horns can be repositioned relative to the target and thereby give a different energy spectrum to the neutrino beam. Given the result from Super-Kamiokande[1] for a Δm^2 around $2.5 \times 10^{-3} eV^2$, the initial configuration of the NUMI Beamline will be for a low energy neutrino beam, which peaks with a neutrino energy around a few GeV. The expected event rates vs. neutrino energy for this configuration is shown in Figure 1 along with the event rates for oscillations. In the low energy configuration, the MINOS far detector should observe about 5000 muon neutrino charged current events in two years with the design intensity assuming no oscillations.

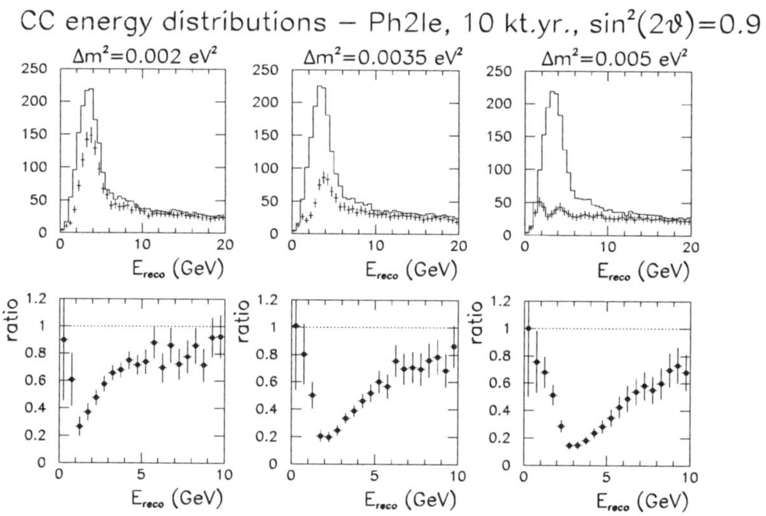

Figure 1. The neutrino energy spectrum for ν_μ charged current events in the MINOS far detector. The solid line shows the result with no oscillations and the points are assuming oscillations with $\sin^2 2\theta = 0.9$ and the given Δm^2. The lower plots show the ratio of oscillations to no oscillations.

The NUMI beam-line construction is currently taking place at Fermilab. The out-fitting of the tunnel and experimental halls was completed in March of 2004. The installation for the beam devices, target and focusing system began at that time. We expect the beam to turn on at the end of 2004 with the first physics data in April of 2005.

3. The Far Detector

The MINOS far detector[5] is a steel and scintillator tracking calorimeter. It consists of 484 planes of 2.54cm thick 8m diameter octagonal steel and a layer of 4cm wide strips of scintillator for a total mass of 5.4 kTons. Each plane records the position of particles traveling through the detector in one view. The two adjacent planes are oriented such that the scintillator strips are rotated 90°, giving us the ability to track particles in three dimensions as they travel along the length of the detector. The light from the scintillator is collected through wavelength shifting fibers running the length of the strip and then into a 16-pixel photo-multiplier tube at each end. To reduce the number of channels used, the light is optically multiplexed such that the light from 8 strips illuminate one pixel in the PMTs. Using the arrival time of the light we obtain a 2.6 ns/plane timing on tracks though out detector allowing us to determine the direction of the particle. Since a large background to our experiment comes from downward going cosmic muons, a shield consisting of two layers of scintillator is placed over the top of the detector allowing us to tag the incoming going muons.

Our detector has a coil through the center to generate a magnetic field within the steel of the detector. The field is created such that muons from the beam neutrino interactions are focused inward allowing a better measurement of their energy. MINOS is thus the first underground detector which will be able to distinguish neutrino and anti-neutrino events. In addition, due to the magnetic field we are able to measure the momentum of escaping muons giving us the ability to have a complete neutrino energy measurement with partially contained events.

This detector was completed in the summer of 2003 and has been taking data since then having observed cosmic rays and a few neutrino events. These data have also been used to commission and calibrate the detector before the beam from Fermilab actually arrives.

4. The Near Detector

The near detector is placed about 1km downstream of the target and 100m underground. It is used to verify the energy spectrum of the neutrino beam so that the event rates at the far detector can be predicted with a lower systematic uncertainty.

The near detector is built in a similar way to the far detector, with 282 planes of steel and 153 planes of scintillator in a squashed 3.8m x 4.8m shape. The scintillator is again readout with the wavelength shifting fiber carrying the light out to a photomultiplier tube. However, due to higher rates in this detector, every fiber is read out by a single channel of a 64 pixel photomultiplier tube. The near detector consists of three sections, the calorimeter area and the spectrometer area. The calorimeter area is fully instrumented having scintillator on each plane. The spectrometer region only has scintillator on every fifth plane. Again the near detector is magnetized such that we have a magnetic field in the target region with the same average field as the the far detector.

The detector planes for this detector were built on the surface before the underground construction was completed and now have started to be brought underground with the completion of the civil construction.

5. The Calibration Detector

In addition to these two detectors in the neutrino beam, MINOS has another detector, the Calibration Detector. This detector has been placed into beamline at the CERN PS to study the response of our detectors to pions, protons, electrons, and muons with a known momenta between 0.5 GeV/c and 10 GeV/c. It consists of 60 planes of 1m x 1m and 2.54cm thick steel and the same 4cm wide scintillator and wavelength shifting fiber. This detector is readout using both the near and far detector electronics and photomultiplier tube configurations. The two configurations are shown to be in good agreement with this data.

Thus we are able to understand the calorimetry response from the hadronic and electromagnetic showers in our detectors. Additionally these data will be used to understand the topological features of the showers in our detector. The data from this detector were taken during runs between 2001 and 2003.

6. The Physics Goals

The physics goals of the MINOS experiment are to demonstrate the oscillation behavior of neutrinos and to provide a precise measurement of the oscillation parameters. One way this goal will be accomplished is by the measurement of the neutrino energy distribution for charged current events. Neutrino oscillations will reduce the observed event rate in specific energy regions corresponding to the Δm^2 by an amount proportional to $\sin^2 2\theta_{23}$. By taking a ratio of the neutrino energy distribution in the far detector with the expectations from no oscillations, we will observe the oscillations causing a dip in the number of observed ν_μ compared with expectations. Figure 2 shows the expected ratio of ν_μ charged current events both 7.4×10^{20} protons on target and 25×10^{20} protons on target assuming a Δm^2 of $2.0 \times 10^{-3} eV^2$ and $\sin^2 2\theta = 1.0$. The figure also shows fits to the Super-Kamiokande results for ν-decay[7] and decoherence[8] for comparison with the oscillation hypothesis. The expectation is to understand the expected event rates to hold the systematic errors in neutrino energy to 5% overall and to within 2% for relative variations between the detectors. This sensitivity will allow us to measure the oscillation parameters with error ellipses as shown on the right in Figure 2.

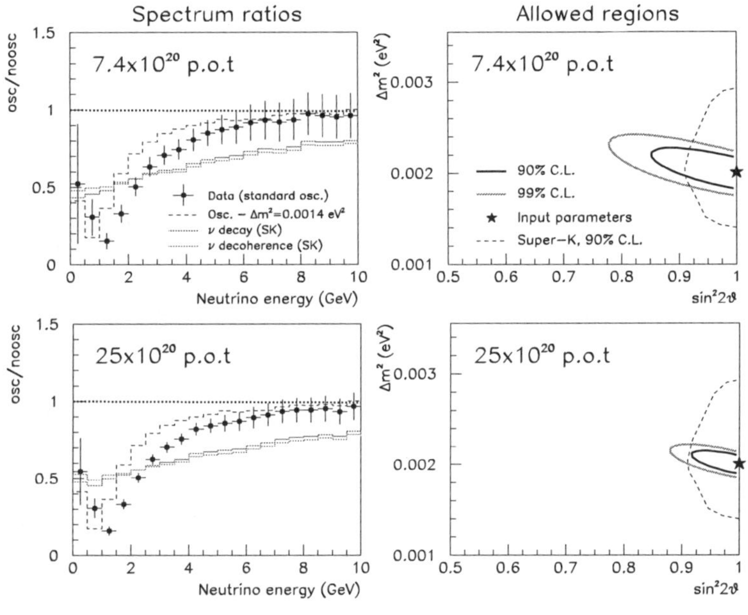

Figure 2. MINOS Oscillation Sensitivity. The plots on the left show the ratio of the number events expected in the MINOS far detector with oscillations compared with no oscillations for both 7.4×10^{20} protons on target and 25×10^{20} protons on target assuming a Δm^2 of $2.0 \times 10^{-3} eV^2$ and $\sin^2 2\theta = 1.0$. The right side shows the resulting error ellipses for the mixing parameters.

In addition to measuring the energy spectrum for charge current ν_μ events, MINOS has sensitivity to the appearance of electron neutrinos in our beam. Depending upon the parameters, MINOS could observe the ν_e appearance in our detector. If observation is not made, an improvement upon the CHOOZ limit[6] will be obtained. Figure 3 shows the expected limits of the MINOS experiment as well as the range of parameter space where observation is possible. Furthermore, MINOS will measure the neutral current event rates in both the near and the far detector giving us and independent measurement of the oscillation parameters. The neutral current measurement will also allowing us to distinguish between a $\nu_\mu - \nu_\tau$ oscillation or $\nu_\mu - \nu_s$ and provide a measurement of the $\nu_\mu - \nu_s$ oscillation fraction. Thus not only will MINOS measure the oscillation parameters but will also give the flavor composition of the mixing.

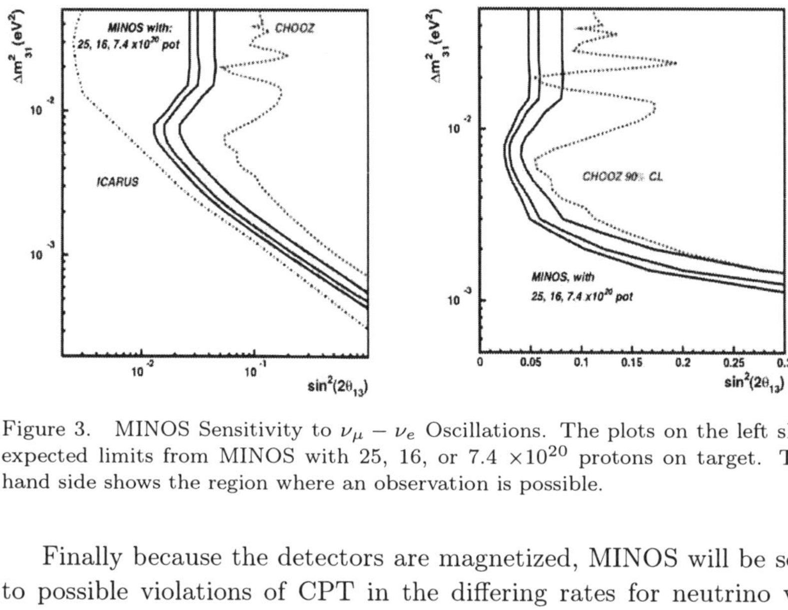

Figure 3. MINOS Sensitivity to $\nu_\mu - \nu_e$ Oscillations. The plots on the left shows the expected limits from MINOS with 25, 16, or 7.4 $\times 10^{20}$ protons on target. The right hand side shows the region where an observation is possible.

Finally because the detectors are magnetized, MINOS will be sensitive to possible violations of CPT in the differing rates for neutrino vs anti-neutrino oscillation behavior. The rate difference can be measured for both atmospheric and beam neutrinos.

The original running plan for the beam is to get 7.4×10^{20} protons on target with four years of MINOS running in the NUMI beamline. A recent request has been for five years of running and 25×10^{20} protons on target. The later request allows improved precision on the measurements made by MINOS.

7. Conclusion

The Far Detector is already taking data in the final configuration for atmospheric neutrinos and cosmic muons. The near detector was built on the surface and is now being transfered and installed underground along with the beamline components for the NUMI beamline. Thus, MINOS is well prepared to start taking data with the neutrino beam as scheduled at the end of 2004. This data will enable us to verify the oscillation hypothesis and accurately measure the parameters of neutrino mixing, Δm^2 and $\sin^2 2\theta_{23}$. Additionally, MINOS should allow us to verify the flavor content of oscillations and measure possible mixing into sterile neutrinos. Finally, the magnetized detectors will allow MINOS to test the behavior for neutrinos and anti-neutrinos independently providing a test for CPT invariance.

Acknowledgments

Funding for MINOS has been provided by the U.S. Department of Energy, Office of Science and by the U.K. Particle Physics and Astronomy Research Council.

References

1. Y. Fukuda *et al.* SuperKamiokande Collaboration, *Phys. Rev. Lett.* **82**, 2644 (1999); *Phys. Rev. Lett.* **85** 3999 (2000). M. Smy at this conference.
2. M. H. Ahn *et al. Phys. Rev. Lett.* **90**, 041801 (2003).
3. S. N. Ahmed *et al.* SNO Collaboration. arXiv:nucl-ex/0309004. J. Maneira at this conference.
4. J. Hylen *et al.*,FERMILAB-TM-2018, Sept. 1997.
5. MINOS Collaboration, "The MINOS Technical Design Report", (1998), Fermilab report NuMI-L-337.
6. M. Appollonio *et al.* CHOOZ Collaboration, *Phys. Lett.* **B466**, 415 (1999).
7. C. Yanagisawa, talk at NOON '03, Kanazawa, Japan.
8. G.L. Fogli, E. Lisi, A. Marrone, and D. Montanino *Phys. Rev.* **D67** 093006 (2003).

ICECUBE: THE CUBIC KILOMETER NEUTRINO TELESCOPE AT THE SOUTH POLE

A. R. FAZELY FOR THE ICECUBE COLLABORATION*

Department of Physics,
Southern University and A&M College,
Baton Rouge, Louisiana 70813
E-mail: fazely@phys.subr.edu

Search for ultra high-energy neutrino induced reactions, as part of a comprehensive probe of the neutrino sky and also investigation of the particle nature of the dark matter, with unique sensitivity to cold dark matter particles are described. We present a description of the design, scientific motivation and goals, performance and status of the IceCube experiment.

1. Introduction

The main motivation for the IceCube experiment [1] is to probe the universe with ultra high energy (UHE) neutrinos and to search for the signature of cold dark matter. The IceCube detector will provide astrophysical and particle physics information, essential to the understanding of the origin of the highest energy cosmic rays as well as a test of the fundamental laws of physics.

The all-particle spectrum, as shown in Figure 1, is dominated by two main features at 3 PeV and at 5 EeV commonly referred to as the "knee" and the "ankle". The spectrum shows a steep drop in the flux of cosmic rays as a function of energy. The slope becomes steeper at the knee and rises at the ankle. Many attempts have been made to explain the drop and increase in the flux at the knee and the ankle, respectively with diverse fortune (see for example Esteban Roulet [2] and references therein).

In the UHE region of the knee, the particle flux drops to $1/m^2-year$ and falls to $1/km^2 - year$ at the ankle. These fluxes are impossible to observe by conventional detectors. The detection of these particles with reasonable

*Full author list is given at the end of the paper

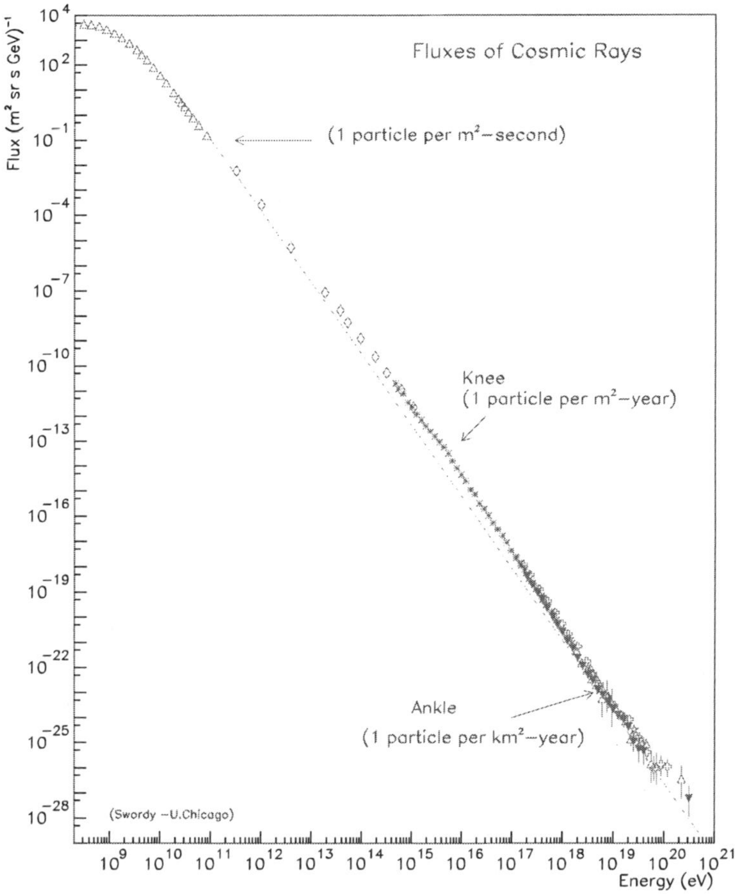

Figure 1. All-particle energy spectrum

statistics will provide the necessary information regarding sources and the nature of these particles. Therefore, at the scale of the IceCube detector, one can begin to perform efficient neutrino detection in the PeV region, above where the "knee" in the all-particle spectrum occurs.

IceCube is designed to search for sources of UHE neutrinos such as Active Galactic Nuclei (AGN), Supernova Remnants (SNR) or micro-quasars, and neutrinos from Gamma Ray Bursts (GRB). The sensitivity of IceCube to astrophysical sources has been studied in reference [4].

Aside from a search for Astrophysical sources of neutrinos, IceCube can also provide answers to a series of questions, related to particle physics, such as search for neutrinos from possible candidates for cold dark matter, weakly interacting massive particles (WIMPs) annihilating in the sun, and magnetic monopoles or other exotic particles such as strange quark matter or Q-balls predicted by SUSY models [5]. Furthermore, IceCube with its cubic kilometer size is able to examine a possible enhancement in neutrino interaction cross sections due to extra dimensions where graviton contributions could increase the total neutrino cross section to the level of hadronic interaction cross sections, of the order of tens of mb[6].

2. IceCube Detector Setup

The IceCube detector layout is shown in Figure 2. IceCube consists of 4800 optical modules (OM) mounted on 80 strings regularly spaced such that each two adjacent strings are 125 m apart. In planar view, IceCube covers an area of approximately 1 km^2. The instrumented part of the string is at a depth of 1,450 to 2,450 m below the surface of the ice. Each PMT string consists of 60 OM's, with OM's equally spaced at a distance of approximately 17 m. The strings are arranged in a hexagonal pattern in planar view. At the ice surface, on the top of each string, a station of the IceTop air shower array will be installed. An IceTop station consists of two ice tanks with a total area of 7 m^2. Two of these tanks have been installed during the 2003-04 season at the South Pole. The IceTop will be operated in coincidence with the in-ice arrays. This provides a veto for the downward going events as well as information on the chemical composition of the cosmic rays up to 10^{18}eV [7]. The AMANDA-II detector [3] will be integrated into IceCube. The present configuration is designed for optimum sensitivity to muon neutrinos in the energy range of 1-100 TeV.

3. The Digital Optical Module

The heart of the Data Acquisition system of IceCube is the Digital Optical Module (DOM). The IceCube DOM shown in Figure 3 contains a 10-inch diameter PMT, HAMAMATSU R-7081. These PMT's provide excellent charge and time resolution. The dynamic range is 200 photo-electrons (pe)/10 nsec, with an integrated dynamic range of more than 2000 pe's. The signal is digitized at the PMT level with a digitization depth of 4 μ-sec.

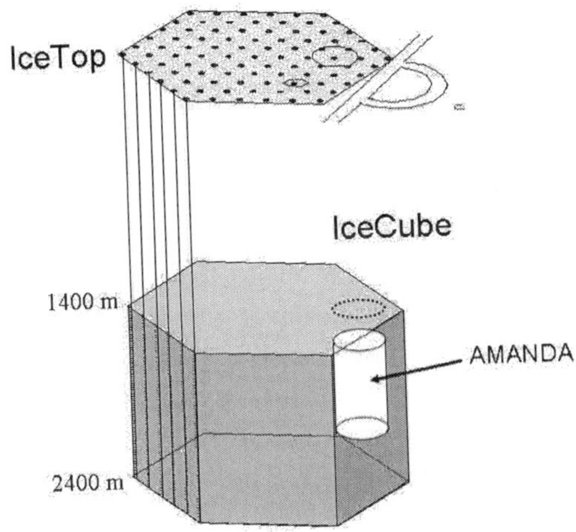

Figure 2. IceCube, IceTop and AMANDA

A single PMT low noise rate of less than 500 Hz, at operating temperatures of the South Pole of -20 to -40 ^{o}C has been achieved with these PMT's. Each PMT and its associated electronics is housed in a glass sphere that can withstand a pressure of 10,000 psi (68,948 kPa). The face of the PMT makes contact with this glass shell through a gel with approximately the same index of refraction as that of the glass.

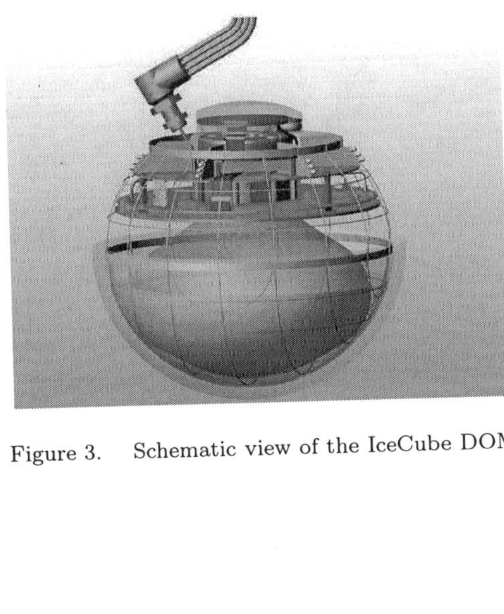

Figure 3. Schematic view of the IceCube DOM

4. IceCube Performance

The main particle detection capability of the IceCube detector is measured in terms of its sensitivity to the detection of muons. Muons are produced in the detector because of charged-current interactions of ν_μ with H and O nuclei in the ice or other nuclei in the ground below the ice. The majority of the downward-going muons in IceCube are due to the interaction of the primary cosmic rays with the atmosphere. Figure 4 summarizes the expected sensitivity to diffuse fluxes as a function of neutrino energy. Solid lines show the expected 90% confidence level (CL) limits for E^{-2} and E^{-1} fluxes, respectively. These calculations assume a period of three years of data accumulation. The dashed lines show the model proposed by Stecker and Salamon describing the photo-hadronic interactions in the AGN core[9]. The dotted line shows the model of Mannheim, Protheroe, and Rachen, estimating neutrino emission from photo-hadronic interactions in AGN jets [10]. Also shown is the GRB estimate of Waxman and Bahcall [12]. Their estimate yields approximately ten GRB events for 1000 GRB's monitored. IceCube performance at higher energies is described by Yoshida, Ishibashi, and Miyamoto. They show that in the 10 - 100 PeV energy range, not only muons, but also τ's survive without decay and would leave detectable signals in horizontal and downward-going events[8].

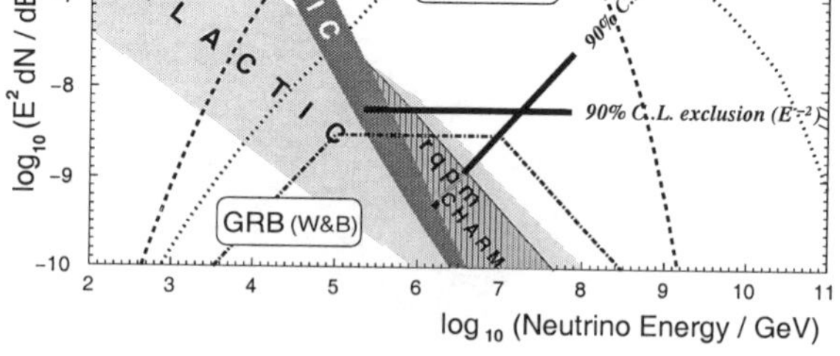

Figure 4. Expected sensitivities of the IceCube detector. See text for explanations.

5. IceCube Deployment

The hot-water drill technology has been developed and perfected over decades as the fastest and the most efficient method for drilling holes in the ice. It turns out that for IceCube hot water also provides the only possible technology. Water is readily available and the water in the hole after deployment of the strings freezes, preserving the optical properties of the ice. The optics for IceCube are well understood from years of experience with AMANDA. Holes with 60 cm diameter will be drilled with 100 ^oC water at a rate of 16 holes per year. The drill system is an evolution of the AMANDA drill called the Enhanced Hot Water Drill (EHWD). The EHWD drill is energy intensive because of the large amount of energy required for the ice-to-water phase transition. The total time required to drill a 2400 m deep hole is about 40 hours. The power consumption for the EHWD will be 5 MW, compared to 2 MW for AMANDA. This, and the larger diameter and the length of the water transporting hoses, will result in 40 hours needed to drill a 2400 m deep hole. Mounting, testing and deployment of a string with 60 DOM's are estimated to take about 20 hours.

6. IceCube Status

The IceCube collaboration includes about 150 scientists from institutions in Belgium, Germany, Japan, Netherlands, New Zealand, Sweden, UK, US and Venezuela. In the US IceCube is funded by the National Science Foundation through a multi-year Major Research Equipment (MRE) grant. The start-up funding for a period of two years from the NSF began in 2002. This year (2004) IceCube has begun its implementation phase as a fully funded NSF, MRE. Furthermore, significant funding for the IceCube project has been approved in Belgium, Germany and Sweden. Full IceCube construction has begun this year and will take 6 years. The first parts of the EHWD's have been shipped to the pole. This year the production and testing of 150 DOM's have been scheduled. The drilling of the first holes is scheduled to begin in January 2005.

Acknowledgments

This work was supported by the National Science Foundation through a subcontract from the University of Wisconsin, Madison.

IceCube Author List

A. Achterberg[26], M. Ackermann[6] J. Ahrens[15], J.N. Bahcall[8], X. Bai[1], R.C. Bay[14], T. Becka[15], K.-H. Becker[2], J. Bergmans[26], D. Berley[16], E. Bernardini[6], D. Bertrand[3], D.Z. Besson[9], E. Blaufuss[16], K. Blundell[19], D.J. Boersma[6], S. Böser[6], C. Bohm[27], O. Botner[25], A. Bouchta[25], O. Bouhali[3], T. Burgess[27], W. Carithers[10], T. Castermans[18], J. Cavin[22], W. Chinowsky[10], D. Chirkin[14], B. Collin[12], J. Conrad[25], J. Cooley[22], D.F. Cowen[12], A. Davour[25], C. De Clercq[28], T. DeYoung[16], P. Desiati[22], R. Ehrlich[16], R.W. Ellsworth[17], P.A. Evenson[1], A.R. Fazely[13], T. Feser[15], T.K. Gaisser[1], J. Gallagher[21], R. Ganugapati[22], H. Geenen[2], A. Goldschmidt[10], J.A. Goodman[16], R.M. Gunasingha[13], A. Hallgren[25], F. Halzen[22], K. Hanson[22], R. Hardtke[22], T. Hauschildt[6], D. Hays[10], K. Helbing[10], M. Hellwig[15], P. Herquet[18], G.C. Hill[22], D. Hooper[19], D. Hubert[28], B. Hughey[22], P.O. Hulth[27], K. Hultqvist[27], S. Hundertmark[27], J. Jacobsen[10], G.S. Japaridze[4], A. Jones[10], A. Karle[22], H. Kawai[5], M. Kestel[12], N. Kitamura[23], R. Koch[23], L. Köpke[15], M. Kowalski[6], J.I. Lamoureux[10], N. Langer[26], H. Leich[6], I. Liubarsky[7], J. Madsen[24], K. Mandli[22], H.S. Matis[10], C.P. McParland[10], T. Messarius[2], P. Mészáros[11,12], Y. Minaeva[27], R.H. Minor[10], P. Miočinović[14], H. Miyamoto[5], R. Morse[22], R. Nahnhauer[6], T. Neunhöffer[15], P. Niessen[28], D.R. Nygren[10], H. Ögelman[22], Ph. Olbrechts[28], S. Patton[10], R. Paulos[22], C. Pérez de los Heros[25], A.C. Pohl[27], J. Pretz[16], P.B. Price[14], G.T. Przybylski[10], K. Rawlins[22], S. Razzaque[11], E. Resconi[6], W. Rhode[2], M. Ribordy[18], S. Richter[22], H.-G. Sander[15], S. Sarkar[19], K. Schinarakis[2], S. Schlenstedt[6], D. Schneider[22], R. Schwarz[22], D. Seckel[1], J. Silk[19], A.J. Smith[16], M. Solarz[14], G.M. Spiczak[24], C. Spiering[6], M. Stamatikos[22], T. Stanev[1], D. Steele[22], P. Steffen[6], T. Stezelberger[10], R.G. Stokstad[10], K.-H. Sulanke[6], G.W. Sullivan[16], T.J. Sumner[7], I. Taboada[20], S. Tilav[1], N. van Eijndhoven[26], W. Wagner[2], C. Walck[27], Y.-R. Wang[22], C.H. Wiebusch[2], C. Wiedemann[27], R. Wischnewski[6], H. Wissing[6], K. Woschnagg[14], S. Yoshida[5]

(1) Bartol Research Institute, University of Delaware, Newark, DE 19716, USA

(2) Fachbereich 8 Physik, BUGH Wuppertal, D-42097 Wuppertal, Germany

(3) Université Libre de Bruxelles, Science Faculty CP230, Boulevard du Trio mphe, B-1050 Brussels, Belgium

(4) CTSPS, Clark-Atlanta University, Atlanta, GA 30314, USA

(5) Dept. of Physics, Chiba University, Chiba 263-8522 Japan

(6) DESY-Zeuthen, D-15738 Zeuthen, Germany

(7) Blackett Laboratory, Imperial College, London SW7 2BW, UK

(8) Institute for Advanced Study, Princeton, NJ 08540, USA

(9) Dept. of Physics and Astronomy, University of Kansas, Lawrence, KS 66045, USA

(10) Lawrence Berkeley National Laboratory, Berkeley, CA 94720, USA

(11) Dept. of Astronomy and Astrophysics, Pennsylvania State University, Univ ersity Park, PA 16802, USA

(12) Dept. of Physics, Pennsylvania State University, University Park, PA 168 02, USA

(13) Dept. of Physics, Southern University, Baton Rouge, LA 70813, USA
(14) Dept. of Physics, University of California, Berkeley, CA 94720, USA
(15) Institute of Physics, University of Mainz, Staudinger Weg 7, D-55099 Mai nz, Germany
(16) Dept. of Physics, University of Maryland, College Park, MD 20742, USA
(17) Dept. of Physics, George Mason University, Fairfax, VA 22030, USA
(18) University of Mons-Hainaut, 7000 Mons, Belgium
(19) Department of Physics, University of Oxford, Oxford, England
(20) Departamento de Física, Universidad Simón Bolívar, Caracas, 1080, Venezuela
(21) Dept. of Astronomy, University of Wisconsin, Madison, WI 53706, USA
(22) Dept. of Physics, University of Wisconsin, Madison, WI 53706, USA
(23) SSEC, University of Wisconsin, Madison, WI 53706, USA
(24) Physics Dept., University of Wisconsin, River Falls, WI 54022, USA
(25) Division of High Energy Physics, Uppsala University, S-75121 Uppsala, Sweden
(26) Faculty of Physics and Astronomy, Utrecht University, NL-3584 CC Utrecht , The Netherlands
(27) Dept. of Physics, Stockholm University, SE-10691 Stockholm, Sweden
(28) Vrije Universiteit Brussel, Dienst ELEM, B-1050 Brussels, Belgium

References

1. http://icecube.wisc.edu
2. E. Roulet, Proceeding of Lepton-Photon Conference,11-16 August 2003, Fermi National Accelerator Laboratory, Batavia, Illinois USA
3. http://amanda.uci.edu/
4. J. Ahrens et al., *Astropart. Phys.***20**, 507 (2004).
5. F. Halzen, astro-ph/0311004.
6. J. Alvarez-Muniz, F.Halzen, T. Han, D. Hooper, *Phys. Rev. Lett.***88** 021301 (2002).
7. C. Spiering for the IceCube Collaboration, astro-ph/0404090.
8. S. Yoshida, R. Ishibashi, and H. Miyamoto, *Phys. Rev.***D69** 103004 (2004).
9. F. W. Stecker and M. H. Salamon, *Space Sci. Rev.***75** 341 (1996).
10. K. Mannheim, R. Protheroe, J. Rachen, *Phys. Rev.***D63** 023003 (2001).
11. G. G. Hill and K. Rawlins, *Astropart. Phys.***19** 393 (2003).
12. E.Waxmann, J.Bahcall, *Phys. Rev.***D59** 023002 (1999).

PLANS FOR EXPERIMENTS TO MEASURE θ_{13}

MAURY GOODMAN
Argonne National Laboratory
Argonne IL 60439, USA
E-mail: maury.goodman@anl.gov

New experiments at accelerators and reactors are being designed to search for a possible non-zero value of the MNS matrix parameter θ_{13}.

1. Neutrino Parameters and the MNS matrix

The neutrino mixing matrix, analogous to the CKM matrix for quarks, is the MNS matrix[1]. In the simplest case, that matrix can be parameterized by three angles and one CP violating phase. With the remarkable recent progress in understanding neutrinos, it is known that at least two of these mixing angles are large, unlike the angles in the quark sector. Despite these large angles, neutrino oscillation analyses are usually based on the assumption that there are only two neutrinos. This simplifies the analyses, and is accurate, but can be confusing.

The three mixing angles are designated θ_{12}, θ_{23} and θ_{13}. Solar neutrino experiments and the KamLAND reactor experiment have measured θ_{12}, also called θ_{sol}. Atmospheric neutrino experiments have been used to measure θ_{23} or θ_{atm}. Long-baseline accelerator neutrino oscillation experiments have been designed to measure this angle. The third mixing angle, θ_{13}, has only been limited to be smaller than $\sin^2(2\theta_{13}) < 0.17$ by the CHOOZ[2] nuclear reactor experiment, at the currently favored value of Δm^2_{atm}. A CP violating phase which appears in the MNS matrix is called δ. Long-baseline neutrino experiments would be sensitive to a non-zero value of θ_{13} by measuring $\nu_\mu \rightarrow \nu_e$ oscillations with a rate smaller than the dominant expected $\nu_\mu \rightarrow \nu_\tau$ oscillations, so this is often called a sub-dominant process.

The current program of long-baseline neutrino experiments includes the K2K experiment in Japan, which has been running since 1998, the NuMI/MINOS program at Fermilab, which will start in 2005, and the CNGS program at CERN which is expected to start in 2006. Accelerator beams consist mostly of muon neutrinos (ν_μ) which are made when pions and Kaons decay. All three experiments expect to measure a change in the number and distribution of ν_μ as a result of $\nu_\mu \to \nu_\tau$ oscillation. In addition, CNGS hopes to measure ν_τ appearance. These ν_τs would be present in all three experiments, but the production of τs in K2K and MINOS is expected to be suppressed by kinematic thresholds.

2. The importance of θ_{13}

I share the view that the single most important task before the neutrino community is to measure a non-zero value for θ_{13} if it exists. While $\sin^2(2\theta_{23})$ is near 1, SNO data supports that $\sin^2(2\theta_{12})$ is approximately 0.7 and is clearly not maximal. Although we don't know what symmetries if any control the values of the MNS mixing angles, it is reasonable to expect that if θ_{12} is not maximal, θ_{13} is probably not exactly zero. The current limit on $\sin^2(2\theta_{13})$ is < 0.17[2]. Some of us suspect that a new search which is sensitive down to 0.02 would have a 90% chance of measuring a non-zero value for θ_{13}!

Such a result would open up new possibilities for future neutrino research. If θ_{13} is not zero, searches for matter effects and CP violation with accelerator neutrinos are feasible, in particular using $\nu_\mu \to \nu_e$ measurements as described below in Equation 3. The observation of matter effects in ν or $\bar{\nu}$ tells us if Δm^2_{31} is positive (normal mass hierarchy) or negative (inverted mass hierarchy). Of even greater interest, the measurement of differences between ν and $\bar{\nu}$ oscillation parameters (after removing any matter effects) is sensitive to CP violating effects in the lepton sector. The measure of CP violation is the Jarlskog invariant, which is proportional to factors which include the product of all mixing angles and Δm^2 values. With the large mixing in the lepton sector, CP violation in the neutrino sector is poised to be 50 times larger than in the quark sector, depending solely on whether θ_{13} is near is current upper limit, or is much smaller.

The probability in vacuum for $\nu_\mu \to \nu_e$ is[3]:

$$
\begin{aligned}
P(\nu_\mu \to \nu_e) &= 2\sin(2\theta_{13}) s_{23} c_{13} s_{12} (s_{12} s_{23} s_{13} - c_{12} c_{23} c_\delta) \sin^2\phi_{32} \\
&+ 2\sin(2\theta_{13}) s_{23} c_{13} c_{12} (c_{12} s_{23} s_{13} + s_{12} c_{23} c_\delta) \sin^2\phi_{31} \\
&- 2\sin(2\theta_{12}) c_{13}^2 [s_{12} c_{12} (s_{13}^2 s_{23}^2 - c_{23}^2) + s_{13} s_{23} c_{23} (s_{12}^2 - c_{12}^2) c_\delta] \sin^2\phi_{21} \\
&+ \frac{1}{2}\sin(2\theta_{12})\sin(2\theta_{13})\sin(2\theta_{23}) c_{13} s_\delta [\sin\phi_{32}\cos\phi_{32} \\
&- \sin\phi_{31}\cos\phi_{31} + \sin\phi_{21}\cos\phi_{21}]
\end{aligned} \tag{1}
$$

where $\phi_{ij} = \Delta m^2_{ij} L/(4E)$ and c_{ij} and s_{ij} refer to the cosine and sine of the mixing angle ij. The first two terms describe the behavior at an L/E corresponding to the large value of Δm^2 and the terms proportional to $\sin\delta$ ($\cos\delta$) are CP odd (even).

In contrast reactor neutrino disappearance at small distances (i.e. L/E $\sim \Delta m^2_{31}$, so L $\sim$ 1 km), the formula is the simpler familiar one:

$$P(\bar{\nu}_e \rightarrow \bar{\nu}_e) \cong 1 - sin^2(2\theta_{13}) sin^2\phi_{32} \tag{2}$$

3. Future Off-Axis Long-Baseline Neutrino Experiments

The search for a few percent $\nu_\mu \rightarrow \nu_e$ in long-baseline experiments is made difficult by three backgrounds. First, there are ν_e in the beam near the percent level from K and μ decays in the beam pipe. Second, when the dominant $\nu_\mu \rightarrow \nu_\tau$ oscillation takes place and a τ is produced by charged current interaction, the τ decays to an electron 17% of the time. Third, for any detector which resembles a calorimeter, some fraction of the neutral current events are going to be indistinguishable from an electron, particularly the single π^0 neutral current events.

The kinematics of neutrino production is such that off the main axis of the neutrino beam, the average neutrino energy is both lower and more peaked toward a single value. The lower energy means that event rates are way down, already a significant limitation for long-baseline neutrino experiments. However, an off-axis narrow-band beam can be used to reduce all three backgrounds. The ν_e in the beam are mostly at high energy, and do not constitute a large background to the expected signal. There are no ν_τ charged current interactions below 4 GeV, so the tau decay background is greatly reduced, and though the neutral current background is not eliminated, it is greatly reduced since there is no high energy tail on the neutrino spectrum. This has led to consideration of new accelerator based neutrino programs in the US and Japan.

3.1. *T2K*

The Japanese Particle Research Center (JPARC) is a new 50 GeV proton synchrotron being built in Tokai. A neutrino beam is being planned.[4] In a first phase the accelerator would operate at 0.77 MW, but an upgrade to 4 MW is being considered. The 22.5 kiloton Super-Kamiokande detector is 295 km away, and the beam could be built to be simultaneously a few degrees off axis to that experiment and to the proposed site for a new 1000 kiloton Hyper-Kamiokande detector in Tochibora. With a 5 year run of JPARC, and the proposed 2 degree off-axis beam, T2K would be able to measure θ_{13} or set a limit on $\sin^2(2\theta_{13}) < 0.006$ at 90% CL for $\delta = 0$. The Hyper-Kamiokande detector would be similar in design to Super-K, using large 50 cm diameter Hamamatsu phototubes.

3.2. *NOνA*

A proposal is being developed for an off-axis experiment using the NuMI beam at Fermilab.[5] Any detector would be built near the surface of the earth, about 10 km away from the center of the NuMI beam. The detector would have a mass of 50 Kilotons, and be sensitive to 1 GeV electron showers. The passive detector is planned to be 7 sheets of 2.5 cm particle board between readout planes. Active detector technologies being considered are resistive plate chambers and liquid and solid scintillator.

The liquid scintillator design is for 14.4 m long multi-cell extrusions of PVC, each containing 32 cells of width 3.75 cm. The cells would be 3 cm thick along the beam direction. A looped fiber would be inserted in each cell and an end-cap would be glued on one end, with a manifold/optical connector assembly at the other. There are no critical tolerances, such as positioning of the fiber. The proposal is under consideration at Fermilab.

3.3. *Brookhaven Wide Band Beam*

An alternative to the off-axis program has been developed at Brookhaven.[6] Using a wide-band beam, a huge detector such as the UNO described in the last section of this paper, and a longer distance to exploit possible matter effects, the Brookhaven idea uses fits to the full energy spectrum to find and measure oscillation parameters.

4. Reactor Experiments

From the discovery of the neutrinos by Reines and Cowan[7] at Savannah River to the evidence for $\bar{\nu}_e$ disappearance at KamLAND[8], reactor neutrino experiments have studied neutrinos in the same way – observation of inverse beta decay with scintillator detectors. Since the signal from a reactor falls with distance L as $1/L^2$, as detectors have moved further away from the reactors over the years, it has become more important to reduce backgrounds. That is achieved with a sufficient overburden, and experiments one kilometer or more away from reactors (Chooz, Palo Verde and KamLAND) have been underground.

The KamLAND experiment measured a 40% disappearance of $\bar{\nu}_e$ presumably associated with the 2nd term in Equation 3:

$$P(\bar{\nu}_e \to \bar{\nu}_e) \cong -\sin^2 2\theta_{13} \sin^2(\frac{\Delta m_{31}^2 L}{4E}) - \cos^4 \theta_{13} \sin^2 2\theta_{12} \sin^2(\frac{\Delta m_{21}^2 L}{4E}) + 1 \tag{3}$$

The Chooz and Palo Verde data put a limit on θ_{13} (through the first term in Equation 3) of $\sin^2 2\theta_{13} < 0.17$. Those experiments could not have had greatly improved sensitivity to θ_{13} because of uncertainties related to knowledge of the flux of neutrinos from the reactors. They were designed to test whether the atmospheric neutrino anomaly might have been due to $\nu_\mu \to \nu_e$ oscillations, and hence were searching for large mixing.

Any new experiment to look for non-zero values of θ_{13} would need the following properties:

- two or more detectors to reduce uncertainties to the reactor flux
- identical detectors to reduce systematic errors
- carefully controlled energy calibration
- low backgrounds and/or reactor-off data

In Equation 3, the values of θ_{12}, Δm_{21}^2 and Δm_{31}^2 are approximately known. In Figure 1, The probability of $\bar{\nu}_e$ disappearance as a function of L/E is plotted with θ_{13} assumed to be near its maximum allowed value. Note that CP violation does not affect a disappearance experiment, and that matter effects can be safely ignored in a reactor experiment. The large variation in P for L/E$>$ 10 km/MeV is the effect seen by KamLAND and solar ν experiments. The much smaller deviations from unity for L/E $<$ 1 km/MeV are the goal for an accurate new reactor experiment.

The optimization of detector distances for such a new experiment is straightforward. The statistical power comes from measuring a deficit of $\bar{\nu}_e$ (up to a few percent) at the far detector, along with a change in the energy spectrum consistent with that deficit. Depending on the value of Δm_{atm}^2, the far detector should be located about 1.7 km away. For higher statistics experiments, the importance of the shape of the spectrum is greater than the rate, and the optimum detector location moves about 30% closer. The optimum locations for detectors is thus sensitive to eventual systematic errors as well as oscillation parameters. But a near detector around 100m and a far detector around 1300 m will be near the optimum. Since the civil construction of laboratories might contribute half or more to the cost of an experiment, site conditions could change the optimization.

A list of possible sites for a new reactor experiment is included in Ta-

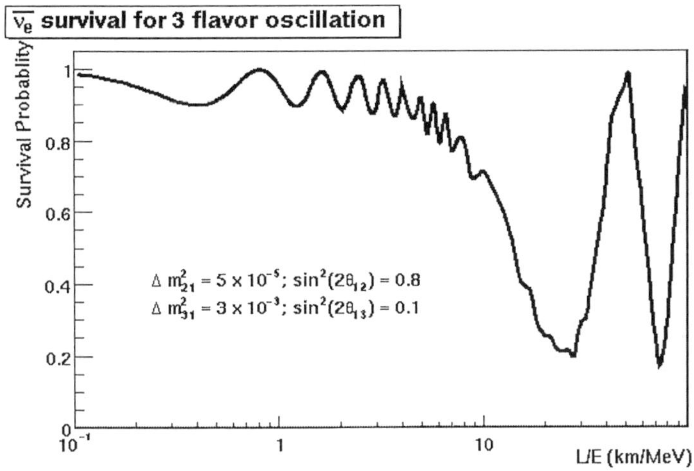

Figure 1. Probability of $\bar{\nu}_e$ disappearance versus L/E for θ_{13} near its current upper limit.

ble 4, along with a tabulation of previous reactor experiment sites.[9] One obvious choice is the site of the 7 GW CHOOZ reactors in France, and a proposal called "Double CHOOZ" is being prepared. A larger 15 ton far detector would be placed in an existing hall 1050 m from the two reactors, and new near detector would be placed close ($\sim 200m$). The existing site is an attraction, but is also a limitation. An even larger detector (50-1000 ton) would be desirable to achieve greater sensitivity on θ_{13}. One of several projects being considered is to use the 7 GW Braidwood Reactors in Illinois, and place two shafts at 200 m and the other at 1800 m, on average from the two reactor cores. Two 25 ton detectors would be built.

A sensitivity of 0.02 in $\sin^2 2\theta_{13}$ can be achieved with as little as 250 ton-Gigawatt-years, while an exposure of 8000 ton-Gigawatt-years may be required to achieve a sensitivity of 0.01.[10] A two or more detector reactor experiment can find a non-zero value for θ_{13} faster and less expensively than an off-axis experiment. It does not face the degeneracies regarding CP parameters and the sign of Δm^2, and hence cannot address those issues. But a measurement of θ_{13} by reactors followed by optimized off-axis experiments would together measure neutrino parameters with much less uncertainty due to degeneracies and correlations.

Reactor	Location	L	Power	Overburden	Mass
		m near/far	GW	MWE	ton
		Previous Reactor Experiments			
Chooz	France	1100	8.5	300	5
Bugey	France	49/95	5.6	16	1/0.5
Palo Verde	Arizona	890	11.6	32	11.3
KamLAND	Japan	$< 180 >$	200 (26)	2700	1000
		Possible sites for New Reactor ν Experiments			
Angra	Brazil	350/1350	4.0	60/600	50/50
Braidwood	Illinois	200/1800	7	250/250	25/50
Double CHOOZ	France	200/1050	7	50/300	10/10
Daya Bay	China	300/1500	11	200/1000	20/40
Diablo Canyon	California	400/1800	7	100/700	25/50
KASKA	Japan	350/1300	24	140/600	8/8
KR2DET	Russia	115/1000	1.5	600/600	45/45

5. UNO nucleon decay project

One of the reasons for the tremendous progress in understanding the neutrino has been the fact that several detectors were built underground to search for nucleon decay. The UNO detector[11] is proposed as a next generation underground water Cerenkov detector that probes nucleon decay beyond the sensitivities of the highly successful Super-Kamiokande detector.[11] The baseline conceptual design of the detector calls for a "Multi-Cubical" design with outer dimensions of 60x60x180 m^3. The detector has three optically independent cubical compartments with corresponding photo-cathode coverage of 10%, 40%, and 10%, respectively. The total (fiducial) mass of the detector is 650 (440) kton, which is about 13 (20) times larger than the Super-K detector. Water Cerenkov technology is the only realistic detector technology available today to allow a search for this decay mode for proton lifetimes up to 10^{35} years. UNO provides other opportunities, such as the ability to observe oscillatory behavior and appearance in the atmospheric neutrinos; precision measurement of temporal changes in the solar neutrino fluxes; supernovae and supernova relic neutrinos; and searches for astrophysical point sources of neutrinos. UNO is an ideal distant detector for a long-baseline neutrino oscillation experiment with neutrino beam energies below about 10 GeV, as envisaged by the Brookhaven working group.

6. APS Neutrino Study

A large number of ideas to study the neutrino sector are being pursued, and they come with a variety of costs and feasibility. Ideas for "neutrino factories" to study θ_{13} are now considered part of the program of the far future. A study in the United States to put the differing ideas for neutrino experiments into a single coherent program is being developed under the auspices of four divisions of the American Physical Society. Six working groups are holding meetings and a report is due in the late summer of 2004. Reactor experiments to see if θ_{13} is non-zero, followed by accelerator projects to measure CP effects if it is, should be a major part of such a program.

References

1. Named to honor Maki, Nakagawa and Sakata.
2. Appollonio et al., Eur.Phys.J. C27 (2003) 331-374.
3. V. Barger et al., Phys. Rev. D63, 113011 (2001), I. Mocioiu and R. Shrock, JHEP 0110, 050 (2001).
4. See Letter of Intent at http://neutrino.kek.jp/jhfnu/loi/loi.v2.030528.pdf and subsequent documents at http://neutrino.kek.jp/jhfnu/.
5. I. Ambats et al., "Proposal to Build an off-axis detector to study numu to nue Oscillations in the NuMI Beamline", March 2004, http://www-off-axis.fnal.gov/
6. M. Diwan et al., hep-ex/0305105.
7. C.L. Cowan et al., Science 124, 103 (1956).
8. Eguchi et al., Phys. Rev. Lett. 90 (2003) 021802.
9. K. Anderson et al., "A New Nuclear Reactor Neutrino Experiment to Measure theta 13", hep-ex/0402041
10. P.Huber et al., hep-ph/0303232.
11. M. Goodman et al., "Physics Potential and Feasibility of UNO", June 2001

LIMITS ON THE NEUTRINO MAGNETIC MOMENT USING 1496 DAYS OF SUPER-KAMIOKANDE-I SOLAR NEUTRINO DATA

D. W. LIU *

Department of Physics and Astronomy
University of California, Irvine
Irvine, CA 92697, USA
E-mail: dwliu@hep.ps.uci.edu

Limits on the neutrino magnetic moment has been obtained by analyzing the energy spectrum distortions using 1496 days Super-Kamiokande-I solar neutrino data. A limit of $\mu_\nu \leq 3.6 \times 10^{-10}$ μ_B at 90% C.L. has been obtained by fitting to the Super-Kamiokande day/night spectra. With additional information from other solar neutrino and KamLAND experiments constraining the oscillation region, a limit of $\mu_\nu \leq 1.1 \times 10^{-10}$ μ_B at 90% C.L. was obtained.

1. Introduction

The interests in the neutrino magnetic moment started with the idea of neutrino. In Pauli's famous letter in which he proposed the existence of neutrino, he suggested that neutrinos were bound in the nuclei by electromagnetic force through neutrino magnetic moments [1]. In the modern theory of particle physics, the Standard Model, neutrinos are massless and do not have magnetic moment due to the left hand nature of weak interaction. Results from recent neutrino experiments [2,3,4,5,6] show strong evidence that neutrinos oscillate and thus must have non-zero masses. Including neutrino mass in the Standard Model, neutrinos can acquire a magnetic moment by the coupling of the left hand and right hand currents through the mass term [7].

$$\mu_\nu = \frac{3G_F e m_\nu}{8\sqrt{2}\pi^2} \approx 3.2 \times 10^{-19}(\frac{m_\nu}{1eV})\mu_B \quad (1)$$

*For the Super- Kamiokande Collaboration

where μ_B is the Bohr magneton. There are various extensions to the Standard Model which can give neutrino large magnetic moments [8].

As the magnetic moments could change the helicity of neutrinos, they are associated with mass eigenstates [9]. The interaction of neutrino mass eigenstates j and k with a magnetic field is characterized by the magnetic moments μ_{jk}. Both diagonal ($j = k$) and off-diagonal ($j \neq k$) moments are possible.

Neutrino magnetic moments have been used to explain the solar neutrino problem [10], *e.g.* spin flavor precession (SFP) [11]. The magnetic field in the Sun could flip the spin of the neutrinos making them either undetectable or detected at reduced rate. Recent results from terrestrial experiment KamLAND [6] ruled out SFP as the solution to the solar neutrino problem [12].

The first experiment to search for the neutrino magnetic moment was done by UCI group using the anti-neutrinos from the reactor [13]. Subsequent experiments of similar type have been carried out over years with limits ranging from 1 $\times 10^{-10}$ μ_B to 4 $\times 10^{-10}$ μ_B [14,15,16]. These limits are several order of magnitude larger than that predicted by theories (1). So the detection of such large neutrino magnetic moment would mean new physics beyond the Standard Model.

Different from the above mentioned experiments, this search for neutrino magnetic moment is done with the high statistics solar neutrino data of Super-Kamiokande-I. The basic idea comes from the fact that if $\mu_\nu \neq 0$, in addition to the usual weak scattering contribution to the neutrino-electron elastic scattering (2), there is an incoherent contribution from magnetic scattering (3). So the existence of neutrino magnetic moment will manifest itself in the energy spectrum of recoiled electrons [9].

$$\left(\frac{d\sigma}{dT}\right)_{WK} = C\left[g_L^2 + g_R^2\,(\,1 - \frac{T}{E_\nu}\,)^2 - g_L g_R \frac{m_e T}{E_\nu^2}\right] \tag{2}$$

where $C = 2G_F^2 m_e/\pi$, $g_L = \sin^2\theta_W + 1/2$ for ν_e, $g_L = \sin^2\theta_W - 1/2$ for ν_μ and ν_τ, and $g_R = \sin^2\theta_W$.

$$\left(\frac{d\sigma}{dT}\right)_{EM} = \mu_\nu^2\,\frac{\pi\alpha_{em}^2}{m_e^2}\,\left(\frac{1}{T} - \frac{1}{E_\nu}\right) \tag{3}$$

where μ_ν is in units of μ_B, E_ν is the neutrino energy, $T = E_e - m_e$, and $T(E_e)$ is the kinetic (total) energy of the recoil electrons.

2. Super-Kamiokande Energy Spectrum

Super-Kamiokande (SK) is a water Cherenkov detector, with a fiducial volume of 22.5 kton, located in the Kamioka mine in Gifu, Japan. Detailed descriptions of the detector can be found elsewhere [17]. SK detects solar neutrinos via the elastic scattering of neutrinos off electrons in the water. Crucial to this search of the neutrino magnetic moment is to measure the recoiled electron energy spectrum with high precision. To achieve this and also minimize energy related systematic effects, the number of hit photomultiplier tubes (PMT) is related to the total electron energy using electrons injected by an electron linear accelerator (LINAC) [18]. The number of hit PMTs in the Monte Carlo simulation of those LINAC electrons is tuned to agree with LINAC data. As a result of this tuning, the systematic uncertainty of the reconstructed energy of electrons between 5 and 20 MeV is less than 0.64%. The uncertainty of the energy resolution is less than 2%. This absolute energy scale is monitored and cross checked by (1) muon decay electrons, (2) spallation products induced by cosmic ray muons, and (3) decay of artificially produced ^{16}N [19]. The data used for this analysis were collected from May 31, 1996 to July 15, 2001 with a livetime of 1496 days. As a real time detector, SK can divide the data sample into day and night data samples which give the day/night spectra.

Figure 1 shows the ratio of SK measured recoil electron energy spectrum and the expected weak scattering spectrum assuming no oscillation. The measured energy spectrum is examined by a χ^2 as in the SK's standard spectrum analysis [20] with the following modifications, 1) considering the oscillation effects on the expected spectrum; 2) using SK day/night energy spectra to take into account the earth effects; 3) including the contribution from magnetic scattering.

$$\chi^2 = \sum_{a=d,n} \sum_{i=1}^{18} \left[\frac{\frac{\alpha}{1+\beta\,\delta_i} \left(W_i^a + \mu_{10}^2 M_i\right) - D_i^a}{\sigma_i^a} \right]^2 + \beta^2 \tag{4}$$

where $W_i^{d,n}$ is the ratio of the oscillated day/night weak scattering spectra to the non-oscillated weak scattering spectrum. The solar neutrino oscillations are approximated by a two-neutrino description with parameters θ (mixing angle) and Δm^2 (difference in mass squared) [3]. M_i is the ratio of the magnetic scattering spectrum to the non-oscillated weak scattering spectrum assuming $\mu_\nu = 10^{-10}\mu_B$. $D_i^{d,n}$ is the ratio of the measured day/night spectra to the non-oscillated weak scattering spectrum. δ_i is the

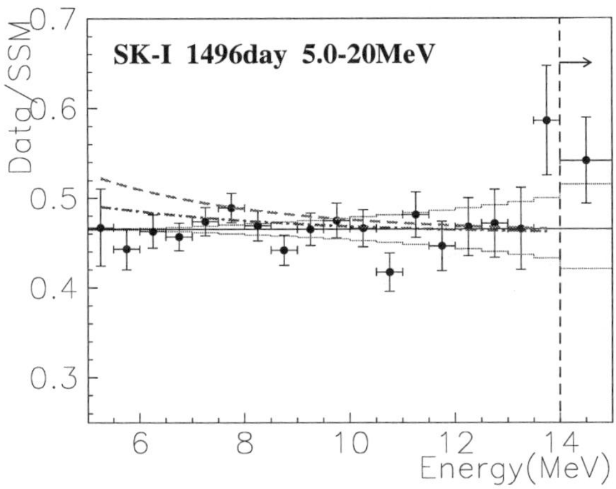

Figure 1. Ratio of SK-I observed recoil electron energy spectrum and the expected non-oscillated weak scattering spectrum. The error bars are the results of the statistical and energy non-correlated systematic errors being added in quadrature. The dotted lines are the correlated systematic errors. The dash-dotted line is the expected oscillated weak scattering spectrum for $\Delta m^2 = 6.6 \times 10^{-5}\text{eV}^2, \tan^2\theta = 0.48$. The dashed line shows the addition of magnetic scattering with $\mu_\nu = 1.1 \times 10^{-10}\mu_B$ on top of the oscillated weak spectrum. (The zero has been suppressed).

energy bin correlated systematic error, $\sigma_i^{d,n}$ is the day/night statistical and uncorrelated systematic errors added quadratically. α is the normalization factor of the measured ^{8}B flux to the expected flux; the ^{8}B flux is not constrained in this analysis. μ_{10}^2 is magnetic moment squared in units of $(10^{-10}\ \mu_B)^2$. β is a parameter used to constrain the variation of correlated systematic errors which come from the uncertainties in the energy scale, resolution and ^{8}B neutrino energy spectrum. Considering neutrinos with only diagonal magnetic moments [9], the survival probability of neutrinos passing through the magnetic field in the Sun is independent of neutrino energy [8]. Thus the shape of the ^{8}B neutrino spectrum will not be changed by the magnetic field in the Sun. In this search we use the SK day/night spectra from 5 to 14 MeV and consider only the ^{8}B solar neutrino flux. Furthermore, we assume $\mu_{\nu 1} = \mu_{\nu 2}$, so the magnetic scattering spectrum would not be changed by neutrino oscillations.

α, β and μ_ν^2 are parameters used to minimize the χ^2 in the oscillation parameter space with the condition $\mu_\nu^2 \geq 0$. As usual, the minimized χ^2 value can be used to exclude certain regions in the oscillation parameter space. Figure 2 shows the excluded regions without and with magnetic scattering. The addition of the magnetic scattering compensate that of weak scattering making some excluded regions allowable now.

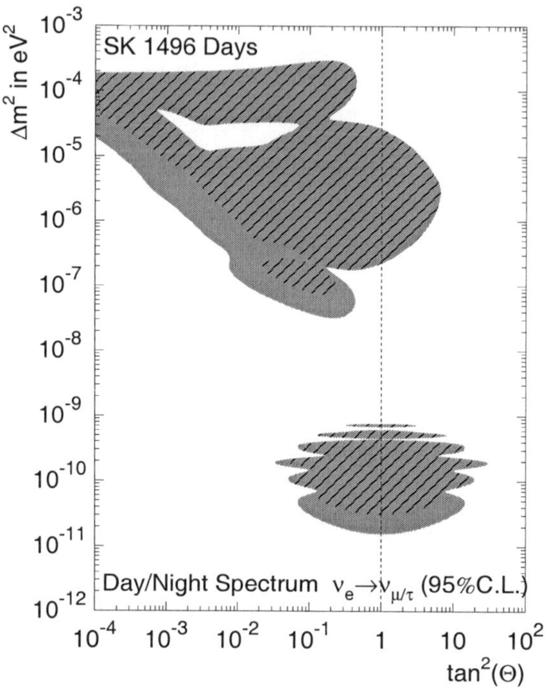

Figure 2. 95% C.L. exclusion region using the SK day/night spectra shape. The shaded area assumes only weak scattering. The hatched region takes into account the contribution from magnetic scattering.

3. Limits On the Neutrino Magnetic Moments

As there is no clear signal of neutrino magnetic moment, an increase of event rates at lower energies, limits on the neutrino magnetic moment have been derived using the following procedure. For each point in the oscillation parameter space, a 90% C.L. limit is obtained using equation (5). Prob is the probability distribution function of $\Delta\chi^2(= \chi^2 - \chi^2_{min})$ which itself is a function of the square of the magnetic moment. The overall limit on the neutrino magnetic moment is then obtained by finding the maximum of the aforementioned limits in the allowed oscillation parameter regions.

Discarding the regions excluded by SK day/night spectra, we found at 90% C.L. $\mu_\nu \leq 3.6 \times 10^{-10}\mu_B$ with the limit at $\Delta m^2 = 3.13 \times 10^{-11}\text{eV}^2$ and $\tan^2\theta = 0.91$ which is in the VAC region.

$$\text{Prob}(\Delta\chi^2(\mu^2 \geq \mu_0^2)) = 0.1 \times \text{Prob}(\Delta\chi^2(\mu^2 \geq 0)) \tag{5}$$

Results from other solar neutrino experiments could help to limit the allowed regions in the oscillation parameter space. Before doing that, we must analyze the effect of neutrino magnetic moments on their measurements. Radiochemical experiments like Homestake [21], SAGE [22] and Gallex/GNO [23] detect solar neutrinos via charged current interactions with nucleons. First neutrinos would not take part in the charge current interaction through magnetic moments, second the neutrino magnetic moment would not interfere their counting of event rates. So their measurements of solar neutrino flux rates are not affected. SNO measure the charged current, neutral current and elastic scattering rates by utilizing their distinctive angular distributions [5]. As mentioned earlier, charged current rates would not be affected. The effects of non-zero magnetic moment on the SNO neutral current interaction have been estimated to be very small [24]. A magnetic moment could change the elastic scattering rates but would not change the angular distribution of the elastic scattering events. Therefore, SNO's charged current and neutral current rates will be essentially unaffected by a non-zero neutrino magnetic moment. The combination of these charged current rates with SNO's neutral current rate and SK's day/night spectra constrains the neutrino oscillation to an area in the large mixing angle (LMA) region as shown in Figure 3 (the area within the dashed lines).

Limiting the search for the neutrino magnetic moment within the region allowed by solar neutrino experiments, we get an upper limit on the neutrino magnetic moment of $\mu_\nu \leq 1.3 \times 10^{-10}\ \mu_B$ at 90% C.L. with the limit at $\Delta m^2 = 2.8 \times 10^{-5}\text{eV}^2, \tan^2\theta = 0.42$.

KamLAND's detection of reactor $\bar{\nu}_e$'s using inverse β-decay interactions [6] would not be affected by a non-zero neutrino magnetic moment. Therefore assuming CPT invariance, the inclusion of the KamLAND results further constrains the neutrino oscillation solutions in the LMA region (the shaded area in Figure 3). This gives a limit on the neutrino magnetic moment at 90% C.L. of $\mu_\nu \leq 1.1 \times 10^{-10}\ \mu_B$ with the limit at $\Delta m^2 = 6.6 \times 10^{-5}\text{eV}^2$ and $\tan^2\theta = 0.48$.

If neutrinos have off-diagonal moments, the magnetic field in the Sun could distort the ^{8}B neutrino flux spectrum, so the limits of neutrino mag-

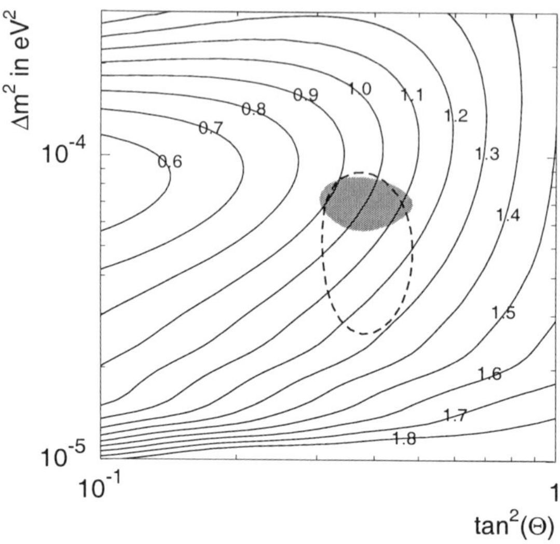

Figure 3. The 90% C.L. μ_ν limit contours (in units of $10^{-10}\mu_B$) and neutrino oscillation allowed regions. The area within the dashed lines is the solar neutrino experiments' allowed region considering both weak and magnetic scattering. The shaded area shows the allowed region for solar experiments plus KamLAND.

netic moment could in principle be affected. But for the LMA region, the effect of the solar magnetic field is negligible [12,25], so the limits on the neutrino magnetic moment in the LMA region would not changed.

4. Conclusion

Limits on the neutrino magnetic moment have been obtained by analyzing the SK day/night energy spectra. The oscillation effects on the shape of the weak scattering spectrum have been taken into account when analyzing energy spectra. A limit of 3.6×10^{-10} μ_B using Super-Kamiokande-I's 1496 days of solar neutrino data is obtained. By constraining the search to only the regions allowed by all neutrino experiments, a limit of 1.1×10^{-10} μ_B is obtained. This result is comparable to the most recent magnetic moment limits from reactor neutrino experiments of 1.3×10^{-10} μ_B (TEXONO) [15] and 1.0×10^{-10} μ_B (MUNU) [16], albeit for neutrinos and not anti-neutrinos.

Acknowledgments

The authors gratefully acknowledge the cooperation of the Kamioka Mining and Smelting Company. Super-Kamiokande has been built and operated from funding by the Japanese Ministry of Education, Culture, Sports, Science and Technology, the U.S. Department of Energy, and the U.S. National Science Foundation. This work was partially supported by the Korean Research Foundation (BK21) and the Korea Ministry of Science and Technology.

References

1. W. Pauli, in K. Winter (Ed.), *Neutrino Physics*, Cambridge University Press, Cambridge, 2000.
2. Y. Fukuda *et al.*, Phys. Rev. Lett, **81**, 1562 (1998);
3. S. Fukuda *et al.*, Phys. Rev. Lett, **86**, 5656, (2001); S. Fukuda *et al.*, Phys. Lett. B **539**, 179, (2002).
4. Q. R. Ahmad *et al.*, Phys. Rev. Lett, **89**, 011301 (2002); Q. R. Ahmad *et al.*, Phys. Rev. Lett, **89**, 011302 (2002).
5. S. N. Ahmed *et al.*, nucl-ex/0309004.
6. K. Eguchi *et al.*, Phys. Rev. Lett. **90**, 021802 (2003).
7. B. W. Lee and R. E. Shrock, Phys. Rev. D **16**, 1444 (1977); W. J. Marciano and A. I. Sanda, Phys. Lett. B **67**, 303 (1977).
8. R. Mohapatra, P. B. Pal, *Massive Neutrino in Physics and Astrophysics*, World Scientific, Singapore, 1998.
9. J. F. Beacom and P. Vogel, Phys. Rev. Lett. **83**, 5222 (1999).
10. J. Schechter, J. W. F. Valle, Phys. Rev. D **24**, 1883 (1981); Erratum-*ibid.* D **25**, 283 (1981); L. B. Okun, M. B. Voloshin and M. I. Vysotsky, Sov. Phys. JETP **64**, 446 (1986).
11. C. S. Lim and W. J. Marciano, Phys. Rev. D **37**, 1368 (1988); E. Kh. Akhmedov, Phys. Lett. B **213**, 64 (1988).
12. J. Barranco *et al.*, Phys. Rev. D **66**, 093009 (2002); hep-ph/0207326 v3.
13. F. Reines, H. S. Gurr and H. W. Sobel, Phys. Rev. Lett. **37**, 315 (1976); P. Vogel and J. Engel, Phys. Rev. D **39**, 3378 (1989).
14. A. I. Derbin *et al.*, JETP Lett. **57**, 768 (1993).
15. H. B. Li *et al.*, Phys. Rev. Lett. **90**, 131802 (2003).
16. Z. Daraktchieva *et al.*, Phys. Lett. B **564**, 190 (2003).
17. Y. Fukuda *et al.*, Nucl. Instrum. Meth. **A501**, 418 (2003).
18. M. Nakahata *et al.*, Nucl. Instrum. Meth. **A421**, 113 (1999).
19. E. Blaufuss *et al.*, Nucl. Instrum. Meth. **A458**, 638 (2001).
20. Y. Fukuda *et al.*, Phys. Rev. Lett. **82**, 2430 (1999).
21. B. T. Cleveland *et al.*, Astrophys. J. **496**, 505 (1998).
22. V. N. Gavrin, Talk given at TAUP' 2003, Sep. 5-9, Seattle, USA
23. E. Bellotti, Talk given at TAUP' 2003, Sep. 5-9, Seattle, USA
24. E. Akhmedov and V. Berezin, Z. Phys. C **54**, 661, (1992); K. Tsuji, S. Nakamura, T. Sato and K. Kobodera, in preparation.
25. O. G. Miranda *et al.*, Phys. Lett. B **521**, 299 (2001); Nucl. Phys. B **595**, 360 (2001); W. Grimus *et al*, Nucl. Phys. B **648**, 376 (2003).

FIRST RESULTS FROM THE SALT PHASE OF SNO

J. C. MANEIRA ON BEHALF OF THE SNO COLLABORATION.
Queen's University
Physics Dep., Stirling Hall
Kingston, Ontario K7L 3N6, Canada
E-mail: maneira@sno.phy.queensu.ca

The Sudbury Neutrino Observatory (SNO) is an underground heavy-water Cherenkov detector designed to detect ^8B solar neutrinos through neutral (NC) and charged (CC) current interactions on deuterons and elastic scattering on electrons. The results from the pure D_2O phase of the experiment confirmed solar model predictions and gave strong evidence for flavor change. In the second phase, 2 tonnes of NaCl were added to the heavy water, in order to enhance the detection of neutral current interactions. This allowed for precision, energy-unconstrained measurements of the solar neutrino flux that exclude maximal flavor mixing at a level of 5σ. The talk will focus both on the characterization of the detector response and the implications of the results on Neutrino Oscillation Physics.

1. Introduction

The discrepancy between measured[1,2,3,4,6,5] and predicted[7,8] solar neutrino fluxes, that lasted for over thirty years, is known as the solar neutrino problem. A possible solution is that the ν_e's produced in the Sun core undergo a flavor change on their way to the Earth, and an elegant mechanism for this is supplied by matter-enhanced oscillations[9,10].

The Sudbury Neutrino Observatory[15] is an underground heavy water Cherenkov detector located in the INCO, Ltd. Creighton mine near Sudbury, Canada, at a depth of about 2 km (6010 m of water equivalent), specifically designed to test the flavor change hypothesis by separately measuring the flux of electron and all flavor solar neutrinos. The active target of the detector, consists of 1000 tonnes of heavy water(D_2O), and is contained in a 12 m diameter transparent acrylic vessel. The Cherenkov photons produced in the D_2O are detected by 9456 8 inch photomultiplier tubes (PMTs) mounted on a 17.8 m diameter stainless steel geodesic sphere and coupled to light concentrators that raise the effective coverage to 55 %. Ultra-pure light water surrounds the acrylic vessel and the PMT support structure, in order to shield the heavy water from high energy γ rays originating in the PMTs and neutrons from the surrounding rock. An additional 91 PMTs are mounted facing outward on the support structure to serve as a cosmic muon veto.

SNO detects ^{8}B solar neutrinos through the processes:

$$\begin{aligned} \nu_e + d &\rightarrow p + p + e^- && \text{(CC)}, \\ \nu_x + d &\rightarrow p + n + \nu_x && \text{(NC)}, \\ \nu_x + e^- &\rightarrow \nu_x + e^- && \text{(ES)}. \end{aligned}$$

The charged current (CC) reaction is sensitive exclusively to ν_e's in the energy range of solar neutrinos. The neutral current reaction (NC) is equally sensitive to all active neutrino flavors ($x = e, \mu, \tau$), but its energy and direction information are lost. The elastic scattering (ES) reaction is sensitive to all electron flavors as well, but with reduced sensitivity to ν_μ and ν_τ.

The counting of the free neutrons from the NC reaction provides a model-independent measurement of the total flux of active ^{8}B solar neutrinos with an energy above the kinematic threshold of 2.2 MeV. In the first phase of SNO, with a pure heavy water target, the neutron was detected by observing the 6.25 MeV γ ray following its capture on a deuteron. In the second phase of SNO, 2 tonnes of salt were added to the D_2O and the NC rate measurement was made by observing an 8.6 MeV γ ray cascade following the neutron capture on ^{35}Cl. This talk presents the published data analysis from the pure D_2O[11,12,13] and salt phases of SNO[14]. In the third phase, the neutrons will be detected by an array of 40 ^{3}He counters, called "Neutral Current Detectors" (NCDs), presently being installed inside the active D_2O volume, in a $1m \times 1m$ grid.

1.1. *Previous Results from the Pure D_2O phase*

The pure D_2O data set was collected in 306.4 livetime days and contained 2928 events, after removal of instrumental and non-Cherenkov like backgrounds and application of the energy and volume cuts. The events were selected above a kinetic energy threshold of 5 MeV and below a fiducial radius of 550 cm.

The data are resolved into contributions from CC, ES, neutrons (NC+background) and low energy Cherenkov background events with the extended maximum likelihood method using probability density functions in kinetic energy, cosine of the angle with respect to the Sun, and radial position. The background rates were constrained to the measured levels. Under the assumption of an undistorted ^{8}B energy spectrum, the solar neutrino fluxes measured with each reaction were found to be, in units of 10^6 cm^{-2}s^{-1}:

$$\begin{aligned}\Phi_{CC} &= 1.76^{+0.06}_{-0.05}(stat.) \pm 0.09(syst.)\\ \Phi_{ES} &= 2.39^{+0.24}_{-0.23}(stat.) \pm 0.12(syst.)\\ \Phi_{NC} &= 5.09^{+0.44}_{-0.43}(stat.)^{+0.46}_{-0.43}(syst.)\end{aligned}$$

When the requirement of an undistorted spectrum is relaxed and only the direction and position variables are used in the signal extraction, the NC flux is found to be:

$$\Phi^{unc}_{NC} = 6.42 \pm 1.57(stat.)^{+0.55}_{-0.58}(syst.)$$

These results showed the presence of a non ν_e active flavor component in the solar neutrino flux, different from zero at a 5.3σ level, demonstrating flavor change. In addition, the NC measurement of the total active ^{8}B solar neutrino flux is in good agreement with solar model predictions.

2. The Salt Phase of SNO

In order to enhance the precision of the NC flux measurement, 2 tonnes of NaCl (salt) were added to the D_2O in May 2001. Neutron capture on ^{35}Cl has a high cross section and results in a multiple γ ray cascade, with a total energy of 8.6 MeV; higher than the 6.25 MeV γ ray from ^{2}H capture. Hence, not only is the neutron capture efficiency higher in the salt phase, but also the neutron events are shifted to higher energies, further away from the low energy background region. In addition, the higher γ multiplicity allows the use PMT hit pattern isotropy to statistically separate between neutrons and single electron events (CC and ES).

The salt data set presented here was collected between July 26, 2001 and October 10, 2002 and corresponds to 254.2 live days.

2.1. *Detector Response*

After removal of instrumental and non-Cherenkov like backgrounds, the times and positions of hit PMTs are used to calculate estimators for the vertex position, the particle direction and energy and the event isotropy. The detector response was calibrated by using a source deployment system capable of reaching most positions in two perpendicular vertical planes inside the acrylic vessel, as well as in vertical axes in the light water volume.

The energy estimator uses an energy normalization obtained with the ^{16}N γ source and an optical model calibrated with a laser diffusing source ("laserball") to estimate the event energy based on its number of hit PMTs (nhit), reconstructed position and direction. The laserball calibration measured an increase in optical attenuation in the D_2O volume throughout the salt phase, as shown in Figure 1a) and the ^{16}N calibration showed a steady drop in the detector response (about $-2\% yr^{-1}$). As shown in Figure 1b), this drift was predicted by Monte Carlo simulations using time-varying D_2O attenuation, so a correction was applied to the energy estimator. The uncertainty in the energy scale was measured with the ^{16}N calibration to be 1.1%.

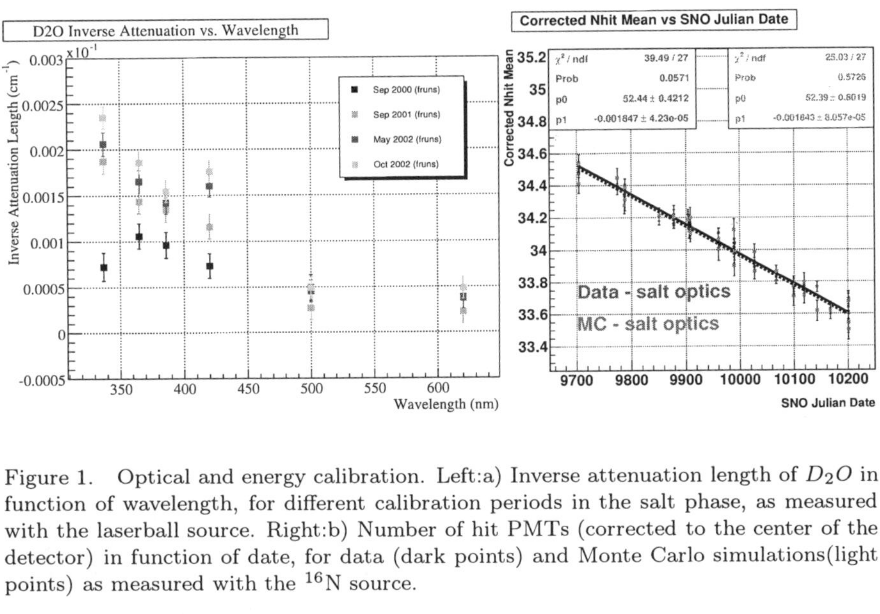

Figure 1. Optical and energy calibration. Left:a) Inverse attenuation length of D_2O in function of wavelength, for different calibration periods in the salt phase, as measured with the laserball source. Right:b) Number of hit PMTs (corrected to the center of the detector) in function of date, for data (dark points) and Monte Carlo simulations(light points) as measured with the ^{16}N source.

The neutron detection efficiency was calibrated with a ^{252}Cf spontaneous fission source and its measured distribution in function of radius, for the analysis kinetic energy threshold of 5.5 MeV, is shown in Figure 2a). The volume-weighted detection efficiency within the analysis fiducial radius of 5.5 m was 0.399 ± 0.012, about a factor of three higher than in the pure D_2O phase.

The angular distributions of the PMT hit pattern of each event was expanded in terms of Legendre polynomials and the combination $\beta_{14} =$

$\beta_1 + \beta_4$ was chosen as an isotropy parameter. Since single Cherenkov ring electron events from CC interactions are less isotropic than multiple ring γ cascades from NC neutrons, β_{14} distributions were used to statistically separate NC and CC events. Figure 2 shows a comparison of data and Monte Carlo β_{14} from the ^{252}Cf (neutrons, multiple γs) and ^{16}N (single γ) sources. The uncertainty in the mean value is 0.87%.

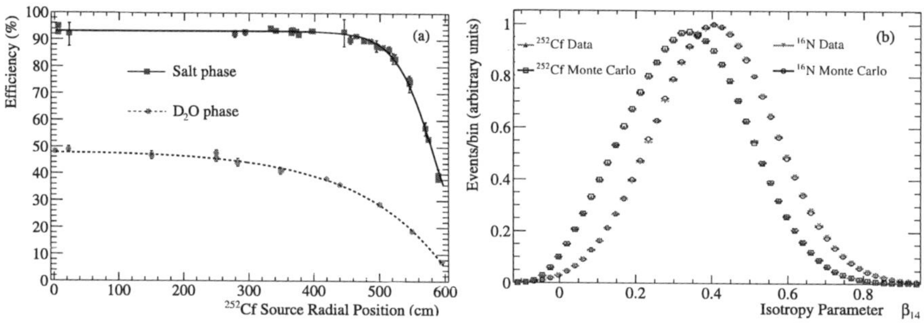

Figure 2. a) Neutron detection efficiency in the salt phase (squares) and in the pure D_2O phase (circles), as measured with a ^{252}Cf source. b) Distributions of the isotropy parameter β_{14} for data and Monte Carlo from the ^{252}Cf and ^{16}N sources.

The dominant backgrounds to the solar neutrino analysis window are deuteron photo-disintegration neutrons due to uranium and thorium chain activity in the D_2O. The average levels of this activity throughout the salt dataset were measured by *ex-situ* and *in-situ* methods, and used to calculate the number of expected neutrons, leading to $74.8^{+24.0}_{-23.6}$ events in the salt dataset. After adding other contributions (like fission, ^{24}Na activation), the total amounts to 113±26 background neutrons. The total contribution from low energy Cherenkov backgrounds was estimated to be lower than 14.7 events (68% C.L.), significantly lower than in the pure D_2Odata analysis due to the higher energy threshold. A class of events characterized by a nearly isotropic light distribution and a reconstructed position close to the acrylic vessel was identified and an upper limit of 5.4 events (68% C.L.) was set for this background. Neutrons produced at the acrylic vessel or in the H_2O can propagate into the fiducial volume, but the enhanced neutron capture efficiency in salt makes the external neutrons readily apparent, so an additional radial distribution function was included in the fits to extract the rate of this component.

2.2. *Solar Neutrino Results*

The maximum likelihood method was applied to extract the fluxes of each neutrino interaction and the external neutron background rate from the salt data set. The likelihood function was constructed with probability density functions of volume-weighted vertex radius ($\rho = (R_{fit}/600)^3$), event angle with respect to the Sun ($cos\theta_{sun}$) and isotropy (β_{14}). As can be seen from the distributions shown in Figure 3, the isotropy parameter allows a good separation of the CC and NC classes of events, making possible an energy-unconstrained and model-independent analysis.

The fitted number of events yield the following fluxes, in units of $10^6 cm^{-2}s^{-1}$:

$$\Phi_{CC} = 1.59^{+0.08}_{-0.07}(stat.)^{+0.06}_{-0.08}(syst.)$$
$$\Phi_{ES} = 2.21^{+0.31}_{-0.26}(stat.) \pm 0.10(syst.)$$
$$\Phi_{NC} = 5.21 \pm 0.27(stat.) \pm 0.38(syst.)$$

The systematic uncertainties are listed in Table II or reference [14].

Adding the constraint of undistorted ^{8}B spectrum for comparison with earlier results yields, in units of $10^6 cm^{-2}s^{-1}$:

$$\Phi_{CC} = 1.70^{\pm}0.07(stat.)^{+0.09}_{-0.10}(syst.)$$
$$\Phi_{ES} = 2.13^{+0.29}_{-0.28}(stat.)^{+0.15}_{-0.08}(syst.)$$
$$\Phi_{NC} = 4.90 \pm 0.24(stat.)^{+0.29}_{-0.27}(syst.)$$

in agreement with the pure D_2O results shown in Section 1.1.

A combined χ^2 analysis of the shape-unconstrained fluxes from the salt phase, the day and night energy spectra from the pure D_2O phase and results from other solar neutrino experiments was carried in order to constrain the allowed regions of the neutrino mixing parameters Δm^2 and $tan^2\theta$. The ^{8}B flux is a free parameter (f_B) in the fit and the hep flux is fixed to $9.3 \times 10^3 cm^{-2}s^{-1}$. This analysis selects the LMA region, as shown in Figure 4a). A global analysis including the results from the reactor neutrino experiment KamLAND[16] further reduces the allowed region (Figure 4b)), and the best fit point is $\Delta m^2 = 7.1^{+1.2}_{-0.6} \times 10^{-5}$, $\theta = 32.5^{+2.4}_{-2.3}$, $f_B = 1.02$ (1σ errors). At 99% C.L., only the lower band of the LMA region is allowed, and maximal mixing is disfavored at a level of 5.4σ.

3. Summary and Prospects

The salt phase of SNO has provided a high precision measurement of the total active ^{8}B solar neutrino flux in a model-independent way, without

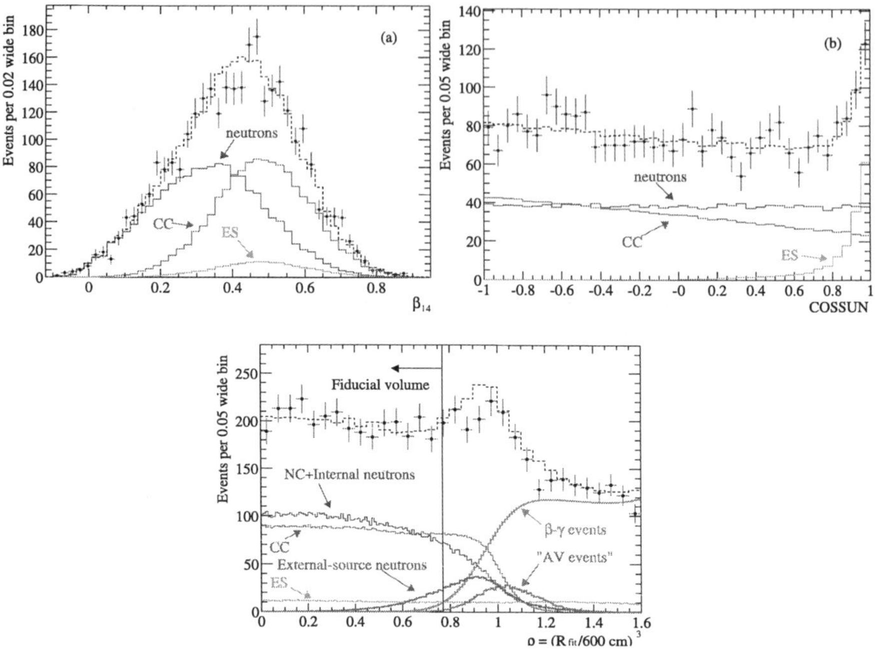

Figure 3. Distribution of (a) β_{14}, (b)$cos\theta_{sun}$ and (bottom plot) volume-weighted radius (ρ) for the selected events. Also shown are the Monte Carlo predictions for CC, ES, NC + internal neutrons and external neutron events scaled to the fit results. The dashed lines represent the summed components.

any assumption on the energy dependence of the neutrino flavor transformation. This has further demonstrated the phenomenon of neutrino flavor transformation and excluded maximal mixing in a scenario of 2-flavor neutrino oscillations.

At the time of this talk, desalination has taken place and detector parameters such as the D_2O attenuation, the absolute energy scale and the neutron capture efficiency have returned to their pre-salt values. Several efforts are ongoing for the deployment and commissioning of the NCD counters in the D_2O volume. The event-by-event neutron counting should provide for a more precise NC flux measurement and CC spectral shape analysis.

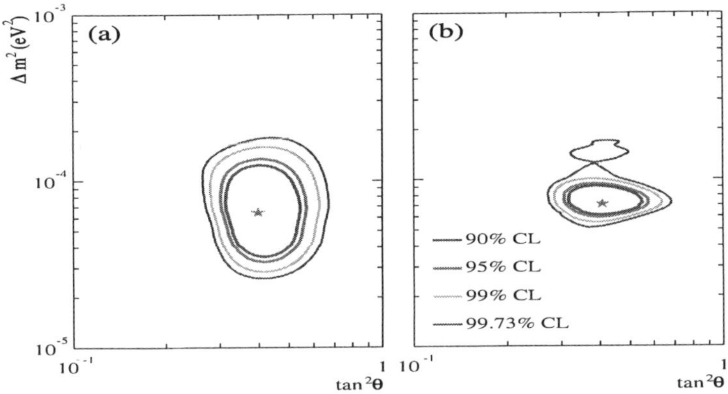

Figure 4. Global neutrino oscillation contours. a) Solar global, b) Solar global + KamLAND.

Acknowledgments

This work is supported by Canada: NSERC, Industry Canada, NRC, Northern Ontario Heritage Fund, Inco, AECL, Ontario Power Generation, HPCVL, CFI; USA: DOE; UK:PPARC. We thank the SNO technical staff for their strong contributions.

References

1. B.T. Cleveland *et al.*, Astrophys. J. **496**, 505 (1998).
2. Y. Fukuda *et al.*, Phys.Rev. Lett. **77**, 1683 (1996).
3. V. Gavrin, 4th Int. Workshop on Low Energy and Solar Neutrinos; Paris, May 19-21, 2003
4. W. Hampel *et al.*, Phys. Lett. **B447**, 127 (1999).
5. T. Kirsten, XXth Int. Conf. on Neutrino Physics and Astrophysics, Munich, May 25-30, 2002; to be published in Nucl. Phys. B Proc. Suppl.
6. S. Fukuda *et al.*, Phys. Lett. **B539**, 179 (2002)
7. J.N. Bahcall, M. H. Pinsonneault, and S. Basu, Astrophys. J. **555**, 990 (2001).
8. A.S. Brun, S. Turck-Chièze, and J.P. Zahn, Astrophys. J. **525**, 1032 (1999); S. Turck-Chièze *et al.*, Ap. J. Lett., v. **555** July 1, 2001.
9. L. Wolfenstein, Phys. Rev. **D17**, 2369-2374 (1978)
10. S.P. Mikheyev, A.Y. Smirnov, Sov. Jour. Nucl. Phys. **42**, 913-917 (1985)
11. Q.R. Ahmad *et al.* (SNO Collaboration), Phys. Rev. Lett. **87**, 071301 (2001).
12. Q.R. Ahmad *et al.* (SNO Collaboration), Phys. Rev. Lett. **89**, 011301 (2002).
13. Q.R. Ahmad *et al.* (SNO Collaboration), Phys. Rev. Lett. **89**, 011302 (2002).
14. S.N. Ahmed *et al.* (SNO Collaboration), e-Print Archive:nucl-ex/0309004, submitted to Phys. Rev. Lett.
15. SNO Collaboration, Nucl. Instr. and Meth. **A449**, 172 (2000).
16. K. Eguchi *et al.*, Phys. Rev. Lett. **90**, 021902(2003)

PRECISION MEASUREMENT OF THE ANOMALOUS MAGNETIC MOMENT OF THE MUON

JONATHAN M. PALEY

For the Muon (g-2) Collaboration
Boston University Physics Department
590 Commonwealth Avenue
Boston, MA 02215, USA
E-mail: jpaley@bu.edu

The present generation of measurements of the anomalous magnetic moment of the muon have reached sub part per million (ppm) precision, at which level they are not only sensitive to electromagnetic and hadronic interactions, but, for the first time, the electroweak interactions. Therefore comparing the experimental results with Standard Model evaluations provides stringent constraints on physics beyond our current model. From the latest measurements at Brookhaven National Laboratory, the anomalous magnetic moment is now known to within 0.5 ppm. These measurements have sparked much discussion in the physics community, as well as a re-examination of the theoretical evaluations of the anomaly. In 2001, measurements were made with negative muons that give a_{μ^-} to 0.7 ppm. I describe the theory, experiment, and analysis techniques used to determine a_{μ^-}.

1. Introduction

Anomalous magnetic moments have provided great insight into the field of subatomic particle physics. In particular, the ultra-precise determination of the anomalous magnetic moment of the electron provides the best determination of the QED coupling constant, α, at $q^2 = 0$. The present measurements of the anomalous magnetic moment of the muon, a_μ, provide a unique and stringent test of the electroweak theory in the Standard Model of particle physics, and are also a sensitive test of physics beyond the Standard Model.

2. Theoretical value of a_μ

The anomalous magnetic moment is defined as $a = (g-2)/2$ where g is the gyromagnetic ratio relating the magnetic moment, $\vec{\mu}_s$ to the spin, $\vec{S}$ of a particle:

$$\vec{\mu}_s = g\frac{e}{2m}\vec{S} \tag{1}$$

The Standard Model calculation of a_μ is typically broken up into parts as $a_\mu^{\rm SM} = a_\mu^{\rm QED} + a_\mu^{\rm Weak} + a_\mu^{\rm Had}$, where the contributions to the anomaly are grouped according to the fundamental interactions involved: QED, Weak, and Hadronic (Had, or QCD). The weak interaction loops contribute about 1.3 ppm to the total value of a_μ, so a measurement of a_μ at the sub-ppm level provides a nice test of the electroweak theory. However, the total uncertainty on $a_\mu^{\rm SM}$, at the level of ~ 0.7 ppm (both the QED and Weak contributions are known to within 0.03 ppm), is dominated by the uncertainty on $a_\mu^{\rm Had}$. This contribution cannot be calculated on first principles and is rather determined via a combination of a dispersion relation and the optical theorem[1]. Hence, the vacuum polarization hadronic states are related to real final states of $e^+e^- \to$ hadrons, and the theoretical uncertainty of a_μ is directly related to the uncertainty on the cross section $\sigma(e^+e^- \to \text{hadrons})$. An indirect determination of the cross section involves using the τ spectral function and the conserved vector current hypothesis (with the appropriate isospin corrections). Unfortunately the pion form factors derived from the e^+e^- and τ data differ by $\sim 10\%$ just above the ρ-resonance energy region. The difference between these two approaches results in a $\sim 3\sigma$ difference between $a_\mu^{\rm Had}(e^+e^-)$ and $a_\mu^{\rm Had}(\tau)$, complicating the interpretation of the measurement of a_μ[2,3].

3. Experiment 821 at BNL

The muon (g-2) experiment at Brookhaven National Laboratory (experiment 821) is based on muon spin precession in a magnetic field. Protons extracted from the Alternating Gradient Synchrotron (AGS) at 24 GeV/c strike a water-cooled nickel target. Pions produced in the proton-nickel collision are collected and transported down a 72 meter-long straight channel in which approximately one third of the pions decay to muons. Forward-going muons are selected, resulting in a highly ($\sim 95\%$) polarized muon beam of momentum ~ 3.09 GeV/c.

Next, these muons are injected into a storage ring with a magnetic field of ~ 1.45 T. The muon beam is focused inside the storage ring using electrostatic quadrupole fields[4]. In practice, the muon beam is injected outside the aperture of the storage ring, and therefore receives a "kick" at approximately 1/4 the betatron length around the ring via a fast, pulsed, non-ferric magnetic kicker[5], placing the beam on a stable orbit inside the aperture.

The muon (g-2) storage ring is a 680 metric ton, continuous C-shaped, super-ferric (superconducting coils with an iron yoke) magnet of diameter 14.2 m. Shimming of the magnet has reduced the higher multipoles to $\lesssim 1$ ppm at the edge of the storage aperture, and has improved the overall homogeneity of the vertical magnetic field inside the storage region such that on average, the field varies by less than 4 ppm across the storage aperture.

Inside the storage ring, the spin vector precesses[6,7] about the momentum vector at a frequency ω_a given as

$$\vec{\omega}_a = \frac{e}{m}\left[a_\mu \vec{B} - \left(a_\mu - \frac{1}{\gamma^2 - 1}\right)(\vec{\beta} \times \vec{E})\right] \tag{2}$$

The above relation assumes $\vec{\beta} \cdot \vec{B} = 0$. The value of γ, and hence the momentum of the muon beam, is chosen such that the $\vec{\beta} \times \vec{E}$ term vanishes. Using this so-called "magic gamma" ($\gamma = \gamma_m \simeq 29.3$), the anomalous magnetic moment of the muon is

$$a_\mu = \frac{m}{e}\frac{\omega_a}{B} \tag{3}$$

The spin difference frequency, ω_a, is determined from the parity-violating decay of the muon: The decay electron is emitted preferentially in the direction of the spin of the muon. Therefore the number of decay electrons above some energy threshold E_t as a function of time is

$$N(t) = N_0(E_t)\exp(-t/\gamma\tau)[1 + A(E_t)\cos(\omega_a t + \phi(E_t))] \tag{4}$$

where the number (N_0), the asymmetry (A) and the phase (ϕ) are all functions of threshold energy. Ideally, the value of ω_a is extracted by fitting the decay electron time spectrum to Eq. 4. The expected relative statistical uncertainty in ω_a is

$$\frac{\delta\omega_a}{\omega_a} \simeq \frac{\sqrt{2}}{\omega_a \gamma\tau A\sqrt{N}} \tag{5}$$

Since both A and N depend on energy-threshold, the energy-threshold is chosen such that the product NA^2 is maximized in order to reduce the statistical uncertainty on ω_a. Typically, the value of E_t that maximizes NA^2 is ~ 1.8 GeV.

Twenty-four electromagnetic calorimeters[8] placed uniformly around the inside of the storage ring measure the time and energy spectra of the decay

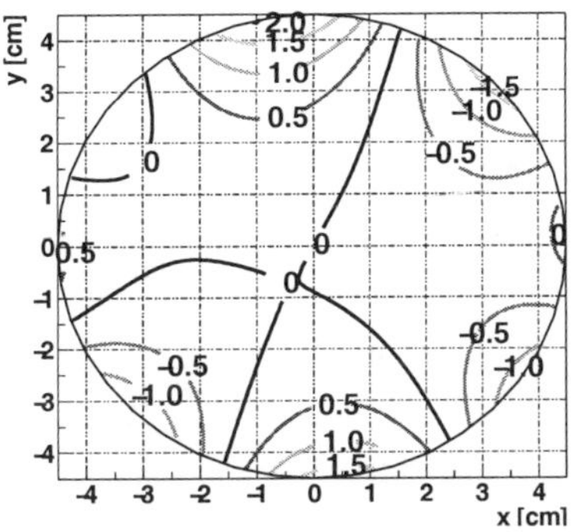

Multipoles (ppm)

	normal	skew
Quad	-0.29	0.15
Sext	-0.71	-0.48
Octu	0.06	0.01
Decu	1.08	0.39

Figure 1. Contour plot of the dipole magnetic field, averaged over azimuth, in the storage aperture. All values are relative differences with respect to the field in the center of the aperture. The contour lines are 0.5 ppm wide.

electrons. The quadrants of the calorimeter are read out by four photomultiplier tubes (PMTs). The analog signal from the calorimeter PMTs is digitized via a 400 MHz waveform digitizer, from which an energy and time are extracted for each decay electron pulse.

The magnetic field of the g-2 storage ring is measured using pulsed NMR, and hence the field measurement is actually that of the Larmor precession frequency (ω_p) of the proton in water in a ~ 1.45 T field. Eq. 3 may be rewritten as

$$a_\mu = \frac{\mathcal{R}}{\lambda - \mathcal{R}}, \quad \text{where} \quad \lambda = \frac{\mu_\mu}{\mu_p}, \quad \text{and} \quad \mathcal{R} = \frac{\omega_a}{\omega_p} \tag{6}$$

λ has been measured using both the hyperfine structure in muonium and the Zeeman transition frequency in muonium[9]. A combination of theory and experiment has determined λ with an uncertainty of 0.03 ppm[11]. The ultimate goal of experiment E821 is to determine the ratio $\mathcal{R}$, and hence a_μ, with a total uncertainty of ~ 0.35 ppm.

4. Magnetic Field Measurement

The magnetic field of the g-2 storage ring is measured every few days during the data collection run via a beam-tube trolley consisting of 17 NMR probes. As the trolley is pulled around the ring on a fixed track inside the ring under vacuum, it measures the field every few millimeters, providing a 3-dimensional map of the magnetic field in terms of $f_p = \omega_p/2\pi$. In between trolley measurements, changes in the field were monitored using NMR probes fixed on the top and the bottom of the storage ring vacuum walls. Furthermore, a feedback loop consisting of a subset of the fixed

Table 1. Systematic error table for the ω_p measurement.

ω_p **Systematic Error Table**	
Source of Error	**Size (ppm)**
Absolute calibration	0.05
Calibration of trolley probes	0.09
Trolley measurements of B_0	0.05
Interpolation with fixed probes	0.07
Uncertainty from muon distribution	0.03
Others†	0.10
Total	0.17

† Includes systematic effects from higher multipoles, trolley temperature and voltage response, and eddy currents from the muon kickers.

probes adjusts the current in the superconducting coils to stabilize the field to within 0.1 ppm.

In calculating the value of a_μ, a weighted average of ω_p over the muon distribution in both space and time must be taken. Furthermore, the trolley probes are calibrated in air with respect to a standard spherical water probe before and after data collection periods. Table 1 lists the systematic uncertainties involved in the measurement of ω_p. The final value of $\langle f_p \rangle = \langle \omega_p / 2\pi \rangle$ for the 2001 data run is 61791400(11) Hz.

5. Spin Precession Frequency Measurement

As mentioned in Section 3, the time spectrum of decay electrons above a threshold energy would ideally be fit to the five-parameter function in Eq. 4:

$$N(t) = N_0 \exp(-t/\gamma\tau)[1 + A\cos(\omega_a t + \phi)] \tag{7}$$

However, because of various background effects, fitting this function to the data results in an unacceptable χ^2/dof by more than 20σ. The largest background effects are coherent betatron oscillations (CBO), pileup (overlapping pulses), gain, and muon losses. All of these backgrounds distort the decay electron time (and energy) spectrum in a time-dependent manner.

The author chose an analysis technique which essentially divides out the exponential decay in the time spectrum. The method[10] involves shifting the times of the decay electrons in one quarter of the data set forward by one half of a g-2 cycle, another quarter of the data set backward by one half of a g-2 cycle, and leaving the rest of the data untouched. The ratio, $r(t)$, defined as

$$r(t) = \frac{N(t+T/2) + N(t-T/2) - 2N(t)}{N(t+T/2) + N(t-T/2) + 2N(t)} \tag{8}$$

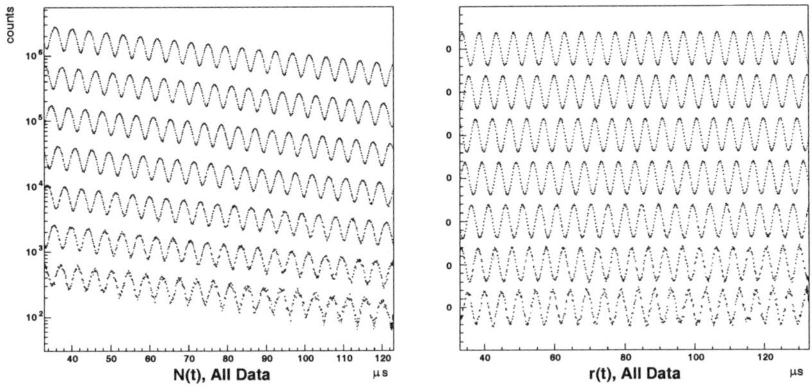

(a) Original spectrum, $N(t)$. (b) Ratio, $r(t)$.

Figure 2. Original $N(t)$ spectrum and ratio function, both produced using all 2001 data. Note that $N(t)$ is plotted on a log vertical scale. The horizontal axis wraps around every 90 μs, starting from 33 μs after injection.

is then formed, converting the ideal five-parameter function (Eq. 4) into a simple three-parameter cosine function (the unknown parameters are an amplitude, frequency (ω_a) and a phase). Fig. 2 shows the original time spectrum on the left, and the ratio, constructed from the same data set, on the right.

The advantage of fitting the ratio, $r(t)$, is an order of magnitude decrease in the sensitivity of ω_a to slow effects, such as gain and muon losses, compared to the traditional approach of fitting the time spectrum itself, $N(t)$. The ratio method still requires pileup and CBO to be dealt with appropriately [12]. In the author's case, pileup is statistically constructed from the data and subtracted from the time spectrum, and the ratio function is modified to account for all known CBO effects. Table 2 lists the systematic uncertainties on ω_a as determined in the author's analysis.

The final value of ω_a is obtained from a simple arithmetic mean of the results from five different ω_a analyses on the 2001 data set [12]. The analyses differ primarily in the applied energy cuts, methods and fit functions used to extract ω_a. The fit results from the analyses have been shown to be in agreement to within the expected deviations. We find $f_a = \omega_a/2\pi =$ 229073.59(15)(5) Hz.

Table 2. Systematic error table for the ω_a measurement.

ω_a **Error Table**	
Source of Error	**Size (ppm)**
Gain	0.1
Pileup	0.05
Coherent Betatron Oscillations	0.08
Muon Losses	0.10
E-field and Pitch Corrections	0.06
Others‡	0.11
Total Syst. Uncertainty	0.21
Statistical Uncertainty	0.69

‡ Includes systematic effects from other beam dynamics effects, AGS background, timing shifts and fit procedure.

6. Results

The value of a_{μ^-} was calculated according to the relation in Eq. 6:

$$a_\mu = \frac{\mathcal{R}}{\lambda - \mathcal{R}} = 11659214(8)(3) \times 10^{-10} \tag{9}$$

where $\mathcal{R}_{\mu^-} = \omega_a/\omega_p = 0.0037072083(26)$ and $\lambda = \mu_\mu/\mu_p = 3.18334539(10)$[11]. The first error is statistical and the second is systematic. The measured value of $\mathcal{R}_{\mu^-}$ is in good agreement with the average of the previous measurements of $\mathcal{R}_{\mu^+} = 0.0037072048(25)$, the difference being $\Delta\mathcal{R} = (3.5 \pm 3.4) \times 10^{-9}$. Assuming charge, parity and time (CPT) symmetry and taking into account correlations between systematic uncertainties between the various data sets, the new average value of $\mathcal{R}_\mu = 0.0037072063(20)$. From this result we obtain the new average value of $a_\mu^{\text{exp}} = 11659208(4)(3) \times 10^{-10}$, a relative uncertainty of 0.5 ppm [12].

Fig. 3 shows a comparison between the experimental values of $a_{\mu^+}^{\text{exp}}$ (weighted average of the 1998, 1999 and 2000 BNL data sets), $a_{\mu^-}^{\text{exp}}$ (from the analysis of the 2001 BNL data set), their weighted average, and two theoretical values of a_μ^{SM}, one based on the e^+e^- hadronic cross section and the other based on the τ spectral function. The differences between the weighted average experimental value and the theoretical value of a_μ are: $\Delta a_\mu(e^+e^-\text{−based}) = (27.0 \pm 10.0) \times 10^{-10}$ and $\Delta a_\mu(\tau\text{−based}) = (12.3 \pm 9.2) \times 10^{-10}$.

Acknowledgments

Experiment E821 at BNL was supported in part by the U.S. Department of Energy, the U.S. National Science Foundation, the German Bundesmin-

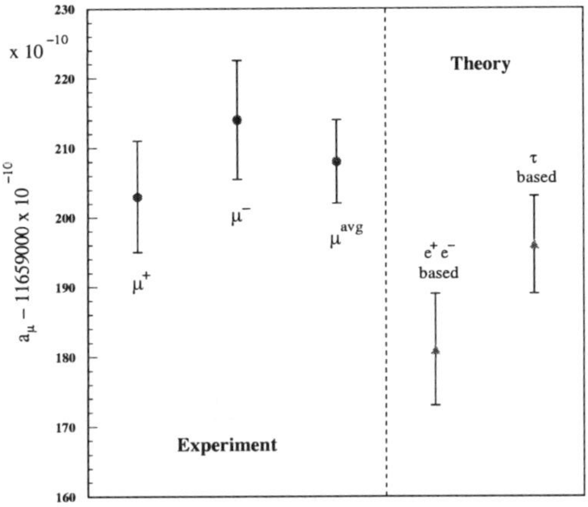

Figure 3. Comparison of experimental and theoretical values of a_μ. The value for $a_\mu^{\text{avg}}(\text{exp})$ assumes CPT and is the weighted average of $a_\mu^+(\text{exp})$ and $a_\mu^-(\text{exp})$.

ister für Bildung und Forschung, the Russian Ministry of Science, and the U.S.-Japan Agreement in High Energy Physics. Funding for this work was provided by the U.S. National Science Foundation's grant to the Boston University Intermediate Energy Group.

References

1. M. Gourdin and E. de Rafael, *Nucl. Phys. B* 10, 667 (1969)
2. M. Davier, *et. al.*, *Eur. Phys. J* C31, 503-510 (2003)
3. S. Ghozzi and F. Jegerlehner, *Z. Phyz.* C67, 585-602 (2003)
4. Y. Semertzidis, *et. al.*, *Nucl. Inst. A* 503, 458-484 (2003).
5. E. Efstathiadis, *et. al.*, *Nucl. Inst. A* 496, 8-25 (2003).
6. L. H. Thomas, *Phil. Mag.* **3**, 1 (1927).
7. J. D. Jackson, *Classical Electrodynamics*, John Wiley & Sons, New York (1975).
8. S. A. Sedykh, *et. al.*, *Nucl. Inst. A* 455, 346-360 (2000).
9. W. Liu, *et. al.*, *Phys. Rev. Lett.* 82, 711 (1999)
10. J. M. Kindem, "The Anomalous Magnetic Moment of the Positive Muon", Ph.D. Thesis, University of Minnesota (1998)
11. K. Hagiwara, *et. al.*, "Review of Particle Physics", *Phys. Rev. D* 66 (2002).
12. G. W. Bennet, *et. al.*, *Phys. Rev. Lett.* 92, 161802 (2004).

MEASUREMENT OF THE STRUCTURE FUNCTIONS g_1^p AND g_1^d WITH THE CLAS AT JEFFERSON LAB

YELENA PROK

University of Virginia, Physics Department, Charlottesville VA 22901, USA
E-mail: yprok@jlab.org

Abstract. Inelastic scattering using polarized nucleon targets and polarized charged lepton beams allows the extraction of the structure functions g_1 and g_2 which provide information on the spin structure of the nucleon. A program designed to study such processes has been underway in Jefferson Lab since 1998. A polarized electron beam, solid polarized NH_3 and ND_3 targets and the CEBAF Large Acceptance Spectrometer (CLAS) in Hall B were used to collect the desired data. 3 billion events were accumulated during the first run, and over 23 billion events were accumulated during the second run. The measurements cover the resonance region with unprecedented detail and add significantly to the DIS data set at low to moderate Q^2 and moderate to high x. The inclusive analysis of the data collected with beam energies of 1.6 GeV and 5.6 GeV has been mostly completed, yielding such important quantities as the photon nucleon asymmetry A_1, the spin structure function g_1, and its first moment Γ_1. These quantities allow us to study various aspects of the nucleon spin structure and to distinguish between different models of the valence spin structure. Measurements at low Q^2 are particularly interesting, as they shed light on the dynamics of transition from the partonic to hadronic degrees of freedom.

REFERENCES

1. Gerasimov S., *Yad. Phys.* **2**, 598 (1965)
2. Drell S. and Hearn A. *Phys. Rev. Lett.* **16**, 908 (1966)
3. Burkert V. and Ioffe B., *Phys. Lett. B* **296**, 223 (1992)
4. Soffer J. and Teryaev O., *Phys. Rev. D* **51**, 25 (1992)
5. Abe K. *et al.*, *Phys. Rev. D* **58**, 112003 (1998)
6. Close F.E. and Melnitchouk W., *hep-ph/0302013* (2003)
7. Isgur N., *Phys.Rev.D* **59**, 034013 (1999)
8. Bloom E. and Gilman F., *Phys. Rev. D* **4**, 2901 (1971)

NEUTRINO EXPERIMENTS WITH SUPER-KAMIOKANDE

M. B. SMY

Department of Physics and Astronomy
University of California, Irvine
Irvine, California 92697-4575
E-mail: msmy@uci.edu

Super-Kamiokande's measurements of the solar and atmospheric neutrino flux yielded first evidence for oscillation of atmospheric ν_μ's into ν_τ's and (together with data from the SNO experiment) first evidence for oscillation of solar ν_e's into ν_μ's or ν_τ's. Data from Super-Kamiokande also constrained the neutrino mixing in both cases to be large. In Super-Kamiokande, the flux and energy spectrum of the first long baseline accelerator ν_μ's were measured, they favor neutrino oscillations consistent with the atmospheric neutrino analysis.

1. Neutrino Oscillation and Super-Kamiokande

Both solar and atmospheric neutrino measurements have yielded significantly smaller interaction rates than expected: the reduction of the solar ν_e flux compared to the standard solar model (SSM [1]) was observed to be between 50% and 65% [2], the atmospheric ν_e flux was found to be the same within uncertainty while the atmospheric ν_μ flux was measured at about 65% of the calculation [3]. Both deficits — known as the solar neutrino problem and the atmospheric anomaly — are explained by neutrino flavor changes caused by oscillations: solar neutrino experiments detect predominantly ν_e and atmospheric experiments are less sensitive to ν_τ interactions.

In case of three neutrinos (and neglecting CP violating phases), the mixing matrix U relating the mass eigenstates $\nu_{1,2,3}$ to the flavor eigenstates $\nu_{e,\mu,\tau}$ can be described phenomenologically using the three mixing angles θ_{12}, θ_{13} and θ_{23}. Using the short notation $s_{ij} = \sin\theta_{ij}, c_{ij} = \cos\theta_{ij}$ the mixing matrix becomes

$$U = \begin{pmatrix} c_{12}c_{13} & s_{12}c_{13} & s_{13} \\ -s_{12}c_{23} - c_{12}s_{23}s_{13} & c_{12}c_{23} - s_{12}s_{23}s_{13} & s_{23}c_{13} \\ s_{12}s_{23} - c_{12}c_{23}s_{13} & -c_{12}s_{23} - s_{12}c_{23}s_{13} & c_{23}c_{13} \end{pmatrix}$$

and the flavor conversion probability for a neutrino of energy E after flight distance L in vacuum is $P_{vac}(\nu_\alpha \longrightarrow \nu_\beta) = \delta_{\alpha\beta} - 4\cdot\sum_{j<k} U_{\alpha j}U_{\beta j}U_{\alpha k}U_{\beta k}O_{jk}$ with $O_{jk} = \sin^2 \frac{\Delta m^2_{jk}L}{4E}$. Atmospheric neutrino data requires [3] a Δm^2_{atm} around 10^{-3}eV^2, while solar neutrino data implies a Δm^2_{sol} below $\approx 10^{-4}\text{eV}^2 \ll \Delta m^2_{\text{atm}}$. For the purpose of this paper, the Δm^2_{sol} is assumed to be Δm^2_{12}. Then, $\Delta m^2_{\text{atm}} \approx \Delta m^2_{13} \approx \Delta m^2_{23}$, $O_{\text{atm}} = \sin^2 \frac{\Delta m^2_{\text{atm}}L}{4E}$, $O_{\text{sol}} = \sin^2 \frac{\Delta m^2_{\text{sol}}L}{4E}$, $\sin\theta_{\text{atm}} = \cos\theta_{13}\sin\theta_{23}$, and $\theta_{\text{sol}} = \theta_{12}$. The atmospheric conversion probabilities are

$$\begin{aligned}
P_{\text{vac}}(\nu_e \longrightarrow \nu_\mu) &\approx \sin^2 2\theta_{13}\sin^2\theta_{23}O_{\text{atm}} \\
P_{\text{vac}}(\nu_\mu \longrightarrow \nu_\tau) &\approx \left(\sin^2 2\theta_{\text{atm}} - \sin^2 2\theta_{13}\sin^2\theta_{23}\right)O_{\text{atm}} \\
P_{\text{vac}}(\nu_\tau \longrightarrow \nu_e) &\approx \sin^2 2\theta_{13}\cos^2\theta_{23}O_{\text{atm}} \\
P_{\text{vac}}(\nu_e \longrightarrow \nu_x) &\approx \sin^2 2\theta_{13}O_{\text{atm}} \\
P_{\text{vac}}(\nu_\mu \longrightarrow \nu_x) &\approx \sin^2 2\theta_{\text{atm}}O_{\text{atm}}
\end{aligned} \tag{1}$$

while the solar ν_e survival probability is

$$P_{\text{vac}}(\nu_e \longrightarrow \nu_e) \approx \cos^4\theta_{13}\left(1 - \sin^2 2\theta_{\text{sol}}O_{\text{sol}}\right) + \sin^4\theta_{13}. \tag{2}$$

Therefore, atmospheric neutrinos yield information about the last column of the mixing matrix and solar neutrinos about the first row. If θ_{13} is small, then both solar and atmospheric neutrinos can be analyzed using a two-neutrino model with effective solar and atmospheric mixing angles.

Matter effects can be taken into account by defining the potential V and replacing $\Delta m^2, \theta$ with the effective parameters $\Delta m^2_{\text{mat}}, \theta_{\text{mat}}$

$$\frac{\left(\Delta m^2_{\text{mat}}\right)^2}{\left(\Delta m^2\right)^2} = \left(\frac{2E\Delta V}{\Delta m^2} + \cos 2\theta\right)^2 + \sin^2 2\theta \qquad \frac{\sin^2 2\theta_{\text{mat}}}{\sin^2 2\theta} = \frac{\Delta m^2}{\Delta m^2_{\text{mat}}} \tag{3}$$

where ΔV is the potential difference between the neutrino flavors. Depending on the sign of ΔV, a resonance can occur [4] (MSW effect) which leads to a large conversion probability, even if the vacuum mixing is small.

Super-Kamiokande (SK) is a cylindrical 50,000 ton water Cherenkov detector. An optical barrier separates an "inner" concentric cylinder (32,000 tons) from the outer "anti-counter". [a] The detector is further described elsewhere [5]. Between May 31st, 1996 and July 15th, 2001 SK exposed the 22,500 tons contained in its fiducial volume (concentric cylinder 2m away from the optical barrier) for 1496 effective days to solar neutrinos and for 1489 effective days to atmospheric neutrinos.

[a] The anti-counter separates "fully contained events" (no entering or exiting charged particles) from "partially contained events" (only exiting charged particles) and "entering events".

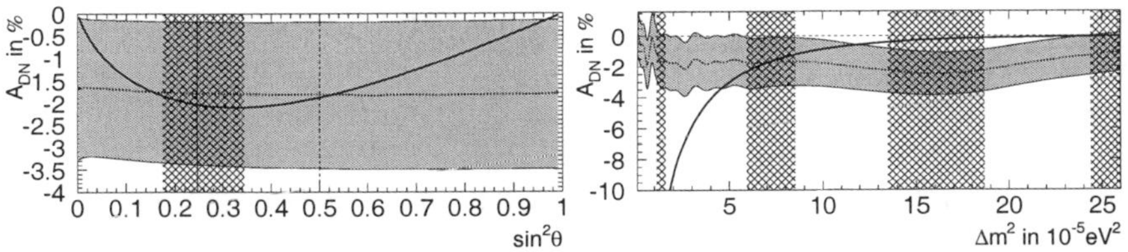

Figure 1. (a) Fitted SK Day/Night Asymmetry as a Function of Mixing. The Δm^2 is $6.3 \times 10^{-5}\text{eV}^2$. The hatched area corresponds to the $\pm 1\sigma$ uncertainty of the ^{8}B flux by Junghans et al [1]. (b) Fitted SK Day/Night Asymmetry as a Function of Δm^2. The hatched area corresponds to the 95% allowed contours reported by the KamLAND collaboration. The gray band is the $\pm 1\sigma$ SK measurement. The solid black line shows the oscillation prediction of the day/night asymmetry.

2. Solar Neutrinos

All solar data[2] are well described with oscillations involving just two neutrino mass eigenstates and imply a mass squared difference between $\Delta m^2 = 3 \times 10^{-5}\text{eV}^2$ and $\Delta m^2 = 1.9 \times 10^{-4}\text{eV}^2$ and a mixing angle between $\tan^2\theta = 0.25$ and $\tan^2\theta = 0.65$ [6]. This region of parameter space is referred to as the Large Mixing Angle solution (LMA). The rate and spectrum of reactor anti-neutrino interactions in the KamLAND experiment [7] are also well reproduced for these mixing angles and some of these Δm^2. Over the Δm^2 range of the LMA, solar ^{8}B neutrinos are ≈100% resonantly converted into the second mass eigenstate by the large matter density inside the sun [4] leading to a ν_e survival probability of $\approx \sin^2\theta$. However, due to the presence of the earth's matter density, the observed probability on earth differs from $\sin^2\theta$ during the night. SK measures the energy, direction, and time of the recoil electron from elastic scattering of solar neutrinos with electrons by detection of the emitted Cherenkov light. Since this process is primarily sensitive to ν_e's, the changing probability induces an apparent dependence of the measured neutrino interaction rate on the solar zenith angle (often a regeneration of ν_e's during the night). SK employs a maximum likelihood fit to the expected solar zenith angle dependence on the neutrino interaction rate [8].

The solar zenith angle θ_z between the solar direction and the vertical direction defines day ($\cos\theta_z < 0$) and night ($\cos\theta_z > 0$). The day/night rate asymmetry is defined as $A_{\text{DN}} = \frac{D-N}{0.5(D+N)}$ where D (N) refers to the average neutrino interaction rate during the day (night). If the neutrino interaction rate during the night varies significantly from the average night rate N, and if the functional form (shape) of this variation is known, the amplitude of this time variation of the rate can be determined more accurately then just calculating A_{DN} from the average rates. These conditions are met for two-neutrino oscillations in the LMA region. In [8] a maximum likelihood fit to the SK data finds a day/night amplitude equivalent to $A_{\text{DN}} = -1.8 \pm 1.6(\text{stat})^{+1.3}_{-1.2}(\text{syst})\%$. The fit assumes $\Delta m^2 = 6.3 \times 10^{-5}\text{eV}^2$ and $\tan^2\theta = 0.55$. The asymmetry calculated from the measured average day and night rates on the other hand is $A_{\text{DN}} = -2.1 \pm 2.0(\text{stat})^{+1.3}_{-1.2}(\text{syst})\%$ [6]. (assuming a step function for the time variation). The dependence of the fitted day/night amplitude on the mixing angle $\sin^2\theta$ is shown in Figure 1(a). Overlaid are the predicted asymmetries and the solar model constraint of the ^{8}B neutrino flux from Junghans et al [1]. The Δm^2 dependence is stronger as can be seen in Figure 1(b). Overlaid are the predicted asymmetries and bands (typically called LMA-0, LMA-I, LMA-II, etc) corresponding to the KamLAND 95% allowed contours: the SK day/night measurement excludes LMA-0, and favors LMA-I.

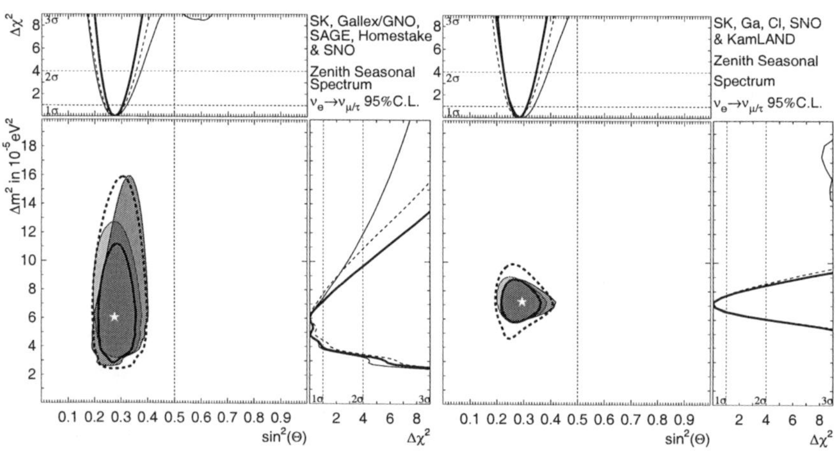

Figure 2. Allowed Regions at 95% C.L. from solar neutrino data (left) and solar neutrino & KamLAND (right) reactor neutrino data. The functions at the top and right of each panel are the marginalized $\Delta\chi^2$ distributions. The dark gray areas/solid lines use the solar neutrino data from Gallex/GNO, SAGE, Homestake & SK the SSM neutrino fluxes (and the Junghans ^{8}B flux constraint), the light gray areas/dashed lines) solar measurements from SK & SNO and no neutrino flux constraints from solar models. The thick solid (dashed) black lines show the 95% C.L. (99.73% C.L.) from all solar exeperiments.

An oscillation analysis of the SK data by itself is found in [8]. It describes the solar zenith angle variation with a likelihood, while the spectrum is fit with a χ^2 method. Figure 2 shows in (dark gray) the allowed regions at 95% C.L. resulting from the combination of experimental data from Gallex/GNO, SAGE, the Homestake experiment and SK. It relies on the ^{8}B flux from Junghans and six low energy neutrino fluxes of the standard solar model [1]. Also shown is a combined fit to SK data, the salt-phase SNO rate measurements, and the SNO day/night asymmetry (which does not rely on any neutrino flux prediction) and a combined fit from all experiments (thick lines). All analyses yield a unique allowed region – the LMA solution – and agree very closely in mixing. Assuming CPT invariance, all three fits are then combined with a binned likelihood analysis [7] of the KamLAND reactor anti-neutrino measurements, the results of which are shown in the right panel. LMA-I is favored over all other solutions by 3.5σ. The oscillation χ^2 is Gaussian; the parameters are determined as $\Delta m^2 = 7.1^{+0.6}_{-0.5} \times 10^{-5}\text{eV}^2$ and $\sin^2\theta = 0.276^{+0.033}_{-0.026}$. At those parameters, the day/night asymmetry is expected to be -1.6% while the amplitude fit to SK data yields $-1.7 \pm 1.6(\text{stat})^{+1.3}_{-1.2}(\text{syst})\%$. [b]

3. Atmospheric Neutrinos and K2K Neutrino Beam

KEK has pointed an almost pure ν_μ beam (from π decays) at SK (K2K experiment). Since the neutrino flux and direction is known, the neutrino energy can be kinematically reconstructed in quasi-elastic neutrino interactions, so both the SK neutrino interaction rate and the neutrino spectrum can be analyzed for neutrino oscillations. The unoscillated beam is measured at KEK with a miniature version of SK, a water Cherenkov detector of about 1,000 tons mass and fine-grained tracking detectors (to separate quasi-elastic neutrino interactions). Atmospheric π's produce both ν_e's and ν_μ's (as well as their anti neutrinos) due to the subsequent μ decay. In charged-current interactions, the flavor of the neutrino can be inferred from the produced lepton. Since the decay photons of neutral pions produce electrons by Compton scattering, a neutral-current enhanced multi-ring sample can be defined. Partially contained events are almost always due to higher energy muons. ν_μ interactions in the rock below SK produce upward-going entering μ's. The energy of those muons can only be estimated statistically

[b]Most other combined oscillation analyses show a 3σ-allowed LMA-II contour due to the use of the SK zenith spectrum χ^2 instead of the likelihood employed in this report.

(a)

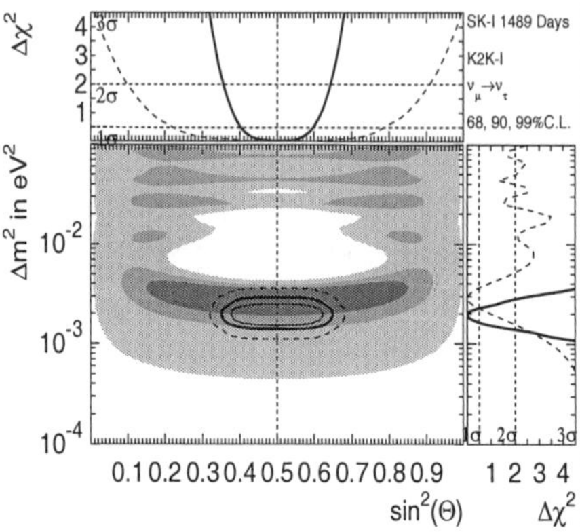

(b)

Mode	best fit parameters	$\Delta\chi^2$
$\nu_\mu \to \nu_\tau$	$\sin^2 2\theta = 1.00$	0.0
	$\Delta m^2 = 2.1 \cdot 10^{-3}\mathrm{eV}^2$	0.0σ
$\nu_\mu \to \nu_e$	$\sin^2 2\theta = 0.97$	110.5
	$\Delta m^2 = 5.1 \cdot 10^{-3}\mathrm{eV}^2$	10.5σ
$\nu_\mu \to \nu_s$	$\sin^2 2\theta = 0.98$	48.9
	$\Delta m^2 = 2.9 \cdot 10^{-3}\mathrm{eV}^2$	7.0σ
LxE	$\sin^2 2\theta = 0.90$	107.8
	α=$5.6 \cdot 10^{-4}/(\mathrm{GeVkm})$	10.4σ
ν_μ decay	$\cos^2\theta = 0.50$	105.6
(short)	$\alpha = 3.7 \cdot 10^{-3}\mathrm{GeV/km}$	10.3σ
ν_μ decay	$\cos^2\theta = 0.33$	20.2
(long)	$\alpha = 1.2 \cdot 10^{-3}\mathrm{GeV/km}$	4.5σ
ν_μ deco-	$\sin^2 2\theta = 0.98$	10.5
herence	$\gamma = 7.3 \cdot 10^{-3}\mathrm{GeV/km}$	3.2σ

Figure 3. (a) Allowed oscillation parameters from SK atmospheric neutrinos for a two-neutrino fit assuming pure $\nu_\mu \leftrightarrow \nu_\tau$ oscillations (lines). Overlaid are the allowed regions from the K2K experiment. The panels on the right hand side and top give the marginalized χ^2: the solid like is from SK atmospheric data, the dashed line from K2K data. (b) Comparison and best-fit parameters of several flavor conversion hypotheses. $\nu_\mu \to \nu_\tau$ is strongly favored by the data.

using Monte Carlo distributions. The average energy is lower for muons that stop inside the detector.

The zenith angle θ of high energy leptons is correlated with the neutrino flight distance L, so neutrino oscillations generate up-down asymmetries in the measured atmospheric flux. While the number of $e-$like events agrees with expectation, the $\mu-$like event rate is significantly less. The muon rate from the KEK beam is also smaller than expected. Downward-going atmospheric neutrino-induced μ-like events agree with the Monte Carlo, upward-going events show the strongest deficit. No such distortion is seen in the multi-ring neutral-current enhanced sample. Figure 3 gives allowed ranges for neutrino mixing and Δm^2 assuming oscillations of ν_μ's into ν_τ's (the best explanation of SK atmospheric data). The regions based on the KEK beam data (SK rate and neutrino spectrum) are also shown. The data prefer maximal mixing and $\Delta m^2 \approx$2–3$\cdot 10^{-3}\mathrm{eV}^2$.

SK data rule out $\nu_\mu \leftrightarrow \nu_e$ oscillations but allow a substantial "ν_e" component (about 30% at 90% C.L.) in the ν_μ oscillations. Figure 4(a) shows the allowed values for θ_{13} as a function of $\Delta m^2_{\mathrm{atm}}$ in a three-neutrino oscillation analysis. Zero mixing for θ_{13} is preferred. In a similar way, the "sterile content" of the oscillations was investigated for $\theta_{13} = 0$ and the Δm^2 hierarchy $\Delta m^2_{\mathrm{LSND}} \approx 1\mathrm{eV}^2 \gg \Delta m^2_{\mathrm{atm}} \approx 10^{-3}\mathrm{eV}^2 \gg \Delta m^2_{\mathrm{sol}} \approx 10^{-4}\mathrm{eV}^2$

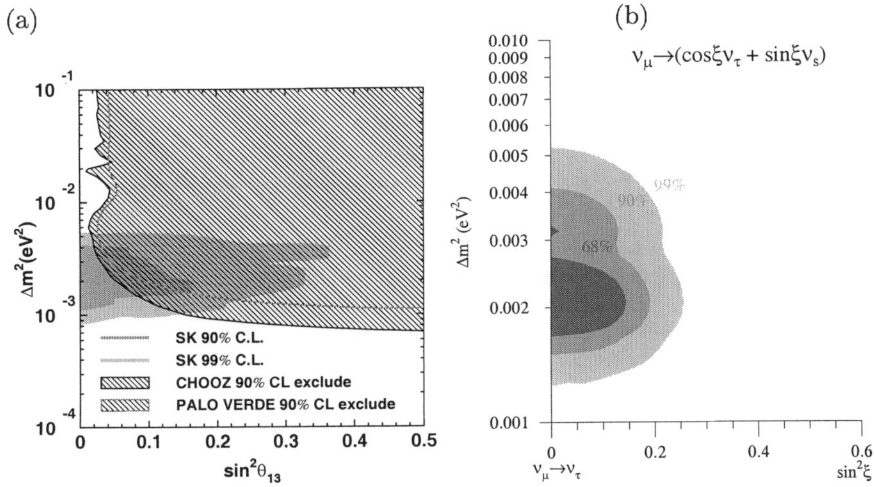

Figure 4. Neutrino Oscillations with Three and Four Neutrinos. (a) Allowed range of the SK atmospheric ν_e disappearance probability compared to the limits for $\overline{\nu_e}$ obtained from reactor experiments. The most stringent limit comes from the CHOOZ experiment. (b) The limit on the sterile content is $\sin^2 \xi < 0.19$ at 90% C.L. Pure $\nu_\mu \leftrightarrow \nu_s$ is disfavored by the absence of an up-down asymmetry in the neutral-current enhanced multi-ring sample and its predicted reduced oscillation probability at high neutrino energy due to matter effects.

(Figure 4): at 90% C.L., the maximum allowed sterile content is 25%. If atmospheric ν_μ indeed oscillate into ν_τ, then about 90 ν_τ charged-current events are expected in the SK data set. The high threshold for this reaction makes it difficult to isolate these events. At this energy, atmospheric neutrinos typically produce a spray of particles, and ring-counting becomes very difficult. Three (statistically correlated) searches were performed. Table 1 summarizes the results[c] of each analysis.

Table 1. Results of Three τ Appearance Searches.

Analysis	Number of τ-Events in Fit	Efficiency	Obs.,	Exp. Signif.
Energy-Flow[c]	79^{+44}_{-40}(stat+syst)	32%	1.8σ	1.9σ
Ring-Counting	145 ± 44(stat)$^{+11}_{-16}$(syst)	44%	3.1σ	2.4σ
Neural Net	99 ± 39(stat)$^{+13}_{-21}$(syst)	51%	2.2σ	2.1σ

[c]The energy-flow analysis uses only 1289 live days of data.

$$U \approx \begin{pmatrix} \frac{\sqrt{3}}{2} & \frac{1}{2} & 0 \\ -\frac{1}{2\sqrt{2}} & \frac{\sqrt{3}}{2\sqrt{2}} & \frac{1}{\sqrt{2}} \\ \frac{1}{2\sqrt{2}} & -\frac{1}{2\sqrt{2}} & \frac{1}{\sqrt{2}} \end{pmatrix}$$

$$\theta_{12} \approx \frac{\pi}{6} \quad \theta_{23} \approx \frac{\pi}{4} \quad \theta_{13} \approx 0$$

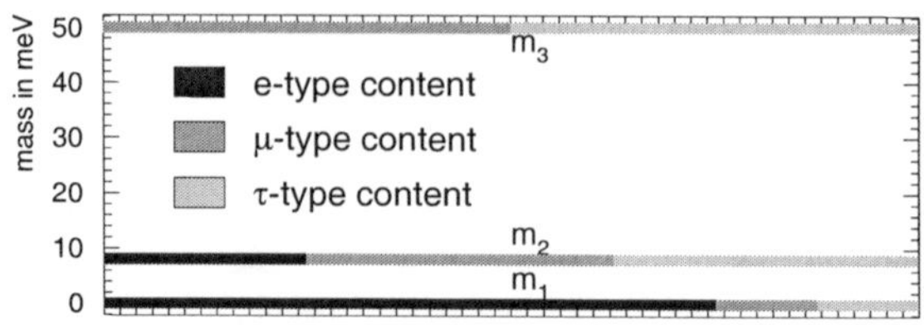

Figure 5. A Possible Leptonic Mixing Matrix U and Neutrino Mass Spectrum. The mixing matrix neglects CP violation and assumes that θ_{13} is indeed zero and the atmospheric mixing is maximal. The mass spectrum additionally assumes, that m_1 is small enough to be negligible and that the masses of the "solar pair" are smaller than the third mass. (hierarchical case).

4. Neutrino Mixing and Masses

In summary, both solar and atmospheric neutrino oscillation parameters are determined. There are no hints for sterile neutrinos in either atmospheric or solar neutrino oscillations, so three neutrinos are sufficient to explain both data sets: Figure 5 shows mass spectrum and flavor content for the hierarchical case ($m_{1,2} \ll m_3$). CP violating phases are neglected. The values of θ_{ij} are the best fits of the atmospheric and solar two-neutrino oscillation analyses.

References

1. J.Bahcall et al., *Astrophys. J.* **555**, 990 (2001); A.Junghans et al., *nucl-ex/0308003*;
2. B.T.Cleveland et al., *Astrophys. J.* **496**, 505 (1998); V.Gavrin, *TAUP2003*; E.Bellotti, *TAUP2003*; S.Fukuda et al., *Phys. Rev. Lett.* **86**, 5651 (2001); S.Fukuda et al., *Phys. Rev. Lett.* **86**, 5656 (2001); Q.R.Ahmad et al., *nucl-ex/0309004*; Q.R.Ahmad et al., *Phys. Rev. Lett.* **89**, 011302 (2002).
3. Y. Fukuda et al., *Phys. Rev. Lett.* **81**, 1562 (1998).
4. S.P.Mikheyev and A.Y.Smirnov, *Sov. Jour. Nucl. Phys.* **42**, 913 (1985); L.Wolfenstein, *Phys. Rev. D* **17**, 2369 (1978).
5. S.Fukuda et al., *Nucl. Instrum. Methods A* **501**, 418 (2003).
6. S.Fukuda et al., *Phys. Lett. B* **539**, 179 (2002).
7. K.Eguchi et al., *Phys. Rev. Lett.* **90**, 021802 (2003); oscillation likelihood from A.Ianni, *J. of Phys. G: Nucl. Part. Phys.* **29**, 2107 (2003).
8. M.B.Smy et al., *hep-ex/0309011*.
9. M. Apollonio, *Phys. Lett. B* **466**, 415 (1999).

MEASUREMENT OF $\sin^2 2\theta_{13}$ BY REACTOR EXPERIMENTS AND ITS SENSITIVITY*

OSAMU YASUDA

Department of Physics, Tokyo Metropolitan University,
1-1 Minami-Osawa, Hachioji, Tokyo 192-0397, Japan
E-mail: yasuda@phys.metro-u.ac.jp

Reactor experiments offer a promising way to determine θ_{13} and are free from parameter degeneracies in neutrino oscillations. It is described how reactor measurements of $\sin^2 2\theta_{13}$ can be improved by a near-far detector complex. The experimental lower bound is derived on the sensitivity to $\sin^2 2\theta_{13} \gtrsim 0.02$ based on the rate analysis, and an idea is given which may enable us to circumvent this bound. It is shown that in the Kashiwazaki-Kariwa plan (KASKA) the sensitivity to $\sin^2 2\theta_{13}$ is approximately 0.02.

1. Introduction

The recent experiments on atmospheric and solar neutrinos and KamLAND have been so successful that we now know the approximate values of the mixing angles and the mass squared differences of the atmospheric and solar neutrino oscillations: $(\sin^2 2\theta_{12}, \Delta m^2_{21} \equiv m^2_2 - m^2_1) \simeq (0.8, 7\times 10^{-5}\text{eV}^2)$ for the solar neutrino and $(\sin^2 2\theta_{23}, |\Delta m^2_{31}| \equiv |m^2_1 - m^2_3|) \simeq (1.0, 2\times 10^{-3}\text{eV}^2)$ for the atmospheric neutrino, where the three flavor framework of neutrino oscillations is assumed. The quantities which are still unknown are the third mixing angle θ_{13}, the sign of the mass squared difference Δm^2_{31} which indicates whether the mass pattern of neutrinos is of normal hierarchy or of inverted one, and the CP phase δ. Among these, θ_{13} is the most important quantity in the near future neutrino experiments.

It has been known that the oscillation parameters θ_{jk}, Δm^2_{jk}, δ cannot be determined uniquely even if the appearance probabilities $P(\nu_\mu \to \nu_e)$ and $P(\bar{\nu}_\mu \to \bar{\nu}_e)$ are measured precisely from a long baseline accelerator

*This work is supported in part by the Grant-in-Aid for Scientific Research in Priority Areas No. 12047222 and No. 13640295, Japan Ministry of Education, Culture, Sports, Science, and Technology.

experiment due to so-called parameter degeneracies, and this problem has to be solved to determine the CP phase in the future long baseline experiments. Among the ideas which have been proposed to solve the problem, combination of a reactor measurement and a long baseline experiment offers a promising possibility (See Ref.[1] and references therein). This combination works because reactor experiments measure the disappearance oscillation probability which depends only on θ_{13} to a good approximation:

$$P(\bar{\nu}_e \to \bar{\nu}_e) \simeq 1 - \sin^2 2\theta_{13} \sin^2\left(\frac{\Delta m^2_{13} L}{4E}\right),$$

where E and L stand for the neutrino energy and the baseline length.

2. Reactor measurements of $\sin^2 2\theta_{13}$ and its sensitivity

To illustrate how a near-far detector complex improves the sensitivity to $\sin^2 2\theta_{13}$ of reactor neutrino experiments, let me discuss the case with a single reactor, one near and one far detectors. To discuss the sensitivity, let me introduce χ^2, which basically indicates whether the difference between the number of events with oscillations and that without oscillations is large enough compared to the total error whose square is the sum of the statistical and systematic errors squared. For simplicity I will mainly discuss the sensitivity in the limit of infinite statistics, i.e., with the systematic errors only. Throughout my talk I will perform the rate analysis only.

Let $m_{\rm n}$ and $m_{\rm f}$ be the number of events measured at the near and far detectors, $t_{\rm n}$ and $t_{\rm f}$ be the theoretical predictions. Then χ^2 is given by

$$\chi^2 = \min_{\alpha' s}\left\{\left[\frac{m_{\rm n} - t_{\rm n}(1+\alpha_{\rm c}+\alpha_{\rm c}^{(\rm r)}+\alpha_{\rm u}^{(r)})}{t_{\rm n}\sigma_{\rm u}}\right]^2 + \left[\frac{m_{\rm f} - t_{\rm f}(1+\alpha_{\rm c}+\alpha_{\rm c}^{(\rm r)}+\alpha_{\rm u}^{(r)})}{t_{\rm f}\sigma_{\rm u}}\right]^2 \right.$$
$$\left. + \left(\frac{\alpha_{\rm c}}{\sigma_{\rm c}}\right)^2 + \left(\frac{\alpha_{\rm c}^{(\rm r)}}{\sigma_{\rm c}^{(\rm r)}}\right)^2 + \left(\frac{\alpha_{\rm c}^{(\rm r)}}{\sigma_{\rm c}^{(\rm r)}}\right)^2\right\}, \qquad (1)$$

where α_u, α_c, $\alpha_c^{(r)}$ and $\sigma_u^{(r)}$ are the variables to introduce the uncorrelated systematic error σ_u of the detectors, the correlated systematic error σ_c of the detector, the correlated systematic error $\sigma_c^{(r)}$ of the flux and the uncorrelated systematic error $\sigma_u^{(r)}$ of the flux, and it is assumed that the uncorrelated errors for the two detectors are the same and are equal to $\sigma_{\rm u}$.

After some calculations[2], I obtain

$$\chi^2 = \left(\frac{m_{\rm n}}{t_{\rm n}} - 1, \frac{m_{\rm f}}{t_{\rm f}} - 1\right) V^{-1} \begin{pmatrix} \dfrac{m_{\rm n}}{t_{\rm n}} - 1 \\ \\ \dfrac{m_{\rm f}}{t_{\rm f}} - 1 \end{pmatrix},$$

where

$$V \equiv \begin{pmatrix} \sigma_{\rm u}^2 + \sigma_{\rm c}^2 + (\sigma_{\rm u}^{(\rm r)})^2 + (\sigma_{\rm c}^{(\rm r)})^2 & \sigma_{\rm c}^2 + (\sigma_{\rm u}^{(\rm r)})^2 + (\sigma_{\rm c}^{(\rm r)})^2 \\ \sigma_{\rm c}^2 + (\sigma_{\rm u}^{(\rm r)})^2 + (\sigma_{\rm c}^{(\rm r)})^2 & \sigma_{\rm u}^2 + \sigma_{\rm c}^2 + (\sigma_{\rm u}^{(\rm r)})^2 + (\sigma_{\rm c}^{(\rm r)})^2 \end{pmatrix}$$

is the covariance matrix. After diagonalizing V, I have

$$\chi^2 = \frac{\left[(m_{\rm n}/t_{\rm n} - 1) + (m_{\rm f}/t_{\rm f} - 1)\right]^2}{4\sigma_{\rm c}^2 + 4(\sigma_{\rm u}^{(\rm r)})^2 + 4(\sigma_{\rm c}^{(\rm r)})^2 + 2\sigma_{\rm u}^2} + \frac{\left[(m_{\rm n}/t_{\rm n} - 1) - (m_{\rm f}/t_{\rm f} - 1)\right]^2}{2\sigma_{\rm u}^2}. \quad (2)$$

The strategy in this talk is to assume no neutrino oscillation for the theoretical predictions t_j (j=n,f) and assume the number of events with oscillations for the measured values m_j (j=n,f) and to examine if a hypothesis with no oscillation is excluded, say at the 90%CL, from the value of χ^2. Hence I have

$$\frac{m_j}{t_j} - 1 = -\sin^2 2\theta_{13} \left\langle \sin^2\left(\frac{\Delta m_{13}^2 L_j}{4E}\right)\right\rangle, \quad (3)$$

where L_j is the distance between the reactor and the near or far (j=n,f) detector, and

$$\left\langle \sin^2\left(\frac{\Delta m_{13}^2 L_j}{4E}\right)\right\rangle \equiv \frac{\displaystyle\int dE\ \epsilon(E) f(E) \sigma(E) \sin^2\left(\frac{\Delta m_{13}^2 L_j}{4E}\right)}{\displaystyle\int dE\ \epsilon(E) f(E) \sigma(E)}.$$

$\epsilon(E)$, $f(E)$, $\sigma(E)$ stand for the detection efficiency, the neutrino flux, and the cross section, respectively. Thus Eq. (2) becomes

$$\chi^2 = \sin^4 2\theta_{13} \left\{ \frac{\left[D(L_{\rm f}) + D(L_{\rm n})\right]^2}{4\sigma_{\rm c}^2 + 4(\sigma_{\rm u}^{(\rm r)})^2 + 4(\sigma_{\rm c}^{(\rm r)})^2 + 2\sigma_{\rm u}^2} + \frac{\left[D(L_{\rm f}) - D(L_{\rm n})\right]^2}{2\sigma_{\rm u}^2} \right\} \quad (4)$$

where (3) was used, and

$$D(L) \equiv \left\langle \sin^2\left(\frac{\Delta m_{13}^2 L}{4E}\right)\right\rangle$$

was defined. The numerical value of $D(L)$ is plotted in Fig.1 as a function of L. Here I adopt the reference values $\sigma_{\rm c} = 0.8\%/\sqrt{2} = 0.6\%$ and

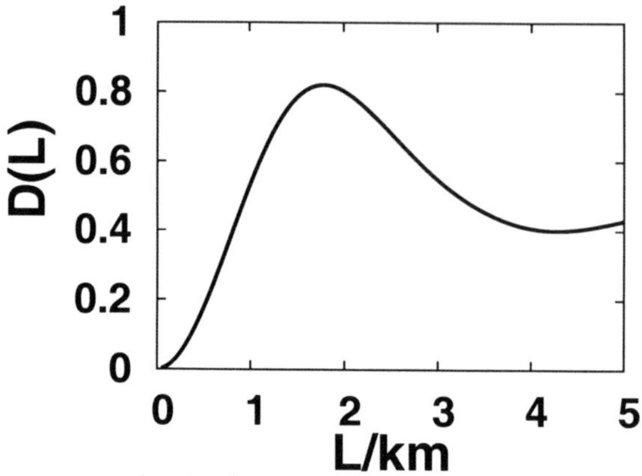

Figure 1. $D(L) \equiv \langle \sin^2\left(\frac{\Delta m^2_{13} L}{4E}\right)\rangle$ as a function of L. $D(L)$ has its maximum value 0.82 at L=1.8km, and approaches to its asymptotic value 0.5 as $L \to \infty$. $|\Delta m^2_{13}| = 2.5 \times 10^{-3}\mathrm{eV}^2$ is used as the reference value.

$\sigma_{\mathrm{u}} = \sqrt{(2.7\%)^2 - (2.1\%)^2 - (0.8\%/\sqrt{2})^2} = 1.6\%$ used in Ref.[1], where basically the reference values were deduced from extrapolation of the previous reactor experiments Bugey and CHOOZ. In the estimation of σ_{c}, I used 2.7% total error and 2.1% error of the flux which are the reference values in the CHOOZ experiment. As for the correlated and uncorrelated errors of the the flux from the reactors, I adopt the same reference values as those used by the KamLAND experiment: $\sigma_{\mathrm{c}}^{(\mathrm{r})} = 2.5\%$, $\sigma_{\mathrm{u}}^{(\mathrm{r})} = 2.3\%$. Putting the present reference values together, I have

$$2\sigma_{\mathrm{u}}^2 = (0.8\%)^2$$
$$4\sigma_{\mathrm{c}}^2 + 4(\sigma_{\mathrm{u}}^{(\mathrm{r})})^2 + 4(\sigma_{\mathrm{c}}^{(\mathrm{r})})^2 + 2\sigma_{\mathrm{u}}^2 = (7.6\%)^2.$$

The contribution from $(4\sigma_{\mathrm{c}}^2 + 4(\sigma_{\mathrm{u}}^{(\mathrm{r})})^2 + 4(\sigma_{\mathrm{c}}^{(\mathrm{r})})^2 + 2\sigma_{\mathrm{u}}^2)^{-1}$ in Eq. (4) is only 1% compared to that from $(2\sigma_{\mathrm{u}}^2)^{-1}$, so virtually this term can be ignored in Eq. (4). Hence χ^2 is given approximately by

$$\chi^2 \simeq \sin^4 2\theta_{13} \frac{[D(L_{\mathrm{f}}) - D(L_{\mathrm{n}})]^2}{2\sigma_{\mathrm{u}}^2}. \qquad (5)$$

Since no oscillation is assumed for the theoretical predictions, the best fit value in $\sin^2 2\theta_{13}$ for the measurement is $\sin^2 2\theta_{13} = 0$. If χ^2 is larger than 2.7, which corresponds to the value at the 90%CL for one degree of freedom, then the hypothesis of no oscillation is excluded at the 90%CL. This implies that the systematic limit on $\sin^2 2\theta_{13}$ at the 90%CL, or the sensitivity in the limit of infinite statistics, is given by

$$\left(\sin^2 2\theta_{13}\right)^{\text{sys only}}_{\text{limit}} \simeq \sqrt{2.7}\frac{\sqrt{2}\sigma_{\rm u}}{D(L_{\rm f})-D(L_{\rm n})}. \tag{6}$$

Eq. (6) tells us that, in order to optimize $\left(\sin^2 2\theta_{13}\right)^{\text{sys only}}_{\text{limit}}$, I have to minimize $D(L_{\rm n}) \equiv \langle\sin^2\left(\Delta m^2_{13}L_{\rm n}/4E\right)\rangle$ and maximize $D(L_{\rm f}) \equiv \langle\sin^2\left(\Delta m^2_{13}L_{\rm f}/4E\right)\rangle$. Since the possible maximum value of $D(L_{\rm f})-D(L_{\rm n})$ is 0.82, which is attained for $L_{\rm f}=1.8$km and $L_{\rm f}=0$, the lower bound of $\sin^2 2\theta_{13}$ at a single reactor experiment can be estimated as:

$$\text{lower bound of }\sin^2 2\theta_{13} \simeq \frac{\sqrt{2.7}\sqrt{2}\sigma_{\rm u}}{0.82} = 2.8\,\sigma_{\rm u}. \tag{7}$$

Eq. (7) indicates that, unless one develops a technology to improve $\sigma_{\rm u}$ significantly compared to the reference value $\sigma_{\rm u}=0.6\%$ assumed in Ref.[1], it is difficult to achieve the sensitivity below $\sin^2 2\theta_{13}=0.016$ in an experiment with one reactor and two detectors. It should be stressed here that the correlated systematic error $\sigma_{\rm c}$ and the errors $\sigma_{\rm u}^{(r)}$, $\sigma_{\rm c}^{(r)}$ of the flux do not appear in the dominant contribution to χ^2, and that it is the uncorrelated systematic error $\sigma_{\rm u}$ divided by the factor $D(L_{\rm f})-D(L_{\rm n})$ that determines the systematic limit on $\sin^2 2\theta_{13}$.

In the ideal case with N pairs of the reactors and the near detectors and one far detector, one can show[2] that Eq. (7) becomes

$$\text{lower bound of }\sin^2 2\theta_{13} \simeq \frac{\sqrt{2.7}\sqrt{1+1/N}\sigma_{\rm u}}{0.82} > 2.0\,\sigma_{\rm u}. \tag{8}$$

Eq.(8) indicates that even in the ideal case the sensitivity can be approximately 0.012 at best, as far as the rate analysis is concerned, no matter how many reactors and detectors there may be, if one adopts $\sigma_{\rm u}=0.6\%$.

In principle there is a way to improve this bound. In the case with one reactor, if one puts M identical detectors at the near site and M identical detectors at the far site, where all these detectors are assumed to have the same uncorrelated systematic error $\sigma_{\rm u}$, then the value of χ^2 is simply multiplied by M. Eqs. (5) and (7) imply that the sensitivity becomes

$$\text{lower bound of }\sin^2 2\theta_{13} \simeq \frac{\sqrt{2.7}\sqrt{2}\sigma_{\rm u}}{0.82\sqrt{M}} = \frac{2.8}{\sqrt{M}}\sigma_{\rm u}.$$

Therefore, it follows theoretically that the more identical detectors one puts, the better sensitivity one gets. Notice that this conclusion is based crucially on the assumption that the uncorrelated systematic error $\sigma_{\rm u}$ of the detectors is independent of the number of the detectors $2M$. This assumption may not be satisfied in general, but if the dependence of $\sigma_{\rm u}$ on M is weaker than $\sqrt{M}$, $\sigma_{\rm u}/\sqrt{M}$ decreases as M increases, and this possibility may give us a way to improve the sensitivity. It is therefore important to examine experimentally the dependence of $\sigma_{\rm u}$ on M.

3. The KASKA plan

The Kashiwazaki-Kariwa nuclear plant consists of two clusters of reactors, and one cluster consists of four reactors while the other consists of three. In the Kashiwazaki-Kariwa (KASKA) plan[3] one near detector is placed near one cluster of the reactors while the other near detector is placed near another cluster. In addition, a far detector is located 1.3 km away from the reactors. In the presence of seven reactors, the total number m_j of the events measured at the j-th detector (j=1 (the first near), 2 (the second near), 3 (the far detector)) is a sum of contributions m_{aj} ($a = 1, \cdots, 7$) from each reactor, and this is also the case for the theoretical predictions t_j and t_{aj} ($a = 1, \cdots, 7$) at the j-th detector. So I have $m_j = \sum_{a=1}^{7} m_{aj}$, $t_j = \sum_{a=1}^{7} t_{aj}$. In the limit of infinite statistics χ^2 is defined as

$$\chi^2 = \min_{\alpha' s} \left\{ \sum_{j=1}^{3} \frac{1}{t_j^2 \sigma_{\rm u}^2} \left[m_j - t_j \left(1 + \alpha_{\rm c} + \alpha_{\rm c}^{(\rm r)} + \sum_{a=1}^{7} \frac{t_{aj}}{t_j} \alpha_{\rm ua}^{(\rm r)} \right) \right]^2 \right.$$
$$\left. + \left(\frac{\alpha_{\rm c}}{\sigma_{\rm c}} \right)^2 + \left(\frac{\alpha_{\rm c}^{(\rm r)}}{\sigma_{\rm c}^{(\rm r)}} \right)^2 + \sum_{a=1}^{7} \left(\frac{\alpha_{\rm ua}^{(\rm r)}}{\sigma_{\rm u}^{(\rm r)}} \right)^2 \right\}.$$

After some calculations, I get

$$\chi^2 = \left(\frac{m_1}{t_1} - 1, \frac{m_2}{t_2} - 1, \frac{m_3}{t_3} - 1 \right) V^{-1} \left(\frac{m_1}{t_1} - 1, \frac{m_2}{t_2} - 1, \frac{m_3}{t_3} - 1 \right)^T,$$

where

$$V_{jk} = \delta_{jk} \sigma_{\rm u}^2 + \sigma_{\rm c}^2 + (\sigma_{\rm c}^{(\rm r)})^2 + (\sigma_{\rm u}^{(\rm r)})^2 \sum_{a=1}^{7} \frac{t_{aj}}{t_j} \frac{t_{ak}}{t_k}. \tag{9}$$

As in the case with one reactor (3), I have

$$\frac{m_j}{t_j} - 1 = - \sin^2 2\theta_{13} \sum_{a=1}^{7} \frac{t_{aj}}{t_j} D(L_{aj}),$$

where L_{aj} is the distance between the a-th reactor (a=1,$\cdots$,7) and the j-th detector (j=1,2,3). Therefore χ^2 is proportional to $\sin^4 2\theta_{13}$:

$$\chi^2 = C \sin^4 2\theta_{13},$$

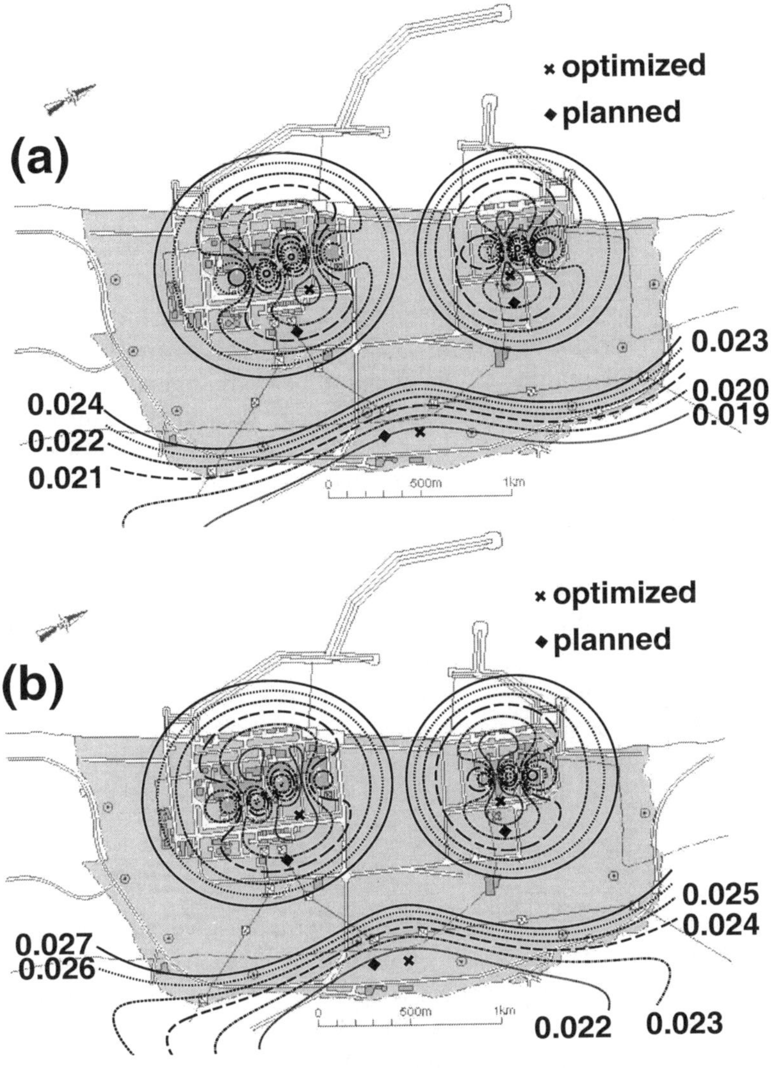

Figure 2. The contour plot of the sensitivity to $\sin^2 2\theta_{13}$ in the limit of infinite statistics (a) and for 20 ton·yr (b) in the KASKA experiment. The optimized and currently planned positions of the detectors are also depicted. When the contour for each detector is plotted, it is assumed that other detectors are located in the optimized positions. The position of the far detector is varied within the campus of the Kashiwazaki-Kariwa nuclear plant. For the data size 20 ton·yr, the value $\sin^2 2\theta_{13} = 0.025$ of the sensitivity seems to be inconsistent with the contour plot in Fig.2(b), which suggests that it is better than $\sin^2 2\theta_{13} = 0.022$ for the planned location of the far detector. However, the contour for the far detector is plotted here on the assumption that the near detectors are both in the optimized position, and therefore this calculation gives the sensitivity better than that for the locations currently planned.

where

$$C \equiv \sum_{j,k=1}^{3} \sum_{a=1}^{7} \frac{t_{aj}}{t_j} D(L_{aj})(V^{-1})_{jk} \sum_{b=1}^{7} \frac{t_{bk}}{t_k} D(L_{bk}).$$

Hence the sensitivity to $\sin^2 2\theta_{13}$ at 90%CL is given by

$$\sin^2 2\theta_{13} = \left(\frac{\chi^2|_{90\%CL}}{C} \right)^{1/2} = \left(\frac{2.7}{C} \right)^{1/2},$$

where it is assumed that the value of $|\Delta m^2_{13}|$ is precisely known and therefore degree of freedom in this analysis is one. The covariance matrix (9) cannot be inverted analytically, so C has to be evaluated numerically. Fig.2 shows the contour plot of the sensitivity to $\sin^2 2\theta_{13}$ in the limit of infinite statistics (a) and for 20 ton·yr (b). Fig.2 indicates that optimization forces the distances between each near detector and the reactors in each cluster be approximately (300±130)m. This results in slightly poorer sensitivity to $\sin^2 2\theta_{13}$ than the hypothetical single reactor case with near and far detectors. This is because in the case with a single reactor, the near detector can be theoretically arbitrarily close to the reactor and the factor $D(L_{\rm f}) - D(L_{\rm n}) \to D(L_{\rm f})$ can in principle be the maximum value 0.82, whereas in the KASKA case, the sensitivity is given by[2]

$$\left(\sin^2 2\theta_{13}\right)^{\rm sys\ only}_{\rm limit} \simeq \frac{\sqrt{2.7}\sqrt{1.04}\,\sigma_{\rm u}}{0.81 \sum_{a=1}^{7} \frac{t_{a3}}{t_j} D(L_{a3}) - 0.31 \sum_{a=1}^{7} \frac{t_{a1}}{t_j} D(L_{a1}) - 0.50 \sum_{a=1}^{7} \frac{t_{a2}}{t_j} D(L_{a2})},$$

and the second and third term in the denominator give small contribution to spoil the sensitivity. From the numerical calculations, one can show that the KASKA experiment has the sensitivity $\sin^2 2\theta_{13} \sim 0.025$ (0.019) with the data size of 20 ton·yr (∞ ton·yr). Since the sensitivity to $\sin^2 2\theta_{13}$ at KASKA is close to the lower bound 0.016 (cf. (7) with $\sigma_{\rm u} = 0.6\%$), the setup of the KASKA plan is not far from the optimum.

References

1. H. Minakata, H. Sugiyama, O. Yasuda, K. Inoue and F. Suekane, Phys. Rev. D **68**, 033017 (2003).
2. H. Sugiyama, O. Yasuda, F. Suekane and G.A. Horton-Smith, to appear.
3. F. Suekane, K. Inoue, T. Araki and K. Jongok, arXiv:hep-ex/0306029.

4 Neutrino Theory

NEUTRINOS, LEPTON FLAVOR AND CP VIOLATION IN A PREDICTIVE SO(10) MODEL*

B. DUTTA AND Y. MIMURA

Department of Physics, University of Regina,
Regina, Saskatchewan S4S 0A2, Canada
E-mails: duttabh@uregina.ca, mimura2y@uregina.ca

R.N. MOHAPATRA

Department of Physics, University of Maryland,
College Park, MD 20742, USA
E-mail: rmohapat@phys.umd.edu

A minimal supersymmetric SO(10) model with one **10** and one **126** Higgs superfield predict all neutrino mixings as well as the solar mass difference squared in agreement with observations. However, the CKM CP phase is constrained to be in the second or third quadrant requiring a significant non-CKM component to CP violation to explain observations. We revisit this issue using type I and II seesaw formula for neutrino masses show that the addition of a nonrenormalizable term restores compatibility with neutrino data and CKM CP violation in both cases. We further find that the MSSM parameter $\tan\beta \geq 30$ in the type I model and lepton flavor violation and lepton electric dipole moments are accessible to proposed experiments in both type I and type II models. We also discuss the unification of the gauge couplings in type I model which requires an intermediate scale.

1. Introduction

The solar and atmospheric neutrino experiment results have now conclusively established that neutrinos have mass and they mix among themselves.A clear outline of the mixing pattern among the three generations of neutrinos has emerged. The two mixing angles responsible for solar and atmospheric neutrino deficits are large, and the third mixing angle corresponding to reactor neutrinos is small [1].

*This work is supported by Natural Sciences and Engineering Research Council of Canada and by the National Science Foundation Grant No. phy-0099544

The light neutrino masses are generated by the seesaw mechanism [2] which requires an extension of the standard model that includes heavy right handed neutrinos. The light neutrino mass matrix is given by

$$M_\nu^{\rm I} = -M_\nu^D M_R^{-1} (M_\nu^D)^T, \tag{1}$$

where M_ν^D is the Dirac neutrino mass matrix and M_R is the right handed Majorana mass matrix.

The above formula is called the type I seesaw formula and the right handed Majorana mass scale, v_R, is almost determined by the mass squared difference needed to explain the atmospheric neutrino result.

We will consider a class of models where this right handed neutrino mass arises from a renormalizable coupling of the form $fNN\Delta_R$, where N is a right handed neutrino, f is a coupling constant and Δ_R is a Higgs field whose vacuum expectation value (vev) gives mass to the right handed neutrino. This feature can be naturally embedded in $SU(2)_L \times SU(2)_R \times U(1)_{B-L}$ or any higher gauge group such as SO(10)), where the Δ_R field is part of an $SU(2)_R$ triplet field. The parity invariance then implies that we also have an $f\nu\nu\Delta_L$ coupling term as a parity partner of the $NN\Delta_R$ coupling. In this class of theories, whenever Δ_R acquires a vev, so does Δ_L and they are related by the formula $\langle \Delta_L \rangle \equiv v_L = \frac{v_w^2}{\lambda v_R}$, where v_w is a weak scale and v_R is the Δ_R vev and λ is a coupling constant in the Higgs potential. The Δ_L vev contributes a separate seesaw suppressed Majorana mass to the neutrino leading to a modified seesaw formula for neutrino masses given below.

$$M_\nu^{\rm II} = M_L - M_\nu^D M_R^{-1} (M_\nu^D)^T, \tag{2}$$

where $M_L = fv_L$ and $M_R = fv_R$. This formula for the neutrino mass matrix is called the type II seesaw formula [3]. In the case where right handed Majorana masses are heavy enough, the second term in the type II seesaw formula can be negligible, and the first term, $M_L = fv_L$, is dominant. We will call this the pure type II seesaw. When both terms are comparable, we will call this the mixed type II seesaw.

Since the experimental data suggests the gauge coupling unification scale in the minimal supersymmetric extension of the Standard Model (MSSM) to be, $M_G \sim 2 \times 10^{16}$ GeV, which is close to the seesaw scale to explain the current neutrino data. The minimal grand unification model for neutrinos is based on the SO(10) group since all standard model fermions and the right handed neutrino fit into the **16**-dimensional representation of SO(10), resulting not only in a complete unification of the quarks and leptons but also yielding possible relations between the quark and lepton mass matrices.

An interesting class of renormalizable SO(10) models with **10** and $\overline{\mathbf{126}}$ Higgs multiplets was proposed in Ref.[4] where it was shown that the model provides a completely realistic description of the charged fermion sector of the standard model and is very predictive in the neutrino sector without any extra symmetry assumptions. Further, in this class of models, the R-parity is an automatic symmetry of the model which guarantees a naturally stable supersymmetric dark matter.

The results and discussions presented in the next sections are based on our recent paper[5].

2. Fitting the neutrino massses and the mixings

The SO(10)-invariant superpotential, includes MSSM Yukawa couplings plus right handed Majorana mass term which are written by using the MSSM superfields:

$$W_Y \supset Y^u_{ij} Q_i U^c_j H_u + Y^d_{ij} Q_i D^c_j H_d + Y^e_{ij} L_i E^c_j H_d + Y^\nu_{ij} L_i N^c_j H_u + \frac{1}{2} f_{ij} N^c_i N^c_j \Delta_R \,, \tag{3}$$

where H_u and H_d are MSSM Higgs doublets which are linear combinations of the SM doublets in **10** and $\overline{\mathbf{126}}$ Higgs multiplets, and Δ_R is part of the $\overline{\mathbf{126}}$ field Δ. We note that the VEV of Δ_R gives right handed neutrino masses and breaks $SU(2)_R \times U(1)_{B-L}$ symmetry down to $U(1)_Y$. As noted the SO(10) Higgs fields, H and Δ, contain two pairs of $SU(2)_L$ Higgs doublets, $H^{u,d}_{10}$ and $H^{u,d}_{126}$, and the MSSM Higgs doublets are linear combinations of two pairs,

$$H_u = \alpha_u H^u_{10} + \beta_u H^u_{126}, \; H_d = \alpha_d H^d_{10} + \beta_d H^d_{126}, \tag{4}$$

where $|\alpha_{u,d}|^2 + |\beta_{u,d}|^2 = 1$. The MSSM Yukawa couplings $Y^{u,d,e,\nu}$ are given by linear combination of h and f, and the fermion mass matrices and Majorana mass matrix for right handed neutrino are given as

$$M_u = (h^* \alpha_u + f^* \beta_u) v_u, \; M_d = (h^* \alpha_d + f^* \beta_d) v_d, \tag{5}$$

$$M_e = (h^* \alpha_d - 3 f^* \beta_d) v_d, \; M^D_\nu = (h^* \alpha_u - 3 f^* \beta_u) v_u, \; M_R = f^* v_R, \tag{6}$$

where $v_{u,d}$ are VEVs of MSSM Higgs doublets and v_R is a VEV of Δ_R. We denote that $v_u = v \sin\beta$ and $v_d = v \cos\beta$, where $v = 174$ GeV. Then we have sumrules for leptonic mass matrices:

$$M_e = c_d M_d + c_u M_u, M^D_\nu = \frac{c_d + 3}{c_u}(M_d - M_e) + M_u, M_R = \frac{M_d - M_e}{4 \beta_d v_d} v_R \,, \tag{7}$$

where

$$c_u = \frac{4\cot\beta}{\alpha_u/\alpha_d - \beta_u/\beta_d}, \quad c_d + 3 = \frac{-4\beta_u/\beta_d}{\alpha_u/\alpha_d - \beta_u/\beta_d}. \tag{8}$$

So if the quark mass matrices M_u and M_d are our input, then leptonic mass matrices are determined by two complex parameters, c_u and c_d, barring the overall scale and the phase of Majorana neutrino mass matrix, M_R. Thus, using the masses of three charged leptons, we predict the neutrino mass matrices in terms of only one real parameter. The light neutrino mass matrix is given by:

$$M_\nu^{\rm II} \propto (M_d - M_e); \tag{9}$$

$$M_\nu^{\rm II\,mixed} \propto (M_d - M_e)(1+\Delta) + 2\frac{c_u}{c_d+3}M_u + \left(\frac{c_u}{c_d+3}\right)^2 M_u(M_d - M_e)^{-1}M_u. \tag{10}$$

$\Delta = 0$ corresponds to type I.

In both pure type II and mixed type II, we can obtain the bi-large mixing structure, for the solar and the atmospheric oscillation. In the bi-large neutrino matrix, the 1-3 mixing is well predicted as $|U_{e3}| \simeq 0.17-0.18$. In the mixed type II (and type I) case, a larger $|c_d|$ is favored to obtain bi-large structure in the neutrino mass matrix. However, fitting of small mass squared ratio at the same time requires a smaller $|c_d|$ (larger strange quark mass) and hence it is hard to achieve a best fit value of the recent combined data analysis. In the pure type II case, the condition of bi-large mixing does not constrain $|c_d|$. However, the electron mass fitting requires the strange mass to be related to the KM phase, and if the KM phase is 60-80 degree, the strange mass can not be large and thus it is hard to achieve the bi-maximal mixing and the small mass squared ratio in the pure type II case as well. If the KM phase is in the second quadrant and especially for the larger value of KM phase, the strange quark mass can be large and the best fit values for neutrino oscillation can be achieved.

In this model, we get the follwoing bound for $\tan\beta$:

$$\min\left(\left|\frac{c_d+3}{c_u}\right|, \left|\frac{1-c_d}{c_u}\right|\right) \le \tan\beta \le \max\left(\left|\frac{c_d+3}{c_u}\right|, \left|\frac{1-c_d}{c_u}\right|\right). \tag{11}$$

In the case where $\hat{\xi} \simeq 1$ which is favored to obtain bi-large structure, $(1-c_d)/c_u \simeq m_t/m_b$, and then we have lower bounds of $\tan\beta$, such as $\tan\beta \gtrsim m_t/m_b|(c_d+3)/(1-c_d)|$. In the case where $c_d \sim -7$ which is favored in the mixed type II (and type I) case, the lower bound of $\tan\beta$ is roughly 30. To obtain small $\tan\beta$ such as $\tan\beta \sim 10$, $c_d \sim -3.5$ is needed. Such small $|c_d|$ can happen only in the case where the KM phase is close to 180 degree in the minimal SO(10) model. The $B - L$ breaking scale has an upper-bound in the type I model.

In order to reproduce the CKM model of CP violation at low energies, we employ corrections to the SO(10) model which originate from non-renormalizable terms in the superpotential. The most simple non-renormalizable term is written including an extra singlet S does not improve the situation.We use the following term

$$\frac{1}{2}\frac{1}{M_P}h'_{ij}\psi_i\psi_j H\Sigma \tag{12}$$

Σ has is a **210** Higgs multiplet which breaks SO(10) down to 2231 symmetry. This nonrenormalizable term can arise from

$$W' = h'\psi\psi\bar{\Delta}' + M\bar{\Delta}'\Delta' + \Delta'\Sigma H \tag{13}$$

The effective theory below the scale M has a nonrenormalizable term of the type in where only **126** part of ΣH field product contributes. If only $(\mathbf{1},\mathbf{1},\mathbf{1})$ and $(\mathbf{1},\mathbf{1},\mathbf{15})$ submultiplets of **210** acquire vev, the effective operator that contributes to fermion masses then has $(\mathbf{2},\mathbf{2},\mathbf{15})$ quantum numbers under G_{224}. This leads to mass formulae for the quarks and leptons of the following type.

$$M_{u,d} = (h^*\alpha_{u,d} + f^*\beta_{u,d})v_{u,d} + h'^*\alpha_{u,d}v_{u,d}\langle\Sigma\rangle/M_P, \tag{14}$$

$$M_{e,\nu^D} = (h^*\alpha_{d,u} - 3f^*\beta_{d,u})v_{d,u} - 3h'^*\alpha_{d,u}v_{d,u}\langle\Sigma\rangle/M_P,\ M_R = f^*v_R \tag{15}$$

where $v_{u,d}$ are VEVs of MSSM Higgs doublets and v_R is VEV of Δ_R. Note also that h' is not a symmetric matrix in general, but we assume that h' is a symmetric matrix like h and f for simplicity.

We can now rewrite the above equations as follows

$$M_e = M_e^0 + \Delta M_e,\ M_R = \frac{M_d - M_e^0}{4\beta_d v_d}v_R \tag{16}$$

$$M_\nu^D = \frac{c_d+3}{c_u}(M_d - M_e^0) + M_u + \frac{1-c_d}{c_u}\Delta M_e \tag{17}$$

where $M_e^0 = c_d M_d + c_u M_u$ and $\Delta M_e = -4h'^*\alpha_d v_d\langle\Sigma\rangle/M_P$. In the mixed type II (and type I) case, we have to increase the (2,2) element while keeping $|c_d|$ to be larger than 5. We find a large solar mixing solution in type I when we switch on the (2,2) element of the ΔM_e. In the pure type II case, we need the (1,1) element of the ΔM_e to give the desired electron mass and we then find a large solar mixing solution with KM phase in the first quadrant.

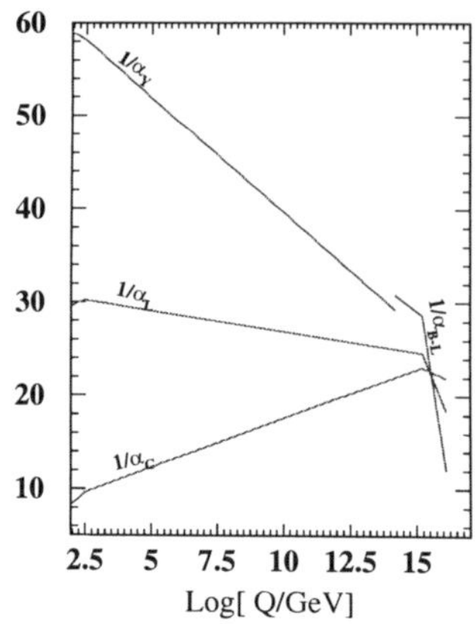

Figure 1. The couplings $1/\alpha_i$ are plotted as a function of scale.

2.1. *Gauge Coupling Unification*

The type I model needs an intermediate scale (v_R) must be below the GUT scale and around 10^{14} GeV. We assume the resulting symmetry breaking chain is of type, $SO(10) \to SU(3)_c \times SU(2)_L \times SU(2)_R \times U(1)_{B-L} \to SU(3)_c \times SU(2)_L \times U(1)_Y$. In addition to the new gauge bosons associated with this new symmetry, the new matter and Higgs that contribute are the following: three RH neutrinos as part of the $SU(2)_R$ lepton doublet; one bidoublet from which the MSSM doublets emerged. In addition we need either the $B-L=2$ triplet pair $(\Delta_L+\bar{\Delta}_L \oplus \Delta_R+\bar{\Delta}_R)$ and $x[(\mathbf{3},\mathbf{3},\mathbf{1},-2/3)+(\bar{\mathbf{3}},\mathbf{3},\mathbf{1},2/3)] \oplus y[(\mathbf{3},\mathbf{1},\mathbf{3},-2/3)+(\bar{\mathbf{3}},\mathbf{3},\mathbf{1},2/3)$ to have successful gauge coupling unification. If we use the Higgs spectrum from Ref.[6] in the context of our model, we find that $x \oplus y$ (from **210**) and $\Delta_L+\bar{\Delta}_L \oplus \Delta_R+\bar{\Delta}_R$ are at about the v_R scale. Using these fields, the Fig. 1 shows the coupling unification in this model and the gauge unification happens at about $10^{15.5}$ GeV.

3. Lepton Flavor Violating Processes and Lepton Electric dipole moments

We now discuss the lepton flavor violating processes e.g. $\mu \to e\gamma$, $\tau \to \mu\gamma$ etc in this model. The supersymmetric contributions for these processes include the neutralino and chargino diagrams.

We work in the basis where the charged lepton masses are diagonal at the highest scale of the theory. We first start with pure type II seesaw model. We assume that v_R (the scale where G_{2231} breaks down to SM) is at the GUT scale. So we have just MSSM and right handed neutrinos below the GUT scale. The right handed masses have hierarchies and therefore get decoupled at different scales. The flavor-violating pieces present in Y_ν induces flavor violations into the charged lepton couplings and into the soft SUSY breaking masses e.g. m^2 terms etc. through the following RGEs:

$$dY_e/dt = \frac{1}{16\pi^2}(Y_\nu Y_\nu^\dagger + \cdots)Y_e \tag{18}$$

$$dm_{LL}^2/dt = \frac{1}{16\pi^2}(Y_\nu Y_\nu^\dagger m_{LL}^2 + m_{LL}^2 Y_\nu Y_\nu^\dagger + \cdots)$$

We use the mSUGRA universal boundary conditions at the GUT scale to draw the figures. In Figs. 2, we show the BR[$\mu \to e\gamma$] as a function of $m_{1/2}$ for different values of A_0. The lightest neutralino is the dark matter candidate in this model and we satisfy the 2σ range of the recent relic density constraint $\Omega_{\rm CDM} = 0.1126^{+0.0087}_{-0.009}$ in the parameter space. When we satisfy the relic density constraint, the m_0 gets determined. For example, m_0 varies between 60-100 GeV for $A_0 = 0$ GeV line in the graph. The Figs also show that the larger $\tan\beta$ has larger lepton flavor violation.

In type I model, the upper bound of v_R, VEV of $\overline{\mathbf{126}}$ which couples to fermions, is determined, for example, v_R can be $\leq 10^{14}$ GeV. Then, G_{2231} symmetry can be maintained between the GUT and the v_R scale. This feature induces larger lepton violation since the right handed neutrino is a part of the doublet which has right handed electrons. The lepton flavor violations are found to be larger in this case and is depicted in Fig. 2(right).The BR[$\tau \to \mu\gamma$] can be as large as 10^{-8} which can be explored in the near future. The electron EDM is plotted in Fig. 3 for the type II and type I.

References

1. J. N. Bahcall, M. C. Gonzalez-Garcia and C. Pena-Garay, *JHEP* **0207**, 054 (2002); V. Barger, D. Marfatia and K. Whisnant, *Int. J. Mod. Phys.E* **12**, 569 (2003).
2. T. Yanagida, in Proceedings of the *Workshop on the Unified Theory and Baryon Number in the Universe* (KEK, Tsukuba, 1979); S. L. Glashow, *Cargese lectures, 1979*; M. Gell-Mann, P. Ramond, and R. Slansky, in *Supergravity*, (North-Holland, Amsterdam, 1979); R. N. Mohapatra and G. Senjanovic, *Phys. Rev. Lett.* **44**, 912 (1980).
3. G. Lazarides, Q. Shafi and C. Wetterich, *Nucl. Phys. B* **181**, 287 (1981); R. N. Mohapatra and G. Senjanovic, *Phys. Rev.D* **23**, 165 (1981).
4. K. S. Babu and R. N. Mohapatra, *Phys. Rev. Lett.* **70**, 2845 (1993).

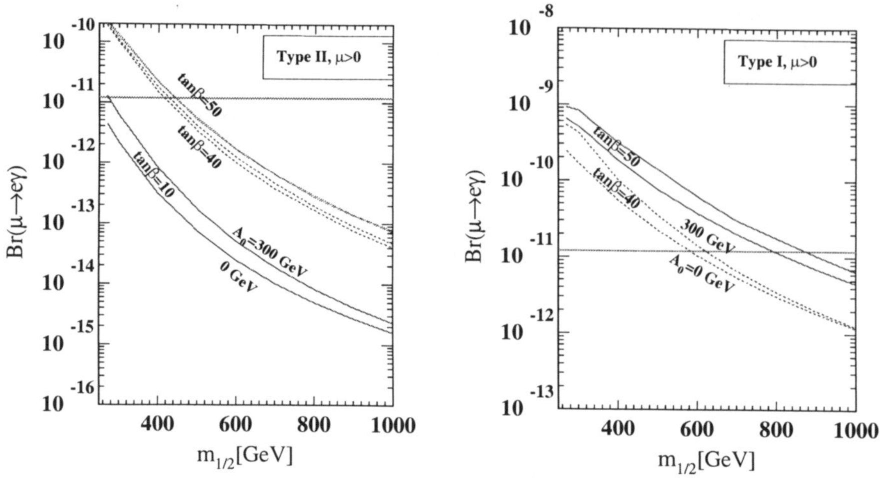

Figure 2. The BR[$\mu \to e\gamma$] is plotted as a function of $m_{1/2}$ for different values A_0 and $\tan\beta$ in pure type II (left) and type I(right) scenarios.

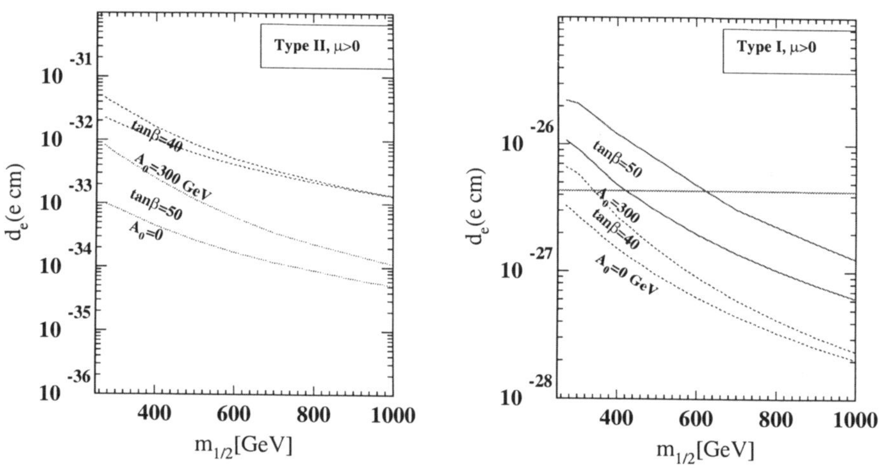

Figure 3. The electron EDM is plotted as a function of $m_{1/2}$ for different values A_0 and $\tan\beta$ in pure type II(left) and type I(right) scenarios.

5. B. Dutta, Y. Mimura and R.N. Mohapatra, hep-ph/0402113, (to appear in *Phys. Rev.* **D**).
6. T. Fukuyama, et al., arXiv:hep-ph/0401213.
7. C. L. Bennett *et al.*, Astrophys. J. Suppl. **148**, 1 (2003).

TESTING THE LEPTOGENESIS MECHANISM OF THE SEESAW MODEL

ELIZABETH JENKINS

Department of Physics,
University of California at San Diego,
9500 Gilman Drive,
La Jolla, CA 92093, USA
E-mail: ejenkins@ucsd.edu

The seesaw theory, the leading theory for particle interactions, provides a viable mechanism for generating the matter-antimatter asymmetry of the universe. Testing the leptogenesis mechanism directly requires measurement of the $d = 6$ operator of the low-energy effective Lagrangian, in addition to the more familiar $d = 5$ operator which generates Majorana masses for the light neutrinos when the electroweak symmetry is spontaneously broken. This important experimental challenge awaits the next generation of particle physicists.

We are at a very interesting crossroads in particle physics. Experiment has now established that neutrinos have mass, and that the neutrino mass eigenstates, ν_1, ν_2 and ν_3 with masses $m_1 < m_2 < m_3$, are not the same as the neutrino flavor eigenstates ν_e, ν_μ and ν_τ which appear in weak doublets with the left-handed charged leptons e^-, μ^- and τ^-. We know quite a bit about the parameters of neutrino mixing from atmospheric and solar neutrino oscillations. What I wish to focus on here is what we do not know about the parameters of the lepton sector.

Our leading candidate theory at high energies is the seesaw model,[1] originally proposed 24 years ago. The seesaw model is the minimal extension of the Standard Model (SM) which adds right-handed neutrinos to its particle content. Including all renormalizable interactions which are allowed by $SU(3) \times SU(2)_L \times U(1)_Y$ gauge symmetry yields the seesaw Lagrangian. Since right-handed neutrinos are gauge singlets, Majorana mass terms for these neutrinos are allowed, and unlike other fermions, they have a mass M which is much larger than the weak scale and which is independent of electroweak symmetry breaking (EWSB). The only interactions of the right-handed neutrinos are Yukawa interactions with the Higgs scalar doublet and the lepton doublets. Thus, the leptonic Lagrangian of the high-energy seesaw theory is given by

$$\begin{aligned}\mathcal{L}_{\text{seesaw}} &= i\,\overline{\ell_L}\,\not{\partial}\ell_L + i\,\overline{e_R}\,\not{\partial}e_R + i\,\overline{N_R}\not{\partial}N_R \\ &\quad -\overline{\ell_L}\,\phi\,Y_e\,e_R - \overline{\ell_L}\,\widetilde{\phi}\,Y_\nu\,N_R - \frac{1}{2}\,\overline{N_R{}^c}\,M\,N_R + \text{h.c.} \qquad (1)\end{aligned}$$

where $\ell_{L\alpha}$ are lepton electroweak doublets, $e_{R\alpha}$ are charged lepton singlets, ϕ is the Higgs doublet and N_R are heavy neutrino singlets. For further explanation of notation, see Ref. 2.

Let me briefly remind you of the attractive features of the seesaw model. First and foremost, the lightness of weakly-interacting neutrinos and the heaviness of sterile neutrinos is explained by the seesaw mechanism. The weakly-interacting neutrinos acquire small Majorana masses after the electroweak symmetry breaks. These Majorana masses are $SU(2)_L$ triplet and require two insertions of the Higgs vacuum expectation value $v/\sqrt{2}$. Thus, the light neutrino Majorana masses are order $(v^2/2M)$ and are suppressed by a factor of $(v/\sqrt{2}M)$ relative to the Dirac masses, proportional to $v/\sqrt{2}$, of the other SM fermions. Hence, the large hierarchy between the light neutrino masses and the rest of the SM fermions can be understood.

A second attractive feature of the seesaw model is that leptogenesis occurs at high energy.[3] The heavy right-handed neutrinos are unstable to decay to lepton and scalar (Higgs) doublets. Interference between tree and loop decay diagrams of the right-handed neutrinos leads to CP violation in the decay processes. Thus, a small lepton asymmetry is generated at high energies in heavy neutrino decay. This leptogenesis is characterized by the CP-violating asymmetry of the lightest right-handed neutrino N_1 with mass M_1, and can be written in terms of the neutrino Yukawa matrix combination $Y_\nu^\dagger Y_\nu$ and the mass ratios of the heavy Majorana neutrinos M_2/M_1 and M_3/M_2.

Other attractive aspects of the seesaw model are that it is a renormalizable high-energy theory, which implies that it has a finite number of high-energy parameters; it is natural in the sense of 't Hooft,[4] i.e. all renormalizable interactions allowed by the SM gauge symmetry are included in the Lagrangian; experiment indicates a seesaw scale M consistent with gauge coupling unification; and it is easy to embed the seesaw model in a grand unified or partially unified theory. With such an embedding, asymmetries of the neutrino field content and of the gauge symmetry of the SM are low-energy artifacts of spontaneous symmetry breakdown of a more symmetric theory.

Although the seesaw model is not the only possible theory of neutrino mass, it is by far the theoretically favored one. A logical alternative to the seesaw model is to add right-handed neutrinos, but to suppose that neutrinos obtain only Dirac masses from Yukawa couplings upon EWSB, just like all the other SM fermions. This supposition requires that lepton number L is a globally conserved quantum number so that Majorana mass terms are forbidden. Such an assumption is unnatural in the technical sense. In this case, in contrast to the seesaw theory, there is no understanding of why light neutrino masses are so much smaller that the masses of all the other SM fermions. Furthermore, there is no leptogenesis at high energy if L is exact, so the matter-antimatter asymmetry of the universe is not addressed. These are strong theoretical reasons for favoring the seesaw model. To summarize: theorists believe that neutrinos are Majorana particles.

In the seesaw model, the lepton asymmetry generated at high energies is thought to be reprocessed by non-perturbative processes at the electroweak scale, resulting in a baryon-antibaryon asymmetry. As is well-known, it is possible to generate a sufficient lepton asymmetry at high energy for the leptogenesis $\rightarrow$ baryogenesis scenario to provide the solution of the matter-antimatter asymmetry of the universe.[5] It is worth pointing out that the magnitude (and sign!) of the baryon-antibaryon asymmetry of the universe is known quite precisely from the WMAP data, so the leptogenesis constraint is significant.

This state of affairs is quite interesting. We have a viable and compelling theory of leptogenesis. It needs to be tested. Can we establish experimentally that leptogenesis *is* the solution of the matter-antimatter asymmetry puzzle? The characteristic energy scale of the seesaw theory is such that we may never build an accelerator which can directly probe its physics. Can we establish the seesaw theory by performing measurements at accessible energies?

At "low" energies, the heavy right-handed neutrinos are not in evidence, so we can measure only the parameters of the low-energy effective theory in which the heavy Majorana neutrinos are integrated out of the theory. The effective Lagrangian has the form

$$\mathcal{L}_{\rm eff} = \mathcal{L}_{SM} + \delta\mathcal{L}_{d=5} + \delta\mathcal{L}_{d=6} + \cdots , \tag{2}$$

where higher dimensional Lagrangian terms with $d > 4$ are suppressed by the power $1/M^{d-4}$ of the seesaw scale. The $d = 5$ and $d = 6$ operators of the low-energy effective seesaw theory are[2]

$$\delta\mathcal{L}_{d=5} = \frac{1}{2}\, c^{d=5}_{\alpha\beta} \left(\overline{\ell_{L\alpha}}^{c}\tilde{\phi}^*\right)\left(\tilde{\phi}^\dagger\, \ell_{L\beta}\right) + \text{h.c.}, \tag{3}$$

$$\delta\mathcal{L}_{d=6} = c^{d=6}_{\alpha\beta} \left(\overline{\ell_{L\alpha}}\tilde{\phi}\right) i\partial\!\!\!/\left(\tilde{\phi}^\dagger \ell_{L\beta}\right), \tag{4}$$

where the coefficients $c^{d=5}$ and $c^{d=6}$ are related to the high-energy parameters of the seesaw theory Y_ν and M by

$$c^{d=5} = Y_\nu^* \,(M^*)^{-1}\, Y_\nu^\dagger\,, \tag{5}$$

$$c^{d=6} = Y_\nu \,(|M|^2)^{-1}\, Y_\nu^\dagger. \tag{6}$$

Measurement of both the $d = 5$ and $d = 6$ operator coefficients suffices for construction of the high-energy seesaw Lagrangian from the low-energy data, when the number of right-handed neutrinos is equal to the number of light fermion generations,[2] which is the most attractive situation theoretically. (The formulae making the connection between the low-energy and high-energy parameters are involved[2] and are suppressed here. The important point is that this connection is possible.)

For three generations of fermions, there are 15 real parameters and 6 phases in the high-energy seesaw Lagrangian. Using chiral symmetries, it is possible to define these physical parameters in a standard high-energy basis[2]

$$Y_e = \frac{\sqrt{2}}{v}\text{diag}(m_e, m_\mu, m_\tau), \qquad M = \text{diag}(M_1, M_2, M_3), \tag{7}$$

$$Y_\nu = \begin{pmatrix} u_{e1}e^{i\Phi_{e1}} & u_{e2}e^{i\Phi_{e2}} & u_{e3}e^{-i(\Phi_{e1}+\Phi_{e2})} \\ u_{\mu1}e^{i\Phi_{\mu1}} & u_{\mu2}e^{i\Phi_{\mu2}} & u_{\mu3}e^{-i(\Phi_{\mu1}+\Phi_{\mu2})} \\ u_{\tau1}e^{i\Phi_{\tau1}} & u_{\tau2}e^{i\Phi_{\tau2}} & u_{\tau3}e^{-i(\Phi_{\tau1}+\Phi_{\tau2})} \end{pmatrix}, \tag{8}$$

with all phases appearing in the Yukawa matrix Y_ν. The low-energy effective theory including the $d = 5$ and $d = 6$ operators also contains 15 real parameters and 6 phases. Three of the real parameters are the charged lepton masses, and are known. The $d = 5$ and $d = 6$ operator coefficients each contain 6 real parameters and 3 phases.

The $d = 5$ operator is the usual Weinberg operator.[6] Upon electroweak symmetry breaking (EWSB), it generates a Majorana mass term for the weakly-interacting light neutrinos with mass matrix

$$m_{\alpha\beta} \equiv -\frac{1}{2}v^2 c^{d=5}_{\alpha\beta} \tag{9}$$

in the flavor basis. Diagonalization of this matrix yields the light neutrino masses m_1, m_2, m_3 and the lepton mixing matrix[7]

$$U_{PMNS} = U_{CKM} \begin{pmatrix} e^{i\phi_1/2} & 0 & 0 \\ 0 & e^{i\phi_2/2} & 0 \\ 0 & 0 & 1 \end{pmatrix}, \tag{10}$$

$$U_{CKM} = \begin{pmatrix} c_{12}c_{13} & s_{12}c_{13} & s_{13}\, e^{-i\delta} \\ -s_{12}c_{23} - c_{12}s_{23}s_{13}\, e^{i\delta} & c_{12}c_{23} - s_{12}s_{23}s_{13}\, e^{i\delta} & s_{23}c_{13} \\ s_{12}s_{23} - c_{12}c_{23}s_{13}\, e^{i\delta} & -c_{12}s_{23} - s_{12}c_{23}s_{13}\, e^{i\delta} & c_{23}c_{13} \end{pmatrix}. \tag{11}$$

Notice that, in contrast to the quark CKM mixing matrix, the lepton PMNS mixing matrix, contains 3 CP-violating phases, since phase redefinitions on the Majorana neutrinos are not allowed.[a] Some of these parameters (θ_{12}, θ_{23}, and the square mass differences $\Delta m^2_{21} \equiv m_2^2 - m_1^2$ and $\Delta m^2_{32} \equiv m_3^2 - m_2^2$) already have been observed in low-energy neutrino oscillations. The presently undetermined parameters, namely s_{13}, δ and the absolute scale of light neutrino mass $\bar{m}$, are the focus of upcoming experiments, and undoubtably will be observed or severely constrained in the coming decade.

The $d = 6$ Lagrangian usually is not considered in the low-energy effective seesaw theory because it is suppressed by an additional power of the seesaw scale. This suppression factor is significant since the seesaw scale is large compared to the weak scale. Upon EWSB, the $d = 6$ operator generates a flavor-nondiagonal $d = 4$ contribution to the kinetic energy term of the light weakly-interacting neutrinos given by the Hermitian matrix

$$\lambda_{\alpha\beta} \equiv \frac{1}{2} v^2 c^{d=6}_{\alpha\beta} \tag{12}$$

in the flavor basis. Normalization of the effective neutrino kinetic energy term and diagonalization of the kinetic energy and Majorana mass terms implies an effective lepton mixing matrix[2]

$$U_{\text{eff}} = \left(\mathbb{I} - \Omega \frac{\lambda}{2} \Omega^\dagger \right) U\,, \tag{13}$$

which is no longer unitary due to effects of the $d = 6$ operator. (Ω represents a rephasing of the charged lepton fields which is defined in Ref. 2. Understanding this subtlety is not important for the discussion here.) Including the $d = 6$ operator, the charged current is given by

$$J^-_\mu \equiv \bar{e}_{L\alpha} \gamma_\mu \left(U_{\text{eff}} \right)_{\alpha i} \nu_i \tag{14}$$

[a] I am working in the convention that the Majorana neutrino fields satisfy $\nu_i^c = \nu_i$.[2]

with lepton mixing matrix U_{eff}, whereas the neutral current is given by

$$J_\mu = \frac{1}{2}\bar{\nu}_i\gamma_\mu \left(U^\dagger_{\text{eff}}\right)_{i\alpha} \left(U_{\text{eff}}\right)_{\alpha j} \nu_j \,, \tag{15}$$

which no longer equals $\frac{1}{2}\bar{\nu}_i\gamma_\mu\nu_i$, since U_{eff} is not unitary.

It turns out that leptogenesis already provides a very interesting and non-trivial constraint on the low-energy parameters $m_{\alpha\beta}$ and $\lambda_{\alpha\beta}$. For example, one can show that leptogenesis vanishes if the λ matrix eigenvalues are degenerate.[2] The leptogenesis constraint also requires nondegenerate M_i and nondegenerate m_i when the heavy neutrinos are strongly hierarchical.

The effects of the $d = 6$ operator at low energy are presumably very tiny. In the seesaw theory, the natural size of λ is of order $(v^2/2M^2)$. For the smallest seesaw scale which seems to be compatible with leptogenesis,[8] $M \sim 10^8$ GeV, $\lambda \sim O(10^{-12})$, but it can be orders of magnitude smaller for larger M. (For example, $M \sim 10^{14}$ GeV implies $\lambda \sim 10^{-24}$.) Probing $\lambda \sim 10^{-5}$ is achievable in next generation experiments.[9] Hence, the experimental situation is challenging in the extreme. Nevertheless, bounding and eventually measuring the $d = 6$ operator seems to be the only way to directly test the leading theory of leptogenesis.

The $d = 6$ operator gives rise to some very interesting physical effects. These include CP-violating asymmetries in neutrino-neutrino oscillations as well as other effects which depend on the phases of the lepton sector.[2,10] Even though experiments in the coming decade will yield in all probability null results, they are nonetheless important to pursue and putting experimental bounds on the $d = 6$ coefficients are of value. For example, by bounding the $d = 6$ coefficients, we can rule out (or in!) more general theories with a scale of new physics of order the TeV scale, or electroweak scale. Experiments which do this need to be part of the future experimental program in high-energy particle physics.

I think it is appropriate to recall the famous remark of Pauli when he first proposed the existence of the neutrino in order to preserve energy and momentum conservation in β decay. He said, "I have done a terrible thing. I have postulated a particle that cannot be detected." While it is true that the neutrino could not be detected when it first was introduced by Pauli, direct experimental evidence for neutrinos was obtained 26 years later. (Interestingly, Pauli was still alive!) This experimental information was crucial for the development of the SM of particle physics. I think we are facing an analogous situation with the seesaw theory. Experiment is very far from the sensitivity required to test leptogenesis. Nevertheless, I believe that someday the experiments will be possible, and that coming generations (if not this generation) will unravel the mystery.

Acknowledgments

I thank the organizers for the invitation to Coral Gables 2003. This work was supported in part by the Department of Energy under grant DE-FG03-97ER40546.

References

1. M. Gell-Mann, P. Ramond and R. Slansky, in *Supergravity*, edited by P. van Nieuwenhuizen and D. Freedman, (North-Holland, 1979), p. 315; T. Yanagida, in *Proceedings of the Workshop on the Unified Theory and the Baryon Number in the Universe*, edited by O. Sawada and A. Sugamoto (KEK Report No. 79-18, Tsukuba, 1979), p. 95; R.N. Mohapatra and G. Senjanović, Phys. Rev. Lett. **44** (1980) 912.
2. A. Broncano, M.B. Gavela and E. Jenkins, Phys. Lett. B **522** (2003) 177 [arXiv:hep-ph/0210271]; Nucl. Phys. B **672** (2003) 163 [arXiv:hep-ph/0307058].
3. M. Fukugita and T. Yanagida, Phys. Lett. **174**B (1986) 45.
4. G. 't Hooft, Cargese Summer Institute (1979) 135.
5. For a recent analysis, see G.F. Giudice, A. Notari, M. Raidal, A. Riotto and A. Strumia, arXiv:hep-ph/0310123.
6. S. Weinberg, Phys. Rev. Lett. **43** (1979) 1571.
7. B. Pontecorvo, *Zh. Eksp. Teor. Fiz.* **33**, 549 (1957) and **34**, 247 (1958); Z. Maki, M. Nakagawa and S. Sakata, Prog. Theor. Phys. **28** (1962) 870.
8. S. Davidson and A. Ibarra, Phys. Lett. B **535** (2002) 25 [arXiv:hep-ph/0202239].
9. A. De Gouvea, G.F. Giudice, A. Strumia, K. Tobe, Nucl. Phys. B **623** (2002) 395 [arXiv:hep-ph/0107156].
10. A. de Gouvea, B. Kayser and R. N. Mohapatra, Phys. Rev. D **67** (2003) 053004 [arXiv:hep-ph/0211394].

TRACING VERY HIGH ENERGY TAU NEUTRINOS FROM COSMOLOGICAL SOURCES IN ICE

J. JONES, I. MOCIOIU, I. SARCEVIC

Department of Physics
University of Arizona
Tucson AZ 85721

M. H. RENO

Department of Physics and Astronomy
University of Iowa
Iowa City IA 52242

Astrophysical sources of ultrahigh energy neutrinos yield tau neutrino fluxes due to neutrino oscillations. We study in detail the contribution of tau neutrinos with energies above 10^6 GeV relative to the contribution of the other flavors. We consider several different initial neutrino fluxes and include tau neutrino regeneration in transit through the Earth and energy loss of charged leptons. We discuss signals of tau neutrinos in detectors such as IceCube, RICE and ANITA.

1. Introduction

The experimental evidence of $\nu_\mu \leftrightarrow \nu_\tau$ neutrino oscillations [1] means that astrophysical sources of muon neutrinos become sources of ν_μ and ν_τ in equal proportions after oscillations over astronomical distances [2]. Although ν_μ and ν_τ have identical interaction cross sections at high energies, signals from $\nu_\tau \rightarrow \tau$ conversions have the potential to contribute differently from ν_μ signals. The τ lepton can decay far from the detector, regenerating ν_τ [3]. This also occurs with μ decays, but electromagnetic energy loss coupled with the long muon lifetime make the ν_μ regeneration from muon decays irrelevant for high energies. A second signal of $\nu_\tau \rightarrow \tau$ is the tau decay itself [4].

We have studied in detail the propagation of all flavors of neutrinos with very high energy ($E \geq 10^6$ GeV) as they traverse the Earth. Because of the high energies attenuation shadows most of the upward-going solid angle at high energies, so we have limited our consideration to nadir angles larger than 80°. We are particularly interested in the contribution from tau neutrinos, produced in oscillations of extragalactic muon neutrinos as they travel large astrophysical distances.

For most astrophysical sources, the neutrinos are produced in pion decays, which determine the flavor ratio $\nu_e : \nu_\mu : \nu_\tau$ to be $1 : 2 : 0$. After propagation over very long distances, neutrino oscillations change this ratio to $1 : 1 : 1$ because of the maximal $\nu_\mu \leftrightarrow \nu_\tau$ mixing. For the GZK flux, ν_e and ν_μ incident fluxes are different because of the additional contributions from $\bar{\nu}_e$ from neutron decay and ν_e from μ^+ decays [5]. Because of this, the flavor ratio at Earth is affected by the full three flavor mixing and is different from $1 : 1 : 1$. Given fluxes at the source $F^0_{\nu_e}$, $F^0_{\nu_\mu}$ and $F^0_{\nu_\tau}$, the fluxes at Earth become:

$$F_{\nu_e} = F^0_{\nu_e} - \frac{1}{4}\sin^2 2\theta_{12}(2F^0_{\nu_e} - F^0_{\nu_\mu} - F^0_{\nu_\tau}) \tag{1}$$

$$F_{\nu_\mu} = F_{\nu_\tau} = \frac{1}{2}(F^0_{\nu_\mu} + F^0_{\nu_\tau}) + \frac{1}{8}\sin^2 2\theta_{12}(2F^0_{\nu_e} - F^0_{\nu_\mu} - F^0_{\nu_\tau}) \tag{2}$$

where θ_{12} is the mixing angle relevant for solar neutrino oscillations. We have assumed that θ_{23}, the mixing angle relevant for atmospheric neutrino oscillations, is maximal and θ_{13} is very small, as shown by reactor experiments, as well as atmospheric and solar data.

Z burst neutrinos[6] from models with ultrahigh energy neutrinos scattering with relic neutrinos to produce Z bosons is another neutrino flux considered below, where neutrino mixing yields flux ratios of $1 : 1 : 1$.

The initial fluxes for GZK and Z burst neutrinos are shown in Fig. 1.

In our propagation of neutrinos and charged leptons through the Earth[7], we have focused on kilometer-sized neutrino detectors, such as ICECUBE [8] and the Radio Ice Cerenkov Experiment (RICE)[9] and on a detector with much larger effective area which uses Antarctic ice as a converter, the Antarctic Impulsive Transient Antenna (ANITA)[10].

2. Neutrino Propagation

Attenuation and regeneration of neutrinos and charged leptons are governed by the following transport equations:

$$\frac{\partial F_{\nu_\tau}(E,X)}{\partial X} = -N_A\sigma^{tot}(E)F_{\nu_\tau}(E,X) + N_A\int_E^\infty dE_y F_{\nu_\tau}(E_y,X)\frac{d\sigma^{NC}}{dE}(E_y,E)$$
$$+ \int_E^\infty dE_y \frac{F_\tau(E,X)}{\lambda_\tau^{dec}}\frac{dn}{dE}(E_y,E) \tag{3}$$

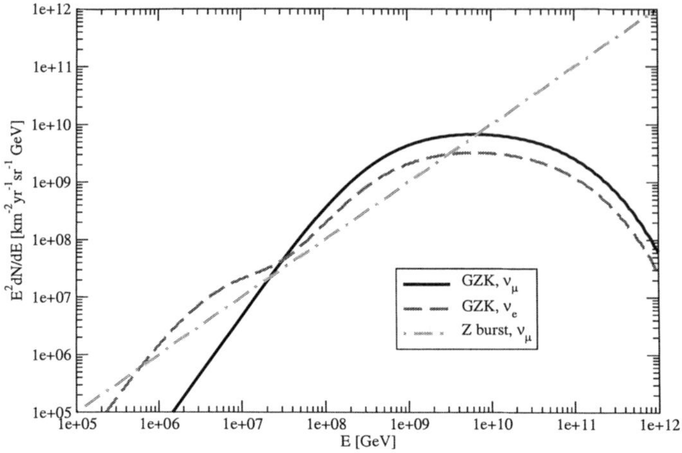

Figure 1. Initial Neutrino Fluxes

$$\frac{\partial F_\tau(E,X)}{\partial X} = -\frac{F_\tau(E,X)}{\lambda_\tau^{dec}(E,X,\theta)} + N_A \int_E^\infty dE_y F_{\nu_\tau}(E_y,X) \frac{d\sigma^{CC}}{dE}(E_y,E) \quad (4)$$

$$-\frac{dE_\tau}{dX} = \alpha + \beta E_\tau \quad (5)$$

Here $F_{\nu_\tau}(E,X) = dN_{\nu_\tau}/dE$ and $F_\tau(E,X) = dN_\tau/dE$ are the differential energy spectra of tau neutrinos and taus respectively, for lepton energy E, at a column depth X in the medium defined by

$$X = \int_0^L \rho(L')dL'. \quad (6)$$

For tau neutrinos, we take into account the attenuation by charged current interactions, the shift in energy due to neutral current interactions and the regeneration from tau decay. For tau leptons we consider their production in charged current ν_τ interactions, their decay, as well as electromagnetic energy loss.

The effective decay length of produced taus does not go above 10^7 cm, even for $E_\tau = 10^{12}$ GeV. This is because electromagnetic energy loss over that distance reduces the tau energy to about 10^8 GeV, at which point the tau is more likely to decay than interact electromagnetically [11].

We have found that the ν_τ flux above 10^8 GeV resembles the ν_μ flux. The lore that the Earth is transparent to tau neutrinos is not applicable in the high energy regime. Tau neutrino pileups at small angles with respect

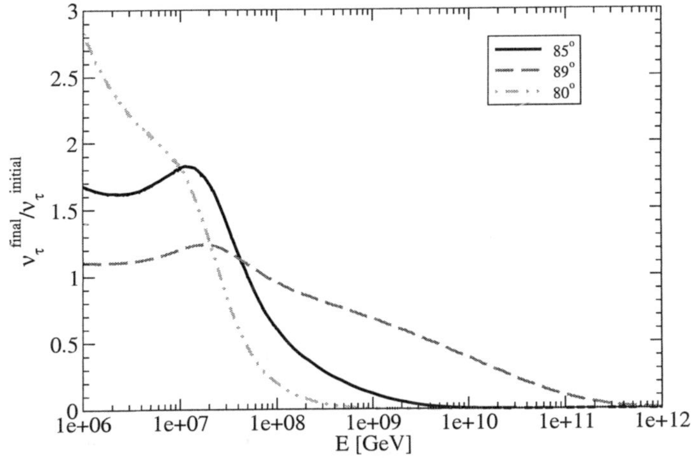

Figure 2. Ratio ν_τ/ν_μ for GZK neutrinos, at nadir angles of 85° and 89°.

to the horizon are significantly damped due to tau electromagnetic energy loss above $E_\tau \sim 10^8$ GeV if the column depth is at least as large as the neutrino interaction length.

At lower energies, $E \leq 10^8$ GeV, regeneration of ν_τ becomes important for trajectories where the other flavors of neutrinos are strongly attenuated, but the ν_τ regeneration is very effective. The regeneration effect depends strongly on the shape of the initial flux and it is larger for flatter fluxes. The enhancement due to regeneration also depends on the amount of material traversed by neutrinos and leptons, i.e. on nadir angle. For GZK neutrinos, we have found that the enhancement peaks between 10^6 and a few$\times 10^7$ GeV depending on trajectory.

Fig. 2 shows the ratio of the tau neutrino flux after propagation to incident tau neutrino flux, for 89°, 85°and 80°. This ratio illustrates a combination of the regeneration of ν_τ due to tau decay and the attenuation of all neutrino fluxes. For 89°, where both the total distance and the density are smaller, the attenuation is less dramatic, and the flux can be significant even at high energy. The regeneration in this case can add about 25% corrections at energies between 10^7 and 10^8 GeV. For 85° the relative enhancement is around 80% and peaked at slightly lower energies, while at 80° it is almost a factor of 3 at low energy. At 80°, however, the flux is very strongly attenuated for energies above a few $\times 10^7$ GeV. It is already clear from here that the total rates will be dominated by the nearly horizontal trajectories that go through a small amount of matter. However, rates can

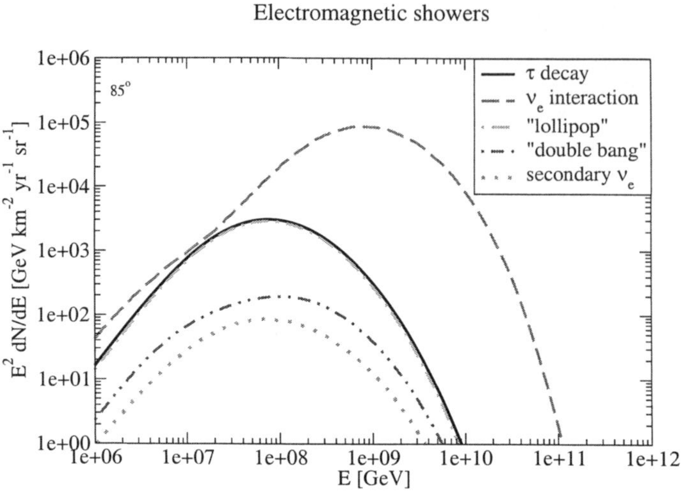

Figure 3. Electromagnetic showers for GZK neutrinos

get significant enhancements at low energies where the regeneration from tau decays adds an important contribution even for longer trajectories.

3. Showers

We have translated the neutrino fluxes and tau lepton fluxes into rates for electromagnetic and hadronic showers at selected angles to see the effect of attenuation, regeneration, and the different energy dependences of the incident fluxes. We have focused on comparing the ν_τ contribution to the ν_e and ν_μ contributions to determine in what range, if any, ν_τ's enhance shower rates. Electromagnetic shower distributions for a nadir angle of 85° are shown in Fig. 3, while Fig. 5 shows hadronic showers.

Fig. 4 shows the ratio of the electromagnetic shower rates at nadir angle 85° in the presence and absence of oscillations for the GZK and Z burst neutrino fluxes (which have a characteristic $1/E$ energy dependence). In absence of oscillations, the only contribution to electromagnetic showers comes from ν_e interactions. In the presence of $\nu_\mu \to \nu_\tau$ oscillations, electromagnetic decays of taus from tau neutrinos add significant contributions to these rates at energies below 10^8 GeV. In the same time, for the GZK flux, $\nu_e \to \nu_{\mu,\tau}$ oscillations reduce the number of ν_e's at low energy, such that below a few $\times 10^6$ GeV there are fewer electromagnetic showers than in the absence of oscillations.

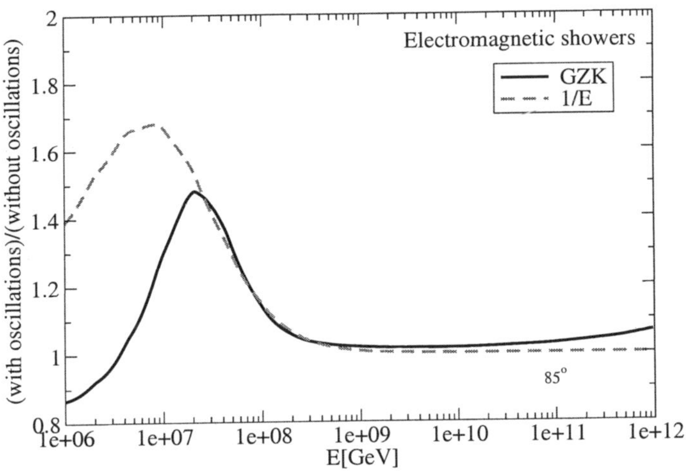

Figure 4. Ratio of electromagnetic shower rates in the presence and absence of $\nu_\mu \to \nu_\tau$ oscillations for GZK and $1/E$ neutrino spectra for nadir angle 85° for a km size detector.

The ν_τ flux enhancements depend on the shape of the initial flux. The electromagnetic showers are more sensitive to this shape than hadronic ones. The relative enhancement in hadronic showers is also smaller than for the electromagnetic showers. This is because for the electromagnetic signal the only contribution in the absence of taus is from electron neutrinos, while for hadrons the tau contribution is compared to a much larger signal, from the interactions of all flavors of neutrinos. We have included contribution from secondary neutrinos, which we find to be relatively small for all fluxes.

For kilometer-sized detectors, at for example a nadir angle of 85°, the maximal enhancement due to ν_τ contribution to electromagnetic shower rates for the GZK flux is about 50% at 3×10^7 GeV, while for a $1/E$ flux, it is even larger, about 70%, at slightly lower energy. These energy ranges are relevant for IceCube, but not for RICE. For energies relevant to RICE, tau neutrinos do not offer any appreciable gain in electromagnetic shower signals compared to $\nu_e \to e$ CC interactions, and they contribute at essentially the same level as ν_μ to hadronic shower rates through NC interactions.

One of the reasons that tau neutrinos do not contribute large signals to kilometer-sized detectors at very high energies is that high energy tau decay lengths are very large, so the probability of a tau decaying in the detector is low. For detectors like ANITA which can sample long trajectories through the ice, one would expect a larger tau neutrino contribution to the signal

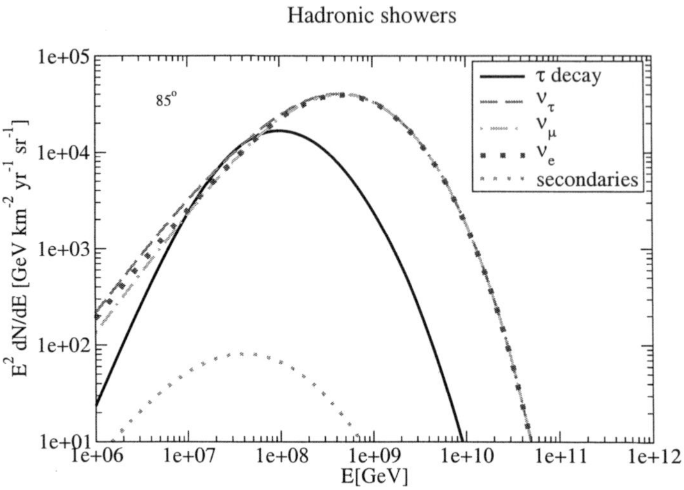

Figure 5. Hadronic showers for GZK neutrinos

from tau decay. Despite the long trajectory (222 km with a maximum depth of 1 km for a neutrino incident at 89° nadir angle) the tau contributions to the electromagnetic shower rate is quite small for fluxes expected to contribute in the ANITA signal. For hadronic showers, the suppression of τ decay to hadrons relative to ν_e NC interaction contributions is about the same as for electromagnetic showers compared to $\nu_e \to e$. The ν_τ contribution to the hadronic shower rate from interactions is about the same as the ν_e contribution. In summary, for ANITA, tau neutrinos do not give any additional signal beyond what one would evaluate based on no regeneration from $\nu_\tau \to \tau \to \nu_\tau$ due to tau electromagnetic energy loss at $E \gtrsim 10^8$ GeV.

Acknowledgments

This work was supported in part by the Department of Energy under contracts DE-FG02-91ER40664, DE-FG02-95ER40906 and DE-FG02-93ER40792.

References

1. Fukuda Y. et al. [Super-Kamiokande Collaboration] 1998, Phys. Rev. Lett., 81, 1562
2. Ahluwalia D.V., Ortiz C.A., Adunas G.Z. 2000, hep-ph/0006092; Athar H., Jezabek M., Yasuda O. 2000, Phys. Rev. D 62, 1033007

3. Halzen F., Saltzberg D. 1998, Phys. Rev. Lett. 81, 4305; Iyer S., Reno M.H., Sarcevic I. 2000, Phys. Rev. D 61, 053003; Iyer Dutta S., Reno M.H., Sarcevic I. 2000, Phys. Rev. D 62, 123001
4. Fargion D. 2002, Ap. J. 570, 909; Feng J.L., Fisher P., Wilczek F., Yu T.M. 2002, Phys. Rev. Lett. 88, 161102
5. R. Engel, D. Seckel, T. Stanev, Phys. Rev. **D64** (2001), 093010.
6. Weiler T.J. 1999, Astropart. Phys. 11, 303; Yoshida S., Sigl G., Lee S. 1998, Phys. Rev. Lett. 81, 5505
7. J. Jones, M.H. Reno and I. Sarcevic: hep-ph/0308042, Phys. Rev. **D69**, 033004.
8. http://icecube.wisc.edu ; Ahrens et al. ICECUBE Collaboration, astro-ph/0305196.
9. Kravchenko I. et al. [RICE Collaboration], astro-ph/0206371
10. Gorham P. et al. [ANITA Collaboration] http://www.ps.uci.edu/~barwick/anitaprop.pdf
11. Iyer Dutta S., Reno M.H., Sarcevic I., Seckel D. 2001, Phys. Rev. D 63, 094020

THE STRANGENESS CHANGING ANTINEUTRINO REACTION $\bar{\nu}_L + P \longrightarrow \Lambda + L^+$

S. L. MINTZ

Physics Department,
Florida International University,
Miami,Florida 33199, USA
E-mail: mintz@helios.fiu.edu

We obtain the cross sections for the reactions $\bar{\nu}_L + p \longrightarrow \Lambda + L^+$ where L is a massive lepton,*i.e.* a muon or a tau lepton. We do this from near threshold to relativistic energies. We further obtain the contributions of various form factors and interference terms to the cross section with a view to discovering if the contributions from hard to observe form factors such as the weak pseudoscalar form factor, F_P, and the weak electric form factor,F_E might be obtained via these reactions. The form factor F_E is particularly interesting as in the more usually observed $p \leftrightarrow n$ transition it is a second class current and forbidden by G-parity. However in the $p \leftrightarrow \Lambda$ transition it is not forbidden and it would be desirable to learn if it is present. Finally we discuss our results and prospects for these experiments.

1. Introduction

Recently we have been studying weak, semileptonic, strangeness changing, reactions of the form $L^- + p \longrightarrow \Lambda + \nu_L$ where L is either an electron, muon or tau lepton. The reason for examining these reactions was due to the interest of a number of experimentalists[1,2] in making use of the electron induced reaction to improve the limit on the electron neutrino mass and to obtain weak strangeness changing transition form factors not previously measured.

The transition matrix element for this process may be written as:

$$< \nu_L \Lambda | H_w | L^- \ p >= \frac{G}{\sqrt{2}} \sin\theta_C \overline{u}_\nu \gamma^\lambda (1-\gamma_5) u_L < \Lambda | J^\dagger_\lambda(0) | p > . \qquad (1)$$

If the quantity $< \Lambda | J^\dagger_\lambda(0) | p >$ were known it would be immediately possible to obtain a differential cross section. We note that here

$$J_\mu(0) = V_\mu(0) - A_\mu(0) \qquad (2)$$

where V_μ and A_μ are the vector and axial vector parts of the weak strangeness changing hadronic current. The matrix elements of the vector and axial vector current may be written as:

$$< \Lambda|V_\mu^\dagger(0)|p >= \bar{u_f}[\gamma_\mu F_V(q^2) + iF_M(q^2)\frac{\sigma_{\mu\nu}q^\nu}{2m_p} - F_S(q^2)\frac{q_\mu}{2m_p}]u_i \quad (3)$$

and

$$< \Lambda|A_\mu^\dagger(0)|p >= \bar{u}_f[\gamma_\mu\gamma_5 F_A(q^2) + \frac{q_\mu\gamma_5 F_P(q^2)}{m_\pi} + \frac{iF_E(q^2)\sigma_{\mu\nu}q^\nu\gamma_5}{2m_p}]u_i \quad (4)$$

where i is the initial particle and f is the final particle, p and Λ respectively. The structure of the particles is contained, of course, in the six form factors,$F_V(q^2)$,$F_M(q^2)$,$F_S(q^2)$,$F_A(q^2)$, $F_P(q^2)$,and $F_E(q^2)$. Thus if we are able to determine these six form factors we can write and evaluate a transition matrix element for the $p \leftrightarrow \Lambda$ transition.

In an electron induced process, two of the form factors, F_P and F_S, are not observable. This is due to the fact that in the transition matrix element squared all terms which contain either a factor of F_S or F_P are proportional to the lepton mass squared. The small size of the electron mass supresses these terms. We should remark that F_S and F_E would describe second class currents in a $p \leftrightarrow n$ transition and would be excluded by G-parity. However this is not true for $p \leftrightarrow \Lambda$ transitions and it would be desirable to know if these form factors are present. Thus in an electron induced process, only F_A, F_V, F_M, and F_E might be observed. In figure 1 we look at the absolute value of the interference terms for the reaction, $e^- + p \longrightarrow \Lambda + \nu_e$. We look at the case for incoming electrons of energy 194.25 MeV which is near threshold. We see that the contribution from the $F_M F_A$ interference term is relatively large (above the 20 percent level) and might be observed in this reaction leading to a value for F_M, the terms containing F_E are considerably smaller and F_E can probably not be acurately determined in this reaction.

We have also studied the case of heavier leptons. In particular we have looked at the reaction $\mu^- + p \longrightarrow \Lambda + \nu_\mu$ from near threshold to a highly relativistic 5.0 GeV. This eliminates the most obvious disadvantage of using electrons as the projectile, namely the small electron mass. Because all terms containing the pseudoscalar form factor,F_P, as well as F_S are proportional to the lepton mass squared, they are unobservable in the electron induced reaction. The lifetime of a muon is short but even near threshold, 0.265 GeV, the muon would achieve a distance of almost 1200 meters from its point of production thus making an experiment possible. In figure 2 we

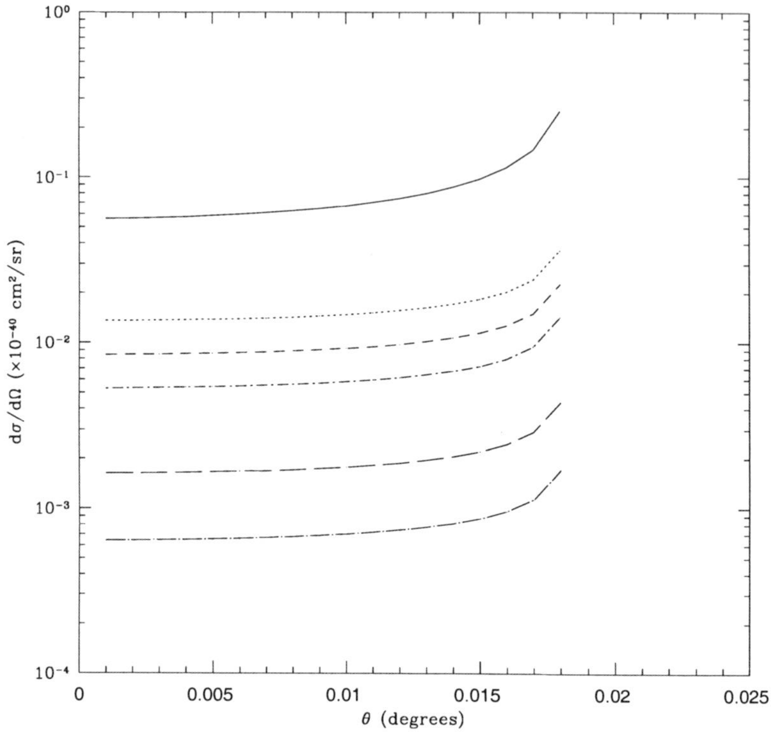

Figure 1. The differential cross section for the reaction $e^- + p \longrightarrow \Lambda + \nu_e$ as a function of outgoing Λ laboratory angle. The solid,dotted,dashed,dot dashed, large dashed and large dot dashed curves are for the total differential cross section, and the absolute value of the $F_A F_M, F_A F_V, F_A F_E, F_V F_M$, and $F_V F_E$ respectively. The incoming electron energy is 194.25 MeV.

show the differential cross section and the individual form factor contributions as well as the absolute value of the important interference terms. We do this for an incident muon of 0.265 GeV which is near threshold.

As can be seen in figure 2, the contributions of the interesting from factors, F_E, F_P, and F_S, are all too small to measure directly. However in this reaction $F_E F_A$, an interference term, might be extracted. This might allow a determination of F_E. . We note that these curves are relatively flat due to the small change in q^2 over this angular range. However it is unlikely that we could extract the pseudoscalar form factor, F_P which would be of great interest for a weak,strangeness changing transition.

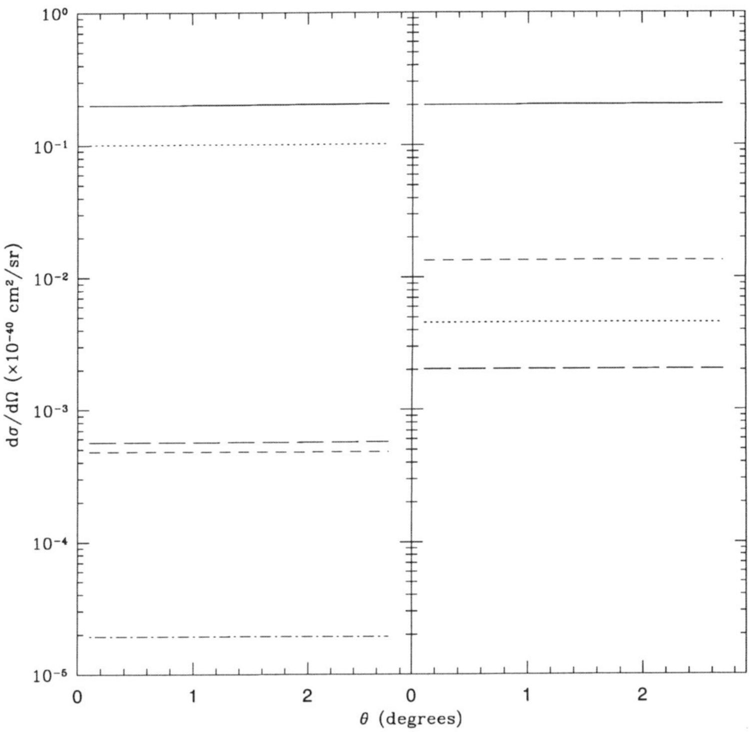

Figure 2. Plot of the differential cross section for the process $\mu^- + p \longrightarrow \Lambda + \nu_\mu$ for an incoming muon of energy 0.265 GeV in the laboratory frame. In the left panel, the solid,dotted, large dashed,dashed and dot dashed curves represent the contributions of the whole differential cross section,F_A^2, F_E^2, F_P^2 and F_S^2. In the right panel the solid,dashed,dotted, and large dashed curves represnt the contributions of the whole differential cross section,and the absolute values of the $F_E F_A$, $F_A F_P$, and $F_E F_V$ respectively.

We therefore consider the tau induced reaction,$\tau^- + p \longrightarrow \Lambda + \nu_\tau$. Because the tau lifetime is very short, only experiments involving highly relativistic tau leptons would be possible. The advantage of using the tau lepton and to a lesser extent, the muon is the large lepton mass which tends to increase contributions containing the pseudoscalar form factor as these contributions are all proportional to the square of the lepton mass.

In figure 3 we plot the interference terms of interest along with the whole differential cross section for the tau induced reaction. Although all of the interference terms are all too small to be observed, we notice that the term $F_A F_P$ containing the pseudoscalar form factor is relatively large. If it were possible to run this reaction near threshold perhaps this interference term would be quite large. However the short lifetime of the tau forbids this. We therefore consider the possibility of using antineutrino reactions of the form $\bar{\nu}_L + p \longrightarrow \Lambda + L^+$. This reaction could be run near threshold for both muon and tau production and perhaps some of the form factors not ordinarily extractable could in fact be determined. We therefore now consider this possibility.

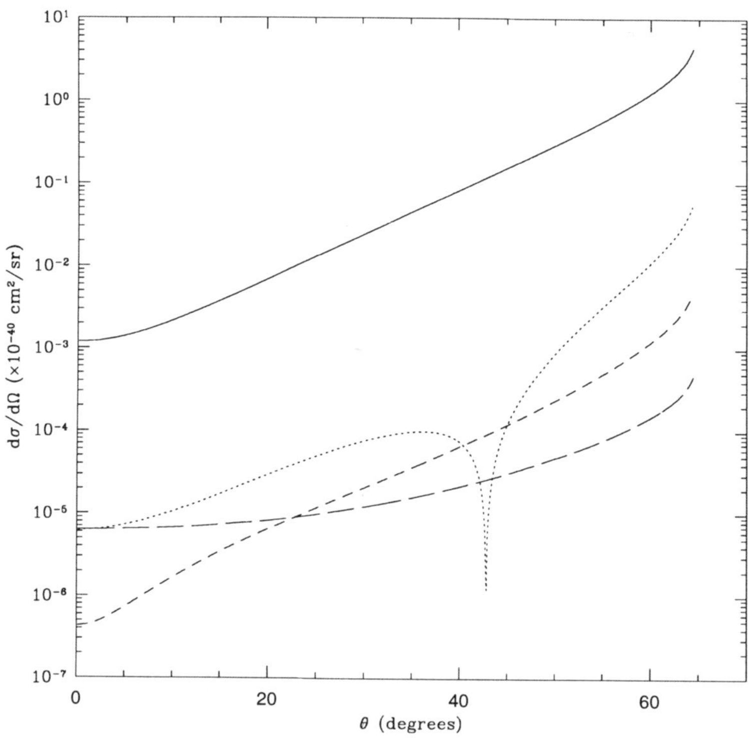

Figure 3. Plot of the differential cross section for the process $\tau^- + p \longrightarrow \Lambda + \nu_\mu$ for an incoming tau of energy 20.0 GeV in the laboratory frame. The solid,large dashed,dotted, and dashed curves represnt the contributions of the whole differential cross section,and the absolute values of the $F_E F_A$, $F_A F_P$, and $F_E F_V$ respectively.

2. Matrix Elements

We now concentrate on the antineutrino induced reaction $\bar{\nu}_L + p \longrightarrow \Lambda + L^+$. The matrix element for this reaction may be written as a first order weak process as:

$$< L^+\Lambda|H_w|\bar{\nu}_L P >= \frac{G}{\sqrt{2}} \cos(\theta_C)\bar{v_\nu}\gamma^\lambda(1-\gamma_5)v_L < \Lambda|J_\lambda(0)|p > . \quad (5)$$

Here the subscript L stands for a massive lepton (electron, muon, or tau). The matrix element of the weak hadronic current is already given by Eqs. (2),(3), and (4). Thus if the form factors are known we can immediately calculate the transition matrix element squared for this process. Using SU(3) relations we have previously obtained the standard form factors describing the weak vector current matrix element[3,4] and we merely quote the results:

$$F_V(q^2) = F_V(0)/(1 - q^2/M_V^2)^2 \quad (6)$$

with $F_V(0) = 1.2247$ and $M_V = .98GeV/c^2$ and

$$F_M(q^2) = F_M(0)/(1 - q^2/M_M^2)^2 \quad (7)$$

with $F_M(0) = 1.793/2m_p$ and $M_M = .71GeV/c^2$. This determines the standard part of the vector current matrix element.

The axial current matrix element is more difficult to obtain but we have already treated it in previous[3,4] work. We quote the results[5,6,7,8].

$$F_A(q^2) = F_A(0)/(1 - q^2/M_A^2)^2 \quad (8)$$

with $M_A = 1.25\ GeV/c^2$ and from Eqs.(11) and (13),$F_A(0) = .8793$. For F_P we use a standard form first given by Nambu, namely

$$F_P(q^2) = -F_A m_k(m_i + m_f)/(q^2 - m_k^2) . \quad (9)$$

Finally we estimate values for F_E and F_S. From a theoretical reference[9] we obtain an estimate for $F_E(0) = .705/2m_p$ in the notation used here. From the same reference $F_S(0) \simeq .344\ F_E(0)$ Making use of our experience that[3] F_E and F_M have similar q^2 dependence and making a similar assumption for F_S, we write:

$$F_E(q^2) = F_E(0)/(1 - q^2/M_M^2)^2 \quad (10)$$

$$F_S(q^2) = F_S(0)/(1 - q^2/M_M^2)^2 \quad (11)$$

where M_M is given in Eq.(12) and $F_E(0)$ and $F_S(0)$ are given above.

Thus we have obtained all of the necessary form factors for evaluating the differential cross section. We give the full transition matrix element squared which now contains contributions from the F_P and F_S form factors which were previously[3] omitted:

$$
\begin{aligned}
|M|^2 = \frac{1}{m_L m_\nu} \Bigg[& \frac{4|F_V|^2}{m_i m_f}[p_f \cdot \nu p_i \cdot L + p_f \cdot L p_i \cdot \nu - L \cdot \nu m_i m_f] + \\
& \frac{4|F_A|^2}{m_f m_i}[p_f \cdot \nu p_i \cdot L + p_f \cdot L p_i \cdot \nu + L \cdot \nu m_i m_f] + \\
& \frac{2|F_M|^2}{m_p^2 m_i m_f}[L \cdot \nu(p_i \cdot L p_f \cdot L + p_i \cdot \nu p_f \cdot \nu + L \cdot \nu m_i m_f) + \\
& + m_L^2(-p_i \cdot \nu p_f \cdot \nu - (1/2) L \cdot \nu p_i \cdot p_f - (3/2) L \cdot \nu m_i m_f)] \\
& - \frac{8 F_V F_A}{m_i m_f}[p_f \cdot \nu p_i \cdot L - p_i \cdot \nu p_f \cdot L + m_L^2 p_i \cdot \nu] + \\
& \frac{4 F_M F_V}{m_p m_f}[L \cdot \nu(p_f \cdot \nu - p_f \cdot L) + m_L^2 (1/2) p_i \cdot \nu - (3/2) L \cdot \nu] + \\
& \frac{4 F_M F_V}{m_p m_i}[L \cdot \nu(p_i \cdot L - p_i \cdot \nu) - (1/2) p_i \cdot \nu] - \\
& \frac{4 F_A F_M}{m_p m_f}[L \cdot \nu(p_f \cdot L + p_f \cdot \nu) - m_L^2 p_f \cdot \nu] - \\
& \frac{4 F_A F_M}{m_p m_i}[L \cdot \nu(p_i \cdot L + p_i \cdot \nu) - m_L^2 p_i \cdot \nu] + \\
& \frac{|F_E|^2}{m_p^2 m_i m_f}[L \cdot \nu(2(p_f \cdot L p_i \cdot L + p_f \cdot \nu p_i \cdot \nu) + q^2 p_f \cdot p_i)] + \\
& \frac{4 F_E F_V}{m_p m_i}[L \cdot \nu(p_i \cdot \nu + p_i \cdot L)] - \\
& \frac{4 F_E F_V}{m_p m_f}[L \cdot \nu(p_f \cdot \nu + p_f \cdot L)] + \\
& \frac{4 F_E F_A}{m_p m_f}[L \cdot \nu(p_f \cdot \nu - p_f \cdot L)] + \\
& \frac{4 F_E F_A}{m_p m_i}[L \cdot \nu(p_i \cdot \nu - p_i \cdot L)] + \\
& \frac{4 F_S^2}{(2m_p)^2 m_i m_f}[m_L^2 L \cdot \nu(p_f \cdot p_i + m_f m_i)] + \\
& \frac{2 F_P^2}{m_i m_f m_k^2}[m_L^2 L \cdot \nu(p_f \cdot p_i - m_f m_i)] + \\
& \frac{8 F_A F_P}{m_i m_f m_k}[m_L^2(p_i \cdot \nu m_f - p_f \cdot \nu m_i)] \Bigg] .
\end{aligned}
\tag{12}
$$

In the above equation we use m_i to denote the mass of the initial hadron,

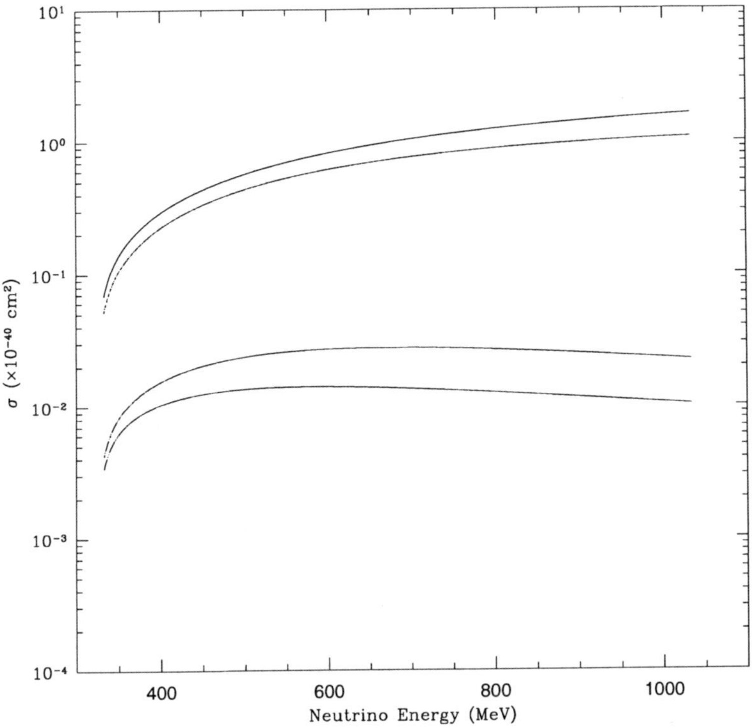

Figure 4. Plot of the cross section for the reaction $\bar{\nu}_\mu + p \longrightarrow \Lambda + L^+$ for muon antineutrino energies from threshold to 1.0 GeV in the laboratory frame. The contributions of the total cross section,the F_A^2, F_AF_E, and F_AF_P respectively are given in order of decreasing size.

here the proton and m_f to denote the final hadron which is the Λ. We use m_p to denote the proton mass which occurs in the definition of current matrix elements so that Eq.(18) may be more readily adapted to other processes. We note that every term in this matrix element squared is proportional to either the neutrino energy,momentum, or mass and to the either the lepton energy, momentum, or mass. The differential cross section can now be calculated by standard methods. The result is:

$$\sigma = \frac{m_\nu m_L m_f}{4\pi E_\nu} \int d\theta |M|^2 \frac{E_L \sin\theta}{m_i + E_\nu(1-\cos\theta)} . \quad (13)$$

3. Results

We can now evaluate the integral, Eq.14. We have so far done this for the muon case but not for the tau case which requires a careful treatment due to its large mass. The result for the muon case may be seen in figure 4. From the figure it is immediately seen that the form factor contributions of interest are too small to be observed. This is a disappointment but we are currently working on the the tau case and hope to have a result soon. This may well turn out to be the most promising possibility as the $F_A F_P$ term was relatively large even well away from threshold. We look forward to completing this work in the near future.

References

1. O.K. Baker,Proposal to the Jefferson Lab PAC,CEBAF(2001).
2. D. Mack, private communication (2002).
3. S.L. Mintz,*Nucl. Phys.* **A 657**,303(1999).
4. S.L. Mintz and M.A. Barnett,*Phys. Rev.* **D 66**,117501(2002).
5. Particle Data Group,*Eur. Phys.J* **C 3**,1(1998).
6. J. Dworkin et al.,*Phys. Rev.* **D 41**,780(1990).
7. J. Wise et al.,*Phys. Lett.* **98 B**,123(1981).
8. M. Bourquin et al.,*Z. Phys.* **C 12**,307(1982).
9. Larry J. Carson,Robert J. Oakes, and Charles R. Wilcox, *Phys. Rev.* **D 37**,3197(1988).
10. W-Y. P. Hwang and E. M. Henley,*Phys. Rev.* **D 38**,798(1988).

A NATURAL FRAMEWORK FOR BI-LARGE NEUTRINO MIXING

STUART RABY*

Department of Physics, The Ohio State University,
174 W. 18th Ave, Columbus, OH 43210, USA
Email: raby@pacific.mps.ohio-state.edu

In this talk I describe a natural framework for bi-large neutrino mixing within the context of two models – 1) a simple generalization of the MSSM and 2) an SO(10) model. Our starting point is the Frampton, Glashow, Yanagida [FGY] neutrino mass ansatz which can easily accomodate bi-large neutrino mixing. The main point of FGY, however, is to obtain a theory of neutrino masses with only one possible CP violating angle. They argue that the sign of the baryon asymmetry of the universe (assuming leptogenesis) is then correlated with CP asymmetries possibly observable in accelerator experiments. Unfortunately, there is a fly in the ointment. It was later shown by Raidal and Strumia [RS] that there is a sign ambiguity which frustrates the above correlation. We note that the Raidal-Strumia ambiguity is resolved in our models.

1. Neutrinos : Masses and Mixing Angles

Let us first summarize the present values of neutrino masses and mixing angles obtained by fitting atmospheric, solar, reactor and accelerator neutrino oscillation data. The atmospheric and solar neutrino masses and mixing angles are given below

- $\Delta m^2_{atm} = |m_3^2 - m_2^2| \approx 3 \times 10^{-3}\ \mathrm{eV}^2$
  $\sin 2\theta_{atm} \approx 1$
- $\Delta m^2_{sol} = |m_2^2 - m_1^2| \approx 7 \times 10^{-5}\ \mathrm{eV}^2$
  $\sin 2\theta_{sol} \leq 1$

*This work is partially supported by doe-er-01545. i also greatly appreciated the hospitality of the organizers.

with an approximate mixing matrix given by

$$\begin{pmatrix} \nu_e \\ \nu_\mu \\ \nu_\tau \end{pmatrix} \approx \begin{pmatrix} c_{sol} & s_{sol} & 0 \\ -s_{sol}/\sqrt{2} & c_{sol}/\sqrt{2} & 1/\sqrt{2} \\ -s_{sol}/\sqrt{2} & c_{sol}/\sqrt{2} & -1/\sqrt{2} \end{pmatrix} \begin{pmatrix} \nu_1 \\ \nu_2 \\ \nu_3 \end{pmatrix}.$$

For this talk, I will assume three light neutrinos. Recall, the observed small neutrino masses can naturally be generated via the See-Saw mechanism. The 3×3 neutrino mass matrix is given by $\sim m_\nu^T \ M_N^{-1} \ m_\nu$, where m_ν (M_N) is the Dirac RL (Majorana RR) mass matrix. Of course, the main problem with neutrino mixing angles is the fact that they are significantly larger than CKM mixing. These large mixing angles may be obtained from one (or a combination) of the following sources – m_ν, M_N; or m_e (the charged lepton mass matrix). In the models presented below, the origin of large neutrino mixing angles is in m_ν.

1.1. *Frampton-Glashow-Yanagida ansatz*

Consider now the FGY [1] ansatz given by

$$\begin{aligned} \mathbf{L} = & \ \left(\nu_e \ a \ + \ \nu_\mu \ a' \ e^{-i\phi/2}\right) \ N_1 \\ & + \ (\nu_\mu \ b \ + \ \nu_\tau \ b') \ N_2 \\ & + \tfrac{1}{2} \ (M_1 \ N_1^2 \ + \ M_2 \ N_2^2) \\ \mathbf{L} \ \equiv & \ \nu \ D^T \ N \ + \ \tfrac{1}{2} \ N \ M_N \ N \end{aligned} \tag{1}$$

where $N_{1,2}$ are the two right-handed (sterile) neutrinos and

$$D^T = \begin{pmatrix} a & 0 \\ a' \ e^{-i\phi/2} & b \\ 0 & b' \end{pmatrix}. \tag{2}$$

This neutrino mass matrix ansatz is expressed in the lepton flavor eigenbasis. The dimensionful parameters a, a', b, b' are chosen to satisfy the relations $b \approx b'$ for maximal atmospheric mixing angle and $a \sim a'$ for a large, but not maximal, solar mixing angle. In addition, due to the zeros in the mass matrix ansatz, there is only one non-vanishing CP violating angle, ϕ. Upon integrating out the heavy sterile neutrinos (N), we obtain the 3×3 FGY light neutrino mass matrix given by

$$\mathbf{M}_{FGY} = D^T \ M_N^{-1} \ D$$

The neutrino mass eigenvalues and small mixing angle are given by

$$m_{\nu_3} \approx 2b^2/M_2 \approx 0.05 \text{ eV} = \sqrt{\Delta m^2_{atm}}$$
$$m_{\nu_2} \approx 2a^2/M_1 \approx 8.4 \times 10^{-3} \text{ eV} = \sqrt{\Delta m^2_{sol}}$$
$$m_{\nu_1} = 0; \quad \theta_{13} \sim m_{\nu_2}/(\sqrt{2}\, m_{\nu_3})$$

1.2. *Raidal and Strumia analysis*

A detailed χ^2 analysis of the FGY ansatz including atmospheric and solar neutrino oscillation data was performed by Raidal and Strumia [2]. Their best fit then makes predictions for $\theta_{13} = 0.078 \pm 0.015$, observable in planned long baseline experiments, and for the effective electron neutrino mass measured in neutrinoless double beta decay, $m_{ee}^{0\nu\beta\beta} = 2.6 \pm 0.4$ meV, which is unobservable in any planned experiment.

In addition, RS perform a detailed analysis of leptogenesis with FGY. They find two successful solutions providing an acceptable cosmological baryon asymmetry [*this is the RS ambiguity*]. One with

$$M_1 \ll M_2; \quad M_1 \approx 10^{11} \text{ GeV}/|\sin\phi| \text{ and } \phi < 0. \tag{3}$$

In this case, the prediction for CP violating neutrino oscillation is given by

$$P(\nu_e \to \nu_\mu) \; < \; P(\nu_\mu \to \nu_e) \equiv P(\bar{\nu}_e \to \bar{\nu}_\mu).$$

In a supersymmetric generalization of FGY, they predict the lepton number flavor violating branching ratios

$$B(\mu \to e\gamma) \approx 2\; r\; 10^{-13};$$
$$B(\tau \to \mu\gamma) \geq 3\; r\; 10^{-12}$$

where $r \approx (\tan\beta/10)^2\ (150 \text{ GeV}/m_{SUSY})^4$.

The other with

$$M_1 \gg M_2; \quad M_2 \approx 10^{12} \text{ GeV}/|\sin\phi| \text{ and } \phi > 0. \tag{4}$$

they find

$$P(\nu_e \to \nu_\mu) \; > \; P(\nu_\mu \to \nu_e) \equiv P(\bar{\nu}_e \to \bar{\nu}_\mu)$$

and

$$B(\tau \to \mu\gamma) \approx 7\; r\; 10^{-11};$$
$$B(\mu \to e\gamma) \geq r\; 10^{-11}.$$

2. "Natural" FGY texture in SUSY with $[\mathrm{SU}(2)\times\mathrm{U}(1)]_{FS}$

We now obtain the FGY texture in a supersymmtric theory with an $[\mathrm{SU}(2)\times\mathrm{U}(1)]_{FS}$ family symmetry protecting the zeros [3]. This is a theory of leptons with the 3 families of electroweak doublets given by $l_i = \begin{pmatrix} \nu_i \\ e_i \end{pmatrix}$. The first two families transform as a doublet under the SU(2) family symmetry with $L_a = l_a,\ a = 1, 2$, and then l_3 is an SU(2) singlet. The superpotential for the neutrino mass sector is given by (we will discuss the charged lepton mass sector next)

$$\begin{aligned} W = \tfrac{H_u}{M} \, (\, L_a \, \phi^a \, N_1 \, + \, L_a \, \tilde{\phi}^a \, N_2 \, + \, l_3 \, \omega \, N_2 \,) \\ + \tfrac{1}{2} \, (\, S_1 \, {N_1}^2 \, + \, S_2 \, {N_2}^2 \,) \end{aligned} \tag{5}$$

where M is a new scale satisfying $M \gg M_Z$. The familon fields $\phi_a,\ \tilde{\phi}_a,\ \omega$ are assumed to get the following vacuum expectation values [VEVs] breaking the family symmetry and generating neutrino masses. We have, in complete generality, $\langle\phi\rangle = \begin{pmatrix} \langle\phi^1\rangle \\ \langle\phi^2\rangle \end{pmatrix}$ and $\langle\tilde{\phi}\rangle = \begin{pmatrix} 0 \\ \langle\tilde{\phi}^2\rangle \end{pmatrix}$. In addition the two right-handed neutrinos $N_i,\ i = 1, 2$ obtain Majorana masses when the SU(2) singlet familon fields get VEVs given by $\langle S_i\rangle = M_i,\ \ i = 1, 2$.

The family symmetry [with specific U(1) charges for all the fields, as discussed in more detail in the paper [3]] is sufficient to make this the most general superpotential consistent with the symmetry. Hence the zeros of the FGY ansatz are obtained, without fine tuning. When the Higgs doublet H_u gets a VEV at the weak scale given by $\langle H_u\rangle = \begin{pmatrix} 0 \\ v\sin\beta/\sqrt{2} \end{pmatrix}$ we finally obtain the FGY neutrino mass matrix with the parameters $a,\ a',\ b,\ b'$ given below.

$$\begin{aligned} a &= v\sin\beta \, \frac{\langle\phi^1\rangle}{\sqrt{2}M}, \quad & a' \, e^{-i\phi/2} &= v\sin\beta \, \frac{\langle\phi^2\rangle}{\sqrt{2}M}, \\ b &= v\sin\beta \, \frac{\langle\tilde{\phi}^2\rangle}{\sqrt{2}M}, \quad & b' &= v\sin\beta \, \frac{\langle\omega\rangle}{\sqrt{2}M}. \end{aligned} \tag{6}$$

Note, however, unlike FGY we must now consider the charged lepton mass matrix which is also constrained by the family symmetry. Until we do this, we cannot be certain that we have obtained the FGY ansatz in the lepton flavor basis.

3. Charged lepton masses

Consider the superpotential for charged leptons given by

$$W_{ch.\ leptons} = \frac{H_d}{M} \left(L_a \, \phi^a \, \bar{e}_1 \; + \; L_a \, \tilde{\phi}^a \, \bar{e}_2 \; + \right.$$
$$\left. l_3 \left(\, \omega \, \bar{e}_2 \; + \; \bar{\omega} \, \bar{e}_3 \, \right) \right) \qquad (7)$$

where the left-handed anti-leptons $\bar{e}_i, \; i = 1, 2, 3$ are SU(2) singlets. When the Higgs H_d gets a VEV given by $\langle H_d \rangle \;=\; \begin{pmatrix} v \cos\beta/\sqrt{2} \\ 0 \end{pmatrix}$ we obtain the charged lepton 3×3 mass matrix below

$$m_l = \begin{pmatrix} \bar{a} & \bar{a}' \, e^{-i\phi/2} & 0 \\ 0 & \bar{b} & \bar{b}' \\ 0 & 0 & \bar{c} \end{pmatrix} \qquad (8)$$

with

$$\bar{a} \;=\; v \cos\beta \, \frac{\langle \phi^1 \rangle}{\sqrt{2}M}, \qquad \bar{a}' \, e^{-i\phi/2} \;=\; v \cos\beta \, \frac{\langle \phi^2 \rangle}{\sqrt{2}M},$$
$$\bar{b} \;=\; v \cos\beta \, \frac{\langle \tilde{\phi}^2 \rangle}{\sqrt{2}M}, \qquad \bar{b}' \;=\; v \cos\beta \, \frac{\langle \omega \rangle}{\sqrt{2}M},$$
$$\bar{c} = v \cos\beta \, \frac{\langle \bar{\omega} \rangle}{\sqrt{2}M}$$

satisfying $\bar{a}, \; \bar{a}' \; \ll \; \bar{b}, \; \bar{b}' \; \ll \; \bar{c}$. Note the parameters $\bar{a}, \; \bar{a}', \; \bar{b}, \; \bar{b}'$ are, up to order one coefficients, the same as $a, \; a', \; b, \; b'$ appearing in the neutrino mass matrix.

The charged lepton mass eigenvalues are approximately given by $m_e \approx \bar{a}, \; m_\mu \; \approx \; \bar{b}, \; m_\tau \; \approx \; \bar{c}$ and the charged lepton mass matrix is diagonalized by the bi-unitary transformation $m_l^{diagonal} \; = U_{\bar{e}}^\dagger \; m_l \; U_e$ with the unitary matrix defining the left-handed mass eigenstates satisfying $U_e \approx Diag(1, e^{i\phi/2}, -e^{i\phi/2})$. Using this matrix we finally obtain the neutrino mass matrix in the flavor basis. We find

$$\mathbf{M} = U_e^T \; [\; D^T \; M_N^{-1} \; D \;] \; U_e \approx \; \mathbf{M}_{FGY} \qquad (9)$$

with

$$D^T \; = \begin{pmatrix} a & 0 \\ a' \, e^{-i\phi/2} & b \\ 0 & b' \end{pmatrix}.$$

Note, the ratios $|a/a'|$ and $|b/b'|$ can be adjusted to accommodate bi-large neutrino mixing.

As a bonus we now see that the ratio of Majorana neutrino masses M_1/M_2 is fixed. We find $(m_e/m_\mu)^2 \approx (\bar{a}/\bar{b})^2 \approx (a/b)^2 \approx (M_1/M_2)(m_{\nu_2}/m_{\nu_3})$. Thus $(M_1/M_2) \sim 10^{-4}$ and the RS ambiguity is resolved!

3.1. *Related issues*

Note, an SU(2) family symmetry in SUSY theories is desirable for completely different reasons. It has been shown that it can ameliorate the SUSY flavor problem [4]. For example, prior to SU(2) symmetry breaking the first and second generation sleptons are degenerate. Hence the off diagonal smuon-selectron mass term necessary for processes such as $\mu \to e\gamma$, seen in the figure below, are suppressed.

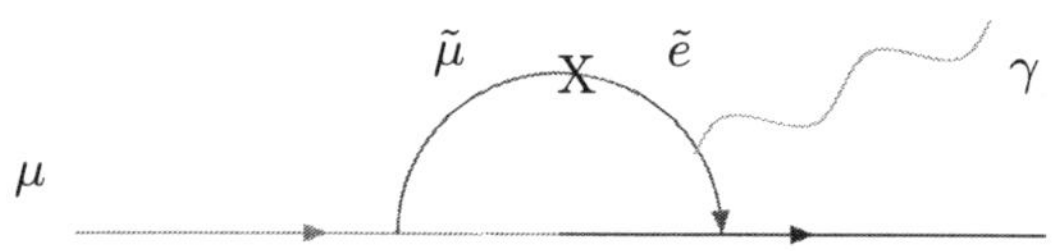

Finally, at the moment we have an effective higher dimensional field theory with new scales $M \sim M_1 \sim M_2 \gg M_Z$. Recall that successful leptogenesis requires $M_1, M_2 > 10^{11}$ GeV. Hence it is reasonable to expect that $M \sim M_{GUT} \approx 3 \times 10^{16}$ GeV. Consider the following virtues of SUSY GUTs. We have

- $M_Z << M_{GUT}$ "Naturally"
- Explains Charge Quantization
- Predicts Gauge Coupling Unification*
- Predicts Yukawa Coupling Unification
- + Family Symmetry $\Longrightarrow$ Hierarchy of Fermion Masses and Protects against large flavor violation
- Neutrino Masses via See - Saw scale $\sim 10^{-3} - 10^{-2}\ M_G \sim M_G^2/M_{Pl}$
- LSP – Dark Matter Candidate, and
- Baryogenesis via Leptogenesis

With all of these virtues it is worth considering embedding our generalized MSSM into an SO(10) SUSY GUT. We consider the following SO(10) SUSY GUT with an $[SU(2) \times U(1)^n]_{FS}$ proposed intially in [5] and analyzed in great detail in [6].

4. SO(10) SUSY GUT × $[SU(2) \times U(1)^n]_{FS}$

The superpotential for the charged fermion sector, including the heavy Froggatt-Nielsen [FG] states $\{ \chi^a, \ \bar{\chi}_a \}$, familon fields $\{ \phi^a, \ \tilde{\phi}^a \}$ and SO(10) adjoint $\{ 45 \}$, is given by

$$\begin{aligned} W \supset\ & 16_3\ 10_H\ 16_3 + 16_a\ 10_H\ \chi^a \\ & + \bar{\chi}_a\ (M_\chi\ \chi^a +\ 45\ \tfrac{\phi^a}{M}\ 16_3 \\ & + 45\ \tfrac{\tilde{\phi}^a \tilde{\phi}^b}{M^2}\ 16_b + A^{ab}\ 16_b) \end{aligned} \tag{10}$$

where we have the three families in 16_a $(a, b = 1, 2)$, 16_3; and the Higgs doublets in 10_H.

In addition, the FG mass M_χ necessarily includes SO(10) breaking VEVs with $M_\chi = M(1 + \alpha X + \beta Y)$. X, Y are the SO(10) breaking VEVs in the adjoint representation of SO(10) with X corresponding to the U(1) in SO(10) which preserves SU(5), and Y the standard weak hypercharge. α, β are arbitrary parameters. Upon integrating out the FG fields we obtain the effective fermion mass operators in the figure. Finally upon giving the familon fields VEVs we obtain the effective Yukawa couplings below.

$$\begin{aligned} Y_u &= \begin{pmatrix} 0 & \epsilon' \rho & -\epsilon\, \xi \\ -\epsilon' \rho & \tilde{\epsilon}\, \rho & -\epsilon \\ \epsilon\, \xi & \epsilon & 1 \end{pmatrix} \lambda \\ Y_d &= \begin{pmatrix} 0 & \epsilon' & -\epsilon\, \xi\, \sigma \\ -\epsilon' & \tilde{\epsilon} & -\epsilon\, \sigma \\ \epsilon\, \xi & \epsilon & 1 \end{pmatrix} \lambda \\ Y_e &= \begin{pmatrix} 0 & -\epsilon' & 3\, \epsilon\, \xi \\ \epsilon' & 3\, \tilde{\epsilon} & 3\, \epsilon \\ -3\, \epsilon\, \xi\, \sigma & -3\, \epsilon\, \sigma & 1 \end{pmatrix} \lambda \end{aligned} \tag{11}$$

with

$$\begin{aligned} \xi &= \langle \phi^1 \rangle / \langle \phi^2 \rangle; \quad \tilde{\epsilon} \propto (\langle \tilde{\phi}^2 \rangle / M)^2; \\ \epsilon &\propto \langle \phi^2 \rangle / M; \quad \epsilon' \sim \langle A^{12} \rangle / M; \\ \sigma &= \frac{1+\alpha}{1-3\alpha}; \quad \rho \sim \beta \ll \alpha. \end{aligned}$$

The small parameters are given by ϵ, $\tilde{\epsilon}$, ϵ'; all other parameters are numbers of order one. All these parameters are fit to the low energy data via a detailed χ^2 analysis.

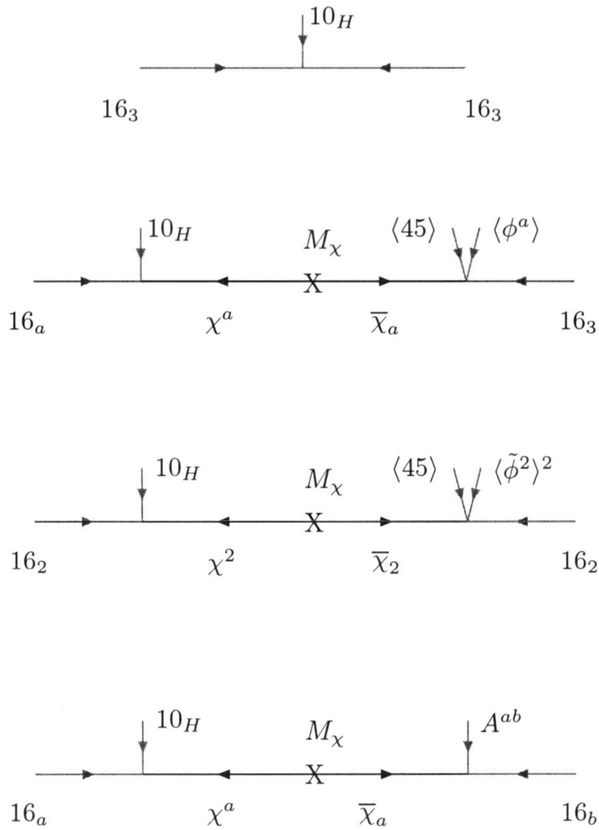

4.1. *Features of the model*

The model has the following nice features. The family hierarchy is obtained via the hierarchy of family symmetry breaking $SU_2 \times U_1 \longrightarrow U_1 \longrightarrow$ nothing where the first (second) breaking is due to the familon VEVs determining the small ratios ϵ, $\tilde{\epsilon}$ (ϵ'). As a result the 3rd family is much heavier than the 2nd family; is then heavier than the 1st family.

We incorporate the following mass patterns – 1) an approximate Georgi - Jarlskog mechanism "naturally" with the familon VEV $\langle 45 \rangle = (B-L)M_G$ giving $m_s \sim \frac{1}{3}m_\mu$ and $m_d \sim 3m_e$; 2) third generation Yukawa unification with $\lambda_t = \lambda_b = \lambda_\tau = \lambda_{\nu_\tau} = \lambda$ @M_G and 3) $\beta \ll \alpha \sim 1$ leads to $m_u < m_d$ even though $m_t \gg m_b$.

We "naturally" satisfy gauge coupling unification and the SU_2 family symmetry suppresses flavor violation such as $\mu \to e\gamma$.

The model has 10 Yukawa parameters (6 real coefficients and 4 phases) to fit 13 fermion masses and mixing angles. Varying these 10 Yukawa parameters [three gauge (α_G, M_G, ϵ_3) and 7 soft SUSY breaking parameters] at M_G we use the two (one) loop RG equations for dimensionless (dimensionful) variables to obtain the χ^2 function at M_Z. All observables at M_Z are evaluated including one loop threshold corrections. The result of the χ^2 analysis, using the code of T. Blazek (see Refs. [7,6]), is given in the Table. The fit is quite good. In particular there are only 10 Yukawa parameters with 16 independent fermion mass and mixing angle observables.

Table 1. χ^2 analysis including 5 precision electroweak and 16 fermion mass and mixing angle observables plus the branching ratio for the process $b \to s\gamma$.

Observable	Data(σ) masses	Theory (in GeV)	Pull
M_Z	91.188 (0.091)	91.21	¡0.50
M_W	80.419 (0.080)	80.40	¡0.50
$G_\mu \cdot 10^5$	1.1664 (0.0012)	1.166	¡0.50
α_{EM}^{-1}	137.04 (0.14)	137.0	¡0.50
$\alpha_s(M_Z)$	0.11720 (0.002)	0.1139	2.65
M_t	174.30 (5.1)	171.3	¡0.50
$m_b(M_b)$	4.220 (0.09)	4.377	3.04
$M_b - M_c$	3.400 (0.2)	3.430	¡ 0.50
$m_c(m_c)$	1.3000 (0.15)	1.212	¡0.50
m_s	0.089 (0.011)	0.100	1.01
m_d/m_s	0.050 (0.015)	0.0751	2.80
Q^{-2}	0.00203 (0.00020)	0.00200	¡ 0.50
M_τ	1.777 (0.0018)	1.777	¡ 0.50
M_μ	0.10566 (0.00011)	.1057	¡ 0.50
$M_e \cdot 10^3$	0.5110 (0.00051)	0.5110	¡ 0.50
V_{us}	0.2230 (0.0040)	0.2213	¡ 0.50
V_{cb}	0.04020 (0.0019)	0.0391	¡ 0.50
V_{ub}/V_{cb}	0.0860 (0.008)	0.0850	¡ 0.50
V_{td}	0.00820 (0.00082)	0.00846	¡0.50
ϵ_K	0.00228 (0.00023)	0.00233	¡ 0.50
$\sin 2\beta$	0.7270 (0.036)	0.6898	1.07
$B(b \to s\gamma) \cdot 10^4$	3.340 (0.38)	3.433	¡ 0.50
TOTAL χ^2	12.16		

4.2. *Bi-large neutrino mixing in $SO_{10} \times [SU_2 \times U_1^n]_{FS}$ model*

Let us now consider neutrino masses in this model. The model has three right-handed neutrinos contained in the fields 16_a, 16_3. Moreover the Dirac neutrino Yukawa matrix is given by

$$Y_\nu = \begin{pmatrix} 0 & -\epsilon' \, \omega & \frac{3}{2} \, \epsilon \, \xi \, \omega \\ \epsilon' \, \omega & 3 \, \tilde{\epsilon} \, \omega & \frac{3}{2} \, \epsilon \, \omega \\ -3 \, \epsilon \, \xi \, \sigma & -3 \, \epsilon \, \sigma & 1 \end{pmatrix} \lambda \tag{12}$$

with $\omega \;=\; 2\,\sigma/(2\,\sigma - 1)$ This then gives the Dirac LR neutrino mass matrix – $m_\nu \equiv Y_\nu \frac{v}{\sqrt{2}} \sin\beta$.

We now add three SO(10) singlet fields $N_i, \; i = 1, 2, 3$ and couple them to neutrinos via the superpotential

$$\begin{aligned} W_{neutrino} = \tfrac{\overline{16}}{M} \left(N_1 \; \tilde{\phi}^a \; 16_a \; + \; N_2 \; \phi^a \; 16_a \right) \\ + \tfrac{\overline{16}}{M} \left(\; N_3 \; \theta \; 16_3 \right) \\ + \tfrac{1}{2} \left(S_1 \; N_1^2 \; + \; S_2 \; N_2^2 \right). \end{aligned} \tag{13}$$

The field $\overline{16}$ (and another field 16) are assumed to obtain VEVs in the right-handed neutrino direction. In fact, these VEVs are necessary (along with the 45 VEVs) to break SO(10) to the standard model. Given $\langle \overline{16} \rangle = V$, $\langle S_i \rangle = M_i, \; i = 1, 2$ and the familon VEVs we obtain the effective neutrino mass terms given by

$$W_\nu^{eff} = \nu \; m_\nu \; \bar{\nu} + \bar{\nu} \; V \; N + \tfrac{1}{2} \; N \; M_N \; N \tag{14}$$

where

$$(V^T)^{-1} = \frac{M}{v_{16}} \begin{pmatrix} -1/(\langle \tilde{\phi}^2 \rangle \; \xi) & 1/\langle \phi^1 \rangle & 0 \\ 1/\langle \tilde{\phi}^2 \rangle & 0 & 0 \\ 0 & 0 & 1/\langle \theta \rangle \end{pmatrix}.$$

Once the heavy $\bar{\nu}$, N fields are integrated out near the GUT scale, we obtain the 3×3 Majorana neutrino mass matrix (in the flavor basis)

$$\mathbf{M} = U_e^T \; [\; m_\nu \; (V^T)^{-1} \; M_N \; V^{-1} \; m_\nu^T \;] \; U_e. \tag{15}$$

Note, defining the 3×2 Dirac neutrino mass matrix

$$D^T \equiv m_\nu \; (V^T)^{-1} \; M_N \begin{pmatrix} 1 & 0 \\ 0 & 1 \\ 0 & 0 \end{pmatrix} = \begin{pmatrix} a & 0 \\ a' & b \\ 0 & b' \end{pmatrix}$$

and the 2×2 right-handed neutrino mass matrix

$$\hat{M}_N \equiv \begin{pmatrix} M_1 & 0 \\ 0 & M_2 \end{pmatrix},$$

we finally obtain the light neutrino mass matrix

$$\mathbf{M} = U_e^T \; [\; D^T \; \hat{M}_N^{-1} \; D \;] \; U_e. \tag{16}$$

Hence we obtain the Frampton, Glashow & Yanagida ansatz without any fine-tuning. In particular, the zeros are exact and fixed by the family symmetry. This may be seen by the intermediate calculation below.

We have

$$
\begin{aligned}
& Y_\nu \;\; (V^T)^{-1} \\
\sim & \begin{pmatrix} 0 & -\epsilon' \; \omega & \frac{3}{2} \; \epsilon \; \xi \; \omega \\ \epsilon' \; \omega & 3 \; \tilde{\epsilon} \; \omega & \frac{3}{2} \; \epsilon \; \omega \\ -3 \; \epsilon \; \xi \; \sigma & -3 \; \epsilon \; \sigma & 1 \end{pmatrix} \\
\times & \begin{pmatrix} -1/(\langle \tilde{\phi}^2 \rangle \; \xi) & 1/\langle \phi^1 \rangle & 0 \\ 1/\langle \tilde{\phi}^2 \rangle & 0 & 0 \\ 0 & 0 & 1/\langle \theta \rangle \end{pmatrix} \\
\sim & \begin{pmatrix} a & 0 & \mathrm{X} \\ a' & b & \mathrm{X} \\ 0 & b' & \mathrm{X} \end{pmatrix}
\end{aligned}
\tag{17}
$$

In addition, bi-large mixing requires significant constraints on the parameters b, b', a, a'. Maximal atmospheric neutrino oscillation requires $b \approx b'$, whereas large mixing for solar neutrino oscillations requires a of order a'. On the other hand, the ratios $|b/b'|$ and $|a/a'|$ are, now, also constrained by charged fermion masses and mixing angles. Without performing a joint χ^2 analysis (including both charged fermions and neutrinos) we can still make the following observations. The fit to charged fermion masses naturally gives $\epsilon' \sim \epsilon \; \xi$. Hence given

$$
\begin{aligned}
b \equiv & \quad \epsilon' \; \omega \; \lambda \; (M_2/\phi^1) \; \frac{M}{v_{16}} \frac{v \sin\beta}{\sqrt{2}} \\
b' \equiv & -3 \; \epsilon \; \xi \; \sigma \; \lambda \; (M_2/\phi^1) \; \frac{M}{v_{16}} \frac{v \sin\beta}{\sqrt{2}} .
\end{aligned}
\tag{18}
$$

we find $|b/b'| = (\epsilon' \; \omega)/(3 \; \epsilon \; \xi \; \sigma) \approx 1$. With regards a, a' we have $\epsilon' \; \xi^{-1} \sim \tilde{\epsilon}$. Moreover

$$
\begin{aligned}
a \equiv & \qquad -\epsilon' \; \omega \; \lambda \; (M_1/\tilde{\phi}^2) \; \frac{M}{v_{16}} \frac{v \sin\beta}{\sqrt{2}} \\
a' \equiv & \; (-\epsilon' \; \xi^{-1} + 3 \; \tilde{\epsilon}) \; \omega \; \lambda \; (M_1/\tilde{\phi}^2) \; \frac{M}{v_{16}} \frac{v \sin\beta}{\sqrt{2}}
\end{aligned}
\tag{19}
$$

gives $|a/a'| = \epsilon'/((-\epsilon' \; \xi^{-1} + 3 \; \tilde{\epsilon})$ which apparently requires fine-tuning of order 1 in 10 to get $a \sim a'$. Clearly in order to test this model, we must perform an extended χ^2 analysis including both charged fermions and neutrinos. Note, there are only two new parameters in the neutrino sector which can be fit to the solar and atmospheric neutrino mass squared differences.

Finally we find $m_{\nu_2}/m_{\nu_3} \approx (m_e/m_\mu)\ (M_1/M_2)\ \tilde{\epsilon}$ and thus $M_1/M_2 \sim 10^3$. Unfortunately, leptogenesis is more complicated in this model since we must first identify the heavy neutrino mass eigenstates and their lepton/Higgs couplings. Moreover there are, in principle, more CP violating phases.

4.3. *Summary*

The Frampton-Glashow-Yanagida neutrino mass matrix ansatz has several nice features. Since it has only one CP violating angle it can, in principle, correlate the sign of the matter - anti-matter asymmetry with CP violating neutrino oscillations. Unfortunately, the Raidal-Strumia analysis uncovered an ambiguity which allows either sign of the CP violating angle for successful leptogenesis, depending on whether the right-handed neutrino masses satisfies $M_1 \ll M_2$ or $M_1 \gg M_2$! In addition the FGY ansatz provides a natural framework for bi-large neutrino mixing!

In this talk we obtained the FGY ansatz in a supersymmetric model with an $[\mathrm{SU}(2) \times \mathrm{U}(1)]_{FS}$ family symmetry. This had two advantages, 1) the FGY ansatz is fixed by the family symmetry and 2) it resolves the RS ambiguity. This is because the same family symmetry acting on neutrinos also constrains the charged fermion masses. As a result, the ratio of right-handed neutrino masses is fixed by the ratio of δm^2 for atmospheric and solar neutrinos and the ratio of charged fermion masses. It is also important to note that we are only able to obtain the normal hierarchy for neutrino masses.

The second model presented in this talk was a SUSY SO(10) model with an $[\mathrm{SU}(2) \times \mathrm{U}(1)^n]_{FS}$ family symmetry. This model is significantly more constrained since it relates all fermion masses and mixing angles. There are only two new arbitrary parameters in the neutrino sector which are fixed by the neutrino mass differences. Nevertheless, we "naturally" obtain the FGY ansatz for neutrinos (in a theory which initially starts with 3 right-handed neutrinos and 3 SO(10) singlet neutrinos). We find a large neutrino mixing angle for atmospheric neutrino oscillations "naturally." On the other hand, an extended χ^2 analysis must be performed to see if we can obtain a large solar mixing angle. If a good fit is obtained for neutrino masses and mixing, it will then be very interesting to analyze CP violation and leptogenesis in this model.

References

1. P. H. Frampton, S. L. Glashow and T. Yanagida, *Phys. Lett.***B548**, 119 (2002).
2. M. Raidal and A. Strumia, *Phys. Lett.***B553**, 72 (2003).
3. S. Raby, *Phys. Lett.***B561**, 119 (2003).
4. M. Dine, R. G. Leigh and A. Kagan, *Phys. Rev.***D48**, 4269 (1993); A. Pomarol and D. Tommasini, *Nucl. Phys.***B466**, 3 (1996); R. Barbieri, G. R. Dvali and L. J. Hall, *Phys. Lett.***B377**, 76 (1996).
5. R. Barbieri, L. J. Hall, S. Raby and A. Romanino, *Nucl. Phys.***B493**, 3 (1997).
6. T. Blazek, S. Raby and K. Tobe, *Phys. Rev.***D60**, 113001 (1999); T. Blazek, S. Raby and K. Tobe, *Phys. Rev.***D62**, 055001 (2000).
7. T. Blazek, M. Carena, S. Raby and C. E. M. Wagner, *Phys. Rev.***D56**, 6919 (1997).

CHASING CHOOZ*

P. RAMOND

We relate the MNS and CKM mixing matrices using ideas from grand unification. We catalog models in terms of the family symmetries of the down quark mass matrices, and emphasize the role of the Cabibbo angle in the lepton mixing matrix. We find a large class of models with an observable CHOOZ angle $\sim \lambda/\sqrt{2}$

1. Introduction

Since 1998 when SuperKamiokande announced incontrovertible evidence of neutrino oscillations, a series of experiments have unravelled most of the fundamental neutrino parameters. Two important pieces of information still await determination: the absolute value of the masses, together with their hierarchy, and the third mixing angle, for which there only exists a limit set by the CHOOZ experiment. From the theoretical side, neutrino masses are hardly surprising in the context of the grand unified[1,2,3,4] generalizations of the Standard Model. While some of us predicted one large mixing angle in the MNS lepton mixing matrix, the recent determination of *two* large mixing angles comes as a surprise, and poses unforeseen theoretical challenges. In the following we offer several remarks as to the theoretical implications of the present data set and note that with a modest infusion of grand unification ideas, it is possible to relate the CKM and MNS matrices. In a wide class of models, the CHOOZ angle is even predicted in terms of the Cabibbo angle.

2. The Data

Five years of neutrino experiments[5,6,7,8,9,10] can be succintly summarized[11]: the MNS lepton mixing matrix

---

*This work was supported in part by the us department of energy under grant de-fg02-97er41029

$$
\mathcal{U}_{MNS} = \begin{pmatrix} \cos\phi_\odot & \sin\phi_\odot & \epsilon \\ -\cos\theta_\oplus\ \sin\phi_\odot & \cos\theta_\oplus\ \cos\phi_\odot & \sin\theta_\oplus \\ \sin\theta_\oplus\ \sin\phi_\odot & -\sin\theta_\oplus\ \cos\phi_\odot & \cos\theta_\oplus \end{pmatrix} ,
$$

has two large angles, one of which may be maximal. The size of the third is limited by the CHOOZ[8] reactor experiment

$$
\sin^2 2\theta_\oplus > 0.85 ; \qquad 0.30 < \tan^2\phi_\odot < 0.65 ; \qquad |\,\epsilon\,|^2 < 0.005 .
$$

It is theoretically intriguing to produce a (3×3) matrix with such angles, and it may be a powerful hint towards the resolution of the riddle of flavor. Oscillation experiments determine only mass differences

$$
\Delta m^2_\odot = |\, m^2_{\nu_1} - m^2_{\nu_2}\,| \sim 7. \times 10^{-5}\ \mathrm{eV}^2
$$

$$
\Delta m^2_\oplus = |\, m^2_{\nu_2} - m^2_{\nu_3}\,| \sim 3. \times 10^{-3}\ \mathrm{eV}^2 ,
$$

but WMAP[12] is the only experiment to set a limit on the absolute value of their masses

$$
\sum_i m_{\nu_i} < .71\ \mathrm{eV}
$$

These allow for three three possible mass patterns[13]:

- Hierarchical with $m_{\nu_1} \prec m_{\nu_2} \prec m_{\nu_3}$
- Inverted with $m_{\nu_3} \prec m_{\nu_2} \simeq m_{\nu_1}$
- Hyperfine with $m_{\nu_1} \simeq m_{\nu_2} \simeq m_{\nu_3}$

In view of the strong theoretical reasons for linking quarks and leptons, the difference of their mixings is quite striking. In the context of the seesaw[14] mechanism, these have possibly fundamental implications as to the structure of the electroweak singlet masses of the right-handed neutrinos. In the following we explore the causes of these differences, and offer some theoretical prediction for the size of the CHOOZ angle.

3. CKM and MNS Mixings

The $\Delta I_{\mathrm{w}} = 1/2$ Higgs doublet breaking of electroweak symmetry yields the quark Yukawa Matrices

$$\mathcal{M}^{(2/3)} = \mathcal{U}_{2/3} \begin{pmatrix} m_u & 0 & 0 \\ 0 & m_c & 0 \\ 0 & 0 & m_t \end{pmatrix} \mathcal{V}^\dagger_{2/3}$$

$$\mathcal{M}^{(-1/3)} = \mathcal{U}_{-1/3} \begin{pmatrix} m_d & 0 & 0 \\ 0 & m_s & 0 \\ 0 & 0 & m_b \end{pmatrix} \mathcal{V}^\dagger_{-1/3}$$

from which the observable quark mixing matrix is deduced.

$$\mathcal{U}_{CKM} = \mathcal{U}^\dagger_{2/3}\mathcal{U}_{-1/3}$$

Experimentally $\mathcal{U}_{CKM}$ is nearly equal to the unit matrix, up to small powers of λ, the Cabibbo angle, allowing for its Wolfenstein[15] expansion which uses the unit matrix as a starting point. The lesson here is that family mixing is roughly the same for charge $2/3$ and $-1/3$ quarks. The charged leptons Yukawa Matrix

$$\mathcal{M}^{(-1)} = \mathcal{U}_{-1} \begin{pmatrix} m_e & 0 & 0 \\ 0 & m_\mu & 0 \\ 0 & 0 & m_\tau \end{pmatrix} \mathcal{V}^\dagger_{-1}$$

has hierarchical mass eigenvalues, as in the quark sector. To explain neutrino masses, it is simplest to add one right-handed neutrino per family. This implies a neutral leptons Yukawa Matrix

$$\mathcal{M}^{(0)}_{Dirac} = \mathcal{U}_0\,\mathcal{D}_0\,\mathcal{V}^\dagger_0 = \mathcal{U}_0 \begin{pmatrix} m_1 & 0 & 0 \\ 0 & m_2 & 0 \\ 0 & 0 & m_3 \end{pmatrix} \mathcal{V}^\dagger_0$$

and a lepton mixing matrix that is similar to the CKM matrix

$$\mathcal{U}_{MNS} = \mathcal{U}^\dagger_{-1}\mathcal{U}_0\,?$$

Can such a matrix be so different from its sister matrix in the quark sector? Most right-thinking theorists think that the right-handed neutrino masses

$$\mathcal{M}^{(0)}_{Majorana} \sim \Delta I_{\rm w} = 0$$

radically change the landscape: unlike those of quarks and charged leptons, the right-handed neutrinos masses are unconstrained by electroweak quantum numbers. They only have lepton number, and since there are very good limits on lepton number conservation, one expects these masses to be very large. This leads to the seesaw[14] mechanism, according to which the neutrino mass matrix is

$$\mathcal{M}^{(0)}_{Seesaw} = \mathcal{M}^{(0)}_{Dirac} \frac{1}{\mathcal{M}^{(0)}_{Majorana}} \mathcal{M}^{(0)\,t}_{Dirac} .$$

We can rewrite it as

$$\mathcal{M}^{(0)}_{Seesaw} = \mathcal{U}_0 \, \mathcal{C} \, \mathcal{U}^t_0 ,$$

where we have introduce the Central Matrix

$$\mathcal{C} \equiv \mathcal{D}_0 \, \mathcal{V}^\dagger_0 \frac{1}{\mathcal{M}^{(0)}_{Majorana}} \mathcal{V}^*_0 \, \mathcal{D}_0 .$$

It is diagonalized by a unitary matrix $\mathcal{F}$

$$\mathcal{C} = \mathcal{F} \mathcal{D}_\nu \mathcal{F}^t ,$$

while the physical neutrino masses are contained in the diagonal $\mathcal{D}_\nu$. This alters the observable MNS lepton mixing matrix to

$$\mathcal{U}_{MNS} = \mathcal{U}^\dagger_{-1} \mathcal{U}_0 \, \mathcal{F} .$$

The seesaw adds to the MNS matrix the extra unitary matrix $\mathcal{F}$. This suggests we put models of neutrino masses in three classes:

- Models of type 0 for which $\mathcal{F}$ contains no large angle,
- Models of type I for which only one large angle is in $\mathcal{F}$,
- Models of type II, for which both large angles reside in $\mathcal{F}$.

Type I models seem more generic to us than type II, since it is natural to have matrices with one, three or no large mixing angles.

4. A Pinch of Grand Unification

If we want to relate the quark and lepton observables, we need to add some theoretical prejudice. Such is not hard to find, as it is made obvious both by anomaly cancellation and the quantum number structures: grand unification. While there is a great deal of uncertainty as to the nature of the grand unified group and the mechanism by which its large symmetries are broken, there is no doubt that such a group is to be found (cosmologically speaking) in our distant past. It is therefore important to study the patterns implied.

Grand Unification relates $\Delta I_{\rm w} = 1/2$ quark and lepton Yukawa matrices. In the simplest case, $SU(5)$, we have

$$\mathcal{M}^{(-1/3)} \sim \mathcal{M}^{(-1)\,t} ,$$

where t is the transpose, while in $SO(10)$ which naturally includes right-handed neutrinos,

$$\mathcal{M}^{(2/3)} \sim \mathcal{M}^{(0)}_{Dirac} .$$

Assuming only these two simple patterns, we infer

$$\mathcal{U}_{-1/3} \sim \mathcal{V}^{*}_{-1} , \qquad \mathcal{U}_{2/3} \sim \mathcal{U}_{0} .$$

Putting these back in the MNS matrix, we find

$$\begin{aligned} \mathcal{U}_{MNS} &= \mathcal{U}^{\dagger}_{-1}\mathcal{U}_{-1/3}\mathcal{U}^{\dagger}_{CKM}\,\mathcal{F} \\ &= \mathcal{V}^{t}_{-1/3}\mathcal{U}_{-1/3}\mathcal{U}^{\dagger}_{CKM}\,\mathcal{F} , \end{aligned}$$

suggesting how the CKM matrix and its Cabibbo expansion enter into the lepton mixing observables. It is interesting to note that it is the structure of the charge $-1/3$ matrix that plays a dominant role in relating quark to lepton mixings.

- If $\mathcal{M}_{-1/3}$ is family-symmetric,

  $$\mathcal{U}_{-1/3} = \mathcal{V}^{*}_{-1/3} ,$$

  and we get

  $$\mathcal{U}_{MNS} = \mathcal{U}^{\dagger}_{CKM}\,\mathcal{F} .$$

  Both large angles must be in $\mathcal{F}$; these are type II models.
- If $\mathcal{M}_{-1/3}$ is not family-symmetric, one or two large angle can come from it, leading to type O and type I models.

Type O models require the two mixing angles among the right-handed down quarks to be uniformly large. Although logically possible, I will not discuss these models in the following.

There is a framework in which type I Models appear naturally. Start from the Yukawa matrices, assuming that the masses a Cabibbo expansion for the mass ratios, as well as the Froggatt-Nielsen[16] scheme for the exponents,

$$\mathcal{M}^{(-1/3)} = \begin{pmatrix} \lambda^4 & \lambda^3 & \lambda^3 \\ \lambda^3 & \lambda^2 & \lambda^2 \\ \lambda & a & b \end{pmatrix} ,$$

and

$$\mathcal{U}_{CKM} = \begin{pmatrix} 1 & \lambda & \lambda^3 \\ \lambda & 1 & \lambda^2 \\ \lambda^3 & \lambda^2 & 1 \end{pmatrix} \qquad \frac{m_s}{m_b} \sim \lambda^2 \qquad \frac{m_d}{m_b} \sim \lambda^4 .$$

It is simplest to consider the no-mixing limit ($\lambda \rightarrow 0$) where

$$\mathcal{M}^{(2/3)} = \begin{pmatrix} 0 & 0 & 0 \\ 0 & 0 & 0 \\ 0 & 0 & m_t \end{pmatrix} , \qquad \mathcal{M}^{(-1/3)} = \begin{pmatrix} 0 & 0 & 0 \\ 0 & 0 & 0 \\ 0 & a & b \end{pmatrix}$$

and

$$\mathcal{U}_{CKM} = \mathcal{U}^{\dagger}_{2/3}\mathcal{U}_{-1/3} = 1$$

Then $SU(5)$ suggest a non-symmetric charge -1 matrix as well,

$$\mathcal{M}^{(-1)} = \begin{pmatrix} 0 & 0 & 0 \\ 0 & 0 & a \\ 0 & 0 & b \end{pmatrix} ,$$

so that

$$\mathcal{U}_{-1} = \begin{pmatrix} 1 & 0 & 0 \\ 0 & \cos\theta & \sin\theta \\ 0 & -\sin\theta & \cos\theta \end{pmatrix} , \qquad \tan\theta = \frac{a}{b} ,$$

contains one unsuppressed mixing angle.

$$\mathcal{U}_{MNS} = \begin{pmatrix} 1 & 0 & 0 \\ 0 & \cos\theta & \sin\theta \\ 0 & -\sin\theta & \cos\theta \end{pmatrix} \mathcal{F} \, .$$

It is natural from this point of view[17,18] to expect unsuppressed atmospheric neutrino mixing, although this approach does not fix the value of the angle.

The relation between the CKM and MNS matrices suggest that the Cabibbo angle plays an important role and should not be neglected. In particular it could very well be that the CHOOZ angle is *solely* a Cabibbo effect. In type I Models, this hypothesis suggests that

$$\mathcal{U}_{MNS} = \begin{pmatrix} 1 & \lambda^{\alpha} & \lambda^{\beta} \\ \lambda^{\alpha} & \cos\theta & \sin\theta \\ \lambda^{\beta} & -\sin\theta & \cos\theta \end{pmatrix} \begin{pmatrix} \cos\phi & \sin\phi & \lambda^{\gamma} \\ -\sin\phi & \cos\phi & \lambda^{\delta} \\ \lambda^{\gamma} & \lambda^{\delta} & 1 \end{pmatrix} ,$$

where the exponents are completely unknown. As a result the CHOOZ angle is given by λ to an unknown power with an unknown prefactor, and our hypothesis is not very useful.

However, in type II models, where the $\Delta I_w = 1/2$ Yukawa matrices are family symmetric, we find a more definitive prediction

$$\mathcal{U}_{MNS} = \begin{pmatrix} 1 & \lambda & \lambda^3 \\ \lambda & 1 & \lambda^2 \\ \lambda^3 & \lambda^2 & 1 \end{pmatrix} \begin{pmatrix} \cos\phi & \sin\phi & \lambda^{\gamma} \\ -\cos\theta\ \sin\phi & \cos\theta\ \cos\phi & \sin\theta \\ \sin\theta\ \sin\phi & -\sin\theta\ \cos\phi & \cos\theta \end{pmatrix} ,$$

so that plausibly

$$\text{CHOOZ angle} \quad \sim \quad \lambda \sin\theta \sim \frac{\lambda}{\sqrt{2}} \, ,$$

assuming $\gamma < 2$. This conclusion is particularly exciting as it can be tested in the foreseeable future; if correct it will also open the way to observable CP violation in neutrino physics!

We end this section with two additional remarks: one is that it will be desirable to fix a Wolfenstein parametrization for the MNS matrix. The second is that there is another class of models where the charge $-1/3$ Yukawa matrix is not family-symmetric. These are Family Cloning[19] models where each family starts with its own gauge group. A tri-chiral order parameter naturally explains

$$SU(3)_1 \times SU(3)_2 \times SU(3)_3 \ \rightarrow \ SU(3)_{1+2+3} \, ,$$

but falls short in unifying the thre weak isospins into one

$$SU(2)_1 \times SU(2)_2 \times SU(2)_3 \ \rightarrow \ SU(2)_{1+2} \times SU(2)_3 \ ,$$

resulting in a Standard Model with lopsided gauge symmetry

$$SU(2)_{1+2} \times SU(2)_3 \times SU(3)_{1+2+3} \ ,$$

yielding asymmetric Yukawa matrices and one large angle in U_{-1}.

5. Large Angles and $\mathcal{F}$

Finally we would like to discuss the theoretical implications of large angles in $\mathcal{F}$. A(3×3) matrix generically contains one or three large angles, but not two. From this point of view type I models are more desirable.

As in our earlier work[20], consider a simplified case with two families. Recall that

$$\mathcal{C} \ \equiv \ \mathcal{D}_0 \, \mathcal{V}_0^\dagger \, \frac{1}{\mathcal{M}^{(0)}_{Majorana}} \, \mathcal{V}_0^* \, \mathcal{D}_0 \ ,$$

and that the $\Delta I_w = 1/2$ Neutral Dirac Mass is hierarchical:

$$\mathcal{D}_0 \ = \ m \begin{pmatrix} a\,\lambda^\alpha & 0 \\ 0 & 1 \end{pmatrix} .$$

If we call M_1 and M_2 the right-handed Majorana masses, simple algebra yields the central matrix

$$\mathcal{C} \ = \ \begin{pmatrix} (\frac{c^2}{M_1} + \frac{s^2}{M_2})\, a^2\, \lambda^{2\alpha} & (\frac{c\,s}{M_1} - \frac{c\,s}{M_2})\, a\, \lambda^\alpha \\ (\frac{c\,s}{M_1} - \frac{c\,s}{M_2})\, a\, \lambda^\alpha & (\frac{s^2}{M_1} + \frac{c^2}{M_2}) \end{pmatrix} ,$$

where s , c are the sine and cosine of a mixing angle.

It is diagonalized by a large mixing angle in one of two cases:

- $\mathcal{C}_{11} \ \sim \ \mathcal{C}_{22} \ \sim \ \mathcal{C}_{12}$. This implies

$$s \ \sim \ b\lambda^\alpha \ , \qquad c \ \sim \ 1 \ ,$$

  and

$$\frac{M_1}{M_2} \sim \lambda^{2\beta} .$$

Hence the Majorana masses must be hierarchical: there is a correlated hierarchy between the $\Delta I_w = 1/2$ and $\Delta I_w = 0$ sectors. Then

$$\mathcal{C} = \lambda^{2\alpha} \frac{m^2}{M_1} \begin{pmatrix} a^2 & a\,b \\ a\,b & b^2 \end{pmatrix} , \quad \beta > \alpha$$

- $\mathcal{C}_{11}\,,\,\mathcal{C}_{22} \prec \mathcal{C}_{12}$. This is the level crossing case. We obtain

$$\lambda^{\alpha} \prec s \preceq \lambda^{\alpha-\beta}$$

so that

$$\mathcal{C} = \frac{\lambda^{\alpha}\, m^2}{\sqrt{-M_1 M_2}} \begin{pmatrix} 0 & a \\ a & 0 \end{pmatrix} ,$$

naturally leading to maximal mixing. Could the right-handed neutrinos of two families be Dirac partners?

6. Conclusions

The neutrino data set presents new theoretical challenges and hopefully hints towards the resolution of the family puzzle. It points to family mixing at very high scales and a hierarchy among the right-handed neutrinos, in accordance with the grand-unified paradigm. This is a welcome feature in terms of modern theories of leptogenesis.

We close with some suggestions to model builders. Go directly to the Planck scale (or just below). There you will find branes awaiting you with Weyl fermions we can right-handed neutrinos. Their interactions and masses are key to understanding the family riddles and symmetries, and even better laboratory neutrino physics opens a window to their masses. Only then should you worry about the Standard Model; and for that you can always use $SO(10)$ or any such model[21,22].

7. Acknowledgements

This is the first Coral Gables Conference without Behram Kursunoglu, and many of us who have attended countless times miss him. I shall always remember him as a charming advocate and tireless promoter of fundamental science. Thanks to his efforts, physicists from all over the world came yearly to South Florida to discuss physics in a relaxed and gracious atmosphere. I also wish to thank Drs A. Datta and L. Everett for useful discussions.

References

1. J.C. Pati and A.Salam, *Phys. Rev. D* **10** (1974) 275.
2. H. Georgi and S. Glashow, *Phys. Rev. Lett.* **32** (1974) 438.
3. H. Georgi, in *Particles and Fields-1974*, edited by C.E.Carlson, AIP Conference Proceedings No. 23 (American Institute of Physics, New York, 1975) p.575; H. Fritzsch and P. Minkowski, *Ann. Phys. NY* **93** (1975) 193.
4. F. Gürsey, P. Ramond, and P. Sikivie, *Phys. Lett. B* **60** (1975) 177.
5. The Super-Kamiokande Collaboration, *Phys. Rev. Lett.* **85** (2000) 3999.
6. The Super-Kamiokande Collaboration, *Phys. Lett. B* **539** (2002) 179.
7. The SNO Collaboration, *Phys. Rev. Lett.* **89** (2002) 011301.
8. M. Apollonio *et al.*, *Phys. Lett. B* **338** (1998) 383; *Phys. Lett. B* **420** (1998) 397.
9. The KamLAND Collaboration, *Phys. Rev. Lett.* **90** (2003) 021802.
10. The K2K Collaboration, *Phys. Rev. Lett.* **90** (2003) 041801.
11. G.L. Fogli *et al.*, *Phys. Rev. D* **66** (2002) 093008; G.L. Fogli *et al.*, *Phys. Rev. D* **67** (2003) 073002.
12. The WMAP Collaboration, astro-ph/0302207; astro-ph/0302209.
13. For a recent review, see A.Yu. Smirnov, *"Neutrino Physics: Open Theoretical Questions"* , Nov 2003. hep-ph/0311259.
14. M. Gell-Mann, P. Ramond, and R. Slansky in Sanibel Talk, CALT-68-709, Feb 1979 , hep-ph/9809459 (retroprint), and in *Supergravity* (North Holland, Amsterdam 1979). T. Yanagida, in *Proceedings of the Workshop on Unified Theory and Baryon Number of the Universe*, KEK, Japan, Feb 1979.
15. L. Wolfenstein, *Phys. Rev. Lett.* **51** (1983) 1945.
16. C. Froggatt and H.B. Nielsen, *Nucl. Phys. B* **147** (1979) 277.
17. N. Irges, S. Lavignac and P. Ramond, *Phys. Rev. D* **58** (1998) 035003.
18. S. King and G. G. Ross *Phys. Lett. B* **57** (2003) 239.
19. Mark J. Bowick and P. Ramond, *Phys. Lett. B* **131** (1983) 367
20. A. Datta, F-S Ling and P. Ramond, NPB **671** (2003) 383. P. Ramond, "*Neutrinos: Windows to Planck Physics*, Hep-ph/0401001.
    Nucl. Phys. B **199** (1982) 223.
21. C. H. Albright and S. M. Barr, *Phys. Lett. B* **461** (1999) 218
22. G. Altarelli, F. Feruglio and I. Masina, JHEP **0011** (2000) 040; W. Buchmüller, Acta Phys.Polon. B32 (2001) 3707-3718; T. Blazek, S. Raby, K. Tobe, *Phys. Rev. D* **62** (2000) 055001.

5 Lorentz Violation, Field Theory and Branes

PROBLEMS WITH GAUGE INVARIANCE IN LORENTZ-VIOLATING QED

B. ALTSCHUL

Department of Physics
Indiana University
Bloomington, IN 47405 USA
E-mail: baltschu@indiana.edu

We consider the radiative generation of a Chern-Simons addition to the effective action for a Lorentz-violating modification of electrodynamics. The coefficient of the induced Chern-Simons term is manifestly finite, but also potentially indeterminate. However, several values have been suggested as having special significance. We discuss two such special values, and it turns out that neither of them can be made consistent with the gauge invariance properties of the theory.

There has been much recent interest in the possibility of adding CPT- and Lorentz-violating terms to the Lagrangians of quantum field theories[1,2,3,4]. These terms may arise from violations of these symmetries at the Planck scale. There are many strong experimental constraints on Lorentz-violating corrections to the standard model, and such corrections must generally be small. However, the subject of Lorentz violation in quantum field theory is still of great theoretical and experimental interest. In particular, we shall find that there is a complicated interaction between Lorentz-violation and the property of gauge invariance; requiring a Lorentz-violating theory to be gauge invariant can have some curious and counterintuitive properties[5].

The simplest perturbatively nontrivial Lorentz-violating correction to the fermion sector of quantum electrodynamics involves the addition of a CPT-violating axial vector term,

$$\mathcal{L}_b = -\bar{\psi}\, \not{b}\gamma_5\psi, \tag{1}$$

to the action. There has been a great deal of interest in the effect of such a term on the radiative corrections[4,6,7,8,9,10,11,12] to the theory. b_μ has the correct C, P, and T symmetries to generate a term of the Chern-Simons form[13,14,15]

$$\mathcal{L}_{CS} = \frac{1}{2}(k_{CS})_\mu \epsilon^{\mu\alpha\beta\gamma} F_{\alpha\beta} A_\gamma \tag{2}$$

through radiative corrections. Astrophysical measurements constrain a physical coefficient $(k_{CS})_\mu$ to be very small[15,16,17]. This raises the question of whether the possible radiative corrections to $(k_{CS})_\mu$ can be reconciled with the physical term's apparently vanishing value without any fine tuning of parameters.

The full answer to this question is closely tied to the question of gauge invariance. However, the path to the answer is fraught with logical pitfalls. The issues involved were first studied by Coleman and Glashow[9]. They argued that the radiative corrections to the Chern-Simons term were necessarily vanishing—a result which would alleviate any paradoxes. Coleman's and Glashow's argument can be understood as follows: At one-loop order, there are two diagrams that can contribute to the Chern-Simons term; these are triangle diagrams, just such as one might encounter in the study of π^0 decay or anomaly cancellation. External photons are attached to two vertices, and the b_μ interaction appears at the third. If the external photons carry momenta p and k, then the Ward identity implies that the amplitude corresponding to the sum of these diagrams must be transverse to both p and k—$\mathcal{M}^{\mu\nu}p_\mu = \mathcal{M}^{\mu\nu}k_\mu = 0$, where $\mathcal{M}^{\mu\nu}$ is the amplitude in question. Differentiating these transversality conditions, we find that $\mathcal{M}^{\mu\nu}$ is necessarily $\mathcal{O}(kp)$. When we set $p = -k$, corresponding to the physical situation, we find that the amplitude must be $\mathcal{O}\left(p^2\right)$. Since the Chern-Simons term is $\mathcal{O}(p)$, it appears that it cannot be generated by radiative corrections.

This seems a simple enough argument; however, there are additional subtleties that call its validity into question. In particular, the argument presupposes that the external photons can carrying different momenta, or, equivalently, that the b_μ vertex can carry a nonvanishing momentum q. So Coleman and Glashow required gauge invariance for all possible momentum exchanges. This is a very natural condition; for example, this is exactly the condition we need to enforce in the similar problem of π^0 decay in order to ensure that we obtain a unique answer. However, unlike a physical pion, the b_μ vertex can only carry zero momentum. If we set $p = -k$ at the outset, then the two transversality conditions are equivalent, and the whole argument breaks down.

This subtlety involving the momentum transfer q is related to the fact that the Lagrange density $\mathcal{L}_{CS} = \frac{1}{2}(k_{CS})_\mu \epsilon^{\mu\alpha\beta\gamma} F_{\alpha\beta} A_\gamma$ is not itself gauge invariant. (This is not such a serious problem as it might seem, for reasons that will shortly become evident.) Requiring transversality for all q is equivalent to requiring that all the Fourier components of $\mathcal{L}_{CS}$ be gauge invariant, which is not the case. However, enforcing gauge invariance only at $q = 0$ merely requires that

$$\int d^4x \, \mathcal{L}_{CS} \, e^{iq\cdot x} = \int d^4x \, \mathcal{L}_{CS} = I_{CS}, \tag{3}$$

where I_{CS} is the Chern-Simons term in the *action*, be gauge invariant. By using integration by parts, we can verify that I_{CS} is indeed gauge invariant. We must conclude therefore that Coleman's and Glashow's argument (which would exclude a Chern-Simons term *a priori*, because it does not have a gauge invariant Lagrange density) does not really apply, and so nonvanishing radiative corrections have not been excluded.

It still remains, however, to perform the actual calculation. Rather than considering only the triangle diagrams, which are valid to $\mathcal{O}(b)$, we shall try to work to all orders in b. In order to do this, we must introduce the full Lagrangian for our theory. The Lagrange density is

$$\mathcal{L} = -\frac{1}{4} F^{\mu\nu} F_{\mu\nu} + \bar{\psi}(i\partial\!\!\!/ - m - eA\!\!\!/ - b\!\!\!/\gamma_5)\psi. \tag{4}$$

We shall primarily be concerned with the $m = 0$ case (although Coleman's and Glashow's argument, as presented above, does not actually apply to the massless case). Considering only massless particles will simplify our calculations and will allow us to construct a clear analogy that illuminates the origins of the difficulties we shall encounter. Although we can eliminate b from the massless Lagrangian by making the chiral transformation

$$\psi \to e^{-i\gamma_5 b\cdot x}\psi, \; \bar{\psi} \to \bar{\psi} e^{-i\gamma_5 b\cdot x}, \tag{5}$$

this transformation is anomalous and does not leave the gauge invariantly regulated fermionic measure invariant[18].

The exact fermion propagator for the massive theory is

$$S(k) = \frac{i}{k\!\!\!/ - m - b\!\!\!/\gamma_5}. \tag{6}$$

When $m = 0$, this is most easily rationalized by breaking it into two terms, corresponding to the two eigenvalues of γ_5. Doing this, we have

$$\begin{aligned} S(k) &= \frac{i}{k\!\!\!/ - b\!\!\!/}\frac{1-\gamma_5}{2} + \frac{i}{k\!\!\!/ + b\!\!\!/}\frac{1+\gamma_5}{2} \\ &= i\left[\frac{k\!\!\!/ - b\!\!\!/}{(k-b)^2}\frac{1-\gamma_5}{2} + \frac{k\!\!\!/ + b\!\!\!/}{(k+b)^2}\frac{1+\gamma_5}{2}\right]. \end{aligned} \tag{7}$$

In the b-exact formulation, all the one-loop diagrams contributing to the photon self-energy (with arbitrary numbers of b_μ insertions) are replaced by a single diagram: the usual $\mathcal{O}\left(e^2\right)$ photon self-energy, with the usual electron propagator replaced by our b-exact expression. So the one-loop self-energy becomes

$$\Pi^{\mu\nu}(p) = -ie^2 \mathrm{tr} \int \frac{d^4k}{(2\pi)^4} \gamma^\mu S(k) \gamma^\nu S(k+p). \tag{8}$$

This may be simplified in the massless case to

$$\Pi^{\mu\nu}(p) = \frac{ie^2}{2} \mathrm{tr} \int \frac{d^4k}{(2\pi)^4} \left[\frac{\gamma^\mu(\not{k}+\not{b})\gamma^\nu(\not{k}'+\not{b}) + \gamma^\mu(\not{k}+\not{b})\gamma^\nu(\not{k}'+\not{b})\gamma_5}{(k+b)^2(k'+b)^2} \right.$$
$$\left. + \frac{\gamma^\mu(\not{k}-\not{b})\gamma^\nu(\not{k}'-\not{b}) - \gamma^\mu(\not{k}-\not{b})\gamma^\nu(\not{k}'-\not{b})\gamma_5}{(k-b)^2(k'-b)^2} \right], \tag{9}$$

where $k' = k+p$. Through an analysis of the structure of (9), we may learn a great deal about the nature of the nonperturbative theory.

We observe that all the information necessary for the calculation of $\Pi^{\mu\nu}(p)$ is contained in the function

$$f_{\alpha\beta}(p,b) = \int \frac{d^4k}{(2\pi)^4} \frac{(k-b)_\alpha (k+p-b)_\beta}{(k-b)^2(k+p-b)^2}. \tag{10}$$

The self-energy involves symmetric and antisymmetric sums $f_{\alpha\beta}(p,b) \pm f_{\alpha\beta}(p,-b)$ contracted with tensors in (μ,ν,α,β). The terms of $f_{\alpha\beta}(p,b)$ that are even in b give rise to contributions to $\Pi^{\mu\nu}(p)$ with different Lorentz structure than the terms that are odd in b, since the odd terms involve a trace over γ_5. However, as a formal object, $f_{\alpha\beta}(p,b)$ still contains everything needed to determine the one-loop self-energy.

The two terms in (9) involving γ_5 contribute to k_{CS} in the effective action. (These terms correspond to the two triangle diagrams.) The two terms are linearly divergent, but their sum is finite. If the theory were defined perturbatively in b, the momentum routings for the two diagrams would be uncorrelated, and we would be free to shift the integration variables in the terms relative to one another; then *any* constant of proportionality between k_{CS} and b could be obtained. So, while the radiatively induced value of $(k_{CS})_\mu$ would be unambiguously finite, it would also be undetermined (and possibly vanishing); its value would depend upon how the theory is regulated.

So we see that this theory has the fascinating property that (when the theory is appropriately defined), the radiatively induced Chern-Simons term may be unambiguously finite, yet entirely undetermined. However, we may hope to avoid this ambiguity in the nonperturbative theory. It was pointed out by Jackiw and Kostelecký[6] that, since there is only a single Feynman diagram in the nonperturbative formalism, there is no longer any freedom to shift the momentum routings relative to one-another; we must regulate the two terms the same way. If we do this, and evaluate the integrals by symmetric integration, we get the definite nonperturbative value of $(k_{CS})_\mu = -\frac{e^2}{16\pi^2}b_\mu$. Since it is the result of a nonperturbative calculation, this value appears to have a special significance.

However, there are gauge-invariance-related problems with this value of $(k_{CS})_\mu$ as well. The nonperturbative viewpoint requires that the same regulator be used for all the terms, regardless of whether they are even or odd in b. Because there is involved only a single Feynman diagram, a truly nonperturbative calculation would involve a single evaluation of $f_{\alpha\beta}(p,b)$ to all orders in b using a unique regularization prescription. We shall therefore use the same methods used by Jackiw and Kostelecký to fix the coefficient of the induced Chern-Simons term to determine the higher-order, CPT-even contributions to the self-energy.

If we shift the integration variable $k \to k - b$ in $f_{\alpha\beta}(p,b)$, then the integrand becomes b-independent. Since the integral is superficially quadratically divergent, the surface term generated by the shift is at most quadratic in b. Therefore, there are no contributions to the self-energy that are higher than second order in b. This result has been previously demonstrated for the b-odd terms, and it might be expected on dimensional grounds for the b-even terms as well, since there is no mass scale in the problem. However, the ease with which it has been demonstrated here shows the usefulness of the propagator (7).

The $\mathcal{O}\left(b^0\right)$ part of $f_{\alpha\beta}(p,b)$ gives the usual QED photon self-energy, and the $\mathcal{O}(b)$ contribution to $\Pi^{\mu\nu}(p)$ just gives the Chern-Simons term. We shall therefore focus our attention on the $\mathcal{O}\left(b^2\right)$ terms.

We shall evaluate the $\mathcal{O}\left(b^2\right)$ terms in $f_{\alpha\beta}(p,b)$ by a direct expansion of the integrand in powers of b. The integral that appears at second order in b is finite when the integration is performed symmetrically. This is the correct prescription for performing the integration in the nonperturbative formalism, because the same property (observer Lorentz invariance) is being used to fix the value of the vacuum polarization at both $\mathcal{O}(b)$ and $\mathcal{O}\left(b^2\right)$. [Of course, the same technique cannot be used to deal with the $\mathcal{O}\left(b^0\right)$ term in the photon self-energy if gauge invariance is to be preserved. We shall set this formal difficulty aside, however, since the $\mathcal{O}\left(b^0\right)$ is necessarily divergent in any regularization scheme and is thus qualitatively different from the higher-order terms.]

We begin our calculation by writing

$$f_{\alpha\beta}(p,b)=\int\frac{d^4k}{(2\pi)^4}\exp\left(-b_\gamma\frac{\partial}{\partial k_\gamma}\right)\frac{k_\alpha}{k^2}\frac{k'_\beta}{(k')^2}. \tag{11}$$

The portion of this expression that is quadratic in b, which we shall denote as $h_{\alpha\beta}(p,b)$, is

$$h_{\alpha\beta}(p,b)=\frac{1}{2}b_\gamma b_\delta\int\frac{d^4k}{(2\pi)^4}\frac{\partial}{\partial k_\gamma}\frac{\partial}{\partial k_\delta}\frac{k_\alpha}{k^2}\frac{k'_\beta}{(k')^2} \tag{12}$$

$$=\frac{1}{96\pi^2}(b_\alpha b_\beta-g_{\alpha\beta}b^2). \tag{13}$$

Transversality of the vacuum polarization requires that $(2p^\alpha g^{\nu\beta}-p^\nu g^{\alpha\beta})h_{\alpha\beta}(p,b)=0$. This condition does not hold; even though $h_{\alpha\beta}(p,b)$ is unambiguously finite, it still violates the Ward identity. Gauge invariance is broken, and we conclude that the nonperturbative regularization scheme is unacceptable. We must therefore discard the special value of $(k_{CS})_\mu$ found by Jackiw and Kostelecký.

To understand how this violation of gauge invariance arises, it is useful to consider an analogy. For a theory with Lagrange density

$$\mathcal{L}'=-\frac{1}{4}F^{\mu\nu}F_{\mu\nu}+\bar{\psi}(i\partial\!\!\!/-eA\!\!\!/-a\!\!\!/)\psi, \tag{14}$$

there is a field rescaling similar to (5),

$$\psi\rightarrow e^{-ia\cdot x}\psi,\ \bar{\psi}\rightarrow e^{ia\cdot x}\bar{\psi}, \tag{15}$$

which eliminates a from the theory. Moreover, this rescaling is not anomalous. However, we may choose not to eliminate a from the Lagrangian. If we consider the theory directly as defined by $\mathcal{L}'$ and attempt to calculate the photon self-energy, we will immediately be led to an evaluation of $f_{\alpha\beta}(p,a)$. The violation of the Ward identity in this instance is clearly an artifact of our unconventional choice of momentum coordinates. A shift in the integration variable $k\rightarrow k-a$ eliminates the problematic term. This shift is precisely equivalent to the field redefinition (15).

Returning to the axial vector theory (4), it seems now that the correct solution to our difficulties would be to shift the origin of the integration in (10) so that $h_{\alpha\beta}(p,b)$ is set to zero. While we believe that this is the physically correct way of regulating this theory, it is inconsistent with the nonperturbative formalism. In the vector theory with a, there arises in the calculation of $\Pi^{\mu\nu}(p)$ only a single term of the form

$f_{\alpha\beta}(p,a)$. However, in the axial vector theory, we encounter the sum $f_{\alpha\beta}(p,b)+f_{\alpha\beta}(p,-b)$. The essence of the nonperturbative formulation is that both terms in (9) must be regulated in the same way. We are allowed only a single shift in the integration variable $k \to k+q$. This shift transforms $f_{\alpha\beta}(p,b)+f_{\alpha\beta}(p,-b) \to f_{\alpha\beta}(p,q+b)+f_{\alpha\beta}(p,q-b) \neq 0$. We see that in order to eliminate the Ward-identity-violating surface term, we must be free to shift the integrations in different terms by different amounts, which is equivalent to defining the theory perturbatively. [Although we are working with a massless theory, we are not allowed to make naive chiral shifts in the integration variable k if we are to retain the special nonperturbative value for $(k_{CS})_\mu$, because the corresponding transformation (5) is anomalous. To properly account for the anomaly, we must use the functional integral formalism. The Fujikawa determinant[18] accompanying the shift then reproduces the correct nonperturbative value of the induced Chern-Simons coefficient; however, there is another, completely undetermined contribution that arises from the ambiguity in the definition of the axial current operator[19]. So it is impossible to shift the integrations separately for the left- and right-handed components of ψ without giving up the uniqueness of the $\mathcal{O}(b)$ result.]

The existence of a surface term that violates gauge invariance at $\mathcal{O}\left(b^2\right)$ is perhaps unsurprising, given what is known about the behavior of the nonperturbative theory at $\mathcal{O}(b)$. In the massless case, the only contribution to the induced Chern-Simons coefficient comes from a surface term. The associated induced Lagrange density is not gauge invariant; however, because the density necessarily involves $\epsilon^{\mu\nu\alpha\beta}$, the Ward identity is preserved and the integrated action remains gauge invariant. There is no such restriction on the form of the induced term at $\mathcal{O}\left(b^2\right)$, and without the protection of a specific structure for the self-energy, the gauge invariance of the action is lost. At each order, the surface term simply violates gauge invariance in the strongest way allowed by its tensor structure.

Finally, we must discuss the possibility of dimensional regularization. A dimensional regulator preserves gauge invariance at all orders in b and sets $h_{\alpha\beta}(p,b)=0$. Moreover, it solves the formal problems associated with the nonperturbative evaluation of the $\mathcal{O}\left(b^0\right)$ contributions, since it allows us to regulate all the terms, even the divergent ones, in the same fashion. However, dimensional regularization also restores the complete ambiguity of the induced Chern-Simons term. The b-odd terms involve γ_5, which does not have a unique extension to $4-\epsilon$ dimensions. Any extension that commutes with $4-n\epsilon$ γ-matrices (for arbitrary n) will have the correct limit as $\epsilon \to 0$, and each extension will give a different result for the Chern-Simons coefficient. Therefore, while a dimensional regularization prescription may be used to implement a completely nonperturbative formulation of the theory, it also renders the coefficient of the induced Chern-Simons term completely undetermined.

We conclude that there seems to be no escape from the ambiguity in the value of the induced Chern-Simons term. Gauge invariance has played a central role in our discussions. In the similar problem of π^0 decay, gauge invariance ensures the uniqueness of the answer. However, in this situation, the enforcement of gauge invariance does exactly the opposite; it ensures that the radiative corrections to the theory are ambiguous. The special values of $(k_{CS})_\mu$ proposed by Coleman and Glashow and by Jackiw and Kostelecký both have problems; the reasoning leading to each value cannot be made compatible with the gauge invariance properties of this theory. This demonstrates that the role of gauge invariance in Lorentz violating theories can be complicated and quite subtle.

References

1. D. Colladay, V. A. Kostelecký, Phys. Rev. D **55**, 6760 (1997).
2. D. Colladay, V. A. Kostelecký, Phys. Rev. D **58**, 116002 (1998).
3. V. A. Kostelecký, R. Lehnert, Phys. Rev. D **63**, 065008 (2001).
4. V. A. Kostelecký, C. D. Lane, A. G. M. Pickering, Phys. Rev. D **65**, 056006 (2002).
5. B. Altschul, hep-th/0311200.
6. R. Jackiw, V. A. Kostelecký, Phys. Rev. Lett. **82**, 3572 (1999).
7. M. Pérez-Victoria, Phys. Rev. Lett. **83**, 2518 (1999).
8. J. M. Chung, Phys. Lett. B **461**, 138 (1999).
9. S. Coleman, S. L. Glashow, Phys. Rev. D **59**, 116008 (1999).
10. J. M. Chung, P. Oh, Phys. Rev. D **60**, 067702 (1999).
11. M. Pérez-Victoria, JHEP **04**, 032 (2001).
12. V. A. Kostelecký, A. G. M. Pickering, Phys. Rev. Lett. **91**, 031801 (2003).
13. R. Jackiw, S. Templeton, Phys. Rev. D **23**, 2291 (1981).
14. J. Schonfeld, Nucl. Phys. B **185**, 157 (1981).
15. S. M. Carroll, G. B. Field, R. Jackiw, Phys. Rev. D **41**, 1231 (1990).
16. S. M. Carroll, G. B. Field, Phys. Rev. Lett. **79**, 2394 (1997).
17. M. Goldhaber, V. Trimble, J. Astrophys. Astron. **17**, 17 (1996).
18. K. Fujikawa, Phys. Rev. Lett. **42**, 1195 (1979).
19. J. M. Chung, Phys. Rev. D **60**, 127901 (1999).

NEW IMPLICATIONS OF LORENTZ VIOLATION

DON COLLADAY

New College of Florida
Sarasota, Florida 34243-2197
E-mail: colladay@ncf.edu

In this proceedings, I summarize two recently discovered theoretical implications that Lorentz violation has on physical systems. First, I discuss new models for neutrino oscillations in which relatively simple combinations of Lorentz-violating parameters can mimic the major features of the current neutrino oscillation data. Second, I will present results on Yang-Mills instantons in Lorentz-violating background fields. An explicit solution is presented for unit winding number in SU(2).

1. Introduction

Enormous success in particle physics has been obtained during the last century by assuming symmetry of the fundamental action under the Lorentz group. Supplementing this with various assumed gauge symmetries and representational content eventually led to the standard model. A key step in constructing the standard model involves spontaneously breaking one of these assumed symmetries as well as relaxing some of the discrete symmetries in the electroweak sector. A natural question arises as to the validity of perfect symmetry under the Lorentz group as well.

In fact, there are theoretical reasons to suspect that Lorentz symmetry breaking may arise naturally in more fundamental theories such as string theory[1] or other attempts at quantum gravity[2]. In addition, there are numerous experimental tests of Lorentz invariance in a variety of sectors[3]. A general framework for including general Lorentz breaking effects into the standard model has been constructed[4,5]. The resulting effective field theory is called the Standard Model Extension (SME). Stability and causality issues as well as generic properties of the dispersion relations have also been studied[6].

2. Lorentz and CPT Violation

For about the past fifteen years, it has been known that miniscule remnant effects that violate Lorentz invariance may arise in a more fundamental theory of nature[1]. In addition, the well known CPT theorem proves that any local, Lorentz invariant quantum field theory must also preserve CPT. In fact, this theorem has been expanded to prove that CPT violation implies Lorentz violation[10], demonstrating that bounds on CPT can be interpreted as bounds on Lorentz violation.

The generic features of such violations may be incorporated into effective field theory using a generic spontaneous symmetry breaking mechanism that is analogous to the conventional Higg's mechanism of the standard model. The crucial difference concerns the field that exhibits a nonzero vacuum expectation value. In conventional Higg's models, the field that is used to break electroweak symmetry is taken as a scalar field in order to preserve Lorentz invariance as well as renormalizability. Consider a generic field theory containing gauge bosons with tensor indices (B^μ for example) with nontrivial couplings to the fermions (terms of the type $B^\mu \overline{\psi} \gamma^5 \gamma_\mu \psi$ for example). A Lorentz covariant potential for the tensor field can induce a nonzero expectation value of the form $\langle B^\mu \rangle$ that will generate Lorentz-violating contributions to the matter sectors.

The SME consists of all possible terms that couple the standard model fields to background tensor fields.[4,5] It is the spirit of the model to be as general as possible so that any experiment that exhibits Lorentz violation in the future can be described in this formalism. The hope is that experimentally identifying specific constants for Lorentz violation that occur in nature may serve as a window to a more fundamental theory. On the theoretical side, the SME is general enough to accommodate any theory that involves Lorentz Violation. For example, it has been argued that any realistic theory of noncommutative geometry must reduce to a subset of the SME[11]. For practical calculations, it is often useful to restrict the couplings to a minimal set that preserves the conventional gauge invariance of the standard model as well as power counting renormalizability. Imposing translational invariance on these couplings yields the minimal SME, useful for quantifying leading order corrections to experiments.

As an example, consider the electron-photon sector. Imposing gauge invariance and restricting to power-counting renormalizable terms in the standard model extension yields a lagrangian of

$$L = \tfrac{1}{2} i \overline{\psi} \Gamma^\nu \overleftrightarrow{D}_\nu \psi - \overline{\psi} M \psi + L_{\text{photon}} \quad , \tag{1}$$

where Γ and M denote

$$\Gamma^{\nu} = \gamma^{\nu} + c^{\mu\nu}\gamma_{\mu} + d^{\mu\nu}\gamma_5\gamma_{\mu} \quad , \tag{2}$$

$$M = m + a_{\mu}\gamma^{\mu} + b_{\mu}\gamma_5\gamma^{\mu} + \frac{1}{2}H^{\mu\nu}\sigma_{\mu\nu} \quad . \tag{3}$$

The parameters a, b, c, d, and H are related to fixed background expectation values of tensor fields. In this sector, stringent bounds on many parameters have been attained. For example, limits on the order $|k| < 10^{-32}$ for photons[12], and $|b_3| < 10^{-24}m_e$ for electrons[13] have been obtained.

Different sectors of the SME have independent parameters for the background fields, therefore the stringent limits in electrodynamics do not rule out potentially large effects in other sectors. For example, as I will discuss next, it may be possible that current experimental data regarding neutrino oscillations can be modeled using Lorentz violating terms, rather than masses.

3. Application to Neutrino Oscillations

The conventional formalism appears to describe much of the current data involving neutrino masses fairly well using mass differences on the order of $\Delta m^2 \sim 10^{-20} GeV^2$. The ratio $\Delta m^2/E^2 \sim 10^{-20}$ happens to be compatible with leading order Planck suppression estimates of the Lorentz-Violation parameters. It is therefore reasonable to ask if these oscillation effects are really a manifestation of Lorentz-violating background fields coupled to neutrinos.

The SME effective hamiltonian for neutrinos and antineutrinos in the presence of Lorentz violation has recently been constructed[5,8]. This model is important as it includes all possible leading order corrections to the neutrino propagators in the presence of Lorentz violation. All previous work on neutrinos in the presence of Lorentz violation has assumed a rotationally invariant subset of the SME (called Fried Chicken models) typically with two neutrino species and nonzero neutrino masses[7]. The general case that includes three neutrino species and allows for violation of rotational symmetry can be expressed using an effective hamiltonian in the active neutrino basis $(\nu_a, \overline{\nu}_a)$, where a represents e, μ, or τ.

$$(h_{eff})_{ab} = |\vec{p}|\delta_{ab}\begin{pmatrix} 1 & 0 \\ 0 & 1 \end{pmatrix} + \frac{1}{2|\vec{p}|}\begin{pmatrix} (\tilde{m})_{ab} & 0 \\ 0 & (\tilde{m})^*_{ab} \end{pmatrix} + \frac{1}{|\vec{p}|}\begin{pmatrix} M_{11} & M_{12} \\ M_{21} & M_{22} \end{pmatrix} , \tag{4}$$

where

$$M_{11} = [(a_L)^{\mu}p_{\mu} - (c_L)^{\mu\nu}p_{\mu}p_{\nu}]_{ab} \quad , \tag{5}$$

$$M_{12} = -i\sqrt{2}p_\mu(\epsilon_+)_\nu[(g^{\mu\nu\sigma}p_\sigma - H^{\mu\nu})C]_{ab} \quad , \tag{6}$$

$$M_{21} = i\sqrt{2}p_\mu(\epsilon_+)^*_\nu[(g^{\mu\nu\sigma}p_\sigma + H^{\mu\nu})C]^*_{ab} \quad , \tag{7}$$

$$M_{22} = [-(a_L)^\mu p_\mu - (c_L)^{\mu\nu}p_\mu p_\nu]^*_{ab} \quad . \tag{8}$$

For illustrative purposes, this form can be restricted to the minimal SME[5] for which only the left-handed neutrino doublet L_a is present. The resulting lagrangian contains the terms

$$L \supset \frac{i}{2}\overline{L}_a\gamma^\mu \overleftrightarrow{D}_\mu L_a - (a_L)_{\mu ab}\overline{L}_a\gamma^\mu L_b + \frac{i}{2}(c_L)_{\mu\nu ab}\overline{L}_a\gamma^\mu \overleftrightarrow{D}_\mu L_b \quad , \tag{9}$$

yielding the effective neutrino hamiltonian

$$(h_{eff})_{ab} = |\vec{p}|\delta_{ab} + \frac{1}{|\vec{p}|}[(a_L)^\mu p_\mu - (c_L)^{\mu\nu}p_\mu p_\nu]_{ab} \quad . \tag{10}$$

Note that c_L (a_L) preserves (violates) CPT. Diagonalization of this matrix yields two momentum dependent eigenvalue differences that govern the neutrino and antineutrino oscillation probabilities.

Some generic features of these oscillation probabilities may be identified by analyzing dimensionless combinations of parameters that appear in the oscillatory function arguments. For the standard massive neutrino case, the relevant ratio is $\Delta m^2 \cdot (L/E)$. The Lorentz violation terms typically contribute $a_L \cdot (L)$ and $c_L \cdot (LE)$ factors in the argument. This means that novel new energy dependences for the oscillations may be attained. In general, there will also be rotationally noninvariant terms contributing to the oscillation arguments. This opens up the possibility for interesting searches for diurnal variations at the Earth's siderial period $\omega \approx 23\ h\ 56m$. A realistic model within the minimal SME that appears consistent with current experimental data is the bicycle model[8]. This model is notable since it consists of a two parameter fit to the currently observed data, while at the same time maintaining the full gauge invariance of the standard model. The bicycle model sets all Lorentz violating parameters to zero, except the rotationally invariant piece of c_L and a single spatial componet of a_L.

Regardless of the specific choice of parameters, there are specific signatures for Lorentz Violation in neutrino oscillations. They are:

- Spectral anomalies (L or L/E oscillation behavior).
- L - E conflicts for experiments in different regions of L - E space that cannot be accommodated using only two mass differences.

- Periodic Variations, such as a diurnal signal.
- Compass asymmetries (effects that cannot be attributed standard physics such as the effect of the Earth's magnetic field on cosmic rays).
- Neutrino-antineutrino mixing.
- Classic CPT test: $P_{\nu_b \to \nu_a} \neq P_{\overline{\nu}_a \to \overline{\nu}_b}$.

Note that the only one of these for which there is a possible signal is the L - E conflict of LSND[14] to be tested by the future data collected by MiniBooNE.

4. Yang-Mills Instantons with Lorentz Violation

Static solutions to pure Yang-Mills theories in four Euclidean dimensions are well known and are called instantons. The pure Yang-Mills sector of the SME contains terms that violate the Lorentz symmetry, but it turns out that many of the properties of instanton solutions remain intact[9]. This result is due to the fact that the instanton solutions rely heavily upon topological arguments as will be discussed in the remainder of this proceedings.

The standard pure Yang-Mills Euclidean action is given by

$$S_0(A) = \frac{1}{2} \int d^4x \; Tr[F^{\mu\nu} F^{\mu\nu}] \quad , \tag{11}$$

where

$$F^{\mu\nu} = \partial^\mu A^\nu - \partial^\nu A^\mu + ig[A^\mu, A^\nu] \quad , \tag{12}$$

is the curvature of the connection A. The topological charge q is defined as

$$q = \frac{g^2}{16\pi^2} \int d^4x Tr \tilde{F}^{\mu\nu} F^{\mu\nu} \quad , \tag{13}$$

where $\tilde{F}^{\mu\nu} = \frac{1}{2}\epsilon^{\mu\nu\alpha\beta} F^{\alpha\beta}$ is the dual of F. A useful identity is: $\frac{1}{4} Tr\tilde{F}F = \partial^\mu X^\mu$, with

$$X^\mu = \frac{1}{4}\epsilon^{\mu\nu\lambda\kappa} Tr(A^\nu F^{\lambda\kappa} - \frac{2}{3} ig A^\nu A^\lambda A^\kappa) \quad . \tag{14}$$

This converts the topological charge integral to a surface integral with the net result that q must be an integer. Note that this argument is independent of the specific form of the action.

The equation of motion for the curvature is

$$[D^\mu, F^{\mu\nu}] = 0 \quad , \tag{15}$$

with a corresponding Bianchi Identity that follows from the definition of F:

$$[D^\mu, \tilde{F}^{\mu\nu}] = 0 \quad . \tag{16}$$

($D_\mu = \partial_\mu + iqA_\mu$ is the usual covariant derivative) This gives a set of nonlinear differential equations for A^μ. A clever argument for solving[15] these equations involves consideration of the inequality

$$\frac{1}{2}\int d^4x Tr(F \mp \tilde{F})^2 \geq 0 \quad . \tag{17}$$

This can be rearranged as

$$S \geq \pm\frac{1}{2}\int d^4x Tr[\tilde{F}^{\mu\nu}F^{\mu\nu}] = \pm\frac{8\pi^2}{g^2}q \quad . \tag{18}$$

The inequality is saturated for $F = \pm\tilde{F}$, implying that self-dual or anti-self-dual curvatures are extremal solutions.

As an example of an explicit self-dual solution, let $q = 1$, with gauge group G = SU(2). The vector potential can be written as

$$A^\mu_{SD} = -\frac{\tau^{\mu\nu}x^\nu}{g(\rho^2 + x^2)} \quad , \tag{19}$$

and the corresponding curvature is

$$F^{\mu\nu}_{SD} = \frac{2\rho^2}{g(\rho^2 + x^2)^2}\tau^{\mu\nu} \quad , \tag{20}$$

where $\tau^{0i} = \sigma^i$ and $\tau^{ij} = \epsilon^{ijk}\sigma^k$ are written in terms of the conventional Pauli sigma matrices. The free parameter ρ controls the instanton size. The anti-self-dual solution ($q = -1$) is obtained using the parity transform of the above solution. Subsequently, all self-dual solutions were classified[16].

Next, the Lorentz-violating case is examined. The quadratic action that preserves gauge invariance is given by

$$S(A) = \frac{1}{2}\int d^4x Tr[(F^{\mu\nu}F^{\mu\nu}) + (k_F)^{\mu\nu\alpha\beta}F^{\mu\nu}F^{\alpha\beta}] \quad , \tag{21}$$

where the parameters k_F are small, constant background fields. Only terms of $O(k_F)$ are kept in the calculations. The first result is that the topological charge q remains integral because the conventional argument is insensitive to the detailed form of the action, provided that gauge invariance is maintained. A modified bound on the action is

$$S \geq \pm\frac{8\pi^2}{g^2}q \pm \frac{1}{4}(k_F^{\mu\nu\alpha\beta} + \tilde{k}_F^{\mu\nu\alpha\beta})\int d^4x Tr\tilde{F}^{\mu\nu}F^{\alpha\beta} \quad , \tag{22}$$

where $\tilde{k}_F^{\mu\nu\alpha\beta} = \frac{1}{4}\epsilon^{\mu\nu\lambda\kappa} k_F^{\lambda\kappa\rho\sigma}\epsilon^{\rho\sigma\alpha\beta}$. It is useful to decompose $k_F = k_F^+ \oplus k_F^-$ according to its duality properties and consider the two cases separately.

For case one ($k_F = -\tilde{k}_F$), the background constants take the form $k_F^{\mu\nu\alpha\beta} = \Lambda^{[\mu[\alpha}\delta^{\nu]\beta]}$, where $\Lambda^{\mu\nu}$ is a symmetric, traceless matrix. The action is then extremal for the modified duality condition

$$F' \simeq \pm \tilde{F}' \quad , \tag{23}$$

where $F'^{\mu\nu} = F^{\mu\nu} + \frac{1}{2}k_F^{\mu\nu\alpha\beta}F^{\alpha\beta}$. Explicit solutions are constructed using $\tilde{x}^\mu = x^\mu + \Lambda^{\mu\nu}x^\nu$, and the vector potential is given by $A^\mu(x) \simeq A^\mu_{SD}(\tilde{x}) + \Lambda^{\mu\nu}A^\nu_{SD}(x)$. These solutions take the form of conventional instantons in skewed coordinates.

For case two ($k_F = \tilde{k}_F$), the background constants are trace free. In this case, the lower bound on S given by (22) varies with δF. This means that the previous modified duality condition fails to generate a solution and the equations of motion must be solved explicitly. This can be done to leading order in k_F by expanding $A = A_{SD} + A_k$, fixing A to be close to the conventional self-dual solution. The equations of motion become

$$[D^\nu_{SD}, [D^\nu_{SD}, A^\mu_k]] + 2ig[F^{\mu\nu}_{SD}, A^\nu_k] = j^\mu_k \quad , \tag{24}$$

where $j^\mu_k \equiv k_F^{\mu\nu\alpha\beta}[D^\nu_{SD}, F^{\alpha\beta}_{SD}]$.

This is a second-order, linear elliptic differential equation. A formal solution can be constructed using the relevant propagator $G(x, y)$:

$$A_k = \int d^4y G(x,y) j_k(y) \tag{25}$$

For the case $q = 1$ with G = SU(2), an explicit solution can be constructed using the following procedure:

- Transform to singular gauge $\rightarrow$ makes fields $\sim O(\rho^2)$.
- To $O(\rho^2)$ in this gauge, can use free propagator

$$G_0(x,y) = \frac{1}{4\pi^2(x-y)^2} \quad .$$

- Gives tensorial structure for general ansatz

$$A^\mu_k = \frac{2\rho^2 x^2}{3g} f(x^2) k_F^{\mu\nu\alpha\beta}\tau^{\alpha(\nu}x^{\beta)} \quad .$$

- Substitute into the full equation of motion with general ρ.

Remarkably this gives a differential equation for $f(x)$, indicating that the tensorial structure is in fact correct to all orders in ρ^2.

5. Conclusion

A general formalism allowing for Lorentz violation (and possible resulting CPT violation)in the neutrino sector has been developed. Possible signals for Lorentz violation include anomalous energy dependence as well as siderial variations. To date, only a tiny subset of the neutrino sector implications have been explored. In addition, it has been shown that instantons can still be classified according to the conventional topological charge.

References

1. V.A. Kostelecký and S. Samuel, Phys. Rev. D **39**, 683 (1989); *ibid.* **40**, 1886 (1989); Phys. Rev. Lett. **63**, 224 (1989); *ibid.* **66**, 1811 (1991); V.A. Kostelecký and R. Potting, Nucl. Phys. B **359**, 545 (1991); Phys. Lett. B **381**, 89 (1996); Phys. Rev. D **63**, 046007 (2001); V.A. Kostelecký, M. Perry, and R. Potting, Phys. Rev. Lett. **84**, 4541 (2000).
2. See, for example: J. Alfaro, H.A. Morales-Técotl, L.F. Urrutia, Phys. Rev. D **66**, 124006 (2002); D. Sudarsky, L. Urrutia, and H. Vecetich, Phys. Rev. Lett. **89**, 231301 (2002); Phys. Rev. D **68**, 024010 (2003).
3. For an overview of recent experimental bounds on Lorentz Violation, see, for example, V.A. Kostelecký, ed.,*CPT and Lorentz Symmetry II*, World Scientific, Singapore, 2002.
4. V.A. Kostelecký and R. Potting, Phys. Rev. D **51**, 3923 (1995).
5. D. Colladay and V.A. Kostelecký, Phys. Rev. D **55**, 6760 (1997); Phys. Rev. D **58**, 116002 (1998).
6. V.A. Kostelecký and R. Lehnert, Phys. Rev. D **68**, 123511 (2003); R. Lehnert, Phys. Rev. D **68**, 085003 (2003).
7. S. Coleman and S.L. Glashow, Phys. Rev. D **59** 116008 (1999); V. Barger, S. Pakvasa, T.J. Weiler, and K. Whisnant, Phys. Rev. Lett. **85**, 5055 (2000); J.N. Bahcall, V. Barger, and D. Marfatia, Phys. Lett. B **534**, 114 (2002).
8. V.A. Kostelecký and M. Mewes, hep/ph-0308300; Phys. Rev. D **69**, 016005 (2004).
9. D. Colladay and P. McDonald, hep-th/0312058.
10. O. Greenberg, Phys. Rev. Lett. **89**, 231602 (2002); Phys. Lett. B **567**, 179 (2003).
11. S. Carroll, *et. al.*, Phys. Rev. Lett. **87**, 141601 (2001).
12. See, for example, V.A. Kostelecký and M. Mewes, Phys. Rev. Lett. **87**, 251304 (2001); Phys. Rev. D **66**, 056005 (2002).
13. See, for example, H. Müller, *et al.*, Phys. Rev. D **68**, 116006 (2003).
14. LSND Collaboration, D. Athanassopoulos *et al.*, Phys. Rev. Lett. **81** 1774 (1998); LSND Collaboration, A. Aguilar *et al.*, Phys. Rev. D **64** 112007 (2001).
15. A. Belavin, A. Polyakov, A. Schwartz, and Y. Tyupkin, Phys. Lett. **59B** 85 (1975).
16. M.F. Atiyah, N.J. Hitchin, V.G. Drinfeld, and Yu.I. Manin, Phys. Lett. **65A** 285 (1978).

EVOLVING BRANES WITH GENERALIZED DYNAMICS

THOMAS CURTRIGHT

Department of Physics, University of Miami
Coral Gables, Florida 33124-8046

Abstract

We discuss evolution of extended data sets in the framework of Nambu mechanics, a framework wherein time development is allowed a liberal interpretation in terms of dynamical generating functions.

1 Prologue

My thanks in advance for your time and attention, and to all the organizers of this year's conference for their hard work on our behalf. Especially, in memoriam, thank you *Behram*, for all the meetings you conducted since I first started attending the Coral Gables conferences in January 1981. These meetings serve as milestones in my own life.

The subject of this talk is Nambu mechanics [1]. Classical and quantum Nambu mechanics are discussed at length in T Curtright and C Zachos [2], and in our two correlated talks at last year's Coral Gables Conference [3, 4]. Moreover, some tentative brane applications are described by each of us individually [5, 6] as well as in our combined University of Cincinnati talks [7]. The present talk is based in large part on the results in those previous papers, with some additional details and embellishments.

Those previous papers provide *extensive* guides to the earlier literature. Please look there for the relevant references. In this talk, for simplicity, I will *not* painstakingly and carefully place citations on the marquee next to most results and equations, nor even in conjunction with some of the modest number of interesting ideas that appear. My apologies to those I might offend by this (a longer version of this talk, with more complete references, is available on-line at http://server.physics.miami.edu/~curtright/talks.html).

The organization of this talk is as follows.

- Brane evolution as motivation
- Simple examples of Nambu mechanics
- Quantum N-brackets: Even and Odd
- Harmonic oscillator on the plane

2 Brane motivator

Suppose

$$dz_j = v_j\left[\boldsymbol{z}\right] d\tau \tag{1}$$

describes a solenoidal flow of points, $\frac{\partial}{\partial z_j} v_j\left[\boldsymbol{z}\right] = 0$. What about the evolution of a continuous collection of points,

$$\boldsymbol{z}\left(\tau, \sigma_1, \sigma_2, \cdots, \sigma_d\right) \ ?$$

The d variables σ_i parameterize the "d-brane" consisting of the points.

Nambu mechanics suggests an interesting answer to this problem. If the flow can be expressed as a Nambu n-bracket $\{\boldsymbol{z}, \cdots\}_{\mathrm{NB}}$ involving the curl of a multitude of $n-1$ invariants I_i, i.e. if

$$\begin{aligned} v_j\left[\boldsymbol{z}\right] &= \varepsilon_{j i_1 \cdots i_{n-1}} \frac{\partial}{\partial z_{i_1}} I_1\left[\boldsymbol{z}\right] \cdots \frac{\partial}{\partial z_{i_{n-1}}} I_{n-1}\left[\boldsymbol{z}\right] \\ &= \varepsilon_{j_1 \cdots j_n} \frac{\partial}{\partial z_{j_1}} z_j \frac{\partial}{\partial z_{j_2}} I_1\left[\boldsymbol{z}\right] \cdots \frac{\partial}{\partial z_{j_n}} I_{n-1}\left[\boldsymbol{z}\right] \\ &\equiv \left\{z_j, I_1\left[\boldsymbol{z}\right], \cdots, I_{n-1}\left[\boldsymbol{z}\right]\right\}_{\mathrm{NB}} \ , \end{aligned} \tag{2}$$

then evolution of the previous d-brane may be implemented through the action

$$\mathbb{A} = \int dz_1 \wedge \cdots \wedge dz_n - d\tau \wedge dI_1 \wedge \cdots \wedge dI_{n-1} \tag{3}$$

provided we also have $d = n-2$. That is, for $n = d+2$, the exact n-form

$$\begin{aligned} \Omega &= \left(dz_1 - v_1\left[\boldsymbol{z}\right] d\tau\right) \wedge \cdots \wedge \left(dz_n - v_n\left[\boldsymbol{z}\right] d\tau\right) \\ &= dz_1 \wedge \cdots \wedge dz_n - d\tau \wedge dI_1 \wedge \cdots \wedge dI_{n-1} \end{aligned} \tag{4}$$

yields a "boundary action" that governs the evolution of the complete d-brane.

$$\mathbb{A} = \int_{M_n} \Omega \tag{5}$$

$$= \int_{\substack{\partial M_n \\ \text{world volume} \\ \text{for } d=n-2 \text{ brane}}} \left[\begin{array}{c} \frac{1}{n}\, \varepsilon^{j_1 \cdots j_n}\, z_{j_1}\, \partial_\tau z_{j_2}\, \partial_{\sigma_1} z_{j_3} \cdots \partial_{\sigma_{n-2}} z_{j_n} \\ + \frac{1}{n-1}\, \varepsilon^{k_1 \cdots k_{n-1}}\, I_{k_1}\, \partial_{\sigma_1} I_{k_2} \cdots \partial_{\sigma_{n-2}} I_{k_{n-1}} \end{array} \right] d\tau d\sigma_1 \cdots d\sigma_{n-2} \ .$$

The d-brane of initial data is evolved as a whole by this action, in concordance with the above individual point-like flows.

3 Some simple examples

As the simplest possible example of Nambu mechanics, consider phase-space for a unit mass free particle on the plane. Time evolution of this system can be expressed either as Hamiltonian flow, generated by a Poisson bracket $\{\cdots\}_{\mathrm{PB}}$, or as Nambu flow, generated by a 4-bracket.

$$\begin{aligned} z_j &\in x, p_x, y, p_y \qquad \text{so} \qquad n = 4. \\ \frac{d}{dt} z_j &= v_j \ \text{ either with} \\ v_j &= \{z_j, H\}_{\mathrm{PB}} \ , \quad H = \left(p_x^2 + p_y^2\right)/2 \ \text{ or with} \\ v_j &= \frac{\partial\left(z_j, L_z, p_x, p_y\right)}{\partial\left(x, p_x, y, p_y\right)} \ , \quad L_z = x p_y - y p_x \end{aligned} \tag{6}$$

The point is, no matter which of these brackets we choose, the *same* phase-space velocity v_j is obtained. Hence, the previous action $\mathbb{A}$ would describe the evolution of a 2-surface or "membrane" of *phase-space* data for this free-particle system.

Other classical bracket systems are nearly as transparent [2]. Although in general, Nambu brackets set different time scales on different dynamical sectors of a theory.

As an illustration of dynamical time scales set through Nambu brackets, consider the two dimensional isotropic oscillator. The $su\,(2)$ algebra of the oscillator (as in Schwinger) may be realized by

$$\begin{aligned} J_x &= \tfrac{1}{2}\left(p_x p_y + x y\right) \\ J_y &= \tfrac{1}{4}\left(p_y^2 + y^2 - p_x^2 - x^2\right) \\ J_z &= \tfrac{1}{2}\left(x p_y - y p_x\right) \end{aligned} \tag{7}$$

Each of these is an invariant under time-evolution as generated by (we have set $m = 1/\omega$)

$$H = \frac{1}{2m}\left(p_x^2 + p_y^2 + x^2 + y^2\right) \tag{8}$$

and they satisfy the usual Poisson brackets

$$\{J_x, J_y\}_{\mathrm{PB}} = J_z \ , \quad \{J_y, J_z\}_{\mathrm{PB}} = J_x \ , \quad \{J_z, J_x\}_{\mathrm{PB}} = J_y \tag{9}$$

Moreover, the *square* of the 2d SHO Hamiltonian is given by the quadratic Casimir

$$J_x^2 + J_y^2 + J_z^2 = \tfrac{1}{16}\left(p_x^2 + p_y^2 + x^2 + y^2\right)^2 = \tfrac{1}{4} m^2 H^2 \tag{10}$$

Now there is a simple $su\,(2)$ bracket identity relating Poisson brackets to Nambu 4-brackets. It follows from nothing but the PB algebra.

For any function $A(x, y, p_x, p_y)$ on the phase-space of the oscillator

$$\left\{A, J_x^2 + J_y^2 + J_z^2\right\}_{\text{PB}} = 2 \times \{A, J_x, J_y, J_z\}_{\text{NB}} \tag{11}$$

Thus evolution through use of this particular 4-bracket corresponds to using the square of the 2d SHO Hamiltonian. Evolution using the Hamiltonian to the first power apparently can *not* be expressed in terms of quantum 4-brackets by using linear combinations of the Js. The best we can do is

$$\begin{aligned}
z_j &\in \; x, p_x, y, p_y \qquad \text{so again} \qquad n = 4. \\
v_j &= \frac{\partial(z_j, J_x, J_y, J_z)}{\partial(x, p_x, y, p_y)} \quad \text{or} \\
\frac{d}{d\tau} z_j &= v_j = \tfrac{1}{8} m^2 \left\{z_j, H^2\right\}_{\text{PB}} \\
&= \tfrac{1}{4} m^2 H \{z_j, H\}_{\text{PB}} \\
&= \tfrac{1}{4} m^2 H \left(\frac{d}{dt} z_j\right)
\end{aligned} \tag{12}$$

Roughly speaking, for the 2d SHO it is not a simple time derivative that nicely fits into a Nambu multi-bracket formalism, but rather it is the derivative $\partial/\partial(t/E)$ for fixed energy levels. This can be written as a 4-bracket.

That is to say, more precisely, we need to re-parameterize the time-energy plane by changing coordinates $(t, E) \to \left(\tau \equiv t/E, F \equiv E^2/2\right)$. This preserves areas on the time-energy plane (and therefore maintains the physics associated with time-energy distributions) since

$$\left\{t/E, E^2/2\right\}_{t,E} = 1 \tag{13}$$

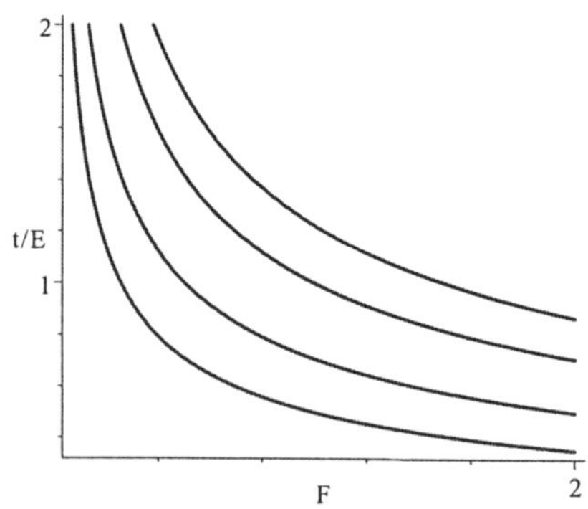

Constant time curves on the $\tau = t/E$, $F = E^2/2$ plane.

More importantly, this parameterization of the time-energy plane dictates that different energy levels evolve according to their own distinct energy-dependent times, as specified by τ. Since the relation between τ and t is invertible, for $E \neq 0$, this is not pathological.

Evolution of SHO extended data and dynamical time scales. We set $m = 1/2$ and rescale $\tau \to 16\tau$ to clean up the numerics, to obtain

$$\frac{d}{d\tau} z = \frac{1}{2} \left\{ z, H^2 \right\}_{\mathrm{PB}} = H \left\{ z, H \right\}_{\mathrm{PB}} , \tag{14}$$

for $z = (x, y, p_x, p_y)$. A representative solution is given by

$$x(E, \tau) = \sqrt{E} \left(\cos(E\tau) + \sin(E\tau) \right) , \quad p_x(E, \tau) = \sqrt{E} \left(-\sin(E\tau) + \cos(E\tau) \right) \tag{15}$$

as opposed to the usual fixed frequency, standard time evolution:

$$x(E, t) = \sqrt{E} \left(\cos(t) + \sin(t) \right) , \quad p_x(E, t) = \sqrt{E} \left(-\sin(t) + \cos(t) \right) . \tag{16}$$

The equivalence map between the single point particle solutions, as given by the relation $\tau = t/E$, is evident.

The geometry of a single phase-space trajectory evolved under τ is indistinguishable from that of a single trajectory evolved under standard time t: Namely it is a circular path centered on the phase-space origin. Only the parameterization scale of the curve is different.

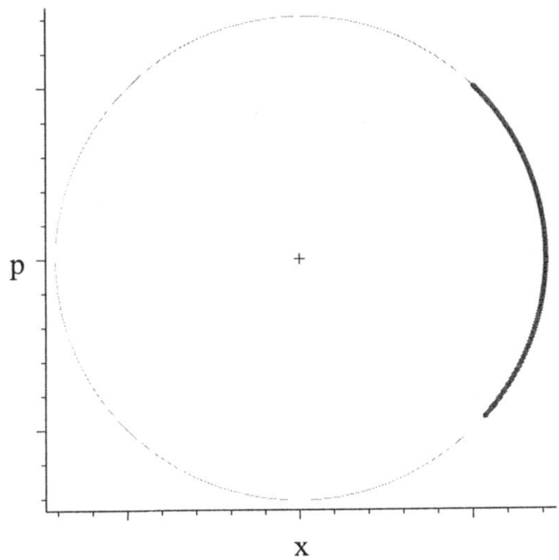

Evolution of a single phase-space point as generated by H, or by $f(H)$ with $f'(H) > 0$.

To see the difference between the two forms of evolution, geometrically, we must compare two or more trajectories. Extended data involving different energies evolve continuously under τ in an amusing way, as indicated in the pictures below.

That is to say, we need to consider the evolution of extended initial data in the phase-space, i.e. a *brane of data.* So consider a straight radial line segment of initial (x, p_x) points, and evolve the entire segment under the action of H^2. The phase-space data undergoes shear! For three $\Delta\tau = 0.5$ steps, we have the following.

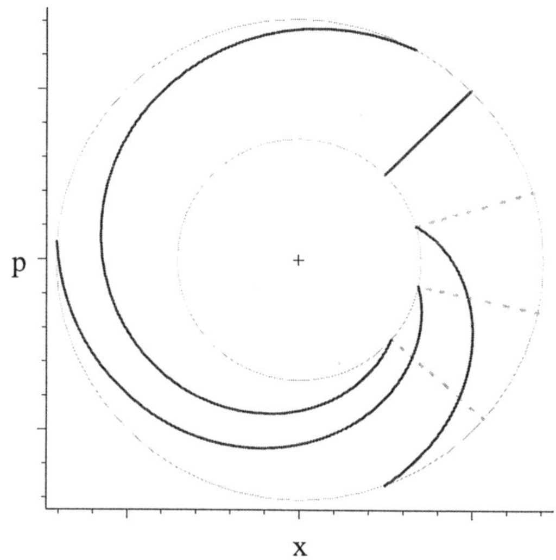

Evolution of extended data under the action of H^2 ($\tau = 0.0, 0.5, 1.0, 1.5$).

(The dotted lines in the Figure are what would be obtained from the initial straight segment if it were evolved under normal time steps using the Hamiltonian H.) Nonetheless, the phase-space τ flow is still that of an incompressible fluid (Liouville's theorem still obtains), as is illustrated in the following Figure, in which a radial wedge of initial data is evolved through one $\Delta\tau = 0.5$ step.

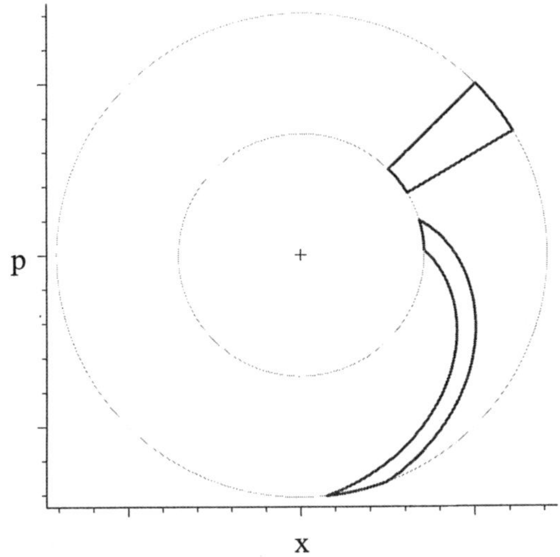

Phase-space volume is preserved under the action of H^2.

4 Quantum brackets

We define quantum Nambu brackets (QNBs) as one of the possibilities originally suggested by Nambu.

$$[A_1, A_2, \cdots, A_k] \equiv \sum_{\substack{\text{all } k! \text{ perms } \{\sigma_1,\sigma_2,\cdots,\sigma_k\} \\ \text{of the indices } \{1,2,\cdots,k\}}} \operatorname{sgn}(\sigma) \; A_{\sigma_1} A_{\sigma_2} \cdots A_{\sigma_k} \, , \tag{17}$$

Alternatively, recursively, there are left- and right-sided resolutions.

$$\begin{aligned} [A_1, \cdots, A_k] &= A_1 [A_2, \cdots, A_k] + \text{ signed permutations} \\ &= [A_1, \cdots, A_{k-1}] A_k + \text{ signed permutations} \end{aligned} \tag{18}$$

Some of the most interesting features of QNBs center around their *not* being derivations, as easily follows from these resolutions. That is

$$[A_1, \cdots, A_k, \; BC\,] - B \; [A_1, \cdots, A_k, C] - [A_1, \cdots, A_k, B] \; C \neq 0 \tag{19}$$

in general. The left-hand side up to $\pm$ is the "derivator": $(-1)^k \Delta_{A_1,\cdots,A_k}(B,C)$. The bracket for fixed $A_1, \cdots, A_k$ is a *derivation* iff $\Delta_{A_1,\cdots,A_k}(B,C) = 0$ for *all* B, C.

Even brackets are even better as they also admit *commutator resolutions.*

$$[A_1, \cdots, A_{2n}] = [A_1, A_2]\,[A_3, A_4]\ \cdots\ [A_{2n-1}, A_{2n}] \\ + \text{ permutations} \tag{20}$$

So, even brackets have direct classical limits as a consequence of the fact that commutators $\Longrightarrow$ Poisson brackets. This we like.

$$\begin{aligned}\frac{1}{n!}\lim_{\hbar\to 0}\left(\frac{1}{i\hbar}\right)^n [A_1, A_2, \cdots, A_{2n}] &= \{A_1, A_2\}_{\mathrm{PB}}\,\{A_3, A_4\}_{\mathrm{PB}}\ \cdots\ \{A_{2n-1}, A_{2n}\}_{\mathrm{PB}} \\ &+ \ \text{permutations} \\ &= \{A_1, A_2, \cdots, A_{2n}\}_{\mathrm{NB}}\ .\end{aligned} \tag{21}$$

Odd quantum brackets, on the other hand, do *not* become just classical odd brackets as $\hbar \to 0$, due to a mismatch in the number of derivatives: a $2n+1$ quantum bracket will only yield $2n$ derivatives in the classical limit, not $2n+1$ derivatives.

Although not usually the case, *sometimes* QNBs are derivations by virtue of algebra of the As. For example, consider a free particle on the plane, again, and use the commutator resolution of the 4-bracket.

$$H = \left(p_x^2 + p_y^2\right)/2\ , \quad L = xp_y - yp_x \tag{22}$$

$$\begin{aligned}[A, L, p_x, p_y] &= \{[A, p_y]\,, [L, p_x]\} - \{[A, p_x]\,, [L, p_y]\} \\ &= i\hbar\,\{[A, p_y]\,, p_y\} + i\hbar\,\{[A, p_x]\,, p_x\} \\ &= 2i\hbar\,[A, H]\end{aligned} \tag{23}$$

This particular example is therefore a derivation.

$$2\,(i\hbar)^2\,\frac{d}{dt}A\,(x, p_x, y, p_y) = [A, L, p_x, p_y] \tag{24}$$

and serves as the simplest example of consistently quantized Nambu mechanics. In this special case, the quantum brackets have the same algebraic structure as their classical limits.

Brackets from fermionic Gaussians The combination of Poisson brackets appearing in the above $\hbar \to 0$ limit-definition of classical Nambu brackets is a well-known structure in the mathematical literature, namely the Pfaffi an.

$$\text{Pfaffian}\left(\{A_i, A_j\}_{\mathrm{PB}}\right) \equiv \sqrt{\det\left(\{A_i, A_j\}_{\mathrm{PB}}\right)}\ . \tag{25}$$

Indeed, the classical Nambu bracket is actually just [8]

$$\{A_1, A_2, \cdots, A_{2n}\}_{\mathrm{NB}} = \mathrm{Pfaffian}\left(\{A_i, A_j\}_{\mathrm{PB}}\right) \ . \tag{26}$$

This may be a useful connection to make, say if it can help to establish all those cases in which the Nambu bracket reduces to a (sum of) single Poisson brackets, as in the $su(2)$ case mentioned earlier, or to establish other such relationships.

Moreover, in a well-defined sense, the commutator resolution above defines a quantum Pfaffi an, i.e. it gives a precise definition when the underlying matrix entries do not commute. The quantum definition of the Nambu bracket that we have chosen, and hence the quantum Pfaffi an, is naturally expressed using fermionic integrals for non-commuting Gaussians.

$$\int \exp\left(-\frac{1}{i\hbar}\sum_{i<j=1}^{2n}\theta_i\left[A_i, A_j\right]\theta_j\right) d\theta_1\cdots d\theta_{2n} = \frac{1}{n!\,(i\hbar)^n}\left[A_1, \cdots, A_{2n}\right] \ , \tag{27}$$

where as usual $\{\theta_i, \theta_j\} = 0$ (actually, a Grassmanian δ_{ij} does not seem to do any harm here), $\int d\theta_i = 0$, and the normalization is $\int \theta_i d\theta_j = \delta_{ij}$. Alternatively, since $\int A(\theta)\, d\theta = \frac{\partial}{\partial\theta} A(\theta)$, we may write this as

$$\frac{\partial}{\partial\theta_1}\cdots\frac{\partial}{\partial\theta_{2n}}\exp\left(-\frac{1}{i\hbar}\sum_{i<j=1}^{2n}\theta_i\left[A_i, A_j\right]\theta_j\right) = \frac{1}{n!\,(i\hbar)^n}\left[A_1, \cdots, A_{2n}\right] \ . \tag{28}$$

Either of these fermionic representations have the correct classical limit, rather transparently, in which limit they become well-known fermionic expressions for $\sqrt{\det}$.

$$\int \exp\left(-\sum_{i<j=1}^{2n}\theta_i\left\{A_i, A_j\right\}_{\mathrm{PB}}\theta_j\right) d\theta_1\cdots d\theta_{2n} = \left\{A_1, \cdots, A_{2n}\right\}_{\mathrm{NB}} \tag{29}$$

Also, the integral representation is perhaps the easiest way to take the $n \to \infty$ limit, such as would be expected to appear in a field theory framework. The sums in the exponential become integrals, and the θ integrations meld into a fermionic functional integral.

$$\int \exp\left(-\frac{1}{2i\hbar}\int d\alpha\int d\beta\;\theta(\alpha)\left[A(\alpha), A(\beta)\right]\theta(\beta)\right)\mathcal{D}\theta \tag{30}$$

Also in this limit, $\int \theta(\alpha)\, d\theta(\beta) = \delta(\alpha - \beta)$. Evaluation and *application* of this fermionic functional integral remains to be carried out, however.

5 Simple harmonic oscillator on the plane

As an elementary example that can *not* be worked out explicitly, consider the SHO on the plane with generators $H_x = \left(p_x^2 + x^2\right)/2$ and $H_y = \left(p_y^2 + y^2\right)/2$ in a 3-bracket. This leads to a propagator with interesting number theoretic underpinnings: It encodes all factorings of integers.

Quantum mechanically [9], data on phase-space is expressed in terms of distributions, usually the so-called Wigner functions (WFs). The WFs in the 2d SHO case are given explicitly by

$$f_{n_x,n_y;k_x,k_y}(x,y,p_x,p_y) \equiv (a_x^* \star)^{n_x} \left(a_y^* \star\right)^{n_y} \frac{f_0(x,y,p_x,p_y)}{\sqrt{n_x! n_y! k_x! k_y!}} (\star a_x)^{k_x} \left(\star a_y\right)^{k_y} , \tag{31}$$

$$a_x \equiv \frac{1}{\sqrt{2\hbar}}(x + ip_x) , \quad a_y \equiv \frac{1}{\sqrt{2\hbar}}(y + ip_y) , \tag{32}$$

$$f_0(x,y,p_x,p_y) = \frac{1}{\pi^2\hbar^2} e^{-\left(x^2+p_x^2+y^2+p_y^2\right)/\hbar} ,$$

$$\star \ \equiv \star_x \star_y = e^{\frac{i\hbar}{2}(\overleftarrow{\partial}_x \overrightarrow{\partial}_{p_x} - \overleftarrow{\partial}_{p_x} \overrightarrow{\partial}_x)} e^{\frac{i\hbar}{2}(\overleftarrow{\partial}_y \overrightarrow{\partial}_{p_y} - \overleftarrow{\partial}_{p_y} \overrightarrow{\partial}_y)} .$$

These states are most easily evolved using particle number densities.

$$\begin{aligned} N_x = a_x^* \star a_x &= \frac{1}{2\hbar}(x - ip_x) \star (x + ip_x) \\ &= \frac{1}{2\hbar}(x - ip_x + \frac{i\hbar}{2}\partial_{p_x} - \frac{\hbar}{2}\partial_x) \star (x + ip_x) \\ &= \frac{1}{2\hbar}\left(x^2 + p_x^2 - \hbar\right) , \end{aligned} \tag{33}$$

That is, $\frac{1}{2}\left(x^2 + p_x^2\right) = \hbar\left(N_x + \frac{1}{2}\right)$, so we are considering an oscillator with $m = 1 = \omega_0$. Similarly

$$N_y = a_y^* \star a_y = \frac{1}{2\hbar}\left(y^2 + p_y^2 - \hbar\right) . \tag{34}$$

The obvious 3-$\star$-bracket to use to evolve the Wigner functions in the Nambu framework is $[f, N_x, N_y]_\star = f \star [N_x, N_y]_\star + N_y \star [f, N_x]_\star + N_x \star [N_y, f]_\star$ or

$$[f, N_x, N_y]_\star = N_y \star f \star N_x - N_x \star f \star N_y . \tag{35}$$

Acting on the $f_{n_x,n_y;k_x,k_y}$, this gives

$$\begin{aligned} \left[f_{n_x,n_y;k_x,k_y}, N_x, N_y\right]_\star &= N_y \star f_{n_x,n_y;k_x,k_y} \star N_x - N_x \star f_{n_x,n_y;k_x,k_y} \star N_y \\ &= (k_x n_y - k_y n_x) f_{n_x,n_y;k_x,k_y} . \end{aligned} \tag{36}$$

Thus the appropriate evolution equations are (restoring the isotropic oscillator frequency, ω_0, but keeping $m = 1$)

$$-i\frac{\partial}{\partial\tau}f_{n_x,n_y;k_x,k_y} = \omega_0\left[f_{n_x,n_y;k_x,k_y}, N_x, N_y\right]_\star = \omega_0\left(k_x n_y - k_y n_x\right)f_{n_x,n_y;k_x,k_y}\,, \tag{37}$$

with solutions

$$f_{n_x,n_y;k_x,k_y}(\tau) = e^{i\omega_0\tau(k_x n_y - k_y n_x)}f_{n_x,n_y;k_x,k_y}\,. \tag{38}$$

The corresponding propagator equation is

$$-i\frac{\partial}{\partial\tau}G_N\left(x,y,p_x,p_y;X,Y,P_X,P_Y;t\right) = \omega_0\left[G_N\left(x,y,p_x,p_y;X,Y,P_X,P_Y;t\right),N_x,N_y\right]_\star\,, \tag{39}$$

with the correctly initialized solution given by

$$\begin{aligned} G_N(x,y,p_x,p_y;X,Y,P_X,P_Y;\tau) &= (2\pi\hbar)^2\times \\ \times\sum_{n_x,n_y,k_x,k_y} & f_{n_x,n_y;k_x,k_y}\left(x,y,p_x,p_y\right)e^{i\omega_0\tau(k_x n_y - k_y n_x)}f_{k_x,k_y;n_x,n_y}\left(X,Y,P_X,P_Y\right)\,. \end{aligned} \tag{40}$$

Now the only issue left is to evaluate the sum.

$$\begin{aligned} &\sum_{n_x,n_y,k_x,k_y} f_{n_x,n_y;k_x,k_y}\left(x,y,p_x,p_y\right)e^{i\omega_0\tau(k_x n_y - k_y n_x)}f_{k_x,k_y;n_x,n_y}\left(X,Y,P_X,P_Y\right) \\ &= \sum_{n_x,n_y,k_x,k_y}\left(a_x^*\star\right)^{n_x}\left(a_y^*\star\right)^{n_y}\frac{f_0\left(x,y,p_x,p_y\right)}{\sqrt{n_x!n_y!k_x!k_y!}}\left(\star a_x\right)^{k_x}\left(\star a_y\right)^{k_y}\times \\ &\times e^{i\omega_0\tau(k_x n_y - k_y n_x)}\left(a_X^*\bigstar\right)^{k_x}\left(a_Y^*\bigstar\right)^{k_y}\frac{f_0\left(X,Y,P_X,P_Y\right)}{\sqrt{k_x!k_y!n_x!n_y!}}\left(\bigstar a_X\right)^{n_x}\left(\bigstar a_Y\right)^{n_y}\,. \end{aligned} \tag{41}$$

A few coherent state tricks here and there and we should be home, modulo the appearance of certain special functions. This is just another generating function for WFs, along the lines of Curtright, Uematsu, and Zachos [10].

We may write two of the sums formally using an ordering convention.

$$\begin{aligned} &\sum_{n_x,n_y}\frac{e^{i\omega_0\tau(k_x n_y - k_y n_x)}}{n_x!n_y!}\left(a_x^*\star\right)^{n_x}\left(a_y^*\star\right)^{n_y}f_0\left(x,y,p_x,p_y\right)\times \\ &\qquad\qquad\times f_0\left(X,Y,P_X,P_Y\right)\left(\bigstar a_X\right)^{n_x}\left(\bigstar a_Y\right)^{n_y} \\ &= \exp\left(e^{-i\omega_0\tau k_y}\underset{\longrightarrow}{a_x^*\star}\,\underset{\longleftarrow}{\bigstar a_X} + e^{i\omega_0\tau k_x}\underset{\longrightarrow}{a_y^*\star}\,\underset{\longleftarrow}{\bigstar a_Y}\right)f_0\left(x,y,p_x,p_y\right)f_0\left(X,Y,P_X,P_Y\right) \end{aligned} \tag{42}$$

Then we have the final two sums, which we cannot evaluate in a closed form, even using the formally ordered series, since we do not know closed form expressions for the sums in the classical limit!

$$\sum_{k_x,k_y} \frac{1}{k_x!k_y!} \exp\left(e^{-i\omega_0\tau k_y} \underset{\longrightarrow}{a_x^*\star} \underset{\longleftarrow}{\bigstar a_X}\right) f_0(x,y,p_x,p_y) (\star a_x)^{k_x} (\star a_y)^{k_y} \times$$
$$\times (a_X^*\bigstar)^{k_x} (a_Y^*\bigstar)^{k_y} f_0(X,Y,P_X,P_Y) \exp\left(e^{i\omega_0\tau k_x} \underset{\longrightarrow}{a_y^*\star} \underset{\longleftarrow}{\bigstar a_Y}\right) \tag{43}$$

As we can see, each of the n and k pairs of sums can be half-way done, leaving the other sum unknown, even in the classical limit.

$$\sum_{k_y,n_x} \frac{e^{i\omega_0\tau(-k_y n_x)}}{k_y!n_x!} (a_x^*\star)^{n_x} f_0(x,y,p_x,p_y) (\star a_y)^{k_y} (a_Y^*\bigstar)^{k_y} f_0(X,Y,P_X,P_Y) (\bigstar a_X)^{n_x}$$
$$= \sum_{k_y} \frac{1}{k_y!} \exp\left(e^{-i\omega_0\tau k_y} \underset{\longrightarrow}{a_x^*\star} \underset{\longleftarrow}{\bigstar a_X}\right) (a_Y^*\bigstar)^{k_y} f_0(X,Y,P_X,P_Y) f_0(x,y,p_x,p_y) (\star a_y)^{k_y}$$
$$= \sum_{n_x} \frac{1}{n_x!} \exp\left(e^{-i\omega_0\tau n_x} \underset{\longrightarrow}{a_Y^*\bigstar} \underset{\longleftarrow}{\star a_y}\right) (a_x^*\star)^{n_x} f_0(x,y,p_x,p_y) f_0(X,Y,P_X,P_Y) (\bigstar a_X)^{n_x}$$
$$\tag{44}$$

It would be marvelous if this reduced to nothing but Dirac deltas enforcing the classical flow. But this quantum miracle does NOT take place. Note that the classical flow is just a modification of the usual Hamiltonian evolution with the time scales for the (x,p_x) and (y,p_y) evolutions set by the values of N_y and $-N_x$, respectively.

$$i\frac{d}{d\tau}x = \omega_0 [x, N_x, N_y]_\star = \omega_0 [x, N_x]_\star \star N_y = \frac{\omega_0}{2\hbar} [x, p_x^2]_\star \star N_y$$
$$= i\omega_0\, p_x\, N_y\ ,$$
$$i\frac{d}{d\tau}p_x = \omega_0 [p_x, N_x, N_y]_\star = \omega_0 [p_x, N_x]_\star \star N_y = \frac{\omega_0}{2\hbar} [p_x, x^2]_\star \star N_y$$
$$= -i\omega_0\, x\, N_y\ , \tag{45}$$
$$i\frac{d}{d\tau}y = \omega_0 [y, N_x, N_y]_\star = -\omega_0 [y, N_y]_\star \star N_x = -\frac{\omega_0}{2\hbar} [y, p_y^2]_\star \star N_x$$
$$= -i\omega_0\, p_y\, N_x\ ,$$
$$i\frac{d}{d\tau}p_y = \omega_0 [p_y, N_x, N_y]_\star = -\omega_0 [p_y, N_y]_\star \star N_x = -\frac{\omega_0}{2\hbar} [p_y, y^2]_\star \star N_x$$
$$= i\omega_0\, y\, N_x\ . \tag{46}$$

Special functions The 3-bracket analysis of the 2d SHO has led to special functions of three variables whose classical limits are defined by the double series

$$F(\alpha,\beta;\gamma)=\sum_{k,n=0}^{\infty}\frac{1}{k!n!}\,\alpha^k\gamma^{kn}\beta^n=F(\beta,\alpha;\gamma)\ . \tag{47}$$

Either of the individual sums may be performed without effort

$$F(\alpha,\beta;\gamma)=\sum_{k=0}^{\infty}\frac{1}{k!}\,\alpha^k e^{\gamma^k\beta}=\sum_{n=0}^{\infty}\frac{1}{n!}\,\beta^n e^{\gamma^n\alpha}\ , \tag{48}$$

but not so easily both sums. If $\alpha,\beta>0$, these series are absolutely convergent for $-1\leq\gamma\leq 1$. When $\alpha,\beta\leq 0$, the series converge absolutely for all $\gamma>0$. Complex values are more interesting. For $\operatorname{Re}\alpha,\operatorname{Re}\beta>0$, the series is convergent if $|\gamma|<1$.

Note the initial γ condition given by the explicit series:

$$F(\alpha,\beta;\gamma=0)=e^\alpha+e^\beta-1\ , \tag{49}$$

as well as the value at $\gamma=1$,

$$F(\alpha,\beta;\gamma=1)=e^{\alpha+\beta}\ . \tag{50}$$

In fact, for γ any root of unity, the sum simplifies since the γ^n terms in the exponential repeat their values. If $\gamma^N=1$, then the sum breaks up into N more elementary series. For example,

$$F(\alpha,\beta;\gamma=-1)=\sum_{n\in\mathbb{Z}^+_{\text{even}}}\frac{1}{n!}\,\beta^n e^{\alpha}+\sum_{n\in\mathbb{Z}^+_{\text{odd}}}\frac{1}{n!}\,\beta^n e^{-\alpha}=e^\alpha\cosh\beta+e^{-\alpha}\sinh\beta\ . \tag{51}$$

There are also the "easy" boundary conditions

$$F(\alpha=0,\beta;\gamma)=e^\beta\ ,\quad F(\alpha,\beta=0;\gamma)=e^\alpha\ . \tag{52}$$

This multi-variable function obeys the 2+1 "heat equation". To see this, take first partial derivatives of (48) to find

$$\frac{\partial}{\partial\alpha}F(\alpha,\beta;\gamma)=F(\alpha,\gamma\beta;\gamma)\ ,\quad \frac{\partial}{\partial\beta}F(\alpha,\beta;\gamma)=F(\gamma\alpha,\beta;\gamma)\ . \tag{53}$$

The remaining first derivative of (47) gives

$$\gamma\frac{\partial}{\partial\gamma}F(\alpha,\beta;\gamma)=\alpha\beta\gamma F(\gamma\alpha,\gamma\beta;\gamma)\ . \tag{54}$$

Then (53) gives $\alpha\beta\gamma F(\gamma\alpha,\gamma\beta;\gamma) = \alpha\beta\frac{\partial}{\partial\beta}F(\alpha,\gamma\beta;\gamma) = \alpha\beta\frac{\partial}{\partial\beta}\frac{\partial}{\partial\alpha}F(\alpha,\beta;\gamma)$, so

$$\gamma\frac{\partial}{\partial\gamma}F(\alpha,\beta;\gamma) = \alpha\frac{\partial}{\partial\alpha}\beta\frac{\partial}{\partial\beta}F(\alpha,\beta;\gamma) \ . \tag{55}$$

Or, in terms of logarithms, the ("light-cone") heat equation

$$\begin{aligned}\frac{\partial}{\partial\ln\gamma}F(\alpha,\beta;\gamma) &= \frac{\partial}{\partial\ln\alpha}\frac{\partial}{\partial\ln\beta}F(\alpha,\beta;\gamma)\\ &= \left(\frac{\partial^2}{\partial(\ln\alpha\beta)^2} - \frac{\partial^2}{\partial(\ln\alpha/\beta)^2}\right)F(\alpha,\beta;\gamma) \ . \end{aligned}\tag{56}$$

A more suggestive way to write all this is to let $x = \ln\alpha$, $y = \ln\beta$, $t = \ln\gamma$, to obtain

$$\begin{aligned}F[x,y;t] &= \sum_{k,n=0}^{\infty}\frac{1}{k!n!}\,e^{kx+ny+nkt}\\ &= \sum_{n=0}^{\infty}\frac{1}{n!}\,\exp\left(ny+e^{x+nt}\right)\\ &= \sum_{n=0}^{\infty}\frac{1}{n!}\,\exp\left(ny+e^{x}\sum_{j=0}^{\infty}\frac{1}{j!}n^jt^j\right) \ , \end{aligned}\tag{57}$$

the latter form suggesting (save for the $1/n!$) that this is a generalization of Jacobi theta functions. The heat equation is now in more traditional form: $\partial F[x,y;t]/\partial t = \partial^2 F[x,y;t]/\partial x\partial y$. Admittedly, this is rather obvious, since *each term* in the series for $F[x,y;t]$ satisfies the heat equation. Still, it follows from the "scale & shift" effects of first-order partials of F:

$$\frac{\partial}{\partial x}F[x,y;t] = e^xF[x,y+t;t] \ , \quad \frac{\partial}{\partial y}F[x,y;t] = e^yF[x+t,y;t] \ , \tag{58}$$

$$\frac{\partial}{\partial t}F[x,y;t] = e^{x+y+t}F[x+t,y+t;t] \ . \tag{59}$$

For a connection with the problem of factoring integers, see the Appendix.

Acknowledgements I thank K Bering, J de Azcárraga, D Fairlie, L Mezincescu, J Morse, Y Nambu, Y Nutku, M Wachs, and C Zachos for discussions related to this talk.

Appendix: The integer-factoring polynomials (IFPs): Generate a set of symmetric polynomials through the sum

$$S(x,y,z) = \sum_{m,n=1}^{\infty} \frac{1}{m!n!} x^m y^n z^{mn} = \sum_{r=1}^{\infty} \frac{1}{r!} z^r p_r(x,y) \ ,$$

$$p_r(x,y) = r! \sum_{k \in d(r)} \frac{1}{k!\,(r/k)!} x^k y^{r/k} \ , \tag{60}$$

where $d(r)$ is the set of all integer divisors of r, including 1 and r. For examples, see the Table below. The various exponents appearing in p_r consist of all ways of factoring the integer r into products of integers, including the cases $r = 1 \times r$ and $r \times 1$. Whenever r is a prime, there are only two terms in the polynomial. This feature of the polynomials may be viewed as a primality test. All polynomials have a common factor of xy. Sometimes the remaining polynomial also factors over the integers, as illustrated by p_6, p_7, p_{10}, and p_{11}. Note the preservation of coeffi cients before and after such factoring, especially in the p_6 and p_{10} cases.

$$\begin{aligned}
p_1 &= xy \ , \\
p_2 &= x^2y + xy^2 = xy\,(x+y) \ , \\
p_3 &= x^3y + xy^3 = xy\left(x^2+y^2\right) \ , \\
p_4 &= x^4y + xy^4 + 6x^2y^2 = xy\left(x^3+6xy+y^3\right) \ , \\
p_5 &= x^5y + xy^5 = xy\left(x^4+y^4\right) \ , \\
p_6 &= x^6y + 60x^3y^2 + 60x^2y^3 + xy^6 = xy\left(x^5+60x^2y+60xy^2+y^5\right) \\
&= xy\,(x+y)\left(x^4 - x^3y + x^2y^2 + 60xy - xy^3 + y^4\right) \ , \\
p_7 &= x^7y + xy^7 = xy\left(x^6+y^6\right) \\
&= xy\left(y^2+x^2\right)\left(y^4 - x^2y^2 + x^4\right) \ , \\
p_8 &= x^8y + 840x^4y^2 + 840x^2y^4 + xy^8 = xy\left(x^7+840x^3y+840xy^3+y^7\right) \ , \\
p_9 &= x^9y + 10\,080x^3y^3 + xy^9 = xy\left(x^8 + 10\,080x^2y^2 + y^8\right) \\
p_{10} &= x^{10}y + 15\,120x^5y^2 + 15\,120x^2y^5 + xy^{10} \\
&= xy\,(x+y)\left(x^2-xy+y^2\right)\left(x^6 - x^3y^3 + 15\,120xy + y^6\right) \\
p_{11} &= x^{11}y + xy^{11} = xy\left(x^{10}+y^{10}\right) \\
&= xy\left(y^2+x^2\right)\left(x^8 - y^2x^6 + y^4x^4 - x^2y^6 + y^8\right)
\end{aligned} \tag{61}$$

The complete theory of such integer-factoring polynomials will have to be discussed elsewhere.

References

[1] Y Nambu, "Generalized Hamiltonian Dynamics" Phys Rev **D7** (1973) 2405-2412.

[2] T L Curtright and C K Zachos, "Classical and quantum Nambu mechanics" Phys Rev **D68** (2003) 085001 [hep-th/0212267].

[3] T L Curtright and C K Zachos, "Quantizing Dirac and Nambu Brackets" talk at the Coral Gables Conference, 11-14 December 2002 [hep-th/0303088].

[4] C K Zachos and T L Curtright, "Deformation Quantization of Nambu Mechanics" talk at the Coral Gables Conference, 11-14 December 2002 [quant-ph/0302106].

[5] C K Zachos, "Membranes and Consistent Quantization of Nambu Mechanics" Phys Lett **B570** (2003) 82-88, talk at the 8th Wigner Symposium, NYC, May 2003 [hep-th/0306222].

[6] T Curtright, "d-Branes in the Stream" [hep-th/0307121].

[7] T Curtright and C Zachos, "Branes, Strings, and Odd Quantum Nambu Brackets" talk contributed to the proceedings of the 3rd International Symposium on Quantum Theory and Symmetries (QTS3), Cincinnati, Ohio, 10-14 Sep 2003 [hep-th/0312048].

[8] K Bering, *private communication.*

[9] C K Zachos, D B Fairlie, and T L Curtright, *Quantum Mechanics in Phase Space*, World Scientific Publishers (2004).

[10] T Curtright, T Uematsu, and C Zachos, "Generating all Wigner functions" J Math Phys **42** (2001) 2396 [hep-th/0011137].

HIDDEN SYMMETRIES IN M-THEORY*

M. J. DUFF

Michigan Center for Theoretical Physics
Randall Laboratory, Department of Physics, University of Michigan
Ann Arbor, MI 48109–1120, USA
E-mail: mduff@umich.edu

Abstract

We focus on D=11 supergravity as the low-energy limit of M-theory and pose the questions: (1) What are the \$D=11\$ symmetries? (2) How many supersymmetries can M-theory vacua preserve?

1 Eleven dimensions

1.1 M-theory

In 1995 it was realized that a non-perturbative unification of the five consistent superstring theories is provided by M-theory, whose low-energy limit is eleven-dimensional supergravity. In addition to strings M-theory involves p-dimensional extended objects, namely the p-branes which couple to the background fields of D=11 supergravity. This resolved the old mystery of why local supersymmetry allows a maximum of eleven dimensions while superstrings stop at ten. Indeed, many of the p-branes were first understood as classical solutions of the supergravity field equations. As a result, supergravity has returned to center stage.

M-theory is regarded by many as the dreamed-of final theory and has accordingly received an enormous amount of attention. It is curious, therefore, that one of the most basic questions of M-theory has until now remained unanswered:

*Research supported in part by DOE Grant DE-FG02-95ER40899.

i) *What are the D=11 symmetries?*

Here we will report some progress.

1.2 The algebra

Eleven is the maximum spacetime dimension in which one can formulate a consistent supersymmetric theory, as was first recognized by Nahm in his classification of supersymmetry algebras. The easiest way to see this is to start in four dimensions and note that one supersymmetry relates states differing by one half unit of helicity. If we now make the reasonable assumption that there be no massless particles with spins greater than two, then we can allow up to a maximum of $N = 8$ supersymmetries taking us from helicity -2 through to helicity $+2$. Since the minimal supersymmetry generator is a Majorana spinor with four off-shell components, this means a total of 32 spinor components. Now in a spacetime with D dimensions and signature $(1, D-1)$, the maximum value of D admitting a 32 component spinor is $D = 11$. (Going to $D = 12$, for example, would require 64 components.) Furthermore, $D = 11$ emerges naturally as the maximum dimension admitting supersymmetric extended objects.

The full D=11 supertranslation algebra is

$$\{Q_\alpha, Q_\beta\} = (C\Gamma^M)_{\alpha\beta} P_M + (C\Gamma_{MN})_{\alpha\beta} Z_{(2)}^{MN} + (C\Gamma_{MNPQR})_{\alpha\beta} Z_{(5)}^{MNPQR}. \quad (1)$$

Note that the total number of algebraically independent charges that could appear on the right hand side is 528. The number actually appearing is

$$11 + 55 + 462 = 528 \quad (2)$$

so the algebra (1) is 'maximally extended'.

1.3 The multiplet

Not long after Nahm's paper, Cremmer, Julia and Scherk realized that supergravity not only permits up to seven extra dimensions but in fact takes its simplest and most elegant form when written in its full eleven-dimensional glory. The unique $D = 11, N = 1$ supermultiplet is comprised of a graviton g_{MN}, a gravitino ψ_M and 3-form gauge field A_{MNP} with 44, 128 and 84 physical degrees of freedom, respectively. The theory may also be formulated in superspace. Ironically, however, these extra dimensions were not at first taken seriously but rather regarded merely as a useful device for deriving supergravities in four dimensions. Indeed $D=4, N=8$ supergravity was first obtained by Cremmer and Julia via the process of *dimensional reduction* i.e. by requiring that all the fields of $D=11, N=1$ supergravity be independent of the extra seven coordinates.

1.4 D=11 supergravity

For future reference we record the bosonic field equations

$$R_{MN} = \frac{1}{12}\left(F_{MPQR}F_{N}{}^{PQR} - \frac{1}{12}g_{MN}F^{PQRS}F_{PQRS}\right) \tag{3}$$

and

$$d*F_{(4)} + \frac{1}{2}F_{(4)} \wedge F_{(4)} = 0, \tag{4}$$

where $F_{(4)} = dA_{(3)}$. The supersymmetry transformation rule of the gravitino reduces in a purely bosonic background to

$$\delta\Psi_M = \mathcal{D}_M\epsilon, \tag{5}$$

where the parameter ϵ is a 32-component anticommuting spinor, and where

$$\mathcal{D}_M = D_M - \frac{1}{288}(\Gamma_M{}^{NPQR} - 8\delta_M^N\Gamma^{PQR})F_{NPQR}, \tag{6}$$

where Γ^A are the $D=11$ Dirac matrices and $\Gamma_{AB} = \Gamma_{[A}\Gamma_{B]}$. Here D_M is the usual Riemannian covariant derivative involving the connection ω_M of the usual structure group Spin(10, 1), the double cover of SO(10, 1),

$$D_M = \partial_M + \frac{1}{4}\omega_M{}^{AB}\Gamma_{AB} \tag{7}$$

2 Hidden spacetime symmetries in D=11

2.1 Spacelike, null and timelike reductions

Long ago, Cremmer and Julia pointed out that, when dimensionally reduced to d dimensions, $D=11$ supergravity exhibits hidden symmetries. For example $E_7(global) \times SU(8)(local)$ when $d=4$ and $E_8(global) \times SO(16)(local)$ when $d=3$.

Cremmer and Julia concentrated on the case where all extra dimensions are spacelike. Here we shall consider timelike and null reductions as well. The global symmetries remain the same but we shall focus on the local symmetries.

In fact, we shall focus particularly on the supercovariant derivative (6) as it appears in the gravitino variation of the dimensionally reduced theory. One finds that, after making a $d/(11-d)$ split, the Lorentz subgroup $G = \mathrm{SO}(d-1,1) \times \mathrm{SO}(11-d)$ can be enlarged to the generalized structure groups $\mathcal{G} = \mathrm{SO}(d-1,1) \times G(spacelike)$, $\mathcal{G} = \mathrm{ISO}(d-1) \times G(null)$ and $\mathcal{G} = \mathrm{SO}(d) \times G(timelike)$ arising in the spacelike, null and timelike dimensional reduction, respectively. As we shall see, these generalized structure groups are the same as the hidden symmetries for $d \geq 3$ but differ for $d < 3$.

First we consider a spacelike dimensional reduction corresponding to a $d/(11-d)$ split. Turning on only d-dimensional scalars, the reduction ansatz is particularly simple

$$g_{MN}^{(11)} = \Big(\Delta^{-1/(d-2)} g_{\mu\nu} 0 0 g_{ij}, \qquad A_{ijk}^{(11)} = \phi_{ijk}, (8)$$

where $\Delta = \det g_{ij}$. For $d \leq 5$, we must also consider the possibility dualizing either $F_{(4)}$ components or (for $d = 3$) Kaluza-Klein vectors to scalars. We will return to such possibilities below. But for now we focus on $d \geq 6$. In this case, a standard dimensional reduction of the $D = 11$ gravitino transformation (5) yields the d-dimensional gravitino transformation

$$\delta\psi_\mu = \hat{D}_\mu \epsilon \tag{9}$$

where

$$\hat{D}_\mu = \partial_\mu + \omega_\mu{}^{\alpha\beta}\gamma_{\alpha\beta} + Q_\mu{}^{ab}\Gamma_{ab} + \tfrac{1}{3!}e^{ia}e^{jb}e^{kc}\partial_\mu\phi_{ijk}\Gamma_{abc}. \tag{10}$$

Here γ_α are $\mathrm{SO}(d-1,1)$ Dirac matrices, while Γ_a are $\mathrm{SO}(11-d)$ Dirac matrices. For completeness, we also note that the d-dimensional dilatinos transform according to

$$\delta\lambda_i = -\tfrac{1}{2}\gamma^\mu[P_{\mu\, ij}\Gamma^j - \tfrac{1}{36}(\Gamma_i{}^{jkl} - 6\delta_i^j\Gamma^{kl})\partial_\mu\phi_{jkl}]\epsilon. \tag{11}$$

In the above, the lower dimensional quantities are related to their $D = 11$ counterparts through

$$\psi_\mu = \Delta^{\frac{1}{4(d-2)}} \left(\Psi_\mu^{(11)} + \frac{1}{d-2} \gamma_\mu \Gamma^i \Psi_i^{(11)} \right), \qquad \lambda_i = \Delta^{\frac{1}{4(d-2)}} \Psi_i^{(11)},$$
$$\epsilon = \Delta^{\frac{1}{4(d-2)}} \epsilon^{(11)},$$
$$Q_\mu^{ab} = e^{i[a} \partial_\mu e_i{}^{b]}, \qquad P_{\mu\, ij} = e^a_{(i} \partial_\mu e_{j)\, a}. \tag{12}$$

This decomposition is suggestive of a generalized structure group with connection given by $\hat{D}_\mu$. However one additional requirement is necessary before declaring this an enlargement of SO$(d-1,1) \times$ SO$(11-d)$, and that is to ensure that the algebra generated by Γ_{ab} and Γ_{abc} closes within itself. Along this line, we note that the commutators of these internal Dirac matrices have the schematic structure

$$[\Gamma^{(2)}, \Gamma^{(2)}] = \Gamma^{(2)}, \qquad [\Gamma^{(2)}, \Gamma^{(3)}] = \Gamma^{(3)}, \qquad [\Gamma^{(3)}, \Gamma^{(3)}] = \Gamma^{(6)} + \Gamma^{(2)}. \tag{13}$$

Here the notation $\Gamma^{(n)}$ indicates the antisymmetric product of n Dirac matrices, and the right hand sides of the commutators only indicate what possible terms may show up. The first commutator above merely indicates that the Γ_{ab} matrices provide a representation of the Riemannian SO$(11-d)$ structure group.

For $d \geq 6$, the internal space is restricted to five or fewer dimensions. In this case, the antisymmetric product $\Gamma^{(6)}$ cannot show up, and the algebra clearly closes on $\Gamma^{(2)}$ and $\Gamma^{(3)}$. Working out the extended structure groups for these cases results in the expected Cremmer and Julia groups listed in the first four lines in the second column of Table 1. A similar analysis follows for $d \leq 5$. However, in this case, we must also dualize an additional set of fields to see the hidden symmetries. For $d = 5$, an additional scalar arises from the dual of $F_{\mu\nu\rho\sigma}$; this yields an addition to (10) of the form $\hat{D}_\mu^{\text{additional}} = \frac{1}{4!} \epsilon_\mu{}^{\nu\rho\sigma\lambda} F_{\mu\nu\rho\sigma} \Gamma_{123456}$. This $\Gamma^{(6)}$ term is precisely what is necessary for the closure of the algebra of (13). Of course, in this case, we must also make note of the additional commutators

$$[\Gamma^{(2)}, \Gamma^{(6)}] = \Gamma^{(6)}, \qquad [\Gamma^{(3)}, \Gamma^{(6)}] = \Gamma^{(7)} + \Gamma^{(3)}, \qquad [\Gamma^{(6)}, \Gamma^{(6)}] = \Gamma^{(10)} + \Gamma^{(6)} + \Gamma^{(2)}. \tag{14}$$

However neither $\Gamma^{(7)}$ nor $\Gamma^{(10)}$ may show up in $d = 5$ for dimensional reasons.

The analysis for $d = 4$ is similar; however here $\hat{D}_\mu^{\text{additional}} = \frac{1}{3!} \epsilon_\mu{}^{\nu\rho\sigma} e^{ia} F_{\nu\rho\sigma i} \Gamma_a \Gamma_{1234567}$. Closure of the algebra on $\Gamma^{(2)}$, $\Gamma^{(3)}$ and $\Gamma^{(6)}$ then follows because, while $\Gamma^{(7)}$ may in principle arise in the middle commutator of (14), it turns out to be kinematically forbidden. For $d = 3$, on the other hand, in additional to a contribution

$\hat{D}_\mu^{\text{additional}} = \frac{1}{2!\cdot 2!}\epsilon_\mu{}^{\nu\rho} e^{ia} e^{jb} F_{\nu\rho ij}\Gamma_{ab}\Gamma_{12345678}$, one must also dualize the Kaluza-Klein vectors $g_\mu{}^i$. Doing so gives rise to a $\Gamma^{(7)}$ in the generalized connection which, in addition to the previously identified terms, completes the internal structure group to SO(16).

The remaining three cases, namely $d = 2$, $d = 1$ and $d = 0$ fall somewhat outside the framework presented above. This is because in these low dimensions the generalized connections $\hat{D}_\mu$ derived via reduction are partially incomplete. For $d = 2$, we find

$$\hat{D}_\mu^{(d=2)} = \partial_\mu + \omega_\mu{}^{\alpha\beta}\gamma_{\alpha\beta} + Q_\mu{}^{ab}\Gamma_{ab} + \tfrac{1}{9}(\delta_\mu^\nu - \tfrac{1}{2}\gamma_\mu{}^\nu) e^{ia} e^{jb} e^{kc}\partial_\nu\phi_{ijk}\Gamma_{abc}, \qquad (15)$$

where $\gamma_{\mu\nu} = -\frac{1}{2}\epsilon_{\mu\nu}(\epsilon^{\alpha\beta}\gamma_{\alpha\beta})$ is necessarily proportional to the two-dimensional chirality matrix. Hence from a two-dimensional point of view, the scalars from the metric enter non-chirally, while the scalars from $F_{(4)}$ enter chirally. Taken together, the generalized connection (15) takes values in $SO(16)_+ \times SO(16)_-$, which we regard as the enlarged structure group. However not all generators are present because of lack of chirality in the term proportional to $Q_\mu{}^{ab}$. Thus at this point the generalized structure group deviates from the hidden symmetry group, which would be an infinite dimensional subgroup of affine E_8. Similarly, for $d = 1$, closure of the derivative $\hat{D}_\mu^{(d=1)}$ results in an enlarged SO(32) structure group. However this is not obviously related to any actual hidden symmetry of the 1/10 split. The $d = 0$ case is subject to the same caveats as the $d = 1$ and $d = 2$ cases: not all group generators are present in the covariant derivative. SL(32, R) requires $\{\Gamma^{(1)}, \Gamma^{(2)}, \Gamma^{(3)}, \Gamma^{(4)}, \Gamma^{(5)}\}$ whereas only $\{\Gamma^{(2)}, \Gamma^{(3)}, \Gamma^{(5)}\}$ appear in the covariant derivative.

Next we consider a timelike reduction for which we simply interchange a time and a space direction in the above analysis. This results in an internal Clifford algebra with signature $(10 - d, 1)$, and yields the extended symmetry groups indicated in the fourth column of Table 1. The same caveats concerning $d = 2, 1, 0$ apply in the timelike case.

Turning finally to the null case, we may replace one of the internal Dirac matrices with Γ_+ (where +, − denote light-cone directions). Since $(\Gamma_+)2 = 0$, this indicates that the extended structure groups for the null case are contractions of the corresponding spacelike (or timelike) groups. In addition, by removing Γ_+ from the set of Dirac matrices, we essentially end up in the case of one fewer compactified dimensions. As a result, the $G(null)$ group in d-dimensions must have a semi-direct product structure involving the $G(spacelike)$ group in $(d+1)$-dimensions. Of course, these groups also contain the original $\mathrm{ISO}(10-d)$ structure group as a subgroup. The resulting generalized structure groups are given in the third column of Table 1. Once again, the same caveats concerning $d = 2, 1, 0$ apply.

2.2 The complete uncompactified D=11 theory

Following Cremmer and Julia's spacelike reduction, the question was then posed: do these symmetries appear magically only after dimensional reduction, or were they already present in the full uncompactified and untruncated $D = 11$ theory? The question was answered by de Wit and Nicolai who made a $d/(11-d)$ split and fixed the gauge by setting to zero the off-diagonal components of the elfbein. They showed that in the resulting field equations the local symmetries are indeed already present, but the global symmetries are not. For example, after making the split $\mathrm{SO}(10,1) \supset \mathrm{SO}(3,1) \times \mathrm{SO}(7)$, we find the enlarged symmetry $\mathrm{SO}(3,1) \times \mathrm{SU}(8)$. There is no global E_7 invariance (although the 70 internal components of the metric and 3-form may nevertheless be assigned to an $E_7/\mathrm{SU}(8)$ coset). Similar results were found for other values of d: in each case the internal subgroup $\mathrm{SO}(11-d)$ gets enlarged to some compact group $G(spacelike)$ while the spacetime subgroup $\mathrm{SO}(d-1,1)$ remains intact[1]. Here we ask instead whether there are hidden *spacetime* symmetries. This is a question that could have been asked long ago, but we suspect that people may have been inhibited by the Coleman-Mandula theorem which forbids combining spacetime and internal symmetries. However, this is a statement about Poincare symmetries of the S-matrix and here we are concerned with Lorentz symmetries of the equations of motion, so there will be no conflict.

The explicit demonstration of $G(spacelike)$ invariance by de Wit and Nicolai is very involved, to say the least. However, the result is quite simple: one finds the same $G(spacelike)$ in the full uncompactified $D = 11$ theory as was already found in the spacelike dimensional reduction of Cremmer and Julia. Here we content ourselves with the educated guess that the same logic applies to $G(timelike)$ and $G(null)$: they are the same as what one finds by timelike and null reduc-

[1] We keep the terminology "spacetime" and "internal" even though no compactification or dimensional reduction is implied.

$d/(11-d)$	$G(spacelike)$	$G(null)$	$G(timelike)$
11/0	$\{1\}$	$\{1\}$	$\{1\}$
10/1	$\{1\}$	$\{1\}$	$\{1\}$
9/2	SO(2)	R	SO(1,1)
8/3	SO(3) × SO(2)	ISO(2) × R	SO(2,1) × SO(1,1)
7/4	SO(5)	$[\mathrm{SO}(3)\times \mathrm{SO}(2)]\ltimes \mathrm{R6}_{(3,2)}$	SO(3,2)
6/5	SO(5) × SO(5)	$\mathrm{SO}(5)\ltimes \mathrm{R}^{10}_{(10)}$	SO(5,C)
5/6	USp(8)	$[\mathrm{SO}(5)\times \mathrm{SO}(5)]\ltimes \mathrm{R}^{16}_{(4,4)}$	USp(4,4)
4/7	SU(8	$\mathrm{USp}(8)\ltimes \mathrm{R}^{27}_{(27)}$	SU*(8)
3/8	SO(16)	$[\mathrm{SU}(8)\times \mathrm{U}(1)]\ltimes \mathrm{R}^{56}_{(28_{1/2},\overline{28}_{-1/2})}$	SO*(16)
2/9	SO(16) × SO(16)	$\mathrm{SO}(16)\ltimes \mathrm{R}^{120}_{(120)}$	SO(16,C)
1/10	SO(32)	$[\mathrm{SO}(16)\times \mathrm{SO}(16)]\ltimes \mathrm{R}^{256}_{(16,16)}$	SO(16,16)
0/11	$\mathrm{SL}(32,R)$	$\mathrm{SL}(32,R)$	$\mathrm{SL}(32,R)$

Table 1: The generalized structure groups are given by $\mathcal{G} = \mathrm{SO}(d-1,1) \times G(spacelike)$, $\mathcal{G} = \mathrm{ISO}(d-1) \times G(null)$ and $\mathcal{G} = \mathrm{SO}(d) \times G(timelike)$.

tion, respectively. The claim that the null and timelike symmetries are present in the full theory and not merely in its dimensional reductions might be proved by repeating the spacelike calculations of DeWit and Nicolai with the appropriate change of Γ matrices. So we propose that, after making a $d/(11-d)$ split, the Lorentz subgroup $G = \mathrm{SO}(d-1,1) \times \mathrm{SO}(11-d)$ can be enlarged to the generalized structure groups $\mathcal{G} = \mathrm{SO}(d-1,1) \times G(spacelike)$, $\mathcal{G} = \mathrm{ISO}(d-1) \times G(null)$ and $\mathcal{G} = \mathrm{SO}(d) \times G(timelike)$. As we have seen, for $d > 2$ the groups $G(spacelike)$, $G(timelike)$ and $G(null)$ are the same as those obtained from dimensional reductions. For the purposes of this section, however, their physical interpretation is very different. They are here proposed as symmetries of the full $D = 11$ equations of motion; there is no compactification involved, whether toroidal or otherwise. (Note that by postulating that the generalized structure groups survive as hidden symmetries of the full uncompactified theory, we avoid the undesirable features associated with compactifications including a timelike direction such as closed timelike curves.)

For $d \leq 2$ it is less clear whether these generalized structure groups are actually hidden symmetries. Yet one might imagine that there exist a yet-to-be-discovered formulation of M-theory in which the $d = 2$ and $d = 1$ symmetries are realized. This would still be in keeping with the apparent need to make a non-covariant split and to make the corresponding gauge choice before the hidden symmetries emerge. A yet bolder conjecture, requiring no non-covariant split or gauge choice since $d = 0$, due to Hull is that there exist a formulation of M-theory with the full SL(32, R). This proposal is nevertheless very attractive since SL(32, R) contains all the groups in Table 1 as subgroups and would thus answer the question of whether all these symmetries are present at the same time. This is an important issue deserving of further study.

Finally, we emphasize that despite the $d/(11-d)$ split these symmetries refer to the full equations of motion and not to any particular background such as product manifolds.

Editor's note: A list of references for the work described in this paper can be found in

M. J. Duff, *Erice lectures on "The status of local supersymmetry"*, hep-th/0403160.

Also see

A. Batrachenko, M. J. Duff, James T. Liu, and W. Y. Wen, *Generalized holonomy of M-theory vacua*, hep-th/0312165.

C. Hull, *Holonomy and Symmetry in M-theory*, hep-th/0305039.

M. J. Duff and James T. Liu, *Hidden Spacetime Symmetries and Generalized Holonomy in M-theory*, hep-th/0303140.

GAUGE/STRING-GRAVITY DUALITY AND FROISSART BOUND

KYUNGSIK KANG

Department of Physics, Brown University
Providence, RI 02912, U.S.A.
e-mail: kang@het.brown.edu

The gauge/string-gravity duality correspondence opened renewed hope and possibility to address some of the fundamental and non-perturbative QCD problems of in particle physics, such as hadron spectrum and Regge behavior of the scattering amplitude at high energies. One of the most fundamental and long-standing problems is the high energy behavior of the total cross-sections. According to a series of exhaustive tests by the COMPETE group, (1) total cross sections have a universal Heisenberg behavior in energy corresponding to the maximal energy behavior allowed by the Froissart bound, i.e., $A + B \ \ln^2(s/s_0)$ with $B \sim 0.32 \ mb$ and $s_0 \sim 34.41 \ GeV^2$ for all reactions, and (2). the factorization relation among $\sigma_{pp,even}, \sigma_{\gamma p}$ and $\sigma_{\gamma\gamma}$ is well satisfied by experiments. I discuss the recent interesting application of the gauge/string-gravity duality of AdS/CFT correspondence with a deformed background metric so as to break the conformal symmetry that lead to the Heisenberg behavior of rising total cross sections, and present some preliminary results on the high energy QCD from Planckian scattering in AdS and black-hole production.

1. Global Description of High Energy Scattering - COMPETE results

The COMPETE group[1] has performed a series of exhaustive tests of the analytic parametrisations for the forward scattering amplitudes against the largest available date at $t = 0$, which includes all measured total cross sections as well as the ratios, ρ, of the real part to imaginary part of the elastic amplitude of $pp, \bar{p}p, \pi^{\pm}p, K^{\pm}p$, and total cross sections of $\gamma p, \gamma\gamma$ and $E^{-}p$. Applying a set of carefully designed criteria for measuring the quality of fits to differentiate the different parametrizations, beyond the usual χ^2/dof, they found $RRPL2_u$ and $RRP_{nf}L2_u$ to be the best analytic amplitude models. Common features of these models are (1) total cross sections have a universal Heisenberg behavior in energy corresponding to the maximal energy behavior allowed by the Froissart bound. i.e.

$B \log^2 (s/s_0)$ for all total cross sections with
$B = 0.315mb, s_0 = 34.41GeV^2$ for $RRPL2_u$ and
$B = 0.315mb, s_0 = 34.03GeV^2$ for $RRP_{nf}L2_u$ irrespectively of degeneracy in Reggeon terms, though favoring non-degeneracy: for example, $B = 0.328mb, s_0 = 49.06GeV^2$ for $(RR)_d PL2_u$.
(2) Total cross sections satisfy the **factorization relation**, $(H_{\gamma p})^2 \to H_{\gamma\gamma} \times H_{pp}$, by the $H = PL2$ terms. Numerically, $\delta = (H_{\gamma p}/H_{pp}) = 0.0031$, in good agreement with the generalized vector dominance.

2. Theoretical Models for Rising Cross Sections

Assuming that the hadron-hadron scattering at high energies as a collision of two flat discs that produce and exchange a pair of mesons in the interaction region of the impact parameter space and that the portion of the energy density that is responsible for the non-renormalizable meson exchange interactions is high enough to create at least a pair of mesons and the portion is exponentially decreasing with the exchanged meson mass and impact distance in analogy to the shock wave process, Heisenberg[2] has argued that the maximum impact distance for which the effective interaction takes place, corresponding to the minimum portion to create a pair of mesons, is $b_{max} = (1/2m)\log(s/s_0), m$ being the exchanged meson pair mass. From this the total cross section is given by

$$\sigma = (\pi/16m_\pi^2)\ log^2\ (s/s_0)$$

which corresponds to a saturating behavior of the **Froissart bound**[3]

$$\sigma_{tot} \le c \log^2 s,$$

in which

$$c \le (\pi/m_\pi^2) = 60mb$$

which is a consequence of the unitarity and positivity of the imaginary part of the scattering amplitudes in the Lehmann ellipse.

The increasing behavior of the total cross sections is a necessary condition for the rigorous proof of the Pomeranchuk theorem[4]. Also one can show from unitarity and analyticity in the form of the derivative dispersion relation that the $s-u$ crossing symmetric, the forward scattering amplitude that saturates the Froissart bound, is of the form $PL2$ from[5].

3. Gauge/Gravity Duality and the Heisenberg Behavior

The gauge/string-gravity duality of the AdS_{d+1}/CFT_d correspondence[6], i.e., the weak coupling gravity of superstring theories in AdS space of $d+1$ dimension is dual to the strong coupling conformal supersymmetric gauge theory in d dimension, has opened the possibility to address the high energy behavior of the scattering amplitude, in particular, the high energy behavior of the total cross sections. But in order to deal with a realistic QCD relevant for the hadron physics, the desired super-gravity solutions must embody the salient features of strong interactions, such as confinement, hadronization with non-zero mass gap and Regge behavior of the hadron scattering amplitude at high energies. On the other hand, the super-string theory contains no mass gap due to zero mass gauge/graviton fields in the string spectra of asymptotic states, and has soft scattering amplitude,

$$A_{string}(s,t) \sim \exp[\frac{-\alpha'}{2}(s\ln s + t\ln t + u\ln u)]$$

to be contrasted to the partonic or hard Regge behavior in gauge theory,

$$A_{qcd}(s,t) \sim s^{2-\frac{n}{2}}$$

Can one find a consistent picture of gauge field properties and Regge amplitude in the strong gauge coupling regime from a suitable string theory via AdS/CFT correspondence?

A suggestion[7] is to deform the Randall-Sundrum type AdS/gravity background metric with an IR cutoff in the holographic radial coordinate r of a space and in particular use a metric of the form,

$$ds^2 = (r/R)^2\eta_{\mu\nu}dx^\mu dx^\nu + (R/r)^2(1-(b/r)^d)^{-1}dr^2 + R^2 ds_Y^2$$

whereby breaking the conformal (and SUSY) symmetry with an IR cut-off at $r_{min} = b$. Here R is the anti-deSitter radius and ds_Y^2 is the metric for 5 (or 6) compact dimensions of 10-d string (or 11-d M-) theory.

One of the most interesting features of this background metric is the warp factor multiplying the $4-d$ flat metric which leads to the holographic relation between r and $4-d$ momentum p. Basically in 10-d string theory, due to the gravitational **Red Shift** for a state localized in the transverse co-ordinate r, $\Delta s = (R/r)\Delta r$ and $p = (r/R)p_s$, so that a state with a characteristic 10-d energy scale, $p_s \sim (1/R)$, corresponds to a 4-d energy scale $p \sim (r/R^2)$, resulting the holographic relation : high energy is large r and low energy is small r. Depending on the structure of the extra $5-d$ or

$6-d$ compact space and the details of the geometry at small r, the precise gauge theory and the breaking of the conformal invariance are different but the high energy QCD is well approximated by this background metric independently of the details of the extra dimensional space at low r near the IR cutoff point $r_{min} \sim \Lambda_{KK} R^2, \Lambda_{KK}$ being the conformal scale, i.e., the mass gap determined by the lightest glue ball (KK mode). In other words, such theory embodies the QCD universalities in the sense that it is approximately conformal in the large momentum region while it has a non-zero mass gap and confinement in the infrared region.

Polchinski and Strassler[8] have argued that with the extra dimensional branes of either warped or non-warped large space-time geometry, the amplitude can be treated essentially as $10-d$ scattering that takes place at a point in AdS in which transverse dimensions are integrated coherently over, so that the soft behavior of the strings would conspire the shape of the bulk wave functions and produce the correct power behavior of the confining gauge theory. In what follows, I will present the analysis of the gauge/gravity following the papers under preparation with H. Nastase[9].

A glueball corresponds to a plane wave state $\psi(r, Y)e^{ix_\mu p^\mu}$ in AdS and scatters with a local proper momentum $p_s(r) = (R/r)p$, i.e., UV shifted in the IR. There is a gauge theory string tension

$$\alpha' = (gN)^{-1/2}\Lambda_{KK}^{-2} \tag{1}$$

and

$$\sqrt{\hat{\alpha}'}p_s = \sqrt{\hat{\alpha}'}p(r_{min}/r) \leq \sqrt{\hat{\alpha}'}p$$

Note that at small r, p_s is larger than the string scale $1/\sqrt{\hat{\alpha}'}$ and A_{string} has the soft behavior.

Gauge theory scattering of glueballs is equated to a scattering inside AdS of the above states via AdS/CFT correspondence, leading to

$$\mathcal{A}(p) = \int drd^5\Omega\sqrt{g}A_{string}(\vec{p})\prod_i \psi_i \tag{2}$$

The integral is damped by the soft behavior of A_{string} at small r and also by the wave functions at large r, since $\sqrt{g} = r^3R^2$, with $\psi \sim Cf(r/r_{min})g(\Omega) \sim C(r/r_{min})^{-\Delta}g(\Omega)$. We get in the scattering region, $\sqrt{\hat{\alpha}'}p \gg 1, r_{scatt} \gg r_{min}$ and

$$\mathcal{A}(p) \sim \int drr^3 \prod_i \left(\frac{r_{min}}{r}\right)^{\Delta_i} A_{string}(pR/r) \sim \left(\frac{\Lambda_{KK}}{p}\right)^{\Sigma\Delta_i - 4} \tag{3}$$

as in QCD. Note this is just from scaling of the amplitude in the scattering region $r_{scatt} \gg r_{min}$ where $10-d$ scattering amplitude for $2 \rightarrow 2$ is dimensionally $g^2(\hat{\alpha})^3 F(p_s\sqrt{\hat{\alpha}'})$. In the Regge limit $0 < -t \ll s$, with $\hat{s} \gg |\hat{t}|, \hat{\alpha}'^{-1}$,

$$\mathcal{A}_{string}(p_x) \sim F\left(p_s\sqrt{\alpha'}\right) \sim (\hat{\alpha}'\hat{s})^{\alpha' t/2+2} \frac{\Gamma(-\hat{\alpha}\hat{t}/4)}{\Gamma(1+\hat{\alpha}'\hat{t}/4)} \tag{4}$$

from the Virasoro-Shapiro amplitude for massless external states ($s+t+u=0$). With this $\mathcal{A}_{string}$, we have, using $\nu = -\hat{\alpha}'\hat{t}$ as integral variable,

$$\mathcal{A} \sim \frac{1}{(\alpha'|t|)^{\Delta/2-3}} \int_0^{\nu_{max}} d\nu \nu^{\Delta/2-3} A_{string}(\nu s/|t|, \nu) \tag{5}$$

since $s/t = \hat{s}/\hat{t}$, and $r = R(\alpha' t/\nu)^{1/2}$.

The Regge behaviour can then be obtained if the dominant contribution of the integral comes from the upper limit, ν_{max}(which comes from r_{min}), which is $\alpha'|t|$, because the saddle point of the integrand is outside of ν_{max},

$$\alpha'|t| < (\Delta-4)/ln(s/|t|),$$

so that the flat space Regge behavior $A \sim (\alpha' s)^{2+\alpha' t/2}$ follows apart from some t factor due to different metrics and wave-function in r. Otherwise, we get

$$\mathcal{A} \sim s^2 |t|^{-\Delta/2} [ln(s/|t|)]^{1-\Delta/2} \tag{6}$$

and the main contribution to the integral comes from again far away from the cutoff, i.e., $r_{scatt} \gg r_{min}$. Note the inverse power of $|t|$ in (6) for any t and sufficiently large s.

Subsequently **Giddings**[10] studied other bulk perspectives and in particular argued that the effect of strong-gravity processes, such as black hole formation, to the high energy behavior of the total cross sections is important in the dual dynamics. Here the key point is that in TeV scale gravity scenario, black holes should be produced once the energy passes the fundamental Planck scale near a few TeV.

Giddings points out that as one increases the energy of the gauge theory scattering, one increases also the relevant energy in string theory. And there are three further (higher) scales (in the case when the string coupling g_s is small but $g_s N$ is large).

The first is the Planck scale

$$M_P \sim g_s^{-1/4}/\sqrt{\hat{\alpha}'},$$

which corresponds in the gauge theory to $\hat{M}_P = g_s^{-1/4}/\sqrt{\alpha'}$, where black holes start to form. Note $\hat{\alpha}' = \Lambda_{KK}^{-2}(g_s N)^{-1/2} = \alpha'$ and $r_{min} = R = 1/\Lambda_{KK}$. The black hole production cross section is approximated by $\sigma \simeq \pi r_H^2 \sim E^{2/7}$ in $10-d$ flat dimension, since we have approximately $10-d$ flat space.
The second scale is where string intermediate states cross over to black hole virtual states,

$$E_c \sim g_s^{-2}/\sqrt{\alpha'} = N^{7/4} M_P/(g_s N)^{7/4} = N^2 \sqrt{\frac{\Lambda_{KK}}{r_{min}}} \frac{1}{(g_s N)^{7/4}}$$

or in the gauge theory

$$\hat{E}_c \sim \frac{N^2 \Lambda_{KK}}{(g_s N)^{7/4}}$$

Here the semi-classical result should be also applicable to the cross section.

The third scale is when the black hole size r_H is comparable to the AdS size R,

$$E \sim M_P^8 r_H^7 \left(M_P^{d-2} r_H^{d-3} \text{ in general}\right) \to E_R = M_P^8 R^7$$

or in the gauge theory $\hat{E}_R = N^2 \Lambda_{KK}$ At this energy, $r_H \sim lnE$ so that $\sigma \sim ln^2 E$ corresponding to the maximal Froissart bound.

Though no proper dynamical theory of black-hole production was used, one may say that Giddings argument may represent some features of scattering in tran-Planckian energy regime. Here two particle scattering at small impact parameter region would induce a large space-time curvature, larger than the fundamental scale, due to a large concentration of energy in a small volume. To a low energy observer, this will appear as a curvature singularity whose horizon will hide the high energy effects from the observer outside. Also it has been known[11] that trapped surfaces do form in the classical trans-Planckian collision of particles. Trapped surface will cause a space-time singularity[12] which will give the horizon censorship. Apparent horizon area can not be smaller than that of the event horizon, which in turn can not be smaller than the black-hole area.

Giddings approximates warped metric by an AdS type

$$ds^2 \approx \frac{R^2}{z^2}(dz^2 + \eta_{\mu\nu} dx^\mu dx^\nu) + R^2 ds_Y^2$$

by changing the variable as $z = R^2/r = Re^{y/R}$. Then $z \in (0, R = r_{min})$ corresponds to $y \in (-\infty, 0)$. In particular, the gauge theory brane is located at the IR cut-off $z = R = r_{min}$. Notice that the confining gauge theory dual is the same as the Randall-Sundrum[13] scenario 2 with the gauge theory living in the IR brane where the warp factor is minimal and with the Planck brane at $z = 0$.

The perturbed AdS metric is

$$ds^2 = (1 + h_{yy})dy^2 + \exp(-2y/R)(\eta_{\mu\nu} + h_{\mu\nu})dx^\mu dx^\nu + R^2 ds_Y^2$$

with the gauge $h_{\mu y} = 0$. In particular, Giddings assumed a black-hole solution due to a static point mass source living in the IR brane and extending to the bulk by putting

$$T_{\mu\nu} = S_{\mu\nu}\delta(y), T_{yy} = T_{y\mu} - 0$$

where $S_{\mu\nu}$ is a static point mass source

$$S_{\mu\nu} = 2m\delta^{d-1}(x)\delta_{\mu 0}\delta_{\nu 0}$$

in the linearized gravitational equations, which he solved by using Neumann's Greens function. The most interesting component of his solution is

$$h_{00} \simeq \frac{k}{RM_P^{d-1}} m(\frac{z}{R})^d \frac{e^{-M_1 r}}{r^{d-3}} \tag{7}$$

either in light or heavy radion region $\mu R \ll 1$ or $\mu R \gg 1$ where M_1 is the mass of the lightest KK mode of graviton, the first pole in the integral expression for h_{00} coming from the first zero of the Bessel function $J_{d/2-1}(qR)$, q^2 being $-p^2$. However for $\mu R \ll 1$ the linear approximation for neglecting brane bending (due to the point mass) is lost before reaching the horizon. So r_H cannot be obtained from this solution even approximately.

But in the heavy radion limit $\mu R \gg 1$, the solution gives an estimate of the horizon size $r_H = r_H(m, y)$ from $h_{00} \sim -1$. In particular, at the IR brane, corresponding to $y = 0$, the black hole size is given by

$$r_H(m, y = 0) \sim \frac{1}{M_1} ln \mid \frac{kmM_1^{d-3}}{RM_P^{d-1}} \mid \sim \frac{1}{M_1} lnm$$

so that

$$r_H(E) \sim lnE$$

and

$$\sigma \sim \frac{1}{M_1^2} ln^2 E$$

In the intermediate case, if $M_1 < M_L$, the horizon forms first, while if $M_1 > M_L$, brane bends first. But even if M_L is smaller, the radion L has the same expression or more precisely LM_L is h_{00} with M_1 and $1/R$ replaced by M_L, which means

$$\sigma \sim \frac{\pi}{M_L^2} ln^2 E$$

The essential point of getting the Froissart-Heisenberg behavior for the cross section is the appearance of the exponential term with the non-zero mass gap. But we find that[9] there is no such exponential in the AdS Aichelburg-Sexl waves[13] for $r \gg R$. In fact it gives just the regular $4-d$ behavior in this limit in Randall-Sundrum[14] for string-like models of Giddings et al[15], prompting us[9] to examine critically how h_{00} is obtained under $r \gg R$ and/or $\exp(r/R) \ll 1$, as well as how the trapped surface in $4-d$ was derived by Eardley and Giddings[16].

Acknowledgment

I would like to thank H. Nastase for the on-going collaborations and enlightening discussions. This talk is the improved version of the earlier presentations which can be found in part in Ref. [17]. This work is supported in part by the U.S. D.o.E. Contract DE-FG-02-91ER40688-Task A, with the Report number Brown-HET-1403,

References

1. J. R. Cudell et al, *Phys. Rev. Lett.* **89**, 201801 (2002), hep-ph/0206172; J. R. Cudell et al, *Phys. Rev.* **D65**, 07024 (2002), hep-ph/ 0107219; See also the COMPETE webpage http://nuclth02.phys.ulg.be/compete.
2. W. Heisenberg, *Z. Phys.* **133** (1952) 65; See also H. G. Dosch, P. Gauron and B. Nicolescu, *Phys. Rev.* **D67** (2003) 077501.
3. M. Froissar, *Phys. Rev.* **23** (1691) 1053; L. Lukaszuk and A. Martin, *Nuovo Cimento* **A52** (1967) 122.
4. Y. Ya. Pomeranchuk, *JETP* **7** (1958) 499.
5. K. Kang and B. Nicolescu, *Phys. Rev.* **D11** (1975) 2461.
6. J. Maldacena, *Adv. Theor. Math. Phys.* **2** (1998) 231; E. Witten, ibid 253; S. S. Gubser, I. R. Klevanov and A. M. Polyakov, *Phys. Lett.* **B428** (1998) 105.
7. E. Witten, *Adv. Theor. Math. Phys.* **2** (1998) 505.
8. J. Polchinski and M. J. Strassler, *Phys. Rev. Lett.* **88** (2002) 03160.

9. K. Kang and H. Nastase, "High Energy QCD from Planckian Scattering in AdS" (in preparation), and "Planckian Scattering Effects ad Black Hole Production in Low M_{pl} Scenarior", (in preparation).
10. S. B. Giddings, *Phys. Rev.* **D67** (2003) 126001, hep-th/0203004.
11. P. D. D'Eath and P. N. Payne, *Phys. Rev.* **D46** (1992) 658; 675; 694.
12. S. W. Hawkings and G. F. R. Ellis, "The Large Scale Structure of Space-Time", Cambridge University Press, 1973,
13. L. Randall and K. Sundrum, *Phys. Rev. Lett.* **83** (1990 4690. See also L. Randall and K. Sundrum, *Phys. Rev. Lett.* **83** (1999) 3370 for one brane model (RS1).
14. P. C. Aichelburg and R. U. Sexl, Gen. Rel and Gravity 2 (1971) 303.
15. S. B. Giddings, S. Kachru and J. Polchinski, *Phys. Rev.* **D66** (2002) 106006.
16. D. M. Eardley and S. B. Giddings, *Phys. Rev.* **D66** (2002) 044011.
17. K. Kang in the 10th International (Blois03) Conference, Helsinki, Finland, June 2003 (*http : //www.physics.helsinki.fi/ blois_03/program.html*); in "Small x and Diffraction 2003 Workshop", FNAL, IL, September 2003 (http://conferences.fnal.gov/smallx/index.html).

LORENTZ VIOLATION AS A QUANTUM-GRAVITY SIGNATURE*

RALF LEHNERT

CENTRA, Departamento de Física, Universidade do Algarve
8000-117 Faro, Portugal
E-mail: rlehnert@ualg.pt

Many theoretical approaches to quantum gravity predict the breakdown of Lorentz symmetry at Planck energies. Kinematical cosmic-ray studies are a sensitive tool in the search for such effects. This talk discusses the construction of test dispersion relations for such analyses.

1. Introduction

One of the most fundamental problems in present-day physics concerns a quantum theory of gravitation. Such a theory is believed to be associated with the Planck scale M_{Pl}, so that quantum-gravity signatures are likely to be suppressed by one or more powers of M_{Pl} at currently attainable energies. The expected minuscule size of potential effects and the absence of a fully realistic and satisfactory underlying theory make quantum-gravity phenomenology particular challenging. A practical approach to overcome this obstacle is to consider the breaking of symmetries that hold exactly in our current fundamental laws, might be violated in approaches to quantum gravity, and can be tested with ultrahigh precision.

Lorentz- and CPT-symmetry breakdown offers a promising tool in this line of investigation.[1] An effective-field-theory framework for the description of such effects at present experimental energy scales is provided by the Standard-Model Extension (SME).[2,3,4] The SME coefficients parametrizing Lorentz and CPT violation can arise in numerous approaches to quantum gravity including strings,[5] spacetime-foam models,[6,7,8] noncommu-

*Funded in part by the Centro Multidisciplinar de Astrofísica (CENTRA) and by the Fundação para a Ciência e a Tecnologia (Portugal) under grant POCTI/FNU/49529/2002.

tative field theory,[9] cosmologically varying scalars,[10,11] random-dynamics models,[12] multiverses,[13] and brane-world scenarios.[14]

The flat-spacetime limit of the SME has provided the basis for numerous experimental and theoretical investigations of Lorentz and CPT breakdown, many of which place tight constraints on SME parameters. Examples of such analyses include ones involving mesons,[15,16,17,18] baryons,[19,20,21] electrons,[22,23,24] photons,[10,25,26,27,28] and muons.[29] We also remark that neutrino-oscillation experiments offer the potential for discovery.[2,30,31]

The quadratic sector of the SME determines the one-particle dispersion relations, which typically exhibit Lorentz-violating modifications,[25,2,4] as expected. It follows that particle-reaction kinematics is altered relative to the conventional case leading to potentially observable shifts in certain reaction thresholds. At energies approaching the Planck scale, such Lorentz-violating effects might be more pronounced, so that kinematical investigations involving ultrahigh-energy cosmic rays (UHECR) can provide a high-sensitivity laboratory for Lorentz-violation searches.

Many recent quantum-gravity investigations outside the context of the SME have exploited the general idea to constrain deviations from conventional dispersion relations.[32,33] Because of the absence of a complete underlying theory, the form of the considered Lorentz-breaking dispersion-relation modifications appears somewhat arbitrary in some of these studies. This talk addresses the question as to whether at least fundamental physics principles that are expected to remain valid in an underlying theory can yield meaningful restrictions on possible dispersion-relation modifications. We primarily consider the principle of coordinate independence and the requirement of compatibility with an effective dynamical framework. We explain why these two conditions should continue to hold in the presence of Lorentz violation regardless of the details of the Planck-scale theory and discuss the ensuing constraints on dispersion relations. Throughout we assume translational invariance and the associated energy-momentum conservation.

The outline of this talk is as follows. Section 2 discusses the need for coordinate independence and its consequences. In Sec. 3, we comment on the compatibility of the modified dispersion relations with dynamics. Some subtle issues regarding the applicability of certain conventional approximations in the Lorentz-violating context are addressed in Sec. 4.

2. Coordinate independence

Although coordinate systems are one of the most important mathematical tools in physics, they do not possess physical reality: the description of a process must remain independent of the choice of coordinates. This allows, for example, different observers, each equipped with a different reference frame, to relate their observations or predictions concerning a given physical system. The principle of coordinate independence is therefore also called observer invariance. Mathematically, observer invariance is usually implemented by choosing a spacetime-manifold description for physical events, tensors or spinors for the representation of observables, and covariant equations as the laws of physics.

It is a common misconception that Lorentz breaking implies the loss of coordinate independence. Such a model would inhibit meaningful physical predictions, so that it is necessary to adhere to observer invariance also in the presence of Lorentz violation. This is by no means a contradiction. Consider, for instance, the conventional motion of a classical point particle of mass m and charge q in an external electromagnetic field $F^{\mu\nu}$ described by

$$m\,\frac{dv^{\mu}}{d\tau} = qF^{\mu\nu}v_{\nu}\ . \tag{1}$$

Here, τ denotes the proper time and v^{μ} the four-velocity of the particle. Note that invariance under rotations of the particle's trajectory is broken by the external nondynamical $F^{\mu\nu}$ resulting in the nonconservation of the charge's angular momentum. However, Eq. (1) is a tensor equation valid in all coordinate systems maintaining coordinate independence. This example also illustrates that transformations of localized dynamical particles and fields (with nondynamical global backgrounds held fixed) must be clearly distinguished from transformations of the coordinates. The former, also called particle transformations, are no longer associated with a symmetry of the model.

In what follows, we consider models that exhibit particle Lorentz violation while maintaining coordinate independence. In many respects, the physical situation is conceptually similar to the external-field case discussed in the previous paragraph. However, in the above example the $F^{\mu\nu}$ background is a local electromagnetic field generated by other charge and current distributions that can in principle be controlled. In the present Lorentz-violating context, such a background is a global property of the effective vacuum outside of experimental control.

We continue our study in a local Minkowski frame. Lorentz-violating dispersion relations are usually taken to be of the form

$$E_0{}^2 - \vec{p}^{\,2} = m^2 + \delta f(E, \vec{p}) \,, \tag{2}$$

where $p^\mu = (E, \vec{p})$ and m are the particle's respective four-momentum and mass, and $\delta f(E, \vec{p})$ describes the Lorentz violation. With our above consideration, the correction δf needs to be observer Lorentz invariant, i.e., it must transform as a scalar under coordinate changes. To obtain a general form of δf, we impose some further mild conditions. First, for small momenta[a] $|\vec{p}| \ll M_{Pl}$ we want to recover the usual dispersion relation with $\delta f = 0$. Second, we want to avoid potential nonlocalities that could arise through the presence of nonpolynomial functions. This yields the ansatz

$$\delta f(E, \vec{p}) = \sum_{n \geq 1} T_{(n)}^{\overbrace{\alpha\beta\,\cdots}^{n \text{ indices}}} \underbrace{p_\alpha p_\beta \,\cdots}_{n \text{ factors}} \,. \tag{3}$$

Here $T_{(n)}^{\alpha\beta\,\cdots}$ is a constant tensor of rank n describing particle Lorentz violation. All the tensor indices $\alpha, \beta \ldots$ are distinct, and each one is properly contracted with a momentum factor, so that each term in the sum is coordinate independent.

This ansatz has various consequences,[33] two of which we mention next: when viewed as a polynomial in E, the modified dispersion relation (2) will in general lift the usual degeneracy between particle, antiparticle, and possible spin-type states. Another implication concerns cases with imposed rotational invariance. Then, the general ansatz (3) does not contain odd powers of $|\vec{p}|$.[33]

3. Effective Field Theory

Kinematical considerations can impose powerful restrictions on particle reactions. However, dispersion-relation constraints can be masked by other effects. For example, a high-energy reaction expected to be suppressed by a modified dispersion relation, could also be prevented by a novel symmetry. Similarly, the presence at high energies of a reaction kinematically forbidden at low energies might perhaps be explained by additional channels due

[a] Here, $\vec{p}$ refers to the components in any frame in which the Earth moves nonrelativistically.

to new undetected particles or the loss of low-energy symmetries. In addition, models of both acceleration mechanisms for UHECRs and atmospheric shower development involve conventional dynamics.

We see that the implementation of general dynamical features both appears necessary for a complete description of UHECR physics and can increase the scope of cosmic-ray analyses. At the same time, it may introduce a certain degree of framework dependence. However, implementing dynamics is tightly constrained by the requirement that known physics must be recovered in certain limits. In what follows, we argue that effective field theory (EFT) is a sensible and general approach for such efforts.

EFTs have proved to be extremely successful in describing diverse physical systems at atomic, nuclear, and elementary-particle scales. Note in particular, that EFT is flexible enough for applications involving discrete condensed-matter backgrounds, situations analogous to those in some quantum-gravity approaches. Moreover, the usual Standard Model itself is normally viewed as an EFT approximating more fundamental physics.

The construction of a suitable EFT can employ a philosophy paralleling that of the dispersion-relation approach in the previous section with the more powerful idea of proceeding at the Lagrangian level: one adds Lorentz-breaking terms $\delta\mathcal{L}$ to the usual Standard-Model Lagrangian $\mathcal{L}_{\mathrm{SM}}$

$$\mathcal{L}_{\mathrm{SME}} = \mathcal{L}_{\mathrm{SM}} + \delta\mathcal{L} \,, \tag{4}$$

where $\mathcal{L}_{\mathrm{SME}}$ denotes the EFT Lagrangian. The correction $\delta\mathcal{L}$ is formed by contracting Standard-Model field operators of unrestricted dimensionality with Lorentz-breaking tensorial parameters (analogous to the $T_{(n)}^{\alpha\beta\,\cdots}$ in Eq. (3)) yielding observer Lorentz scalars. This general EFT for Lorentz violation is the SME mentioned in the introduction. Its apparent generality makes it difficult and perhaps even impossible to find some other effective theory for Lorentz breaking containing the Standard Model with dynamics significantly different from the SME.

4. Additional considerations

In this section, we point out that care is required when approximating Lorentz-violating dispersion relations. This is best demonstrated by a specific example.

Consider the modified dispersion relation

$$E^2 - \vec{p}^{\,2} = m^2 + \frac{E\vec{p}^{\,2}}{M} \,, \tag{5}$$

where M is some high-energy scale such M_{Pl}. The usual quantum-field reinterpretation of the negative-energy solutions yields the respective exact particle and antiparticle energies E_+ and E_-:

$$E_{\pm}(\vec{p}) = \sqrt{\frac{\vec{p}^{\,4}}{4M^2} + \vec{p}^{\,2} + m^2} \pm \frac{\vec{p}^{\,2}}{2M} \,. \tag{6}$$

If one instead employs the ultrarelativistic approximation $E \simeq |\vec{p}|$ in the correction term, as is sometimes done in such kinematical studies, one obtains $\delta f \simeq |\vec{p}|^3/M$. After the reinterpretation, we now have

$$E_{\pm}(\vec{p}) = \sqrt{\frac{|\vec{p}|^{\,3}}{M} + \vec{p}^{\,2} + m^2} \,. \tag{7}$$

Next, we look at photon decay into an electron-positron pair[b] $\gamma \rightarrow e^+ + e^-$ assuming a dispersion relation of the type (5) for both the leptons and the photon. For each particle, we take $M = -M_{Pl}$, and for the photon we set $m = 0$. Using the exact expression for the particle energies (6), one has to consider two distinct incoming photon states γ_+ and γ_-, where the subscripts correspond to those of the particle energy in Eq. (6). One can then show[33] that the decay $\gamma_- \rightarrow e^+ + e^-$ is allowed above a certain threshold. If the observed value $m = 0.511$ MeV for the electron and positron masses is used, the numerically determined threshold value for the incoming photon three-momentum is $|\vec{p}_{\rm min}| \simeq 7.21$ TeV. If one instead employs the approximate particle energies (7), photon decay is forbidden throughout the validity range of Eq. (7).

Considering the TeV threshold scale for this decay and the Planck suppression of the correction δf, the ultrarelativistic approximation is indeed excellent if one is interested in the particle energy only. However, threshold analyses are based on exact energy-momentum conservation and can thus be sensitive to the slightest deviations. Even in a conventional photon decay, the ultrarelativistic approximation renders the lepton momenta lightlike, which seemingly permits the decay in forward direction. In the present case, $E \simeq |\vec{p}|$ introduces an additional degeneracy into the problem. As a result, the approximate solution is spacelike, whereas the exact expression (6) determines both a timelike and a spacelike branch. It is the presence of the timelike momenta that permits the decay.

[b]This process is kinematically forbidden in conventional physics

References

1. For an overview see, e.g., *CPT and Lorentz Symmetry II*, edited by V.A. Kostelecký (World Scientific, Singapore, 2002).
2. D. Colladay and V.A. Kostelecký, Phys. Rev. D **55**, 6760 (1997); **58**, 116002 (1998).
3. V.A. Kostelecký, Phys. Rev. D, in press (hep-th/0312310).
4. V.A. Kostelecký and R. Lehnert, Phys. Rev. D **63**, 065008 (2001); R. Lehnert, hep-ph/0401084.
5. V.A. Kostelecký and S. Samuel, Phys. Rev. D **39**, 683 (1989); Phys. Rev. Lett. **66**, 1811 (1991); V.A. Kostelecký and R. Potting, Nucl. Phys. B **359**, 545 (1991); Phys. Lett. B **381**, 89 (1996); Phys. Rev. D **63**, 046007 (2001); V.A. Kostelecký, M. Perry, and R. Potting, Phys. Rev. Lett. **84**, 4541 (2000).
6. R. Gambini and J. Pullin, in Ref. 1; J. Alfaro *et al.*, Phys. Rev. D **66**, 124006 (2002); D. Sudarsky *et al.*, Phys. Rev. Lett. **89**, 231301 (2002); Phys. Rev. D **68**, 024010 (2003); G. Amelino-Camelia, Mod. Phys. Lett. **A18**, 1073 (2003); R. Myers and M. Pospelov, Phys. Rev. Lett. **90**, 211601 (2003); N.E. Mavromatos, Nucl. Instrum. Meth. B **214**, 1 (2004).
7. See, however, C.N. Kozameh and M.F. Parisi, gr-qc/0310014.
8. C. Adam and F.R. Klinkhamer, Nucl. Phys. B **607**, 247 (2001); F.R. Klinkhamer and C. Rupp, hep-th/0312032; F.R. Klinkhamer and G.E. Volovik, hep-th/0403037.
9. S.M. Carroll *et al.*, Phys. Rev. Lett. **87**, 141601 (2001); Z. Guralnik *et al.*, Phys. Lett. B **517**, 450 (2001); C.E. Carlson, C.D. Carone, and R.F. Lebed, Phys. Lett. B **518**, 201 (2001); A. Anisimov *et al.*, Phys. Rev. D **65**, 085032 (2002); I. Mocioiu, M. Pospelov, and R. Roiban, Phys. Rev. D **65**, 107702 (2002); J.L. Hewett, F.J. Petriello, and T.G. Rizzo, Phys. Rev. D **66**, 036001 (2002).
10. V.A. Kostelecký, R. Lehnert, and M.J. Perry, Phys. Rev. D **68**, 123511 (2003); O. Bertolami, R. Lehnert, R. Potting and A. Ribeiro, Phys. Rev. D **69**, 083513 (2004); R. Lehnert, hep-ph/0312093.
11. N. Arkani-Hamed *et al.*, hep-th/0312099.
12. C.D. Froggatt and H.B. Nielsen, hep-ph/0211106.
13. J.D. Bjorken, Phys. Rev. D **67**, 043508 (2003).
14. C.P. Burgess *et al.*, JHEP **0203**, 043 (2002); A.R. Frey, JHEP **0304**, 012 (2003); J. Cline and L. Valcárcel, hep-ph/0312245.
15. KTeV Collaboration, H. Nguyen, in Ref. 1; OPAL Collaboration, R. Ackerstaff *et al.*, Z. Phys. C **76**, 401 (1997); DELPHI Collaboration, M. Feindt *et al.*, preprint DELPHI 97-98 CONF 80 (1997); BELLE Collaboration, K. Abe *et al.*, Phys. Rev. Lett. **86**, 3228 (2001); BaBar Collaboration, B. Aubert *et al.*, hep-ex/0303043; FOCUS Collaboration, J.M. Link *et al.*, Phys. Lett. B **556**, 7 (2003).
16. V.A. Kostelecký and R. Potting, Phys. Rev. D **51**, 3923 (1995).
17. D. Colladay and V.A. Kostelecký, Phys. Lett. B **344**, 259 (1995); Phys. Rev. D **52**, 6224 (1995); Phys. Lett. B **511**, 209 (2001); V.A. Kostelecký and R. Van Kooten, Phys. Rev. D **54**, 5585 (1996); O. Bertolami *et al.*, Phys. Lett.

B **395**, 178 (1997); N. Isgur *et al.*, Phys. Lett. B **515**, 333 (2001).

18. V.A. Kostelecký, Phys. Rev. Lett. **80**, 1818 (1998); Phys. Rev. D **61**, 016002 (2000); Phys. Rev. D **64**, 076001 (2001).
19. L.R. Hunter *et al.*, in *CPT and Lorentz Symmetry*, edited by V.A. Kostelecký (World Scientific, Singapore, 1999). D. Bear *et al.*, Phys. Rev. Lett. **85**, 5038 (2000); D.F. Phillips *et al.*, Phys. Rev. D **63**, 111101 (2001); M.A. Humphrey *et al.*, Phys. Rev. A **68**, 063807 (2003); Phys. Rev. A **62**, 063405 (2000); V.A. Kostelecký and C.D. Lane, Phys. Rev. D **60**, 116010 (1999); J. Math. Phys. **40**, 6245 (1999).
20. R. Bluhm *et al.*, Phys. Rev. Lett. **88**, 090801 (2002); Phys. Rev. D **68**, 125008 (2003).
21. F. Canè *et al.*, physics/0309070.
22. H. Dehmelt *et al.*, Phys. Rev. Lett. **83**, 4694 (1999); R. Mittleman *et al.*, Phys. Rev. Lett. **83**, 2116 (1999); G. Gabrielse *et al.*, Phys. Rev. Lett. **82**, 3198 (1999); R. Bluhm *et al.*, Phys. Rev. Lett. **82**, 2254 (1999); Phys. Rev. Lett. **79**, 1432 (1997); Phys. Rev. D **57**, 3932 (1998).
23. B. Heckel, in Ref. 1; L.-S. Hou, W.-T. Ni, and Y.-C.M. Li, Phys. Rev. Lett. **90**, 201101 (2003); R. Bluhm and V.A. Kostelecký, Phys. Rev. Lett. **84**, 1381 (2000).
24. H. Müller *et al.*, Phys. Rev. D **68**, 116006 (2003).
25. S.M. Carroll, G.B. Field, and R. Jackiw, Phys. Rev. D **41**, 1231 (1990); M.P. Haugan and T.F. Kauffmann, Phys. Rev. D **52**, 3168 (1995); V.A. Kostelecký and M. Mewes, Phys. Rev. Lett. **87**, 251304 (2001).
26. R. Jackiw and V.A. Kostelecký, Phys. Rev. Lett. **82**, 3572 (1999); H. Müller *et al.*, Phys. Rev. D **67**, 056006 (2003); V.A. Kostelecký and A.G.M. Pickering, Phys. Rev. Lett. **91**, 031801 (2003); G.M. Shore, Contemp. Phys. **44**, 503 (2003); B. Altschul, hep-th/0311200; D. Colladay and P. McDonald, hep-th/0312058; R. Lehnert and R. Potting, to appaer.
27. J. Lipa *et al.*, Phys. Rev. Lett. **90**, 060403 (2003); H. Müller *et al.*, Phys. Rev. Lett. **91**, 020401 (2003); P. Wolf *et al.*, gr-qc/0401017.
28. V.A. Kostelecký and M. Mewes, Phys. Rev. D **66**, 056005 (2002).
29. V.W. Hughes *et al.*, Phys. Rev. Lett. **87**, 111804 (2001); R. Bluhm *et al.*, Phys. Rev. Lett. **84**, 1098 (2000).
30. S. Coleman and S.L. Glashow, Phys. Rev. D **59**, 116008 (1999); V. Barger *et al.*, Phys. Rev. Lett. **85**, 5055 (2000); J.N. Bahcall, V. Barger, and D. Marfatia, Phys. Lett. B **534**, 114 (2002); V.A. Kostelecký and M. Mewes, hep-ph/0308300; S. Choubey and S.F. King, hep-ph/0311326.
31. V.A. Kostelecký and M. Mewes, Phys. Rev. D **69**, 016005 (2004).
32. See, e.g., W. Kluźniak, Astropart. Phys. **11**, 117 (1999); astro-ph/9905308; T. Kifune, Astrophys. J. **518**, L21 (1999); O. Bertolami and C.S. Carvalho, Phys. Rev. D **61**, 103002 (2000); T. Jacobson, S. Liberati, and D. Mattingly, Phys. Rev. D **66**, 081302 (2002); T.J. Konopka and S.A. Major, New J. Phys. **4**, 57 (2002); H. Vankov and T. Stanev, Phys. Lett. **B** 538, 251 (2002); M. Jankiewicz *et al.*, hep-ph/0312221; O. Gagnon and G.D. Moore, hep-ph/0404196.
33. R. Lehnert, Phys. Rev. D **68**, 085003 (2003); hep-ph/0402133.

GENERALIZED HOLONOMY IN M-THEORY *

J. T. LIU
Michigan Center for Theoretical Physics
Randall Laboratory, Department of Physics, University of Michigan
Ann Arbor, MI 48109–1120, USA
E-mail: jimliu@umich.edu

This is Generalized holonomy.

1. Introduction

While much recent progress has been made in understanding the nature of M-theory, must still remains rather mysterious. In particular, despite many conjectures, it is still not obvious what the fundamental degrees of freedom truly are that describe M-theory. Absent such a complete description, one may hope to at least discern important aspects of the theory indirectly through its symmetries. In this sense, it has of course long been expected that M-theory has a large and as yet unexplored set of hidden symmetries beyond the obvious ones such as Poincare in a flat background. As part of an exploration of hidden symmetries, we have introduced the notion of generalized holonomy as a tool for classification of all supersymmetric vacua of M-theory[1,2,3]. Here I present a brief introduction to generalized holonomy and give some elementary examples.

In principle, one would like to work directly in M-theory. However, without a better understanding, in practice we focus on eleven-dimensional supergravity as a particularly interesting corner of M-theory. The supergravity theory contains a metric g_{MN}, a three-form potential $C_{(3)}$ with four-form field strength $F_{(4)} = dA_{(3)}$ and the gravitino Ψ_M. The eleven-dimensional supergravity contains 32 real supercharges. Hence the number n of supersymmetries preserved by a particular vacuum must be an integer

*This work is supported in part by the us department of energy under grant de-fg02-95er40899

$0 \leq n \leq 32$. Our goals are thus to: 1) determine what values of n are allowed for supergravity vacua, and 2) for the allowed values of n, classify (or construct) the corresponding vacua.

2. Holonomy and Generalized Holonomy

Counting the number of preserved supersymmetries, n, is equivalent to counting the Killing spinors ϵ satisfying

$$\mathcal{D}_M \epsilon = 0, \tag{2.1}$$

where the generalized covariant derivative is given by

$$\mathcal{D}_M = D_M - \tfrac{1}{288}(\Gamma_M{}^{NPQR} - 8\delta_M^N \Gamma^{PQR})F_{NPQR}. \tag{2.2}$$

In vacua with vanishing $F_{(4)}$, the Killing spinors satisfy the integrability condition

$$[D_M, D_N]\epsilon = \tfrac{1}{4}R_{MN}{}^{AB}\Gamma_{AB}\epsilon = 0, \tag{2.3}$$

The subgroup of Spin(10, 1) generated by this linear combination of Spin(10, 1) generators Γ_{AB} corresponds to the Riemannian holonomy group H of the spin connection ω_M. The number of supersymmetries, n, is then given by the number of singlets appearing in the decomposition of the 32 of Spin(10, 1) under H. This is of course the well-known connection between holonomy and supersymmetry.

In Euclidean signature, the holonomy groups have been classified [4]. In Lorentzian signature, much less is known but the question of which subgroups H of Spin(10, 1) leave a spinor invariant has been answered [5]. On the other hand, vacua with non-vanishing $F_{(4)}$ have not been addressed using the language of holonomy, at least until now.

For non-vanishing $F_{(4)}$, the supercovariant derivative has the form of a derivative with a generalized connection

$$\mathcal{D}_M = \partial_M + \tfrac{1}{4}\Omega_M, \tag{2.4}$$

where

$$\Omega_M = \omega_M^{AB}\Gamma_{AB} - \tfrac{1}{72}(\Gamma_M{}^{NPQR} - 8\delta_M^N \Gamma^{PQR})F_{NPQR}. \tag{2.5}$$

This generalized connection takes its values in the even part of the eleven-dimensional Clifford algebra, namely SL(32, $\mathbb{R}$). Note, however, that some generators are missing from the covariant derivative, as not all combinations of the Dirac matrices arise.

Integrability of the supercovariant derivative gives rise to a condition

$$[\mathcal{D}_M, \mathcal{D}_N]\epsilon = \tfrac{1}{4}\mathcal{R}_{MN}\epsilon = 0, \tag{2.6}$$

where

$$\mathcal{R}_{MN} = 2(\partial_{[M}\Omega_{N]} + \Omega_{[M}\Omega_{N]}). \tag{2.7}$$

As put forth in [1], the number of preserved supersymmetries is now given by the number of singlets appearing in the decomposition of the 32 of $\mathcal{G}$ under $\mathcal{H} \subset \mathcal{G}$ where $\mathcal{G}$ is the generalized structure group and $\mathcal{H}$ is the generalized holonomy group that $\mathcal{R}_{MN}$ takes its values in. Discussions of generalized holonomy may also be found in [6,7].

In subsequent papers by Hull [8] and Papadopoulos and Tsimpis [9] it was shown that $\mathcal{G}$ may be as large as the Clifford algebra, $\mathrm{SL}(32,\mathbb{R})$. As a result, a vacuum of eleven-dimensional supergravity admits precisely n Killing spinors iff

$$\mathrm{SL}(31-n,\mathbb{R}) \ltimes (n+1)\mathbb{R}^{(31-n)} \not\supseteq \mathcal{H} \subseteq \mathrm{SL}(32-n,\mathbb{R}) \ltimes n\mathbb{R}^{(32-n)}, \tag{2.8}$$

i.e. the generalized holonomy is contained in $\mathrm{SL}(32-n,\mathbb{R}) \ltimes n\mathbb{R}^{(32-n)}$ but is not contained in $\mathrm{SL}(31-n,\mathbb{R}) \ltimes (n+1)\mathbb{R}^{(31-n)}$.

3. Generalized holonomy for $n = 16$ solutions

We now turn to a generalized holonomy analysis of some basic supergravity solutions. Starting with the maximally supersymmetric backgrounds ($n = 32$), namely $\mathbb{E}^{1,10}$, $\mathrm{AdS}_7 \times S^4$, $\mathrm{AdS}_4 \times S^7$ and Hpp, it should be clear that they all have trivial generalized holonomy, in accord with (2.8). However, only flat space may be described by (trivial) Riemannian holonomy.

Somewhat more interesting to consider are the four basic objects of M-theory preserving half of the supersymmetries (corresponding to $n = 16$). These are the M5-brane, M2-brane, M-wave (MW) and the Kaluza-Klein monopole (MK). The latter two have $F_{(4)} = 0$ and may be categorized using ordinary Riemannian holonomy, with $H \subset \mathrm{SO}(10,1)$. We now look at these in turn.

3.1. *The M5-brane*

The familiar supergravity M5-brane solution [10] may be written in isotropic coordinates as

$$\begin{aligned} ds^2 &= H_5^{-1/3}dx_\mu^2 + H_5^{2/3}d\vec{y}^{\,2}, \\ F_{ijkl} &= \epsilon_{ijklm}\partial_m H_5, \end{aligned} \tag{3.1}$$

where $H_5(\vec{y})$ is harmonic in the six-dimensional transverse space spanned by $\{y^i\}$, and $\epsilon_{ijklm} = \pm 1$. While the transverse space only needs to be Ricci flat, we take it to be $\mathbb{E}^5$, so as not to further break the supersymmetry.

A simple computation of the generalized covariant derivative on this background yields

$$\begin{aligned} \mathcal{D}_\mu &= \partial_\mu - \tfrac{1}{6}\Gamma_{\bar{\mu}}{}^{\bar{i}} P_5^+ H^{-3/2}\partial_i H, \\ \mathcal{D}_i &= \partial_i + \tfrac{1}{3}\Gamma_{\bar{i}}{}^{\bar{j}} P_5^+ \partial_j \ln H - \tfrac{1}{2}\Gamma^{(5)}\partial_i \ln H. \end{aligned} \tag{3.2}$$

Here, $P_5^\pm = \frac{1}{2}(1 \pm \Gamma^{(5)})$ is the standard 1/2-BPS projection for the M5-brane, where $\Gamma^{(5)} = \frac{1}{5!}\epsilon_{ijklm}\Gamma^{\bar{i}\bar{j}\bar{k}\bar{l}\bar{m}}$. All quantities with bars indicate tangent space indices. To obtain the generalized holonomy of the M5-brane, we compute the commutator of supercovariant derivatives, (2.6). The result is

$$\begin{aligned} \mathcal{R}_{\mu\nu} &= 0, \\ \mathcal{R}_{ij} &= -\tfrac{8}{9}\Gamma_{\bar{i}\bar{j}} P_5^+ (\partial_k \ln H)^2 + \tfrac{8}{3}\Gamma_{[\bar{j}}{}^{\bar{k}} P_5^+ (\partial_i \partial^k \ln H - \tfrac{8}{3}\partial_{i]} \ln H \partial^k \ln H), \\ \mathcal{R}_{\mu i} &= H^{-1/2}[\tfrac{2}{3}\Gamma_{\bar{\mu}\bar{j}} P_5^+ (\partial_i \partial^j \ln H - \tfrac{2}{3}\partial_i \ln H \partial_j \ln H) + \tfrac{2}{9}\Gamma_{\bar{\mu}\bar{i}} P_5^+ (\partial_j \ln H)^2]. \end{aligned} \tag{3.3}$$

Note that the purely longitudinal holonomy is trivial, as the M5-brane is flat along the longitudinal directions.

For the transverse holonomy, the only combination of Dirac matrices entering $\mathcal{R}_{ij}$ is given by $\Gamma_{\bar{i}\bar{j}} P_5^+$. Defining a set of Hermitian generators $T_{ij} = -\frac{i}{2}\Gamma_{\bar{i}\bar{j}} P_5^+$, it is easily seen that they generate the usual SO(5) algebra. As a result, the transverse holonomy is simply $SO(5)_+$, where the + refers to the sign of the M5-projection. The mixed commutator $\mathcal{R}_{\mu i}$, on the other hand, introduces an additional set of Dirac matrices, $K_{\mu i} = \Gamma_{\bar{\mu}\bar{i}} P_5^+$. Since $\Gamma_{\bar{\mu}} P_5^+ = P_5^- \Gamma_{\bar{\mu}}$, it is clear that the $K_{\mu i}$ generators commute among themselves. On the other hand, commuting $K_{\mu i}$ with the $SO(5)_+$ generators T_{ij} yield the additional combinations $K_\mu = \Gamma_{\bar{\mu}} P_5^+$ and $K_{\mu ij} = \Gamma_{\bar{\mu}\bar{i}\bar{j}} P_5^+$. Picking a set of Cartan generators T_{12} and T_{34} for $SO(5)_+$, we may see that the complete set $\{K_\mu, K_{\mu i}, K_{\mu ij}\}$ has weights $\pm 1/2$. As a result, they transform as a set of 4-dimensional spinor representations of $SO(5)_+$. We conclude that the generalized holonomy of the M5-brane is

$$\mathcal{H}_{\mathrm{M5}} = SO(5)_+ \ltimes 6\mathbb{R}^{4(4)}. \tag{3.4}$$

3.2. *The M2-brane*

Turning next to the M2-brane, the analysis of its generalized holonomy is rather similar to that of the M5-brane. We start with the supergravity solution, which may be written as [6]

$$
\begin{aligned}
ds^2 &= H_2^{-2/3} dx_\mu^2 + H_2^{1/3} d\vec{y}^{\,2}, \\
F_{\mu\nu\rho i} &= \epsilon_{\mu\nu\rho} \partial_i (1/H_2).
\end{aligned}
\tag{3.5}
$$

A similar examination of the commutator of supercovariant derivatives for this solution indicates the presence of both compact generators $T_{ij} = -\frac{i}{2}\Gamma_{\bar{i}\bar{j}} P_2^+$ and non-compact ones $K_{\mu i} = \Gamma_{\bar{\mu}\bar{i}} P_2^+$. Here, $P_2^\pm = \frac{1}{2}(1 \pm \Gamma^{(2)})$ where $\Gamma^{(2)} = \frac{1}{3!}\epsilon_{\mu\nu\rho}\Gamma^{\bar{\mu}\bar{\nu}\bar{\rho}}$ is the M2-brane projection. Furthermore, the coordinates on (3.5) correspond to a 3/8 longitudinal/transverse split. Hence the transverse holonomy in this case is $\mathrm{SO}(8)_+$.

To obtain the generalized holonomy group $\mathcal{H}_{M2}$, we must first close the algebra formed by T_{ij} and $K_{\mu i}$. Upon doing so, we find the additional generators $K_{\mu ijk} = \Gamma_{\bar{\mu}\bar{i}\bar{j}\bar{k}} P_2^+$. As in the M5 case, we may see that the set $\{K_{\mu i}, K_{\mu ijk}\}$ form eight-dimensional representations of $\mathrm{SO}(8)_+$. However, some care must be taken in identifying these representations as the 8_v, 8_s or 8_c (up to an overall automorphism due to triality). The resulting M2 generalized holonomy is given by

$$
\mathcal{H}_{\mathrm{M2}} = \mathrm{SO}(8)_+ \ltimes 12\mathbb{R}^{2(8_s)}.
\tag{3.6}
$$

Additional details on this identification may be found in [2].

3.3. *The M-wave*

We now turn to the pure geometry solutions. The wave (MW) is given by [11]

$$
ds^2 = 2\, dx^+\, dx^- + K\, dx^{+\,2} + d\vec{y}^{\,2},
\tag{3.7}
$$

where $K(\vec{y})$ is harmonic on the nine-dimensional Euclidean transverse space $\mathbb{E}^9$. In a vielbein basis $e^+ = dx^+$, $e^- = dx^- + \frac{1}{2}K\, dx^+$, $e^i = dy^i$, the only non-vanishing component of the spin connection is given by $\omega^{+i} = \frac{1}{2}\partial_i K\, e^+$. Thus the gravitational covariant derivative acting on ϵ is given by

$$
D_+ = \partial_+ + \tfrac{1}{4}\partial_i K \Gamma_- \Gamma_i, \qquad D_- = \partial_-, \qquad D_i = \partial_i.
\tag{3.8}
$$

Note that the metric is given by $ds^2 = 2e^+e^- + e^ie^i$, so that light cone indices are raised and lowered as, *e.g.*, $\Gamma_- = \Gamma^+$ in tangent space.

The only non-vanishing commutator of covariant derivatives is given by

$$
M_{+i} = -\tfrac{1}{4}\partial_i\partial_j K \Gamma_- \Gamma_i,
\tag{3.9}
$$

so we may identify the generalized holonomy generators as $T^i = \Gamma_-\Gamma_i$. Since $\Gamma_-^2 = 0$, these nine generators are mutually commuting, and the MW generalized holonomy is

$$\mathcal{H}_{\rm MW} = \mathbb{R}^9. \tag{3.10}$$

In addition to being a subgroup of $SL(16,\mathbb{R}) \ltimes 16\mathbb{R}^{16}$, this may also be viewed as a subgroup of ISO(9) appropriate to backgrounds with a null Killing vector.

3.4. *The M-monopole*

The final basic M-theory object we consider is the Kaluza-Klein monopole, which is given by the Euclidean Taub-NUT solution [12]

$$ds^2 = dx_\mu^2 + H(dr^2 + r^2 d\Omega_2^2) + H^{-1}(dz - q\cos\theta\, d\phi)^2, \tag{3.11}$$

where $d\Omega_2^2 = d\theta^2 + \sin^2\theta\, d\phi^2$ and $H = 1 + q/r$. As is well known, this space is Ricci flat and hyper-Kähler, and so has $Sp(1) \simeq SU(2)$ holonomy. Since this solution does not involve $F_{(4)}$, its generalized holonomy is similarly SU(2)

$$\mathcal{H}_{\rm MK} = SU(2). \tag{3.12}$$

4. Some $n = 8$ examples

We now turn to intersecting configurations preserving fewer supersymmetries [13,14,15]. While large classes of intersecting brane solutions and configurations involving to branes at angles have been constructed, we will only examine some of the simple cases of orthogonal intersections yielding $n = 8$.

4.1. *Branes with a KK-monopole*

It is well known that the basic supergravity p-brane solutions are not restricted to having only flat Euclidean transverse spaces. In particular, the M5 and M2 solutions of (3.1) and (3.5) demand only that the transverse space spanned by $\{\vec{y}\}$ is Ricci flat. A simple possibility, thus, is to replace $\mathbb{E}^4$ with a Taub-NUT configuration in four of the transverse directions to the brane. For the M5 case, the resulting M5/MK solution has the form [16]

$$ds^2 = H_5^{-1/3} dx_\mu^2 + H_5^{2/3}[dy^2 + H_6(dr^2 + r^2 d\Omega_2^2) + H_6^{-1}(dz - q_6\cos\theta\, d\phi)^2]. \tag{4.1}$$

Here, the M5-brane is delocalized along the y direction, so the harmonic functions have the form $H_5 = 1 + q_5/r$ and $H_6 = 1 + q_6/r$. This represents the lifting of a NS5/D6 configuration to eleven dimensions.

Noting that four of the five transverse directions is replaced by a Taub-NUT space, the corresponding Riemannian holonomy is contained in the SO(5) tangent space group in the sense of SU(2) ⊂ SO(4) ⊂ SO(5). The embedding of the self-dual connection in SO(4) leads to explicit SU(2) generators $T_{ab}^{(\mathrm{MK})} = -\frac{i}{2}\Gamma_{\bar{a}\bar{b}}P_K^+$ where $P_K^{\pm} = \frac{1}{2}(1 \pm \Gamma^{1234})$ and $a, b, \ldots = 1, \ldots, 4$. On the other hand, as shown in section 3.1, the $SO(5)_+$ generalized holonomy of M5 in the transverse directions involve the P_5^+ projection, and is generated by $T_{ij}^{(\mathrm{M5})} = -\frac{i}{2}\Gamma_{\bar{i}\bar{j}}P_5^+$, where $i, j, \ldots = 1, \ldots 5$. As a result, the transverse holonomy of this M5/MK configuration arises as the closure of $T_{ij}^{(\mathrm{M5})}$ and $T_{ab}^{(\mathrm{MK})}$.

It is straightforward to see that the combination of $T_{ij}^{(\mathrm{M5})}$ and $T_{ab}^{(\mathrm{MK})}$ gives rise to precisely $\mathrm{SO}(5)_+ \times \mathrm{SU}(2)_-$ where ± refers to the embedding inside the $\hat{D}$ structure group $\mathrm{SO}(5)_+ \times \mathrm{SO}(5)_-$ for a 6/5 split. The additional M5 mixed commutator generators $\{K_\mu, K_{\mu\, i}, K_{\mu\, ij}\}$ now transform under both $\mathrm{SO}(5)_+$ and $\mathrm{SU}(2)_-$. Working out the weights of these generators under $\mathrm{SO}(5)_+ \times \mathrm{SU}(2)_-$ demonstrates that the generalized holonomy of this M5/MK configuration is

$$\mathcal{H}_{\mathrm{M5/MK}} = [\mathrm{SO}(5)_+ \times \mathrm{SU}(2)_-] \ltimes 6\mathbb{R}^{2(4,1)+(4,2)}. \tag{4.2}$$

For the M2-brane, the eight-dimensional transverse space may be given a hyper-Kähler metric [16], which is generically of holonomy Sp(2). However, we only consider the product of two independent Taub-NUT spaces, with holonomy Sp(1) × Sp(1). Provided both are oriented properly with the M2, this yields a single additional halving of the supersymmetries, leading to $n = 8$. The transverse holonomy of this solution corresponds to the embedding SO(8) × SU(2) × SU(2) ⊂ SO(8) × SO(4) × SO(4) ⊂ SO(8) × SO(8) ⊂ SO(16), where SO(16) is the $\hat{D}$ structure group corresponding to a 3/8 split. The complete generalized holonomy group is

$$\mathcal{H}_{\mathrm{M2/MK/MK}} = [\mathrm{SO}(8) \times \mathrm{SU}(2) \times \mathrm{SU}(2) \ltimes 3\mathbb{R}^{(8_s,2,2)}] \ltimes 6\mathbb{R}^{2(8_s,1,1)}. \tag{4.3}$$

With only a single Taub-NUT space, the generalized holonomy is instead

$$\mathcal{H}_{\mathrm{M2/MK}} = [\mathrm{SO}(8) \times \mathrm{SU}(2) \ltimes 3\mathbb{R}^{2(8_s,2)}] \ltimes 6\mathbb{R}^{2(8_s,1,1)}. \tag{4.4}$$

4.2. *Branes with a wave*

For solutions with an extended longitudinal space, it is possible to turn on a wave in a null direction along the brane. We consider the M2/MW

Table 1. Some generalized holonomies. For $n = 16$, we have $\mathcal{H} \subseteq \mathrm{SL}(16,\mathbb{R}) \ltimes 16\mathbb{R}^{16}$, while for $n = 8$, it is instead $\mathcal{H} \subseteq \mathrm{SL}(24,\mathbb{R}) \ltimes 8\mathbb{R}^{24}$.

n	Background	Generalized holonomy
32	$\mathbb{E}^{1,10}$, $\mathrm{AdS}_7 \times S^4$ $\mathrm{AdS}_4 \times S^7$, Hpp	$\{1\}$
18,...,26	plane waves	$\mathbb{R}^9$
16	M5	$\mathrm{SO}(5) \ltimes 6\mathbb{R}^{4(4)}$
16	M2	$\mathrm{SO}(8) \ltimes 12\mathbb{R}^{2(8_s)}$
16	MW	$\mathbb{R}^9$
16	MK	$\mathrm{SU}(2)$
8	M5/MK	$[\mathrm{SO}(5) \times \mathrm{SU}(2)] \ltimes 6\mathbb{R}^{2(4,1)+(4,2)}$
8	M2/MK/MK	$[\mathrm{SO}(8) \times \mathrm{SU}(2) \times \mathrm{SU}(2) \ltimes 3\mathbb{R}^{(8_s,2,2)}] \ltimes 6\mathbb{R}^{2(8_s,1,1)}$
8	M2/MK	$[\mathrm{SO}(8) \times \mathrm{SU}(2) \ltimes 3\mathbb{R}^{2(8_s,2)}] \ltimes 6\mathbb{R}^{2(8_s,1,1)}$
8	M2/MW	$[\mathrm{SO}(8) \times \mathrm{SL}(16,\mathbb{R}) \ltimes \mathbb{R}^{(8,16)}] \ltimes 8\mathbb{R}^{(8,1)+(1,16)}$
8	M5/MW	$[\mathrm{SO}(5) \times \mathrm{SU}^*(8) \ltimes 4\mathbb{R}^{(4,8)}] \ltimes 8\mathbb{R}^{2(4,1)+2(1,8)}$
8	MW/MK	$\mathbb{R}^5 \times (\mathrm{SU}(2) \ltimes \mathbb{R}^{2(2)})$
8	M2/M5	$\mathrm{SL}(24,\mathbb{R}) \ltimes 8\mathbb{R}^{24}$

combination, where the supergravity solution is given by [14]

$$ds^2 = H_2^{-2/3}(2dx^+dx^- + K\,dx^{+\,2} + dz^2) + H_2^{1/3}d\vec{y}^{\,2},$$
$$F_{+-zi} = \partial_i \frac{1}{H_2}. \tag{4.5}$$

Here, both K and H_2 are harmonic on the eight-dimensional overall transverse space; the wave is delocalized along the z direction.

Taking commutators of the supercovariant derivatives, the generalized holonomy algebra is formed by the closure of the MW algebra, generated by $T_i^{(MW)} = \Gamma_-\Gamma_{\bar{i}}$, and the M2 algebra, generated by $T_{ij}^{M2} = -\frac{i}{2}\Gamma_{\bar{i}\bar{j}}P_2^+$ and $K_{\mu i}^{M2} = \Gamma_{\bar{\mu}\bar{i}}P_2^+$ where μ denotes one of the longitudinal coordinates, $+$, $-$ or z. Here P_2^+ is the M2 projection $P_2^\pm = \frac{1}{2}(1 \pm \Gamma^{(2)})$, where $\Gamma^{(2)} = \Gamma^{+-\bar{z}}$. Working out the commutators is straightforward although somewhat tedious. The resulting generalized holonomy group turns out to be [2]

$$\mathcal{H}_{\mathrm{M2/MW}} = [\mathrm{SO}(8) \times \mathrm{SL}(16,\mathbb{R}) \ltimes \mathbb{R}^{(8,16)}] \ltimes 8\mathbb{R}^{(8,1)+(1,16)}. \tag{4.6}$$

4.3. *Other examples*

Some additional backgrounds, including plane wave solutions, were investigated in [2]. Here we summarize the findings in Table 1

Acknowledgments

This work was done in collaboration with A. Batrachenko, M. J. Duff and W. Y. Wen.

References

1. M. J. Duff and J. T. Liu, *Hidden spacetime symmetries and generalized holonomy in M-theory*, Nucl. Phys. B **674** (2003) 217 [hep-th/0303140].
2. A. Batrachenko, M. J. Duff, J. T. Liu and W. Y. Wen, *Generalized holonomy of M-theory vacua*, arXiv:hep-th/0312165.
3. A. Batrachenko and W. Y. Wen, *Generalized holonomy of supergravities with 8 real supercharges*, arXiv:hep-th/0402141.
4. M. Berger, *Sur les groupes d'holonomie homogene des varietes a connexion affine et des varietes riemanniennes*, Bull. Soc. Math. France **83** (1955) 225.
5. R. L. Bryant, *Pseudo-Riemannian metrics with parallel spinor fields and non-vanishing Ricci tensor*, math.DG/0004073.
6. M. J. Duff and K. Stelle, *Multimembrane solutions of $d = 11$ supergravity*, Phys. Lett. B **253** (1991) 113.
7. M. J. Duff, *M-theory on manifolds of G_2 holonomy: the first twenty years*, hep-th/0201062.
8. C. Hull, *Holonomy and symmetry in M-theory*, hep-th/0305039.
9. G. Papadopoulos and D. Tsimpis, *The holonomy of the supercovariant connection and Killing spinors*, JHEP **0307** (2003) 018 [hep-th/0306117].
10. R. Gueven, *Black p-brane solutions of $D = 11$ supergravity theory*, Phys. Lett. B **276** (1992) 49.
11. C. M. Hull, *Exact pp-wave solutions of 11-dimensional supergravity*, Phys. Lett. B **139** (1984) 39.
12. S. K. Han and I. G. Koh, *$N = 4$ supersymmetry remaining in Kaluza-Klein monopole background in $D = 11$ supergravity*, Phys. Rev. D **31** (1985) 2503.
13. G. Papadopoulos and P. K. Townsend, *Intersecting M-branes*, Phys. Lett. B **380** (1996) 273 [hep-th/9603087].
14. A. A. Tseytlin, *Harmonic superpositions of M-branes*, Nucl. Phys. B **475** (1996) 149 [hep-th/9604035].
15. J. Gauntlett, D. Kastor and J. Traschen, *Overlapping branes in M-theory*, Nucl. Phys. B **478** (1996) 544 [hep-th/9604179].
16. J. P. Gauntlett, G. W. Gibbons, G. Papadopoulos and P. K. Townsend, *Hyper-Kaehler manifolds and multiply intersecting branes*, Nucl. Phys. B **500** (1997) 133 [hep-th/9702202].

SOFT BREAKING IN SUSY, STRING AND INTERSECTING D BRANE MODELS

PRAN NATH

Department of Physics, Northeastern Univesity,
Boston, Massachusetts 02115, USA
E-mail: nath@neu.edu

A brief review of recent developments on soft breaking in SUSY, string and intersecting D brane models is given. The constraints on soft breaking parameters from WMAP data are discussed. A discussion of the constraints of modular invariance on soft breaking within heterotic string models is given. It is shown that the radiative electroweak symmetry breaking leads to a more stringent constraint on soft parameters here than in SUGRA models. Finally, a brief review of recent developments on soft breaking in the framework of intersecting D brane models is given. The issue of gauge coupling unification within intersecting D brane models is briefly discussed.

This paper is in celebration of Paul Frampton's Sixtieth Birthday, and in memory of Behram Kursunoglu.

1. Introduction

In this paper we will discuss soft breaking in SUSY, strings and intersecting D brane models and the implications of recent WMAP data[1,2] in constraining the soft breaking parameters. The implications of the WMAP data are important enough that we will discuss these as the first item in Sec.2 as the results have important implications for the discovery of supersymmetry. Next in Sec.3 we will discuss the implications of modular invariance on soft breaking in strings in a generic way where we also discuss the constraints of radiative breaking of the electroweak symmetry in string models. We will show that radiative electroweak symmetry breaking puts sever constraints on μ and $\alpha_{strring}$[3]. (where μ is the Higgs mixing parameter which appears in the superpotential in the form $\mu H_1 H_2$). In Sec.4 we discuss soft breaking in intersecting D brane models[4] and then comment on gauge coupling unification regarding the conditions under which unification can occur in intersecting brane D model. Finally, in Sec.5 we end with some conclusions.

2. Constraints on Soft Breaking from WMAP

Soon after the accurate WMAP data became available the implications of this data for constraining the parameter space of soft breaking were discussed. In this analysis the hyperbolic branch of radiative breaking of the electroweak symmetry[5] plays an important. The origin of this branch can be seen by examining the equation for the radiative electroweak symmetry breaking equation. Thus one of the radiative symmetry breaking equation reads $C_1^2 m_0^2 + C_3 m_{1/2}'^2 + C_2' A_0^2 + \Delta\mu_{loop}^2 = \mu^2 + \frac{1}{2}M_Z^2$, where m_0 is the universal scalar mass, $m'_{\frac{1}{2}}$ is determined by $m_{\frac{1}{2}}$ which is the universal gaugino mass, and A_0 is the universal trilinear coupling all taken at the GUT scale, and where C_1, C_2' etc are determined in terms of gauge and Yukawa couplings. For the case when the loop correction to μ is small, one typically finds that the co-efficients C_1, C_2' etc are all positive and the soft parameters lie on the surface of an ellipsoid. However, for the case when the loop correction to μ^2 is large, one finds that C_1 can be considered to be turning negative converting the ellipsoid into a hyperboloid. On the hyperboloid m_0 and $m_{1/2}$ can get very large for a fixed μ. Now as $m_{1/2}$ gets very large, one finds that the gaugino masses can become much larger than μ and thus the lowest mass supersymmetric particle (lsp) and the next to the lowest mass supersymmetric particle (nlsp) and the light chargino become degenerate with mass essentially μ. In this case the squarks could lie in the several TeV region and beyond the reach of even the LHC. The question then is if supersymmetry will be hidden from us. Quite remarkably it turns out that even in this case it is possible to discover SUSY via the direct detection of dark matter. We note that a part of the hyperbolic branch (HB) for low values of $m_{1/2}$ is the so called focus point (FP)[6] region.

We turn now to a discussion of dark matter. The most recent WMAP data[1,2] puts a very severe constraint on the relic density of cold dark matter[7,8,9]. This constraint then defines the region of the parameter space where SUSY may be discovered. One of these is the region where relic density constraints can be satisfied by coannihilation between neutralinos and staus. This is the conventional region. The second region deep on HB/FP is the one where $m_{1/2} >> |\mu|$ and here one may have coannihilation of a very different type, i.e., coannihilation involving $\chi_1^0\chi_1^{\pm}$, $\chi_1^0\chi_2^0$, and $\chi_1^0\chi_2^0$. Surprisingly one finds that coannihilation among these particles allows for a satisfaction of the relic density constraints even with m_0 and $m_{1/2}$ lying in the region of several TeV. Even more remarkable is the fact that the neutralino-proton scattering cross-sections lie within reach of the future dark matter detectors such as GENIUS[10] and ZEPLIN[11].

3. Modular Invariance and Soft Breaking

Next we discuss soft breaking in a generic heterotic string model (for a sample of models based on heterotic strings, see Ref.[12]) where modular invariance has proved to be a useful tool[13,14] (For a recent of discussion of phenomenology based on modular invariance see Ref.[15,16]). For the purposes of the discussion we put aside all of the difficult questions regarding SUSY breaking in strings[17] such as stabilization of VEVs and picking the right vacuum state. The main purpose of this analysis is to exhibit certain features of soft breaking in string models which are pretty generic. Thus we consider generic heterotic superstring compactification on orbifolds without going into any specific details of the models. Effectively the only constraint that we use is the T-duality symmetry. Specifically, we use a generic $SL(2, Z)$ subgroup of modular invariance associated with the large radius- small radius symmetry. The scalar potential in the 4-dimensional theory depends on the dilaton field S and on the moduli fields T_i and the potential is invariant under the modular transformations $T_i \to T_i' = (a_iT_i - ib_i)(ic_iT_i + d_i)^{-1}$, $(a_id_i - b_ic_i) = 1$, $(a_i, b_i, c_i, d_i \in Z; i = 1, 2, 3)$. Following the supergravity techniques we assume that the superpotential consists of the visible sector and a hidden sector where supersymmetry is broken. In the hidden sector we assume that the superpotential is a product of Dedekind functions of an appropriate weight and a modular invariant function $F(S, T)$.

The analysis of soft breaking under the constraints of modular invariance shows[3] that the the soft parameters A_0 and B_0 which are coefficients of the trilinear and bilinear terms in V_{soft} depend on the quantity $e^{-D/2}/\sqrt{f_\alpha}$ where $D = -ln(S + \bar{S})$, and $f_\alpha = \prod_{i=1}^{3}(T_i + \bar{T}_i)$, and the subscript α implies that we are evaluating the moduli at one of the self dual points corresponding to $T_i = (1, e^{i\pi/6})$. The above leads to values of $f_\alpha = 8, 4\sqrt{3}, 6, 3\sqrt{3}$ and these imply degeneracy of allowed vacua. The factor $e^{-D/2}$ also has important implications as discussed below. A similar analysis holds for the gaugino masses[3]. Next we discuss briefly the radiative electro-weak symmetry breaking constraints in string models. There are two electroweak symmetry breaking constraints. These are $\mu^2 = -\frac{1}{2}M_Z^2 + (m_{h_1}^2 - m_{h_2}^2 \tan^2\beta)/(\tan^2\beta - 1)$ and $\sin 2\beta = -2\mu B/(m_{h_1}^2 + m_{h_2}^2 + 2\mu^2)$ where $\tan\beta = \langle H_2 \rangle / \langle H_1 \rangle$ and where H_2 gives mass to the up quark and H_1 gives mass to the down quark and the lepton. Now in SUGRA models[18] one uses the first constraint to determine μ. However, in string models μ is in principle determined. Thus the first equation puts a strong consistency check on string theory. The second equation can be viewed as a new way to determine the string constant.This is so because as pointed out already B_0 is proportional to $e^{-D/2}$ which is proportional to $1/g_{string}$. Using the string value of B_0 and the second electroweak symmetry breaking equation one can then determine α_{string} in terms of low energy parameters.

4. Soft Breaking in Intersecting D Brane Models

There has been considerable interest recently in building models based on Type II strings with D-branes[19,20,21,22]. These models are interesting since they contain chiral fermions. Specifically massless matter fields such as quarks, leptons and Higgs lie on the intersection of the branes and appear as bifundamental representations. In intersecting brane models at least four stacks of branes are needed to get the Standard Model. The branes wrap cycles on the tori and for the a-th type brane the stack number and the wrapping numbers (n_a^i, m_a^i) for ith torus are constrained by the RR tadpole cancellation conditions. The number of massless generations correspond to the number of times the branes intersect in the compact space and the intersection is thus determined in terms of the wrapping numbers n and m so that $I_{ab} = \prod_{m=1}^{3}\left(\mathbf{n}_a^{(m)}\mathbf{m}_b^{(m)} - \mathbf{n}_b^{(m)}\mathbf{m}_a^{(m)}\right)$, and the condition for preserving supersymmetry is given by $\phi_a^{(1)} + \phi_a^{(2)} + \phi_a^{(3)} = 2\pi$, $\phi_a^{(m)} = \tan^{-1}(\frac{F_a^{(m)}}{Re(T_m)})$, $F_a^{(m)} = (\frac{m_a^{(m)}}{n_a^{(m)}})$ where we are following the notation of Ref.[4] to which the reader is referred to for full details. The computation of the soft breaking requires knowledge of the Kahler potential for the bifundamentals and in Ref.[4] it was proposed that it take the following form $\tilde{K}_{\alpha\bar{\beta}}^{[ab]} = \delta_{\alpha\bar{\beta}}(s+\bar{s})^{\frac{\nu_{ab}}{2}-1}\prod_{m=1}^{3}(t_m+\bar{t}_{\bar{m}})^{\nu_{ab}^{(m)}-\frac{\nu_{ab}}{2}}(u_m+\bar{u}_{\bar{m}})^{\nu_{ab}^{(m)}-\frac{\nu_{ab}}{2}}$ where $\nu_{ab}^{(m)}$ are the shift vectors. They are related to angles between stacks a and b so that $\nu_{ab}^{(m)} = (1/\pi)\arccos\left(1 + \mathbf{F}_a^{(m)}\mathbf{F}_b^{(m)}\Re(T_m)^{-2}/\prod_{c=a,b}\sqrt{1 + (\mathbf{F}_c^{(m)}\Re(T_m)^{-2}}\right)$, where $\nu_{ab} = \sum_m \nu_{ab}^{(m)} = 1 \text{ or } 2 \bmod 2Z$. From the above the Kahler potential for the parallel brane case[23] D9-D5 and D5-D5 sectors can be easily recovered on using the following shift vectors: (i) for D_9D_5 case one uses $\nu_{ab}^m = (\frac{1}{2}, \frac{1}{2}, 1)$, and (ii) for the D_5D_5 case one uses $\nu_{ab}^m = (\frac{1}{2}, \frac{1}{2}, 0)$.

With the above ingredients one can carry out an analysis of soft breaking in D brane models[4] using the same procedure as in SUGRA models[18] (for other approaches see, Refs.[24,25]). The analysis of soft breaking does not address the difficult task of supersymmetry breaking in the intersecting brane models and how one may stabilize the VEVs and achieve vanishing vacuum energy. Rather we will use the standard parametrization procedure using the visible and hidden sector techniques of SUGRA models. Using this procedure one obtains quite general results for soft breaking in terms of the wrapping numbers, moduli and shift vectors which are left arbitrary. The formulae obtained are valid for a wide class of models. The soft breaking obtained for D brane models also contain sources of CP violation, specifically a total of seven CP phases appear. These are the phases of the dilation and six phases associated with the phases of the t and u moduli. Further, these soft breaking results limit to the results found for the parallel brane cases of D_9D_5 and D_5D_5. Thus it should be possible to satisfy the edm constraints using the cancellation mechanism[26] here as in the case of the parallel D brane case. An application of the general formulae were made for specific models[20,27] which is necessary in carrying out the phenomenology of such models. There are some patterns of symmetry breaking that emerge in these analyses and such patterns can in principle be tested in accelerator experiments. Of course, there are no fully realistic D brane models so the examples are meant only as illustrative..

Next me discuss the issue of unification of the gauge couplings in the intersecting D brane models. We begin by recalling that in MSSM gauge coupling unification occurs with MSSM spectrum at a scale of $M_{GUT} = 2\times 10^{16}$ GeV and one finds the relation $\alpha_2 = \alpha_3 = (5/3)\alpha_Y$. In intersecting brane models unification of gauge coupling constants is not a generic feature[28]. However, one may ask under what conditions such a unification might occur. One interesting observation is that one may choose brane configurations such that one has the relation $1/\alpha_Y = (2/3)/\alpha_3 + 1/\alpha_2$. The further constraint $\alpha_2 = \alpha_3$ then leads to the unification of the gauge coupling constants. So the question then is how can a relation such as $\alpha_2 = \alpha_3$ come about. This issue is tied very deeply with the constraint of how the satisfaction of supersymmetry is achieved in the intersecting brane models. To illustrate this phenomenon let us start with the formula for the gauge couplings for the intersecting brane case. Here one has $\alpha_a = (M_{st}/4\sqrt{2\pi}M_{pl})(Re(T_1)Re(T_2)Re(T_3))^{\frac{1}{2}}/\prod_{i=1}^{3}(Re(T)_i^2 n_a^{i2} + m_a^{i2})^{\frac{1}{2}}$. Clearly then the unification depends on the value of the moduli $Re(T_m)$. The moduli $Re(T_i)$ are constrained by the condition that preserves supersymmetry. It is instructive to write this condition so that $F_a^1/Re(T_1) + F_a^2/Re(T_2) + F_a^3/Re(T_3) = F_a^1 F_a^2 F_a^3/Re(T_1)Re(T_2)Re(T_3)$. There are three distinct cases. These are (Case A): Here all fluxes $F_a^m \neq 0$ and all $Re(T_m)$

are determined by the constraint of preserving SUSY. Thus the unification of gauge coupling constants can only be accidental in this case. (CaseB): Here one has one flux $F_a^m = 0$ and the ratios $Re(T_1) : Re(T_2) : Re(T_3)$ are determined but not their absolute value. Again in this case the unification of gauge coupling can only be accidental. (Case C): Here one flux $F_a^m = 0$ and only one ratio $Re(T_i) : Re(T_j)$ is determined while $Re(T_k)$ is arbitrary. In this case equality of $\alpha_2(M_X)$ and $\alpha_3(M_X)$ can be arranged and the unification of gauge couplings can occur. Assuming that the constraint of case C holds one can get an estimate of the sizes. Thus one finds in this case $\alpha_G = M_{st}/4\sqrt{2\pi}M_{pl}(Ref(T_1))^{\frac{3}{2}}f_G$, where f_G is the factor that depends on the wrapping numbers. A value $M_{st} = 2 \times 10^{16}$ GeV, and $Re(T_i)f_G^{\frac{2}{3}} \simeq 6$ leads to the MSSM value $\alpha_G \simeq \frac{1}{24}$.

5. Conclusion

We summarize now the results. In Sec.2 we showed that soft breaking is highly constrained by the WMAP data. However, quite surprisingly the analysis of relic density in mSUGRA shows that there is a wide range of m_0 and $m_{\frac{1}{2}}$ where the WMAP constraints are satisfied. Specifically it is shown that the WMAP relic density constraints can be satisfied in the HB/FP region where m_0 tends to be rather large and in fact large enough that it may lie outside side the reach of the Large Hadron Collider (LHC). However, an analysis of the neutralino-proton cross section shows that this region would still be accessible to the dark matter detectors with sensitivities reaching up to $\sigma_{\chi p} \sim 10^{-47} cm^2$. In Sec.3 we analysed the constraints of modular invariance on soft breaking in string models. Radiative electroweak symmetry breaking in string models was also analysed. It was shown that radiative breaking constraints in string models are far more stringent than previously realized. Finally we discussed soft breaking in intersecting brane models in Sec.4 and it was noted that interesting patterns of soft breaking arise on intersecting D branes which could be discernible in experiment. The issue of gauge coupling unification was also discussed and it was pointed out that the gauge coupling unification can be arranged in intersecting brane models when one of the fluxes vanished, i.e., $F_a^m = 0$ (for each a) and only one ratio $Re(T_i) : Re(T_j)$ is determined by the constraint that preserves SUSY. Otherwise the unification will have to be totally accidental.

Acknowledgments

The work reported in this paper is based on recent collaborative works with several colleagues, specifically with Utpal Chattopadhyay, Achille Corsetti, Tarek Ibrahim, Boris Kors, and Tomasz Taylor. The author thanks them for enjoyable collaborations. This research was supported in part by NSF grant PHY-0139967.

References

1. C. L. Bennett *et al.*, *First Year Wilkinson Microwave Anisotropy Probe (WMAP) Observations: Preliminary Maps and Basic Results*, astro-ph/0302207.
2. D. N. Spergel *et al.*, "First Year Wilkinson Microwave Anisotropy Probe (WMAP) Observations: Determination of Cosmological Parameters," *Astrophys. J. Suppl.* **148**, 175 (2003) [arXiv:astro-ph/0302209].
3. P. Nath and T. R. Taylor, *Phys. Lett.* **B548**, 77 (2002) [hep-ph/0209282].
4. B. Kors and P. Nath, *Nucl. Phys.* **B681**, 77 (2004) [arXiv:hep-th/0309167].
5. K.L. Chan, U. Chattopadhyay and P. Nath, *Phys. Rev.* **D58**, 0960041998.
6. J. L. Feng, K. T. Matchev and T. Moroi, *Phys. Rev.* **D61**, 075005 (2000).
7. U. Chattopadhyay, A. Corsetti and P. Nath, *Phys. Rev.* **D68**, 035005 (2003) [hep-ph/0303201].
8. J. R. Ellis, K. A. Olive, Y. Santoso and V. C. Spanos, *Phys. Lett.* **B565**, 176 (2003).
9. H. Baer and C. Balazs, *JCAP* **0305**, 006 (2003); H. Baer, C. Balazs, A. Belyaev, T. Krupovnickas and X. Tata, *JHEP* **0306**, 054 (2003); A. B. Lahanas and D. V. Nanopoulos, *Phys. Lett.* **B568**, 55 (2003); H. Baer, T. Krupovnickas and X. Tata, *JHEP* **0307**, 020 (2003); A. B. Lahanas, N. E. Mavromatos and D. V. Nanopoulos, *Int. J. Mod. Phys.***D12**, 1529 (2003).
10. H.V. Klapdor-Kleingrothaus, et.al., "GENIUS, A Supersensitive Germanium Detector System for Rare Events: Proposal", MPI-H-V26-1999, hep-ph/9910205.
11. D. Cline *et al.*, "A Wimp Detector With Two-Phase Xenon," *Astropart. Phys.* **12**, 373 (2000).
12. For a sample of heterotic string model building see, A. Font, L. E. Ibanez, H. P. Nilles and F. Quevedo, *Nucl. Phys.* **B307**, 109 (1988) [Erratum-ibid. **B310**, 764 (1988)]; I. Antoniadis, J. R. Ellis, J. S. Hagelin and D. V. Nanopoulos, *Phys. Lett.* **B194**, 231 (1987); B. R. Greene, K. H. Kirklin, P. J. Miron and G. G. Ross, *Nucl. Phys.* **B292**, 606 (1987); P. Nath and R. Arnowitt, *Phys. Rev.* **D39**, 2006 (1989); A. E. Faraggi, *Phys. Lett.* **B278**, 131 (1992); S. Chaudhuri, S. W. Chung, G. Hockney and J. Lykken, *Nucl. Phys.* **B456**, 89 (1995); Z. Kakushadze, G. Shiu, S. H. H. Tye and Y. Vtorov-Karevsky, *Int. J. Mod. Phys.* **A13**, 2551 (1998)
13. S. Ferrara, N. Magnoli, T. R. Taylor and G. Veneziano, *Phys. Lett.* **B245**, 409 (1990); A. Font, L. E. Ibanez, D. Lüst and F. Quevedo, *Phys. Lett.* **B245**, 401 (1990); H. P. Nilles and M. Olechowski, *Phys. Lett.* **B248**, 268(1990);

P. Binetruy and M. K. Gaillard, *Phys. Lett.* **B253**, 119 (1991); M. Cvetic, A. Font, L. E. Ibanez, D. Lüst and F. Quevedo, *Nucl. Phys.* **B361**, 194 (1991).
14. A. Brignole, L. E. Ibanez, C. Munoz and C. Scheich, *Z. Phys.* **C74**, 157 (1997); B. de Carlos, J. A. Casas and C. Munoz, *Nucl. Phys.* **B399**, 623 (1993); A. Brignole, L. E. Ibanez and C. Munoz, *Phys. Lett.* **B387**, 769 (1996).
15. G. Kane, J. Lykken, S. Mrenna, B. D. Nelson, L. T. Wang and T. T. Wang, *Phys. Rev.* **D67**, 045008 (2003).
16. U. Chattopadhyay and P. Nath, in progress..
17. H. P. Nilles, *Phys. Lett.* **B115**, 193 (1982); S. Ferrara, L. Girardello and H. P. Nilles, *Phys. Lett.* **B125**, 457 (1983); M. Dine, R. Rohm, N. Seiberg and E. Witten, *Phys. Lett.* **B156**, 55 (1985); C. Kounnas and M. Porrati, *Phys. Lett.* **B191**, 91 (1987).
18. A.H. Chamseddine, R. Arnowitt and P. Nath, *Phys. Rev. Lett.* **49**, 9701982; R. Barbieri, S. Ferrara and C.A. Savoy, *Phys. Lett.* **B119**, 3431982; L. Hall, J. Lykken, and S. Weinberg, *Phys. Rev.* **D27**, 2359(1983): P. Nath, R. Arnowitt and A.H. Chamseddine, *Nucl. Phys.* **B227**, 121(1983). For a recent review see, P. Nath, "Twenty years of SUGRA," arXiv:hep-ph/0307123.
19. R. Blumenhagen, L. Görlich and B. Körs, *Nucl. Phys.* **B569**, 209 (2000); *JHEP* **0001**, 040 (2000); hep-th/0002146; G. Pradisi, *Nucl. Phys.* **B575**, 134 (2000); C. Angelantonj, I. Antoniadis, E. Dudas and A. Sagnotti, *Phys. Lett.* **B489**, 223 (2000); S. Förste, G. Honecker and R. Schreyer, *Nucl. Phys.* **B593**, 127 (2001); *JHEP* **0106**, 004 (2001); C. Angelantonj and A. Sagnotti, R. Blumenhagen, B. Körs, D. Lüst and T. Ott, *Nucl. Phys.* **B616**, 3 (2001); R. Blumenhagen, V. Braun, B. Körs and D. Lüst, *JHEP* **0207**, 026 (2002); R. Blumenhagen, L. Görlich and T. Ott, *JHEP* **0301**, 021 (2003); G. Honecker, *Nucl. Phys.* **B666**, 175 (2003);. M. Larosa and G. Pradisi, *Nucl. Phys.* **B667**, 261 (2003); C. Kokorelis, *JHEP* **0209**, 029 (2002).
20. M. Cvetic, G. Shiu and A. M. Uranga, *Phys. Rev. Lett.* **87**, 201801 (2001) [hep-th/0107143]; *Nucl. Phys.* **B615**, 3 (2001)
21. M. Berkooz, M. R. Douglas and R. G. Leigh, *Nucl. Phys.* **B480**, 265 (1996); D. Cremades, L. E. Ibanez and F. Marchesano, *JHEP* **0207**, 009 (2002); M. Cvetic, P. Langacker and G. Shiu, *Phys. Rev.* **D66**, 066004 (2002); M. Cvetic, P. Langacker and G. Shiu, *Nucl. Phys.* **B642**, 139 (2002); D. Cremades, L. E. Ibanez and F. Marchesano, *JHEP* **0307**, 038 (2003)
22. For a review see, C. Angelantonj and A. Sagnotti, *Phys. Rept.* **371**, 1 (2002)
23. L. E. Ibanez, C. Munoz and S. Rigolin, *Nucl. Phys.* **B553**, 43 (1999).
24. M. Grana, *Phys. Rev.* **D67**, 066006 (2003).
25. M. Grana, T. W. Grimm, H. Jockers and J. Louis, arXiv:hep-th/0312232.
26. T. Ibrahim and P. Nath, *Phys. Lett.* **B418**, 98 (1998); *Phys. Rev.* **D61**, 093004 (2000); M. Brhlik, L. L. Everett, G. L. Kane and J. Lykken, *Phys. Rev. Lett.* **83**, 2124; (1999); *Phys. Rev.* **D62**, 035005 (2000)
27. M. Cvetic, P. Langacker and J. Wang, *Phys. Rev.***D68**, 046002 (2003) .
28. R. Blumenhagen, D. Lüst and S. Stieberger, *JHEP* **0307**, 036 (2003)

SOME RELATIONS BETWEEN TWISTED K-THEORY AND E8 GAUGE THEORY

H. SATI
Department of Mathematics
The University of Adelaide SA 5005 Australia
email: hsati@maths.adelaide.edu.au

Abstract

Recently, Diaconescu, Moore and Witten provided a nontrivial link between K-theory and M-theory, by deriving the partition function of the Ramond-Ramond fields of Type IIA string theory from an E8 gauge theory in eleven dimensions. We give some relations between twisted K-theory and M-theory by adapting the method of Diaconescu-Moore-Witten and Moore-Saulina. In particular, we construct the twisted K-theory torus which defines the partition function, and also discuss the problem from the E8 loop group picture, in which the Dixmier-Douady class is the Neveu-Schwarz field. In the process of doing this, we encounter some mathematics that is new to the physics literature. In particular, the eta differential form, which is the generalization of the eta invariant, arises naturally in this context. We conclude with several open problems in mathematics and string theory.

My talk was based on joint work with V. Mathai[1].

References

[1] Varghese Mathai and Hisham Sati, "Some Relations between Twisted K-theory and E8 Gauge Theory," JHEP 0403 (2004) 016 [arXiv hep-th/0312033].

MATTER FIELDS ON A TWO-SHEETED SPACE-TIME*

KAMESHWAR C. WALI
Physics Department, Syracuse University
Syracuse, NY 13244-1130 U.S.A.
email: wali@physics.syr.edu

Abstract

As a prelude to the standard model in curved space time, we present a model of a left- and a right- chiral field living on a two-sheeted space-time. The resulting action functionals include novel interactions due to extended gravity.

1 Introductory Remarks

It is widely recognized that the classical continuum picture of space-time is untenable at all scales. The twin pillars of modern physics, namely, 1). General Relativity, in which the dynamics of the classical continuum of space-time is coupled to the dynamics of the matter moving within it and 2). Quantum Field Theory, with rules of quantization to be applied, in principles, to all degrees of freedom including gravity, are found to be incompatible. Besides each pillar has its own structural fractures, such as space-time singularities in the case of general relativity and infinities and divergences in the case of quantum field theories. An ideal unified theory demands the quantization of space, time and matter all treated on the same footing. A continuum space-time at all scales seems to be inadequate for such an ideal theory. Recent ideas based both on string theory and approaches to quantum gravity suggest strongly that space-time structure at the Planck scale may be discrete with noncommutative coordinates. The continuum space-time of classical physics, it is hoped, will emerge in certain limiting regimes, just as the classical behavior of quantum systems emerges in an appropriate limit.

At the present time we do not have a precisely defined noncommutative space that meets the above requirements. However, in recent years, Connes

*This report is based on the work done in collaboration with Nguyen Ai Viet, Department of High-Energy Physics, Center for Theoretical Physics,Institute of Physics, Hanoi, Vietnam

has proposed an alternate approach to the study of the structure of space-time, based on noncommutative geometry (NCG) [1, 2]. It has given rise to the description of the Standard Model with a geometrical interpretation of the Higgs field on the same footing as the gauge field. Spontaneous symmetry breaking follows as a natural consequence. It enables one, in principle, to calculate some arbitrary parameters of the Standard Model, such as the Weinberg angle, the top quark and Higgs masses.

The main motivation for the present work is to extend Connes' formalism to include gravity. As a prelude to the full standard model, we study a simplified model consisting of a left- and a right- handed chiral spinor fields along with their respective abelian gauge fields on a two-sheeted space-time. The two-sheeted space-time can be considered as a discretized version of the Kaluza-Klein theory in which the internal space of a circle is replaced by two discrete points. In this brief report, I summarize the basics of the formalism and the summary of results. For details and references to related work, the reader is referred to the published paper [4].

2 Non-commutative differential geometry and the spectral triple on a two-sheeted space-time

In his formalism, Connes bypasses the precise specification of the manifold as the starting point. Instead, he formulates its description in terms of an associative and involutive algebra, commutative or noncommutative. One may think of this as a generalization of the well-known theorem due to Gelfand [1, 3], which states that the classical topological space based on a continuum can be completely recovered from the abelian algebra of smooth functions. The starting point in his formalism is what is called a spectral triple [Connes] : 1) an involutive, unital algebra of functions $\mathcal{A}$, 2) Hilbert space $\mathcal{H}$, and 3) a self-adjoint operator called Dirac operator $\mathcal{D}$. The existence of a finite basis for the module of differential forms is understood. Also, the the algebra $\mathcal{A}$ of generalized functions is now not necessarily commutative.

Intuitively, two sheeted space-time can be viewed as an extension of the physical four-dimensional space-time manifold M by a discrete, internal space $\{a, b\}$, a discretized version of Kaluza-Klein theory where the circle is replaced by two discrete points in the fifth dimension.

The spectral triple:

Algebra $\mathcal{A}$ is the algebra of smooth function $C^\infty(M, \mathbf{C})$ tensored with the complex-valued functions on the set $\{a, b\}$:

$$\mathcal{A} = C^\infty(M) \otimes (\mathbf{C} \oplus \mathbf{C}) \cong C^\infty(M, \mathbf{C}) \oplus C^\infty(M, \mathbf{C}). \tag{2.1}$$

Hilbert space $\mathcal{H}$ is the space of square-integrable sections of a spinor bundle,

$$\mathcal{H} \equiv L^2(M,S) \oplus L^2(M,S). \tag{2.2}$$

Dirac operator $\mathcal{D}$ is the generalized self-adjoint operator

$$D = \Gamma^M D_M, \quad M = 0,1,2,3,5 \tag{2.3}$$

with the generalization of the gamma matrices γ^μ given by

$$\Gamma^\mu \equiv \begin{pmatrix} \gamma^\mu & 0 \\ 0 & \gamma^\mu \end{pmatrix}, \qquad \Gamma^5 \equiv \begin{pmatrix} 0 & \gamma^5 \\ \gamma^5 & 0 \end{pmatrix}. \tag{2.4}$$

Generalized space-time derivatives are given by

$$D_\mu \equiv \begin{pmatrix} \partial_\mu & 0 \\ 0 & \partial_\mu \end{pmatrix}. \tag{2.5}$$

These derivatives act along each of the space-time sheets. The derivative between the two sheets is given as

$$D_5 \equiv \sigma \bar{D}_5, \tag{2.6}$$

where

$$\bar{D}_5 \equiv \begin{pmatrix} 0 & m \\ m & 0 \end{pmatrix}, \qquad \sigma \equiv \begin{pmatrix} 0 & 1 \\ 1 & 0 \end{pmatrix}. \tag{2.7}$$

The parameter m has the dimension of mass in order to give the fifth component of space-time the same dimension as the other components. The action of the operator DF on F is defined to be

$$D(F) \equiv [D,F] = \Gamma^M (DF)_M = \Gamma^M D_M F, \tag{2.8}$$

We define the generalized wedge product as the fully anti-symmetric product that truncates all the "junk" forms in a trivial way, namely,

$$\Gamma^M \wedge \Gamma^N = \frac{1}{2}(\Gamma^M\Gamma^N - \Gamma^N\Gamma^M). \tag{2.9}$$

except when in Eqn.(2.9), $M, N = 5$,in which case

$$\Gamma^5 \wedge \Gamma^5 = 1. \tag{2.10}$$

With the spectral triple given, it is straightforward to construct the algebra of forms in the two-sheeted space-time. The exterior derivative of any odd form is its anti-commutator with the operator D as a consequence of the Leibnitz rule. Hence the exterior derivative DU of the generalized one-form $U = \Gamma^M U_M$ is

$$DU = \{D, U\} = \Gamma^M \wedge \Gamma^N (DU)_{MN}, \tag{2.11}$$

2.1 Riemannian geometry in two-sheeted space-time

The Equivalence Principle extended to the two-sheeted space-time requires a locally orthonormal basis. Continuing to work with the Γ-representation, the locally orthonormal basis is represented by Γ^A

$$\Gamma^a \equiv \begin{pmatrix} \gamma^a & 0 \\ 0 & \gamma^a \end{pmatrix}, \qquad \Gamma^{\dot 5} \equiv \begin{pmatrix} 0 & \gamma^5 \\ \gamma^5 & 0 \end{pmatrix}, \tag{2.12}$$

where γ^a and γ^5 are the usual flat Dirac matrices. By choosing γ^5 we have specialized to the two sheets of space-time of chiral spinors. We shall use a $\dot 5$ index in the orthonormal basis to distinguish it from 5 in the general case.

The inner product is defined as the trace

$$<\Gamma^A, \Gamma^B> \;=\; Tr(\Gamma^A \Gamma^B) \;=\; 2\eta^{AB}.\mathbf{1} \tag{2.13}$$

The curvilinear basis will be represented by $\Gamma^M(x)$. To obtain $\Gamma^M(x)$ and define the metric, we postulate the generalized vielbeins $E^A_M(x)$, as the following diagonal matrix of zero-forms:

$$E^\mu_a(x) \equiv \begin{pmatrix} e^\mu_{1a}(x) & 0 \\ 0 & e^\mu_{2a}(x) \end{pmatrix}, \qquad E^\mu_{\dot 5}(x) \equiv 0 ,$$

$$E^5_a(x) \equiv -\begin{pmatrix} a_{1a} & 0 \\ 0 & a_{2a} \end{pmatrix} \equiv -A_a(x) = -E^\mu_a A_\mu ,$$

$$E^5_{\dot 5}(x) \equiv \begin{pmatrix} \phi_1^{-1}(x) & 0 \\ 0 & \phi_2^{-1}(x) \end{pmatrix} \equiv \Phi^{-1}(x) , \tag{2.14}$$

where $e^\mu_{1,2a}(x)$ are two different vielbeins on the two sheets of space-time. Similarly, $a_{1,2}(x)$ and $\phi_{1,2}(x)$ are respectively vector and scalar fields.

The vielbeins E^M_A are invertible giving,

$$E^a_\mu(x) \equiv \begin{pmatrix} e^a_{1\mu}(x) & 0 \\ 0 & e^a_{2\mu}(x) \end{pmatrix}, \qquad E^a_5(x) \equiv 0 ,$$

$$E^{\dot 5}_\mu(x) \equiv A_\mu(x)\Phi(x) , \qquad E^{\dot 5}_5(x) \equiv \begin{pmatrix} \phi_1(x) & 0 \\ 0 & \phi_2(x) \end{pmatrix} \equiv \Phi(x) . \tag{2.15}$$

Then, the $\Gamma^M(x)$ matrices are given as $\Gamma^M(x) = \Gamma^A E^M_A(x)$,

$$\Gamma^\mu(x) \equiv \begin{pmatrix} \gamma^a e^\mu_{1a}(x) & 0 \\ 0 & \gamma^a e^\mu_{2a}(x) \end{pmatrix}, \qquad \Gamma^5(x) \equiv \begin{pmatrix} -a_1(x) & \gamma^5\phi_2^{-1}(x) \\ \gamma^5\phi_1^{-1}(x) & -a_2(x) \end{pmatrix}, \tag{2.16}$$

where $a_{1,2}(x) = \gamma^a a_{1,2a}$ are two gauge connection one-forms on the two different sheets.

Using the inner product defined as the trace on the Hilbert space, we can calculate the representation of the metric tensor. "Metric" field components involve a pair of conventional vierbeins and, in addition, a pair of vector and a pair of scalar field.

3 Physical content of the gravity sector in two-sheeted space-time

The essential point is how to impose a consistent system of constraints to solve the generalized Cartan structure equations without over-constraining the metric. To this end, we find the following set of minimal condtions:

i) Spacetime torsion free condition

$$T_{aBC} = 0, \tag{3.1}$$

ii) Metric compatibility condition

$$\tilde{\Omega}^{*}_{AB} = -\Omega_{BA} \tag{3.2}$$

iii) "Reality" condition

$$\tilde{\Omega}^{*}_{AB} = \Omega_{AB} \tag{3.3}$$

iv) An additional condition on the connections

$$\Omega_{AB\dot{5}} = 0 \tag{3.4}$$

These conditions are sufficient to express the non-vanishing connection coefficients and torsion components in terms of metric fields. Also, more importantly, they provide Lagrangians with proper kinetic terms for all the metric component fields. .

Hence,to summarize, it is to be noted that the gravity sector of the theory contains $e^{\mu}_{a}(x), e^{a}_{\mu}(x)$, the tensor field $v^{\mu}_{\nu}(x)$, the vector and scalar fields $a_{\pm\mu}(x), \Phi_{\pm}(x)$. In the final Lagrangian, the tensor field $v^{\mu\nu}$ will be a mass eigenstate. Thus, one component of each pair of tensor, vector and scalar fields has zero mass, while the other is massive. The special case in which the components on the second sheet are identical to the ones on the first sheet, reproduces exactly the Kaluza-Klein zero mode theory.

4 Matter fields in curved two-sheeted space-time

The matter field Lagrangian consists of a gauge sector and a fermionic sector. The gauge sector contains two Abelian gauge fields and two Higgs scalar fields as part of a genaralized one-form. The fermionic sector consists of a left-chiral spinor on one sheet and a right-chiral spinor on the other.

4.1 The gauge sector

The generalized one-form:

$$U = B(x) = \Gamma^M(x)B_M(x),$$

where

$$B_\mu(x) = \begin{pmatrix} b_{1\mu}(x) & 0 \\ 0 & b_{2\mu}(x) \end{pmatrix} \quad , \quad B_5(x) = \begin{pmatrix} h_1(x) & 0 \\ 0 & h_2(x) \end{pmatrix}. \tag{4.1}$$

Then, the curvature or field strengths are given by the two-form (henceforth the x-dependence will be understood),

$$G = DB + B \wedge B, \tag{4.2}$$

In terms of the above components, the Lagrangian is given as follows:

$$\mathcal{L}_G = -\frac{1}{4}Tr < G, G >= -\frac{1}{4}Tr(2\tilde{G}_{ab}G^{ba} + 4\tilde{G}_{a\dot{5}}G^{\dot{5}a} + \tilde{G}_{\dot{5}\dot{5}}G^{\dot{5}\dot{5}}) \ , \tag{4.3}$$

where the trace is over the two dimensional matrices.

A straight-forward calculation results in the Lagrangian

$$\mathcal{G}_G = \mathcal{L}_{G1} + \mathrm{L}_{G2} + \mathcal{L}_{G3} + \mathcal{L}_{G4}, \tag{4.4}$$

where $\mathcal{L}_{G1}$ and $\mathcal{L}_{G2}$ contain the kinetic terms for the gauge and Higgs fields respectively; $\mathcal{L}_{G3}$ has the quartic Higgs potential of the correct form for spontaneous symmetry breaking and $\mathcal{L}_{G4}$ contains various interaction terms between the gauge sector vector and scalar fields and those from the gravity sector.

4.2 Fermion sector

With two chiral spinor fields Ψ_L on one sheet of space-time and Ψ_R on the other, the fermionic Lagrangian has the standard form,

$$\mathcal{L}_F \quad = \quad i\bar{\Psi}\Gamma^A(E_A^M(D_M + iB_M) + \tfrac{i}{4}\Gamma^B\Gamma^C\Omega_{BCA})\Psi \tag{4.5}$$

where

$$\Psi = \begin{pmatrix} \psi_L \\ \psi_R \end{pmatrix},$$

$$\psi_L = \frac{(1+\gamma_5)}{2}\psi \quad , \quad \psi_R = \frac{(1-\gamma_5)}{2}\psi, \tag{4.6}$$

ψ is the 4-component Dirac spinor and Ω_{BCA} are the non-vanishing connection coefficients expressed in terms of the metric fields. Consequently,this generalized Lagrangian includes gravitational interactions with the matter fields.

5 Physical implications, Summary and Conclusions

The formalism provides formal expressions for the action functionals in terms of formal fields as elements of an algebra. They need to be redefined by rescaling using the dimensional parameters, G_N ,the Newtonian constant, g_V,a new gravitational constant associated with torsion, g , the dimensionless gauge coupling and finally, m ,the parameter having the dimension of mass. Then, the requirement that the kinetic terms of each field have the standard coefficients determines the masses and coupling constants in terms of the above mentioned four parameters. On the formal side, the redefinition of the fields leads to a subalgebra of the original algebra where the fields on the two sheets are complex conjugates of each other.

To summarize the main results, one finds that the gravitational sector consists of the standard Einstein-Hilbert action, accompanied by a Fierz-Pauli Lagrangian for a massive tensor field with mass m. The mass of the vector field accompanying the zero mass vector field is $m/\sqrt{2}$. It also has a massless Brans-Dicke scalar field and massive scalar field. Among the various interaction terms, there is a scalar field potential with a cosmological constant term.

The matter field Lagrangians have the expected spontaneous symmetry breaking leading to a zero mass gauge boson accompanied by a massive one with mass $\sqrt{2}m$. More importantly, there are parity violating interactions involving matter fields and the metric component massive tensor, vector and scalar of the extended gravity sector. In the context of the full standard model, these interactions could be CP-violating and thus provide the much needed CP violation in the early universe to explain baryon-anti-baryon asymmetry.

Thus from the point of view of physics, the formalism provides a rich and complex structure with new interactions, linear and non-linear. With the extended gravity,it has cosmological implications and suggests many avenues for further research

References

[1] A. Connes, *Non-Commutative Geometry*, [Academic Press, 1994].

[2] A. Connes, "Essay on Physics and Non-Commutative Geometry", in D.G. Quillen, G.B. Segal, S.T. Tsou editors, *The Interface of Mathematics and Particle Physics*, [Clarendon Press, Oxford, 1990].

[3] G.F. Simmons, *Introduction to Topology and Modern Analysis*, [Krieger, 1983].

[4] Nguyen Ai Viet and Kameswar C. Wali, Phys. Rev. D **67**, 1240299 (2003).

6 Kurşunoğlu Memorial Tributes

MEMORIAL SESSION FOR BEHRAM KURSUNOGLU

PAUL FRAMPTON

Department of Physics and Astronomy
University of North Carolina at Chapel Hill
Chapel Hill, NC 27599-3255
E-mail: frampton@physics.unc.edu

Behram *loved* physics and was very concerned about maintaining the highest standards in physics, especially at these Coral Gables conferences.

I first met Behram in 1972, the first time attending the Coral Gables conference. But I really got to know hime as a friend only since the Coral Gables meetings restarted in 1992. We spoke often on the phone at length about speakers and topics.

Earlier this month, on returning from a few months in Europe, I was moved to hear on the phone message machine in my Chapel Hill office a long and enthusiastic message from Behram, very much alive giving ideas about Coral Gables 2003.

Behram's research was well summarized by Philip Mannheim on Thursday morning. He pursued a line, the unification of gravity with electromagnetism, which is one of the most difficult in theoretical physics. From what Phillip said, the value of Behram's research remains hard to evaluate. It's unfortunate that he died while this research was still ongoing. Only time will tell.

Behram was fascinating also when he discussed the giants of theoretical physics all of whom he had met and knew including Einstein, Dirac, Schroedinger, Oppenheimer, Tellet and others.

Behram was always a gentleman of great grace and style - both in his interactions with other people and in his dress. He should have been here today wearing his stetson hat and cane, making certain everyone had a great time. We all miss him.

On behalf of the organizing committee of the Coral Gables conference I would like to present a plaque to Sevda. It reads

IN FOND MEMORY OF PROFESSOR BEHRAM N. KURSUNOGLU
March 14, 1922 (Turkey) - October 25, 2003 (Florida).
THEORETICAL PHYSICIST, ORGANIZER AND CONVENER OF
32 CONFERENCES IN FLORIDA SINCE 1964.
COLLEAGUE AND DISCIPLE OF ALBERT EINSTEIN,
PAUL A.M. DIRAC, AND ERWIN SCHROEDINGER.

FOR MY FATHER

ISMET BIL KURSUNOGLU
6200 Leonardo Street
Coral Gables, Florida, 33146

My family and I would like to thank you all for coming here today to pay tribute to my father, Behram Kursunoglu. I am especially thankful to all of you who overcame extraordinary distance and other complications to attend today's event. I would like this day to be remembered from the perspective of a celebration, and to review his accomplishments, his love of you, and life itself.

First and most importantly I would like to recognize the valiant, complete, and tireless love and devotion of my mother to my father. Sevda's constant efforts set out the canvas against which all of his achievements grew – as great as those achievements were, we all realize that to him his greatest achievement was the joy of meeting, marrying and spending his entire life with Sevda. He loved his wife – he adored his wife. He was miserable when she wasn't at his side beyond his 8 hour working day at the university. I know how much she will miss him, but she certainly will have nothing but fond memories of their time together. Their accomplishments are many, but having shared so much love for one another, throughout it all, in the end, that love was the only thing that mattered.

Behram was a remarkable man. The constant joy of seeing and being with his family, along with the joy of his physics was the essence of this man. In this regard we are all so thankful to all the efforts made by Ayda and Jay, Michael and Lauren, in visiting, helping and aiding him for all these years – he loved them all. There was also the joy of seeing Sevil and Johan during their visits to Coral Gables, Fort Lauderdale, and on their international trips; and his special delight in visiting in person and on the phone with Alexa and Lucas whom he so much enjoyed and who continuously enriched his life. He also enjoyed all the visits and travels and talks with me and my wife Carole who considered him as her second father.

To all of you who accompanied my father over the years to various gatherings, luncheons, meetings, parties and all of the social events in Miami, around this great nation and all over the globe, I want to let you know how much he enjoyed each and every one of those relationships and opportunities, no matter how brief or limited. He loved each and every one of you. He loved life.

I and my family are most grateful to my father for realizing his vision of coming to this great country. Imagine starting from his north eastern Turkish Village of Chikura and eventually migrating to live in one of the greatest countries in the world, in the capacity of and having mastered one of the most complex career subjects that one could imagine.

Through it all he had the opportunity to link up with some of the great physicists of the last century. His success is closely linked with, and mimics many others of you who have also made the journey to settle in this great country. He loved America and was a proud American. But let's not forget, that at the same time he managed to remain true to his roots and had a very special love for his mother country, Turkey. He had a profound respect and gathered huge inspiration from Kemal Ataturk's vision of a society based on reason, fairness, equality of the sexes, and most importantly, for promoting a society steeped in education and knowledge of the arts and sciences.

On the home front and from a kid's perspective it was a joy growing up. Lots and lots of stimulation – imagine being surrounded by terms and ideas such as "Unified Field Theory," the expanding universe, super strings, and black holes. Who needed Harry Potter? I was living in and experiencing a Harry Potter story in real-life. There were lots of very good memories – a montage of wonderful memories that now and throughout my life continue to suddenly appear from the recesses in my mind. For example the physics picnics at Mattheson Hammock, or us piling into the Center for Theoretical Studies' station-wagon for the numerous trips to the beautiful Florida Everglades where we would go on walks with my father, Paul Dirac, and with Bob Kelly as our guide.

Then there were the rides across country often taking the latest new car, which in the early years was a Chevrolet, then later Oldsmobiles, and finally moving up in the later years to driving the two Cadillacs that he enjoyed immensely. The cross-country trips included visits to the various National labs. I recall vividly driving on the then craggy and elevated approach road to Los Alamos only to re-visit that same road with him 40 years later and find to our surprise that it had become a super highway.

They were all very exciting times for the kids, his colleagues, and the parents alike. I began to realize that the car was a key tool, in that it was used by my father in ferrying me around to visit and play with my friends, including hundreds of trips to visit my best friend Jeff Benjamin for sleep-overs, launching of our model rockets, and many other wonderful experiences.

There where many other adventures such as taking tea at the offices of the Center for Theoretical Studies where one would see my father or Paul Dirac surrounded by chalk boards filled with amazing mathematics.

There were all of the trans-Atlantic journeys including on-board high seas adventures on the Michaelangelo, or Behram's doing the impossible and re-creating a Czarist era reception dinner of plenty in what was otherwise a very dreary Moscow of 1987. We would wander Istanbul's Kapale charcha and I recalled vendors hounding him in German but with them recoiling, realizing their error after having mistaken him for a German business magnate, only to have him respond to them in exactly their dialect of Turkish – don't forget that after all he was also a supreme comic and mime.

It was a whirl-wind life, picking up a new suit at a Saville Row shop, or dining with Carole, myself, and my mother at Effez in London; the many, many trips to the Foils book store in London that he so loved to visit. Locally we recall the often too numerous trips to Sears and the OK Feed Store, and the various horticultural distributors in the area, eventually turning what started out as a fairly sparse home garden into what is now a veritable tropical jungle. My earliest recollect is going with him to the Coconut Grove Bank to deposit his salary checks, and the swimming lessons at Venetian pool. He was of course full of wonderful stories many of which were conducted in the voices that would exactly imitate the luminaries in the world of physics – including his famous stories and imitations of Eugene Wigner, Edward Teller, and Paul Dirac, not to mention a very decent rendering of his various Russian colleagues. All of this was done within a framework of humor and wonderful story telling, that only Behram could marshal.

He enjoyed recounting stories from Hodja – the classical 13th century Turkish philosopher who lived in middle Anatolia. One story goes once a villager saw Hodja sitting by a very large lake – throwing large quantities of yeast into it – the villager said "Hodja what are you doing?" Hodja remarked "I am making yogurt for the whole city." The villager remarked "Hodja that is crazy you'll never be able to make yogurt that way!" To which Hodja replied "Maybe so but imagine if it works!" Behram's philosophy paralleled that of Hodja in that he was an eternal optimist.

My most fond memory from my youth was very simply the great joy that occurred everyday at 5 pm when I would hear the garage door knowing that it heralded my father's return home from the University. I would run with shear joy to meet him at the garage door entrance to our home, and then experiencing even more pleasure when I was old enough to carry in this briefcase for him. As with all kids who are fortunate enough to have a father, their fathers always appear larger than life, and our case was no different.

This last month, November 19, 2003, will mark the 50th anniversary of his meeting with Professor Albert Einstein. My father always believed in seeking out and associating with the top people in his field. He also always insisted in hearing the news from the horse's mouth. So with that mind-set, on that day he went to Einstein's home in Princeton continuing his search for an answer to the toughest questions in physics. He very much enjoyed that meeting and certainly was most impressed that Einstein indicated time alone would tell which of their theories was right.

Behram was of course most excited with the sudden realization that there might have been two horses at that meeting. None the less that one meeting provided the critical mass to propel his already inquisitive mind for the rest of his life.

Another equally important meeting took place 60 years ago, at the first meeting of his dear friend Nejat Veziroglu in Turkey. Imagine, Prof. Veziroglu has known Behram longer that any of us. His friendship was so much appreciated and was characteristic of the many other great friendships that Behram made through the course of his lifetime. When we look out we see many of those friends here today. Your cumulative experiences and your kindness in sharing them with me and my family have been of great comfort – indeed your historical relationships with my father are so vivid in my mind that they cumulatively result in a reassembly of all the myriad complex events of his life into what is a very good living memory of the man. That is of great comfort to me and my family, and I thank you for that.

If we could look just bit a beyond this beautiful area, we would see the beautiful University of Miami main campus where my father was given the opportunity to pursue his studies of the fundamental sciences. I certainly don't have the time today to detail all of his life's works and all of the achievements that he realized while at the University, but suffice it to say there where many. I would refer you to an excellent and extensive summary of his life works compiled by his life long colleague Professor Arnold Perlmutter, as well as a retrospective by

Professor Philip Mannheim. (See the following two contributions in these proceedings.)

I am deeply grateful to all those who made his career, while at the University of Miami, so rewarding and successful. In particular I would like to thank Dr. Henry King Stanford who had the presence of mind to recognize my father's ability and to help him in realizing his dreams. Certainly since those pioneering days, the University continues to demonstrate excellence and growth in many areas, exceeding any one person's predictions. I am so grateful that my father was able to play a role in creating that vision and for the opportunities afforded to us through my family's long relationship with the University of Miami.

My father at heart was very simple, yet to some he appeared extraordinarily complex and demanding. But these tendencies could be instantly soothed. For example by offering him a simple pot of tea and some roasted filbert nuts you could have a friend for life. By the way, I am very thankful to those of you who were the sources of those Leblebi nuts that you would bring him. I can think of no other food substance that he enjoyed more.

Many such observations, stories, and a compilation of his life's work have been captured in a popular book that my father has written called "The Ascent of Gravity" (subtitled Adventures of a Physicist). It will be published sometime in the next year and you can refer to it to learn more about the thoughts, ideas, and experiences of this great man.

In the end, Behram Kursunoglu was all about striving to push the envelope in the process of discovery. He was always moving forward to unlock those issues in science that have escaped explanation for a very long time. I am certain, knowing the scope of his intellectual capability, that history will prove that he was right in his theory. What I also know is that the hope, inspiration, discoveries, and contributions from my father along with his colleagues will continue to profoundly shape and improve the world we live in.

To the students and young people, all of whom my father admired and enjoyed talking to, take measure in what my father has done. And realize that it is actually quite simple, for you only have to learn to imagine the unimaginable, or to create a solution where there appears to be none, and be willing to challenge conventional thought. But most importantly never give up and never lose sight of your dreams. By the application of this simple formula Behram showed us that you too can move mountains and turn lakes into yogurt.

THE WORK OF BEHRAM KURSUNOGLU

PHILIP D. MANNHEIM*
Department of Physics
University of Connecticut
Storrs, CT 06269, USA
E-mail: mannheim@uconnvm.uconn.edu

Talk presented at the 2003 Coral Gables conference in honor and appreciation of the work of Professor Behram Kursunoglu, general relativist extraordinaire and founder of the Coral Gables series of conferences, whose untimely death occurred shortly before the 2003 conference.

Professor Behram Kursunoglu died shortly before the 2003 Coral Gables conference, the latest in a long line of Coral Gables conferences which Behram had been organizing since 1964. This particular conference had originally been planned to honor Professor Paul Frampton on the occasion of his sixtieth birthday, but with Behram's untimely passing the conference additionally became a memorial to him, to his work and to his life.

At the conference Behram had been scheduled to speak on the topic "Will Einsteinian theories continue to dominate twenty-first century physics?". And even though I had not had an opportunity to discuss his talk with him, I believe I know what he was going to talk about. Specifically, a few years earlier at a previous Coral Gables conference at the turn of the millennium, Behram had commented to me about the fact that Time magazine had named Albert Einstein to be the man of the twentieth century. Behram remarked to me that in another hundred years Time magazine would declare Einstein to be the man of the twenty-first century as well. Behram felt that up to now we had only just begun to skim the surface of the implications of general relativity, and that during the upcoming century general relativity would prove to be the foundation from which even greater and deeper physics was going to arise.

*This work has been supported in part by the U. S. Department of Energy (D.O.E.) grant no. DE-FG02-92ER40716.00.

It should come as no surprise that Behram thought this way about Einstein, since Behram's own career had focused on general relativity ever since Behram's graduate student days at the University of Cambridge in the early 1950s, with general relativity being at the very center of Behram's research during his entire professional career. Behram's own particular interest had been on a geometric unification of gravitation first with electromagnetism and subsequently with all of the other fundamental forces, an issue to which Einstein himself devoted much of his own career; and to establish the context of Behram's own work, it is useful to recall the main ways in which such a unification by metrication has historically been sought.

Characteristic of attempts to achieve unification through metrication was the need to find some generalization of general relativity, akin to the way general relativity had itself generalized special relativity. Specifically, while special relativity provided a single comprehensive framework for treating both electromagnetism and Newton's laws of motion (by modifying the latter), special relativity did not encompass gravity, with Newton's law of gravitation not being invariant under Lorentz transformations. The remedy for this was not to come up with a gravitational law of force which (like the Lorentz electromagnetic law of force) would be Lorentz invariant, but to instead enlarge the set of invariances to be obeyed by nature, with general coordinate invariance replacing Lorentz invariance, and with the metric $g_{\mu\nu}(x)$ emerging as the gravitational field.

The first attempt at unification through metrication was developed by Weyl[1] as early as 1918, only two years after Einstein's development of general relativity. Weyl introduced a local transformation he referred to as a "gauge" transformation under which the metric and the electromagnetic field would transform as

$$g_{\mu\nu}(x) \rightarrow e^{2\alpha(x)} g_{\mu\nu}(x) \tag{1}$$

$$A_\mu(x) \rightarrow A_\mu(x) - e\partial_\mu \alpha(x) \quad , \tag{2}$$

with gravitation and electromagnetism thus unifying by sharing a common $\alpha(x)$. Given such a joint transformation, in addition Weyl departed from the Riemannian geometry of general relativity and replaced it by a new geometry, "Weyl Geometry", in which the connection was generalized to the A_μ-dependent

$$\hat{\Gamma}^\lambda_{\mu\nu} = \frac{1}{2} g^{\lambda\sigma} \left(\partial_\mu g_{\sigma\nu} + \partial_\nu g_{\sigma\mu} - \partial_\sigma g_{\mu\nu}\right) + \frac{1}{e} g^{\lambda\sigma} \left(g_{\sigma\nu} A_\mu + g_{\sigma\mu} A_\nu - g_{\mu\nu} A_\sigma\right) \quad , \tag{3}$$

so that instead of being zero, the covariant derivative of the metric would instead be given by

$$g^{\mu\nu}{}_{;\nu} = \partial_\nu g^{\mu\nu} + \hat{\Gamma}^{\mu}_{\nu\sigma} g^{\sigma\nu} + \hat{\Gamma}^{\nu}_{\nu\sigma} g^{\sigma\mu} = \frac{2A^\mu}{e} \quad ; \tag{4}$$

with Eq. (4) having the remarkable property of being invariant under the joint Eqs. (1) and (2) (a transformation under which $\hat{\Gamma}^{\lambda}_{\mu\nu}$ transforms into itself). With the covariant derivative of a tensor transforming as a vector, Weyl was thus able to connect $g^{\mu\nu}$ and A^μ in an intricate geometrical fashion, though at the price of departing from Riemannian geometry. As a theory Weyl's theory has two noteworthy aspects. First, it introduced the notion of a gauge transformation, with its usage in Eq. (1) entailing a change in the magnitude of $g^{\mu\nu}$ and thus of its size or gauge. (In the parlance of modern gauge theories, it is a change in the (complex) phase of a field which is now known as a gauge transformation, with Eq. (2) continuing to be known as a gauge transformation, but with Eq. (1) now being referred to as a scale transformation.) Second, if imposed as a symmetry, such a scale invariance would require all mass parameters to be zero identically, and thus exclude any fundamental Newton constant G_N. With it thus possessing neither the Einstein equations or Riemannian geometry, Weyl's theory was not so much a generalization of Einstein gravity, but rather a fairly substantial departure from it, and has not been followed in the literature.[a]

The second attempt at unification through metrication is the now very familiar work of Kaluza and Klein based on a role for spacetime dimensions beyond four, a program which later evolved into superstring theory. In the embryonic work of Kaluza and Klein themselves spacetime was envisaged as being 5-dimensional, with its thus 15-dimensional gravitational field decomposing into a 10-component $g_{\mu\nu}$ which was to serve as the standard 4-dimensional spacetime gravitational field, a 4-component $g_{5\mu} = A_\mu$ to serve as the standard electromagnetic field, thus leaving over a 1-component $g_{55} = \phi$ which would transform as a 4-dimensional spacetime scalar, a field

[a]There is however, a less radical version of Weyl's theory in which the invariance of Eq. (1) is retained in a geometry which is a strictly Riemannian one in which $g^{\mu\nu}{}_{;\nu} = 0$. Such a theory is known as conformal gravity and has been pursued by many authors as a candidate gravitational theory, one in which mass scales and G_N are induced dynamically. Such a theory has actually been one of my own research interests, one which Behram had graciously invited me to report on at several Coral Gables conferences, with the theory being found capable of readily addressing the dark matter and dark energy problems which currently challenge the standard Newton-Einstein gravitational theory.

which is now identified as a Higgs field. With such Kaluza-Klein theories being based on the 5-dimensional Einstein equations, they immediately possessed a fundamental G_N, albeit one which was (unlike the role played by e in Eq. (4) above) not connected to any of the intrinsic structure of the theory. However, the theory did have one nice feature in regard to mass scales, namely the compactification of the fifth of the five dimensions would lead to a compactification radius and thus to a dynamical mass scale for particles; though if the compactification radius is taken to be given by the Planck length scale $(\hbar G_N/c^3)^{1/2}$ associated with G_N, such extra dimensions would be way too miniscule to be detectable.[b]

The third attempt at unification through metrication is the non-symmetric gravity program followed by Behram himself[4,5],[c] an approach also seriously considered at one point by both Einstein[6,7] and Schrödinger[8], and also worked on more recently by Professor John Moffat an attendee at the 2003 Coral Gables conference.[d] On noting that without symmetrization, a general 4-dimensional spacetime rank two tensor $\hat{g}_{\mu\nu}$ would contain 10 symmetric and 6 anti-symmetric components, we see that in the 16 components of $\hat{g}_{\mu\nu}$ there is precisely the number of degrees required for a 10-component symmetric gravitational $g_{\mu\nu}$ and a 6-component anti-symmetric electromagnetic $F_{\mu\nu}$. Unification of gravity with electromagnetism can thus be achieved by working in a geometry in which the full $\hat{g}_{\mu\nu}$ metric is not required to be symmetric. To decompose the full $\hat{g}_{\mu\nu}$ into its symmetric and anti-symmetric pieces one can thus set

$$\hat{g}^{\mu\nu} = g^{\mu\nu} + \frac{F^{\mu\nu}}{q} \tag{5}$$

where q is an appropriate parameter. What makes this decomposition so interesting is that with $g_{\mu\nu}$ being dimensionless, the parameter q has to have the same dimension as the electromagnetic field strength, viz.

[b]While quite interested in the Kaluza-Klein theory because of its unification aspects, it is of interest to note that Einstein nonetheless cautioned[2] that "Among the considerations which question this theory stands in the first place: It is anomalous to replace the four-dimensional continuum by a five-dimensional one and then subsequently to tie up artificially one of these five dimensions in order to account for the fact that it does not manifest itself". It is thus of contemporary interest to note that with the recent advent of the large extra dimension, brane-localized gravity program of Randall and Sundrum[3] this concern of Einstein may finally have been addressed.

[c]While Behram concentrated on non-symmetric unification throughout his entire career, at the 2003 Coral Gables conference Dr. John Brandenburg reported on some recent work done in conjunction with Behram on unification through higher dimensions.

[d]The paper of Moffat and Boal[9] provides reference to Behram's early work on the subject.

(mass/length/time^2)$^{1/2}$. Consequently, and unlike Weyl's approach, a unification based an a non-symmetric $\hat{g}_{\mu\nu}$ requires the presence of an intrinsic dimensionful parameter from the outset; and on setting

$$q = \frac{c^2}{r_0(2G_N)^{1/2}} \ , \tag{6}$$

we see that there has to be an intrinsic length scale, r_0, in the theory, one defined above purely in terms of classical quantities. Moreover, in a unification which is to include quantum mechanics as well, this length scale could be identified as the fundamental Planck scale via

$$r_0 = \left(\frac{\hbar G_N}{c^3}\right)^{1/2} \ , \tag{7}$$

though in his work Behram considered other options for the value of r_0, even including the pure imaginary ones allowed of r_0 and q since their overall phases were only fixed by the requirement that the product $r_0^2 q^2$ be identified as the real and positive $c^4/2G_N$. In a unification based on Eq. (5) then, not only do we unify gravitation with electromagnetism, we additionally see an intrinsic and indispensable role for Newton's constant from the very outset.

To see how things work in the theory it is convenient to first consider a theory with just the regular symmetric $g_{\mu\nu}$ and Lagrangian

$$L = g^{1/2}\left[g^{\mu\nu}R_{\mu\nu} + \frac{G_N}{c^4}\phi^{\mu\nu}(\phi_{\mu\nu} - 2F_{\mu\nu})\right] \ , \tag{8}$$

where $F_{\mu\nu}$ is the usual electromagnetic tensor and $\phi_{\mu\nu}$ is an auxiliary antisymmetric field which is not writable as the covariant curl of a vector. Euler-Lagrange variation of this theory yields

$$\begin{aligned} \partial_\mu(g^{1/2}\phi^{\mu\nu}) &= 0 \ , \quad F_{\mu\nu} = \phi_{\mu\nu} \ , \quad \partial_\sigma\phi_{\mu\nu} + \partial_\nu\phi_{\sigma\mu} + \partial_\mu\phi_{\nu\sigma} = 0 \ , \\ R_{\mu\nu} - \frac{1}{2}g_{\mu\nu}R &= \frac{2G_N}{c^4}T_{\mu\nu} \ , \quad T_{\mu\nu} = \frac{1}{4}g_{\mu\nu}\phi^{\alpha\beta}\phi_{\alpha\beta} - \phi_{\mu\alpha}\phi^{\alpha}{}_{\nu} \ , \end{aligned} \tag{9}$$

to thus yield as solution the standard Einstein-Maxwell equations of motion. Now because both $g^{\mu\nu}F_{\mu\nu}$ and $\phi^{\mu\nu}R_{\mu\nu}$ are kinematically zero, the Lagrangian of Eq. (9) can be replaced by the equivalent Lagrangian

$$L = g^{1/2}\left[\left(g^{\mu\nu} + \frac{1}{q}\phi^{\mu\nu}\right)\left(R_{\mu\nu} - \frac{1}{r_0^2 q}F_{\mu\nu}\right) + \frac{1}{2r_0^2q^2}\phi^{\mu\nu}\phi_{\mu\nu}\right] \ , \tag{10}$$

with the replacing of $g^{\mu\nu}$ by $g^{\mu\nu} + \phi^{\mu\nu}/q$ thus still allowing one to recover the standard Einstein-Maxwell equations of motion in an ordinary $g^{\mu\nu}$ symmetric geometry.

While the above analysis shows that it is permissible to replace $g^{\mu\nu}$ by $g^{\mu\nu} + \phi^{\mu\nu}/q$, the resulting equations of motion do not put electromagnetism and gravity on a completely equivalent footing, since even while the Maxwell tensor serves as the source of the Einstein tensor, gravity does not serve as a source for electromagnetism. To remedy this Behram departed from ordinary Einstein gravity with its symmetric $g_{\mu\nu}$, and instead went to a 16-component $\hat{g}_{\mu\nu}$, a theory in which tensor manipulations would then have to be defined anew, in much the same manner as Weyl had to define the A_μ-dependent $\hat{\Gamma}^\lambda_{\mu\nu}$ of Eq. (3) in his theory. To facilitate the construction of the needed geometry, Behram went beyond the standard $e^{\mu a}$ vierbein formulation of gravity introduced by Weyl in 1928 (in which $g^{\mu\nu} = \eta_{ab}e^{\mu a}e^{\nu b}$) by introducing a second vierbein field $f^{\mu a}$. In terms of this second vierbein the electromagnetic tensor can then be introduced as

$$F^{\mu\nu} = \eta_{ab}\left(e^{\mu a}f^{\nu b} - e^{\nu a}f^{\mu b}\right) \quad , \tag{11}$$

while the contravariant $\hat{g}^{\mu\nu}$ can be introduced as

$$\hat{g}^{\mu\nu} = \frac{1}{\sqrt{2}}\left[\eta_{ab}e^{\mu a}e^{\nu b} + \frac{1}{q^2}\eta_{ab}f^{\mu a}f^{\nu b} + \frac{1}{q}\eta_{ab}\left(e^{\mu a}f^{\nu b} - e^{\nu a}f^{\mu b}\right)\right] \quad , \tag{12}$$

with its covariant counterpart having symmetric and anti-symmetric parts which are once and for all defined as

$$\hat{g}_{\mu\nu} = g_{\mu\nu} + \frac{1}{q}\phi_{\mu\nu} \quad . \tag{13}$$

The two vierbeins thus define the theory. As well as serve as the building blocks of the theory, these two vierbeins additionally provide the theory with a road to symmetry, since the $SO(3,1)$ invariance of $g^{\mu\nu}$ to Lorentz transformations in the $e^{\mu a}$ space is not only augmented by an $S0(2)$ invariance of the quantity on the right-hand side of Eq. (11), the combining of the two vierbeins into one 8-component vector allows it to serve as the fundamental representation of an 8-dimensional symplectic group. In this way Behram had thus developed a starting point for an incorporation of the symmetries of elementary particle physics (the use of complex r_0 and q even allows for a connection to unitary groups), an issue he pursued relentlessly throughout his life.

In Behram's generalized theory there is a generalized connection $\Gamma^\lambda_{\mu\nu}$, one not symmetric in its two covariant indices, and it is with respect to this connection that $\hat{g}_{\mu\nu;\rho}$ is to vanish according to

$$\hat{g}_{\mu\nu;\rho} = \partial_\rho \hat{g}_{\mu\nu} - \Gamma^\lambda_{\mu\rho}\hat{g}_{\lambda\nu} - \Gamma^\lambda_{\rho\nu}\hat{g}_{\mu\lambda} = 0 \quad , \tag{14}$$

and it is with respect to this connection that the Riemann tensor is to be constructed as

$$\hat{R}^{\sigma}{}_{\mu\nu\rho} = -\partial_\rho \Gamma^{\sigma}_{\mu\nu} + \partial_\nu \Gamma^{\sigma}_{\mu\rho} - \Gamma^{\lambda}_{\mu\nu}\Gamma^{\sigma}_{\lambda\rho} + \Gamma^{\lambda}_{\mu\rho}\Gamma^{\sigma}_{\lambda\nu} \quad . \tag{15}$$

For such a Riemann tensor the Ricci tensor $\hat{R}_{\mu\nu}$ would possess both a symmetric ($R_{\mu\nu}$) and an anti-symmetric ($\bar{R}_{\mu\nu}$) piece. Given this structure Behram then introduced a generalized Lagrangian ($\hat{g} = \det\hat{g}_{\mu\nu}$, $g = \det g_{\mu\nu}$)

$$L = \hat{g}^{1/2}\hat{g}^{\mu\nu}\left(\hat{R}_{\mu\nu} - \frac{1}{r_0^2 q}F_{\mu\nu}\right) + \frac{2}{r_0^2}\left(\hat{g}^{1/2} - g^{1/2}\right) \quad . \tag{16}$$

Its Euler-Lagrange variation leads to the same Einstein equation for the symmetric $R_{\mu\nu}$ as given before in Eq. (9), but with the $F_{\mu\nu} = \phi_{\mu\nu}$ and $\partial_\nu(g^{1/2}\phi^{\mu\nu}) = 0$ equations being replaced by

$$\begin{aligned} \phi_{\mu\nu} &= F_{\mu\nu} - r_0^2 \bar{R}_{\mu\nu} \ , \\ \left[g^{1/2}\phi^{\nu}{}_{\mu}\right]_{||\nu} &= 0 \ , \end{aligned} \tag{17}$$

where the $||$ symbol denotes a very particular geometric derivative[5] that Behram was able to construct in his theory. As such Eq. (17) thus provides the intricate connection between electromagnetism and geometry which Behram had sought.[e]

As well as work on unification through metrication, Behram will also be remembered for the Coral Gables Conference Series which have been occurring over a span of almost forty years. The very first of these conferences took place in 1964[f] at which Behram presented a paper "A new symmetry group for elementary particles". As a conference organizer Behram was indefatigable in continuing this conference series over the years, while also never flagging as a researcher. Indeed, he continued to make original presentations at the Coral Gables conferences right up to the end. Thus he presented "Recent developments in gravitational theory and an intrinsic

[e]As noted in Pais's book on Einstein[10] (a book which also references Behram's work), one of Einstein's motivations for turning to non-symmetric unified theories was to try to find a theory whose particle-like solutions would be singularity-free. Interestingly, Behram was able to show[5] that the solutions to the theory based on Eq. (16) were singularity-free.

[f]This was an auspicious year for me personally as it was my first year in graduate school at the Weizmann Institute, and even now I can still recall the excitement of that period when Professors Yuval Ne'eman, Harry Lipkin and (frequent Weizmann visitor and Coral Gables Conference Series co-founder) Sydney Meshkov would report on the Coral Gables conferences.

cosmological parameter" at the 2000 Coral Gables conference,[g] "New Facts on the nature of gravitational force and non-linear oscillations of space" at the 2001 Coral Gables conference,[h] and "The layered structure of the universe" at the 2002 Coral Gables conference.

Beyond physics, Behram also established and ran the Global Foundation which organized meetings on and concerned itself with the great social and international issues of the day. He was an urbane and cultured human being, one who can truly be called a Renaissance man. During his lifetime Behram championed many causes, but the one which I suspect may have given him the most pleasure was the eventual coalescence of elementary particle physics and gravitation, something he presciently first advocated more than fifty years ago, and patiently continued to persistently champion ever thereafter.

References

1. H. Weyl, *Sitzungsber. Preuss. Akad. Wiss.* 465 (1918).
2. A. Einstein, *Science* **74**, 438 (1931).
3. L. Randall and R. Sundrum, *Phys. Rev. Lett.* **83**, 4690 (1999).
4. B. N. Kursunoglu, *Phys. Rev.* **88**, 1369 (1952)
5. B. N. Kursunoglu, *Jour. Phys. Essays* **4**, 439 (1991).
6. A. Einstein, *Sitzungsber. Preuss. Akad. Wiss.* 414 (1925).
7. A. Einstein, *Can. J. Math.* **2**, 120 (1950).
8. E. Schrödinger, *Space-Time Structure*, Cambridge University Press, Cambridge UK (1950).
9. J. W. Moffat and D. H. Boal, *Phys. Rev. D* **11**, 1375 (1975).
10. A. Pais, *Subtle is the Lord: The Science and the Life of Albert Einstein*, Oxford University Press, Oxford UK (1982).

[g] It is of interest to note that the article which appeared immediately following Behram's paper in the 2000 proceedings was given by none other than the sixtieth birthday honoree of the 2003 conference Professor Paul Frampton who in 2000 had presented a paper "Quintessence and cosmic microwave background". As usual, here was Paul working creatively on one of the hot topics of the day, and I personally wish him many productive years to come.

[h] It is of interest to note that the article which appeared immediately following Behram's paper in the 2001 proceedings was given by Behram's long-time associate and Coral Gables Conference Series co-founder Professor Arnold Perlmutter who had presented an application of Behram's theory: "A conjecture on the existence of attractive and repulsive gravitational forces in the generalized theory of gravitation". There is perhaps no higher compliment that one can pay a colleague than the one Arnold thus gave to Behram by working on his ideas.

THE LIFE AND WORK OF BEHRAM KURSUNOGLU

ARNOLD PERLMUTTER
Department of Physics, University of Miami
Coral Gables, Florida, 33124

Behram Kursunoglu was born in a small town near Trabzon, close to Turkey's eastern border with the Caucasus, on March 14, 1922 (Einstein's birth date). When he was about eleven years old, the Governor General of that region addressed his class and asked them to state the principles of the Turkish state. Several students quoted the Koran, but Behram stood up and recited the essence of Kemal Ataturk's speech to the nation. Upon hearing that, the Governor General ordered that a suit be made for Behram, of the best blue serge material.

He graduated from the University of Istanbul in 1945, and then went on scholarship to Edinburgh University, where he studied under Max Born and Nicholas Kemmer, receiving his B Sc in 1949. He then attended Cambridge University where he studied with P A M Dirac, A Papapetrou, H Bondi and others, and received his PhD. in 1952. Among his classmates were Abdus Salam, John Ward, Denis Sciama, and Angas Hurst. He was a post-doctoral fellow under Hans Bethe at Cornell University during 1952-1954. He then helped found and served as an adviser to the Turkish Atomic Energy Commission, in 1957. Ultimately, he came to the University of Miami, in 1958.

Behram's first published research was based on his PhD thesis in 1952, which was his proposal to modify the Einstein-Schrödinger unified field theories. Around this time, and later on, he corresponded frequently with Einstein and Erwin Schrödinger. He worked throughout most of the rest of his life on the physical and mathematical aspects of his theory, despite some detours in the intervening years.

During his stay at Cornell he used Tamm-Dancoff methods to study the nature of the nuclear forces, and also to study bremsstrahlung due to high energy protons. His most prescient work during this period was the proposal, in 1953, to make use of high energy electron scattering to study the distribution of charge in protons and neutrons. This was the experimental technique used by Robert Hofstadter during 1956 – 1960, who used electrons of up to 600 MeV in energy to study the charge distribution of the protons and many of the heavier nuclei, thereby also inferring the charge distributions of the neutron.

In 1962 Behram wrote a classic text, "Modern Quantum Theory," to the acclaim of many critics, starting with Werner Heisenberg. During the 1960's he also wrote a large number of papers on diffusion in relativistic plasmas, and beginning with the First Coral Gables Conference on Symmetry Principles at High Energy, in 1964, he introduced a new wave equation and a new symmetry group for elementary particles, work that he continued until the 1970's.

But still, Behram's long meeting with Albert Einstein in 1953 was a high point in his life, after which he never lost sight of his principal ambition, which was to create a viable unified field theory. Some of the consequences of his work were the prediction, in 1959, of the existence of four kinds of neutrinos, and in 1975, his proposal that neutrinos have mass. The theory, which was based on a non-symmetric complex field of the form $g_{\mu\nu} + iq^{-1}\Phi_{\mu\nu}$, where the antisymmetric tensor, $\Phi_{\mu\nu}$, differs from the ordinary electromagnetic tensor by not being expressible as the curl of a vector potential, and q is a very large electric field. By means of the correspondence principle, there exists a fundamental length such that $r_0^{\ 2} q^2 = \dfrac{G}{2c^4}$, and this theory reduces to Einstein's 1916 theory when r_0 becomes zero. He was then led to his depiction of an elementary particle, which he named "orbiton," as comprised of an infinite number of spherical layers of magnetic charge separated by layers of electric charge, such that the total magnetic charge vanishes and the total electric charge is zero or one. Hence, there are no free magnetic monopoles in his theory. More recently, he demonstrated that his theory predicts four different kinds of gravitational forces – long range attractive (conventional), long range repulsive (accounting for accelerated expansion), short range attractive (nuclear), and short range repulsive (weak).

Behram was tireless in his devotion to the study of fundamental aspects of the universe, and of the social and political ramifications of science. The thirty-two famous Coral Gables Conferences on high energy physics and cosmology, including this one, the twenty-five conferences throughout the world on the production and use of energy, his innovative course on Nuclear War/Nuclear Peace and associated seminars for college instructors, along with his creation of the Center for Theoretical Studies at the University of Miami, all bear testimony to his tenacity in the service of his ideals. We are all beneficiaries of his effectiveness.

Some representative papers from his collected works are listed below.

1. B. Kursunoglu, "Velocity of light in generalized theory of gravitation," Phys. Rev. D **14**, 1518-1523 (1976).

2. B. Kursunoglu, "Consequences of nonlinearity in the generalized theory of gravitation," Phys. Rev. D **13**, 1538-1560 (1976)

3. B. Kursunoglu, "Gravitation and magnetic charge," Phys. Rev. D **9**, 2723-2745 (1974); Erratum, Phys. Rev. D **12**, 1850 (1975)

4. B. Kursunoglu, "Theory of Relativistic Supermultiplets. II. Periodicities in Hadron Spectroscopy," Phys. Rev. D **2**, 717-735 (1970)

5. B. Kursunoglu, "Theory of Relativistic Supermultiplets. I. Baryon Spectroscopy," Phys. Rev. D **1**, 1115-1132 (1970)

6. B. Kursunoglu, "Dynamical Theory of Hadrons and Leptons," Phys. Rev. **167**, 1452-1461 (1968)

7. B. Kursunoglu and C. A. Hurst, "Space-Time and Origin of Internal Symmetries," Phys. Rev. **155**, 1663-1671 (1967)

8. B. Kursunoglu, "New Symmetry Group for Elementary Particles. I. Generalization of Lorentz Group Via Electrodynamics," Phys. Rev. **135**, B761-B768 (1964)

9. B. Kursunoglu, "Brownian Motion in a Magnetic Field," Phys. Rev. **132**, 21-26 (1963)

10. B. Kursunoglu, "Proton Bremsstrahlung," Phys. Rev. **105**, 1846-1853 (1957)

11. B. Kursunoglu, "Transformation of Relativistic Wave Equations," Phys. Rev. **101**, 1419-1424 (1956)

12. B. Kursunoglu, "Tamm-Dancoff Methods and Nuclear Forces," Phys. Rev. **96**, 1690-1701 (1954)

13. B. Kursunoglu, "Derivation and Renormalization of the Tamm-Dancoff Equations," Phys. Rev. **92**, 1069-1070 (1953)

14. B. Kursunoglu, "Expectations from a Unified Field Theory," Phys. Rev. **92**, 506-507 (1953)

15. G. Rickayzen and B. Kursunoglu, "Unified Field Theory and Born-Infeld Electrodynamics," Phys. Rev. **89**, 522-523 (1953)

16. B. Kursunoglu, "Gravitation and Electrodynamics," Phys. Rev. **88**, 1369-1379 (1952)

17. B. Kursunoglu, "On Einstein's Unified Field Theory," Phys. Rev. **82**, 289-290 (1951)

THE KURSUNOGLUS

LADY BLANKA ROSENSTIEL, O.S.J.
American Institute of Polish Culture, Inc.
1440 79th Street Causeway, Suite 117
Miami, Florida 33141

Some reminiscences on the life of Professor Behram Kursunoglu and his family.

The Kursunoglu family and I met a long time ago. Going back to the late sixties, together with my husband Lewis and Henry K. Stanford, former President of the University of Miami, we often discussed the wonderful program for High Energy Physics at the Center for Theoretical Studies that Behram organized and directed at the University. Behram was a man of vision, always full of ideas, with a willingness to accomplish them. He possessed great enthusiasm for new possibilities.

In January of 1976 we let Behram have our beautiful yacht, the *Gallant Lady*, to entertain some Russian Scientists who came to participate in his physics conference. After that event there were many occasions when, after the conferences at the Center, I had the pleasure (my husband Lewis passed away) to host at my home the entire group of Nobel Laureates (Dirac, Wigner, Lamb) as well as Edward Teller and many other physicists who came from all over the world to attend the conferences. When not moored at the "Casa Rosenstiel" our yacht was used for cruising during cocktail parties or lunchtime in Biscayne Bay to show off the magic of Miami from the waterways.

All my many friends were delighted to meet so many illustrious personages. We thought there would be no "normal conversation" with them, and that they would be difficult to entertain. On the contrary, when the pianist started to play there was singing and dancing – this was real fun!

Throughout many years, up to 1995, when I sold my home, my place was always open to the Kursunoglus and their friends. Thereafter, Behram and Sevda were always inviting me and my brother Waldemar J. Dowiak to the after-conference parties that were arranged at various places. I was very much flattered when Behram asked me to deliver a speech on the contribution of Polish scientists to the development of world science during one of the dinners. My speech was very much applauded when I talked about Mikolaj Kopernik (Copernicus), Marie Sklodowska Curie, and other world renowned Polish scientists whose contributions shine as guiding lights even to this day.

Beside the many "official" involvements, Behram, Sevda and their children became very dear friends and we spent a lot of time together at various parties, Polonaise Balls, weddings, and also during trips to Washington D.C. On one of these trips, we were invited to the Turkish Embassy for afternoon tea. Behram introduced me to the Ambassador as the person who had organized Polonia in the U.S. through two organizations which I had founded: The American Institute of Polish Culture and the Chopin Foundation of the United States. Behram suggested that the Ambassador seek my advise on how to organize the Turkish population in America! Though I did appreciate Behram's introduction, I had to admit that the Polonian group was not an easy task to organize, no less difficult that other ethnic groups in the U.S.

Soon thereafter, I joined the Ataturk Society in Washington, which sponsored special Ataturk Awards to illustrious people. One of them was Behram Kursunoglu. Although he was already suffering from Parkinson disease, Behram delivered a most moving and memorable acceptance speech.

I shall always remember Behram. He was an outstanding human being – distinguished, serious in his profession, a great bon vivant, and fun in private gatherings. He will be greatly missed by all his friends.

The Kursunoglus will always remain in my heart as an exemplary family – kind, understanding, loving and caring, and just plainly a beautiful people I was, and am, privileged to know.